MW00511069

The Physics and Applications of

High Brightness Electron Beams

Proceedings of the ICFA Workshop

The Physics and Applications of
High Brightness Electron Beams

Chia Laguna, Sardinia 1–6 July 2002

editors

James Rosenzweig
Department of Physics & Astronomy,
UCLA, CA, USA

Gil Travish
Department of Physics & Astronomy,
UCLA, CA, USA

Luca Serafini
INFN Milan, Italy

World Scientific

NEW JERSEY • LONDON • SINGAPORE • SHANGHAI • HONG KONG • TAIPEI • BANGALORE

Published by

World Scientific Publishing Co. Pte. Ltd.
5 Toh Tuck Link, Singapore 596224
USA office: Suite 202, 1060 Main Street, River Edge, NJ 07661
UK office: 57 Shelton Street, Covent Garden, London WC2H 9HE

British Library Cataloguing-in-Publication Data
A catalogue record for this book is available from the British Library.

THE PHYSICS AND APPLICATIONS OF HIGH BRIGHTNESS ELECTRON BEAMS

Copyright © 2003 by World Scientific Publishing Co. Pte. Ltd.
All rights reserved. This book, or parts thereof, may not be reproduced in any form or by any means, electronic or mechanical, including photocopying, recording or any information storage and retrieval system now known or to be invented, without written permission from the Publisher.

For photocopying of material in this volume, please pay a copying fee through the Copyright Clearance Center, Inc., 222 Rosewood Drive, Danvers, MA 01923, USA. In this case permission to photocopy is not required from the publisher.

ISBN 981-238-726-9

Printed in Singapore by World Scientific Printers (S) Pte Ltd

PREFACE: THE PHYSICS AND APPLICATIONS OF HIGH BRIGHTNESS ELECTRON BEAMS

J. B. ROSENZWEIG, G. TRAVISH

UCLA Dept. of Physics and Astronomy
405 Hilgard Ave., Los Angeles, CA 90095-1547
E-mail: rosen@physics.ucla.edu

L. SERAFINI

Istituto Nazionale di Fisica Nucleare
Sezione di Milano
Via Celoria 16, 20133 Milano, Italy

The ICFA-sponsored (ICFA Panel on Beam Dynamics , ICFA Panel on Advanced Accelerators) workshop on "The Physics and Applications of High Brightness Beams" was held on July 1-7, 2002, in Chia Laguna, Sardinia, Italy. This workshop, which had a fine international representation from the fields of intense electron sources, free-electron lasers, advanced accelerators, ultra-fast laser-plasma, beam-plasma and laser-beam physics, has an interesting heritage that is worth discussing, to place these proceedings in the proper perspective.

This workshop summarized in these proceedings represents the merging of two ICFA-sponsored series, one on high brightness beams, and the other on free-electron laser physics. The most recent high brightness beam workshop was entitled "The Physics of High Brightness Beams", endorsed by the ICFA Panel or Novel and Advanced Accelerators, and was held in Los Angeles in November 1999. It concentrated on the emerging physics of intense beam systems, and was notable for the exciting synthesis of methods and concepts between electron beam and ion beams (the proceedings of this workshop, which document this synthesis well, is available from World Scientific[1]). On applications side, the final installment "Arcidosso Series" of workshops, held in Sept. 2001, was entitled "The Physics of, and Physics With, the X-ray Free-Electron Laser". This title foreshadows the merging of discussions between the physics of the enabling device (in this case the FEL), with the examination of the physics of the enabled applications.

In order to merge these two traditions, some decisions needed to be made. The most difficult was the realization that the immensely profitable inclusion of ion beam physics in the scope of the workshop had to be foregone. This was motivated obviously by the electron beam-application flavor of the Arcidosso series, and was underlined by the recent rapid progress in electron beam source and application physics. In addition, with the workshop site in Sardinia, the present Italian initiatives in high brightness electron beams and FELs provided the final push needed to concentrate the workshop solely on electron beams.

The applications discussed at the workshop ranged from self-amplified spontaneous FELs, to Thomson scattering light sources, and also to advanced accelerators based on plasmas and/or lasers. These applications have many physical characteristics in common with both the beams that are required to in the diverse schemes, and also with each other. We note first that the intense beams required for the applications display a wealth of behavior — with aspects of high-field acceleration, plasma and radiative processes — that is indeed reminiscent of the applications themselves.

One unifying way of describing the fields involved in the processes of both beam creation and application is the normalized potential amplitude In the context of the rf photoinjector one defines a single parameter $\alpha = eE_{rf} / k_{rf} m_e c^2$ that controls the longitudinal dynamics in the rf gun; in the context FELs, one introduces the transverse wiggle amplitude parameter $a_u = eB_u / k_u m_e c^2$; in laser-plasma interactions, the quantity $v_{osc} / c = eE_l / \omega_l m_e c$ measures the normalized transverse momentum (not velocity!) of the electron oscillation in the laser field; in Thomson-scattering sources, one has in analogy with FEL and laser-plasma interaction a wiggle amplitude parameter given by $a_l = eE_l / \omega_l m_e c$; in plasma accelerations, the so-called "wave-breaking" field normalizes the self-fields of a plasma wave by $\tilde{E} = E / E_{WB} = eE_l / m_e c \omega_p$. While all of these parameters have different physical implications in their respective scenarios, the fact that all approach unity means that in all cases one must consider *relativistic oscillatory motion.*

Further, in creating and manipulating intense electron beams, even in the presence of very large fields, one must aim to have extremely cold beam distributions. These distributions, despite their near-vanishing phase space volume, are very far from familiar thermal descriptions, and display intricate correlations, arising *e.g.* from emittance compensation, or chicane compression, in six-dimensional phase space. Both the present of very large collective fields, and the related phase space structure make these beam systems qualitatively different than those encountered in most high-energy accelerators. One must keep these characteristics in mind when modeling and measuring the beams, and also in using the beams in applications, as FELs, Thomson scattering devices, and plasma wake-field accelerators, are a few phenomena that can be mentioned which are critically dependent on the details of the beam's phase space, and not on simple rms descriptions of the beam. It is clear from recent experiments in these areas that one must provide a detailed ("start-to-end") simulation model of any high-brightness beam driven measurement, in order to have hope of extracting the most meaningful physics understanding from such experiments. Similar considerations apply when designing the next generation of applications, which include, critically, the X-ray SASE FEL. The workshop presented a number of discussions of mature start-to-end simulations of detailed experimental measurements of FEL, and collective effects in pulse compression. This proliferation also provides clear evidence of the tight coupling of beam and application physics at the levels of design and experimental interpretation.

This state of affairs led further explorations of the similarities between the high brightness beam-generation systems and the subsequent applications in the areas of physics, measurements and technology. These similarities, and their exploitation, formed a core theme of the Chia Laguna workshop. For example, one of the most exciting recent topics arising in the high-brightness beam field is that of the coherent synchrotron radiation (CSR) instability during chicane compression. This instability has more than a passing resemblance to the SASES FEL instability. In looking at this new CSR instability in an old way, one may exploit the existing, powerful tools of FEL theory such as a normalized analysis. In fact a dimensionless gain parameter, analogous the FEL ρ can be defined for the CSR instability, but instead of being small, it is much greater than unity; these dimensionless gain parameters can thus be used to illuminate the similarities and differences between the two instabilities[2].

In the rapidly changing field of high-brightness beams, one is pushed to examine old ideas in new contexts. An excellent example of this is the idea of velocity bunching a beam emitted from the photoinjector[3]. This is an old scheme that has not yet been applied to pre-bunched high brightness beams, but it is one that may also help evade problems induced by the CSR instability, as the beam is not bent during bunching.

The similarities between electron source and application become completely blurred in the case of the plasma-based source. Such schemes rely on very intense pulses of electrons that are captured out of the background plasma by a wave in the plasma itself. The fortuitous scaling of such sources with experimental parameters indicates that they may in the future compete with high brightness beams from photoinjectors in, e.g., light source applications[4,5].

In the course of the Chia Laguna workshop, the advanced technical challenges in both the beam and applications were discussed. Particular attention was paid to measurements. As pointed out above, one needs extremely detailed knowledge of the beam phase space to understand advanced beam-based applications; this knowledge must presently come from both simulations and measurements. These cutting edge measurements must deal with unprecedented small phase volumes, and must further be pushed to sub-picosecond resolution. The present "tool-box" for making these measurements was vigorously evaluated during the workshop.

The workshop would not have been a success without its working group leaders (T. Limberg, H-D. Nuhn, J. Schmerge, M. Uesaka) and their scientific secretaries. Further debts are owed to the Grand Hotel Chia Laguna, the workshop local organization (M. Laraneta, S. Angius, R. Centioni, S. Giromini, and M. Pistoni), the program committee (C. Pellegrini, Giuseppe Dattoli, Paul

Emma, Bill Fawley, Massimo Ferrario, John Galayda, Kwang-Je Kim, Patrick O'Shea, and Siegfried Schreiber, Gennady Stupakov, M. Uesaka), and the organizing committee (J. Rosenzweig, Luca Serafini, W. Barletta, Sergio Bertolucci, Ilan Ben Zvi Hans Braun, Swapan Chattopadhyay, Max Cornacchia, Klaus Floettmann, Stephen Milton, Luigi Palumbo, and Alberto Renieri). Thanks are due to the generous institutions who contributed financial support of the workshop (ANL-APS, DESY, ENEA, INFN, JLAB, LBNL, Northern Illinois University – NICADD, Sinchrotrone Trieste, SLAC UCLA, University of Tokyo, Dipartimento di Energetica dell' Univ. di Roma), with particular gratitude being expressed to the US DOE, Office of High Energy Physics for supporting the cost of publishing these proceedings. This long list of supporting institutions indicates the wide range of the laboratories and universities that are connected by interest in the subjects covered by the Chia Laguna workshop.

It is hoped that these proceedings give some flavor of the workshop plenary presentations and discussions, and also that they provide a unique reference to the status of the relevant research fields as of this writing, a snap-shot of the high-brightness beam physics community as it moves quickly into the future.

James Rosenzweig
Los Angeles, March 2003

Luca Serafini
Milan, March 2003

Gil Travish
Los Angeles, March 2003

---

[1] "The Physics of High Brightness Beams", Ed. J.B. Rosenzweig and L. Serafini (World Scientific, Singapore, 2000).
[2] "Comparison of the Coherent Radiation-induced Microbunching Instability in an FEL and a Magnetic Chicane" S. Reiche and J.B. Rosenzweig, these proceedings.
[3] "Recent advances and novel ideas for high brightness electron beams production based on photo-injectors", M. Ferrario, *et al.*, these proceedings.
[4] "Plasma Density Transition Trapping as a Possible High-Brightness Electron Beam Source," M. Thompson, *et al.,* these proceedings.
[5] "A laser-driven accelerator and Thomson x-ray source*", Donald Umstadter, *et al.,* these proceedings.

Contents

WORKING GROUP A

High Brightness Beam Production and Characterization

SUMMARY FOR WORKING GROUP A ON HIGH BRIGHTNESS BEAM PRODUCTION

J.F. SCHMERGE
Stanford Linear Accelerator Center
2575 Sand Hill Rd,
Menlo Park, CA 94025, USA
E-mail: Schmerge@slac.stanford.edu

Working group A was devoted to high brightness beam production and characterization. The presentations and discussions could be categorized as cathode physics, new photoinjector designs, computational modeling of high brightness beams, and new experimental methods and results. Several novel injector and cathode designs were presented. However, a standard 1.5 cell rf photoinjector is still the most common source for high brightness beams. New experimental results and techniques were presented and thoroughly discussed. The brightest beam produced in a rf photoinjector published at the time of the workshop is approximately 2 10^{14} A/(m-rad)^2 at Sumitomo Heavy Industries in Japan with 1 nC of charge, a 9 ps FWHM long laser pulse and a normalized transverse emittance of 1.2 μm. The emittance was achieved by utilizing a temporally flat laser pulse which decreased the emittance by an estimated factor of 2 from the beam produced with a Gaussian pulse shape with an identical pulse length.

1. Introduction

Working group A was devoted to high brightness electron beam production. More than 25 people attended the group sessions during which 16 talks were presented. The balance of the time was spent in discussions on methods to create higher brightness beams and other related issues. The presentations and discussion topics could be categorized as follows; cathode physics, new photoinjector designs, computational modeling of high brightness beams, and new experimental methods and results.

In addition there were several talks during the plenary sessions related to high brightness beam production. In one of the plenary sessions P. Piot presented a summary talk on the status of current high brightness sources. Among the graphs he presented is Figure 1 which shows the beam brightness plotted as a function of injector frequency with the laboratory conducting the experiment labeled for each point. Two points of specific interest are the Sumitomo Heavy Industries (SHI) [1] and the ELSA 2 [2] experiments which both achieved normalized transverse emittances of 1 μm with 1 nC of charge. The low emittance in the SHI experiment was achieved by utilizing a 10 ps FWHM temporally flat laser pulse which decreased the emittance by an estimated factor of two from the beam produced with a Gaussian pulse shape with an identical length. More details of this experiment are discussed in section 5.1. The ELSA 2 experiment used a laser with Gaussian pulse shape but with a comparatively long pulse length of 60 ps due to the low rf frequency of 144 MHz. Both experiments have achieved the nominal emittance requirement of 1 μm with 1 nC of charge desired by many proposed light sources such as SASE FELs [3] and energy recovery linac based sources [4]. These are believed to be the first sources meeting the desired emittance specification.

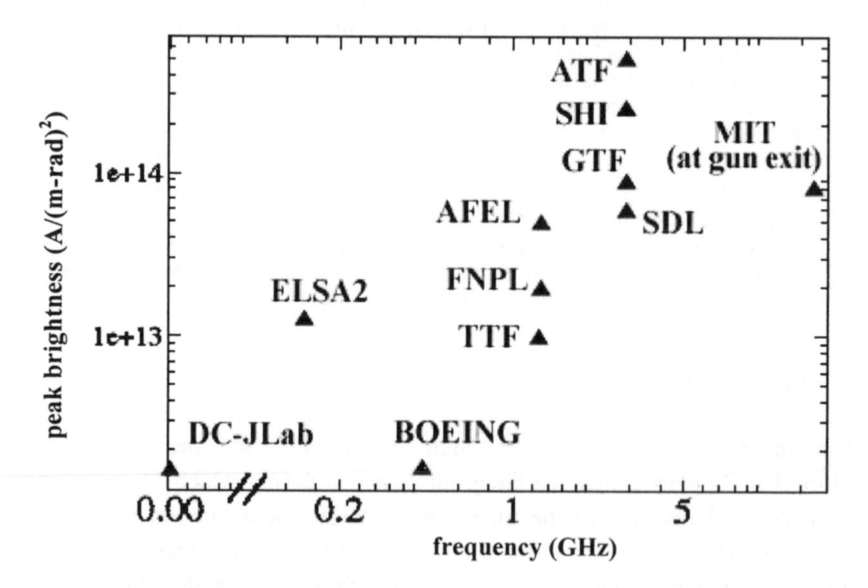

Figure 1. The plot shows peak beam brightness versus injector frequency. The brightest beams to date have been produced in S-band rf guns. The three brightest beams have all been produced with electrically identical 1.6 cell rf guns. The plot was provided by P. Piot.

Photoinjectors are currently the most common source used for high brightness beam production. One of the most popular designs is the BNL/SLAC/UCLA S-band gun with 1.6 cells. Electrically identical copies of this gun have been used to produce the three brightest beams plotted in Figure 1. However, while 2856 MHz appears to be the most common frequency, there is no experimental evidence that it is the optimal frequency. In fact many of the novel injector designs are pushing to either very low frequency such as pulsed DC guns or extremely high frequency such as laser wakefield accelerator based injectors. It was the consensus of the group that the new technologies will require many years of development before they can exceed the performance from an rf photoinjector. However, it should be pointed out that after more than a decade of existence, rf photoinjectors are just now approaching the peak brightness predicted by simulations. Thus even though the new injector designs will require years of development, they can and should be pursued. Nonetheless it appears that the best available source for high brightness beams in the near future is likely to be a standard rf photoinjector.

Despite the abundance of rf photoinjectors, few have been able to meet the peak brightness predicted by simulation. The experimentalists in the group held the laser beam quality primarily responsible for the electron beam not achieving the optimal performance suggested by simulation. The best simulation results are always achieved by assuming an ideal electron beam. Thus the laser beam illuminating the cathode must have a flat-top profile in both the longitudinal and transverse dimensions in order to produce a uniform density cylinder of charge and achieve the peak brightness from a photoinjector. In addition the shot to shot fluctuations such as energy, position and timing need to be minimized in order to optimize the electron beam brightness.

While laser experts can easily produce one of the aforementioned laser qualities, it is much more difficult to produce all the desired qualities simultaneously with minimal jitter. It was the consensus of the group that to date the optimal rf photoinjector drive laser has yet to be designed and built. The desired laser is certainly not commercially available. Too often photoinjector based accelerators do not properly prioritize the drive laser system and consequently the appropriate cost and effort is not spent on the laser design and construction. The result is often accelerator physicists attempting to improve the performance of the drive laser in order to improve the electron beam quality.

Cathode emission uniformity also limits the experimentally achieved beam brightness. Most groups address this problem with so called laser cleaning where the laser fluence is increased to the point that surface contaminants can be vaporized. Surface contaminants produce local variations in the work function and thus once they are removed the cathode emission exhibits less spatial variation. This procedure needs to be repeated on a regular basis to maintain the quantum efficiency and uniformity. Of course the ideal method would be to measure the electron distribution exiting the cathode and modify the laser spatial intensity to produce a uniform electron beam. While this idea has been around for many years it has yet to be experimentally implemented on an rf photoinjector. All the limiting factors described above must be addressed in order to achieve the peak available brightness from a standard rf photoinjector.

2. Cathode Physics

A physics limitations of beam brightness produced by a conventional rf photoinjector is the thermal emittance produced at the cathode. The total emittance exiting the injector is given approximately by the quadratic sum of the thermal emittance and the emittance produced in the injector due to time varying rf forces, space charge and other effects in the emittance compensation process. Measurement of the thermal emittance from a Cu cathode [5] resulted in an emittance of 0.6 μm per mm radius of the cathode which is approximately a factor of 2 larger than theoretical predictions. The total emittance in recent experiments have come close to the thermal emittance limit and therefore no substantial reductions in emittance can be achieved without reducing the thermal emittance. The group consensus was that the thermal emittance is rapidly becoming a beam brightness limiting factor. The emission process needs to be better understood to potentially reduce the thermal emittance. The penetration depth, laser polarization and surface roughness may all affect the thermal emittance. Unfortunately, to the author's knowledge, no experiments are in progress to directly measure the thermal emittance from photoinjector photo-cathodes.

A novel cathode type discussed at the workshop was needle cathodes. A needle cathode has a very small but sharp tip which greatly enhances the rf field and consequently concentrates the emission at the tip from either field emission or Schottky enhanced photo-emission. Figure 2 provided by J. Lewellen shows the on axis field in a 1.5 cell rf gun with a needle cathode. The field enhancement at the tip of a 600 μm diameter needle with 200 μm flat-top radius is a factor of three above the field with a standard flat cathode. The benefit of a needle cathode is low emittance due to the small source size but the small source size also produces relatively low charge. The simulations predict an emittance of 0.1 μm with 20 pC of charge and a 4 ps flat-top pulse shape. The brightness from the needle is higher than a beam produced in a conventional photo-cathode since the emittance falls faster than the square root of charge. The thermal emittance from the needle cathode is assumed negligible due to the small source size. M. Uesaka showed a Tungsten Needle photo-cathode experiment in a DC gun with a measured QE of 3%. There are plans to install the needle into an rf gun with calculations of over 100 A peak current with a 10 μm needle.

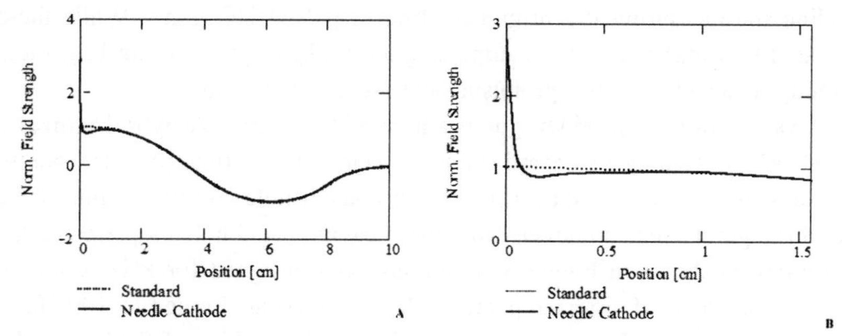

Figure 2. The on axis field versus position is plotted for both a needle and standard cathode. The axis has been expanded in (B) and it can be seen that the field at the needle tip is enhanced by a factor of three over a standard cathode. The plot was provided by J. Lewellen.

A fair amount of discussion was spent on cathodes but very few presentations were made on the subject. Most of the groups represented used metal cathodes but several groups utilize semiconductor cathodes. Metals still appear the most common choice often because of rf breakdown concerns, vacuum considerations and lifetime. M. Useaka reported a group in Japan is working on a metal transmission cathode where the laser is incident upstream of the cathode but no details were available. Nonetheless the group realized the work is interesting since it would allow for normal incidence laser operation without the need for a mirror installed near the electron beam creating undesirable wakefields. This can also be accomplished with grazing incidence laser operation. However, the laser optic corrections necessary to compensate for the elliptic beam and time dependent laser arrival time across the cathode often prevent groups from operating at grazing incidence.

3. Novel Injector Design

Pulsed DC electron guns [6] were one type of novel injector discussed. J. Luiten gave a plenary talk on the status of such guns. The pulsed DC guns described uses a photocathode inside a DC gun with GV/m fields for ns pulse durations to prevent breakdown. The electrons are accelerated over milimeter distances into the MeV energy range and exit through a hole in the anode. The pulsed DC gun allows higher gradients to be achieved in the gun and eliminates the time dependent rf focussing effects in conventional rf guns. It does not completely eliminate the time dependent space charge term although the higher gradient helps reduce its effect. Simulations indicate the possibility of producing sub-micron emittance beams and several groups are now working on building and measuring the brightness from a pulsed DC gun. While these guns are likely many years from surpassing the rf photoinjector beam brightness currently achieved they are a promising and exciting new source.

A variant of the pulsed DC gun is a hybrid DC rf gun. A hybrid utilizes a pulsed DC gun feeding a standard rf photoinjector. The normal photocathode in the photoinjector is replaced by a plate with a small hole and the output of the pulsed DC gun is injected directly into the rf cavity. A. Zholents presented the hybrid design shown in Figure 3 which has been designed for kHz repetition frequencies. Standard room temperature photoinjectors are limited to 100 Hz or lower repetition rates due to cooling limitations from the high rf fields used for acceleration. However, with a pulsed DC gun at the front end, the rf fields can be substantially decreased eliminating the cooling problem and still accelerate the beam to a few MeV energy exiting the hybrid gun. Preliminary simulations indicate this can be accomplished with minimal effect on the beam brightness.

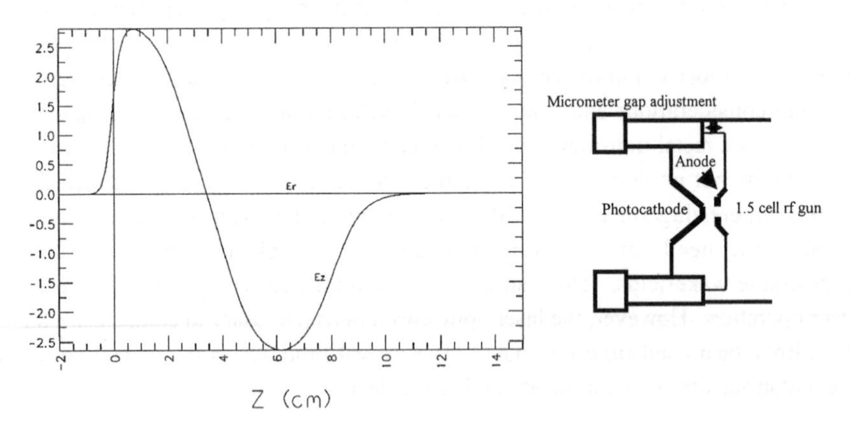

Figure 3. The on axis field as a function of longitudinal position for a Hybrid DC-rf gun is shown. A conceptual schematic for the photo-cathode and pulsed DC gun anode are also shown with an adjustable cathode position. The DC anode is immediately followed by a 1.5 cell S-band rf gun with a small hole in the standard cathode plate for electron injection from the pulsed DC gun. The figure was provided by A. Zholents.

As mentioned earlier S-band guns are not the only frequency of operation for rf photoinjectors. Lower frequency injectors have had success at producing high brightness beams such as ELSA 2 and the AFEL multicell rf gun [7]. These injectors suffer from undesireable effects such as limited gradient due to rf breakdown, increased thermal emittance, long pulses often requiring additional bunch compression and subsequent problems associated with compression such as CSR. However, they are easy to fabricate, require relatively low solenoid field strengths and the laser pointing and timing jitter requirements are easier to meet due to the larger beam size.

Higher frequency rf guns have the ability to reach higher gradients, but generate lower charge per bunch, are more difficult to fabricate, require higher solenoid fields for emittance compensation and require more stringent requriements on the laser such as timing jitter and pointing stability. Several X-band rf guns have been fabricated and tested [8-9]. At even higher frequency in the optical regime are laser wake-field based injectors. These injectors were not discussed in the injector group but were included in working group D which covered laser acceleration schemes. From the discussion it was not clear at which rf frequency the beam brightness is maximized.

A flat beam photoinjector design was presented by S. Lidia. The salient differences in the described flat beam injector and a standard photoinjector is the imposition of a solenoid field at the cathode, and a skew quadrupole triplet immediately following the booster cavity. The large correlation in the 4D xy phase space of the beam as it leaves the solenoid field is translated into a large ratio between the x and y emittances with the removal of the cross correlation (xy', x'y) terms. Figure 4 shows a schematic of the beamline and some recent

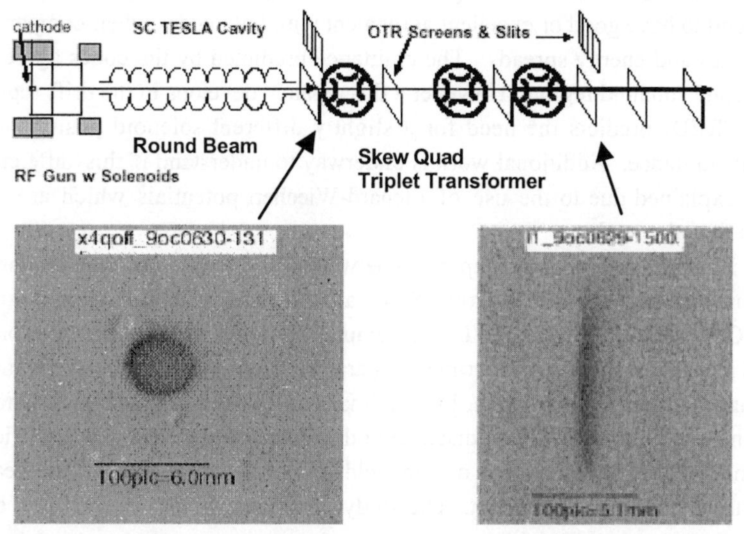

Figure 4. A schematic of the beamline for generating a flat beam is shown along with images of the electron beam before and after converting it to a flat beam with the skew quadrupole triplet. The figure was provided by S. Lidia.

experimental results of flat beam production at the A0 photoinjector facility at Fermilab. The source is intended for driving a recirculating linac with standard bending magnets and insertion devices to produce femto-second scale synchrotron radiation. The production of ultrafast radiation pulses involves a two-stage compression process. The first stage of compression compresses the electron beam from 20ps to 2ps and the second stage compresses the x-ray pulse from 2ps to 50-100fs in an asymmetric Bragg mirror pair. The last stage of compression requires a flat electron beam with low vertical emittance (<0.4 μm) and high horizontal emittance (~20 μm). Recent measurements with beams in the charge range of 200-300 pC and laser pulse lengths from 10-34 ps FWHM yield an emittance approximately 1-2 μm vertically and 30-50 μm horizontally.

4. Computational Modeling Issues

Several presentations were made on simulation codes. M. Quattromini talked about the code TREDI. TREDI is a fully 3D macroparticle Monte Carlo code devoted to the simulation of electron beams through rf guns, linacs, magnets and other beamline components where self fields are accounted for by means of Lienard-Wiechert retarded potentials. The code has been compared against PARMELA and HOMDYN using the SPARC project injector parameters. It was found to have good or excellent agreement with both codes when comparing beam sizes and energy spreads. The emittance predicted by the codes agree in the gun and solenoid region. However a discrepancy develops in the drift region where TREDI predicts the need for a slightly different solenoid position for optimal emittance. Additional work is underway to understand if this difference can be explained due to the use of Lienard-Wiechert potentials which are not used in the other codes.

C. Limborg presented a comparison between a slice emittance experiment at the Source Development Laboratory (SDL) at BNL and PARMELA simulation [10]. Comparison of all three Twiss parameters shows good agreement once careful measurements of all experimental parameters are input in the simulation including the thermal emittance, longitudinal and transverse profiles, solenoid map, and gun field map. It was discovered that decreasing the on axis gun field in the half cell by 20% from the design field was necessary to predict the beam sizes measured in the experiment. The study demonstrates that simulations can match the experiment if all the experimental parameters are carefully measured. Figure 5 shows the simulated and measured slice Twiss parameters at the end of the SDL linac for two different solenoid current values.

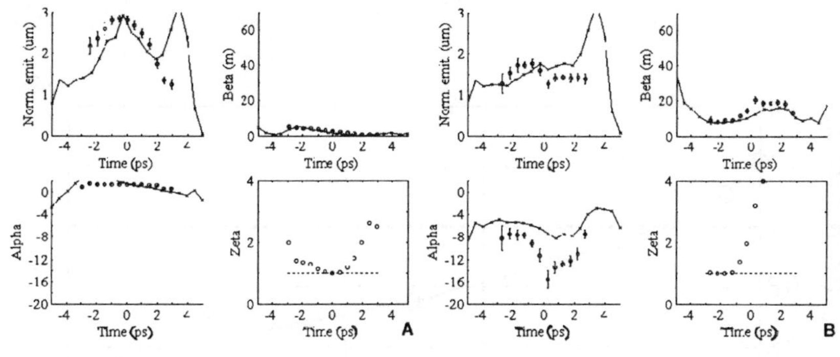

Figure 5. The simulated and measured Twiss parameters as a function of time are shown. The solenoid field is set at 98 A in (A) and 104 A in (B). Simulations are shown as solid lines and measured data as points. Zeta is commonly called the mismatch parameter. The figure was provided by C. Limborg.

C. Limborg also presented a PARMELA simulation of an emittance measurement experiment at the Gun Test Facility at SLAC. Typically, simulations report the emittance as the result of the calculation of the emittance from the full 6-D phase space distributions. However, typical experiments determine the emittance from measuring beam size only. In this experiment the rms beam sizes were determined by projecting a beam image in one plane and then truncating the wings at 5% of the peak value to minimize the effect in the data analysis of non-beam-related pixels. The Twiss parameters are then fit to the resulting rms beam sizes with the measured beamline parameters. In the simulation the beam sizes were computed identically but with different wing truncations values and the Twiss parameters fit to the results just like the experiment. Figure 6 shows a plot of the emittance as a function of the wing truncation level with 10,000 particles simulated in PARMELA and the particle number artificially increased during post processing to 100,000 particles to improve the statistics. As can be seen from the figure the total beam emittance is significantly underestimated with truncations greater than 2%. However, in real experiments the noise level is often above 2% and is the reason typical experiments use truncations as high as 10%. Thus, measured emittances can be less than the value computed by simulation. The study points out that care must be taken when comparing only the emittance value between simulation and experiment. A more accurate method is to compare all the Twiss parameters to be sure the simulation accurately describes the experiment since the emittance parameter by itself can be misleading. The study also demonstrates the need for a large dynamic range in the electron beam imaging system to avoid truncation errors.

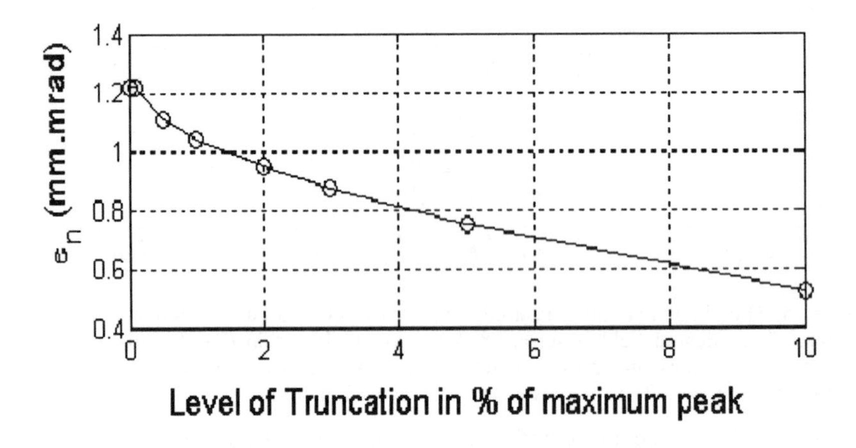

Figure 6. The computed emittance as a function of wing truncation level is shown for a typical set of beam parameters. The emittance computed by PARMELA from the full 6-D phase space distribution is 1.3 μm. The figure is provided by C. Limborg.

K. Floettmann pointed out that the phase space emittance is not constant in a drift under certain circumstances. The phase space beam emittance is the area of the ellipse in momentum-size space while trace space emittance is the area of the ellipse in divergence-size space. The two quantities are only exactly equal in a beam with no energy spread. With large energy spreads one finds that the emittance can vary significantly in a drift due to the development of the fan like or bow-tie phase space distributions. Figure 7 shows the phase space emittance and trace space emittance as a function of distance along the beamline. At various positions along the beam-line correlated energy spreads are added or removed and the effect on the two emittances can be observed. As can be seen in the figure the two emittances are not always identical and can be substantially different depending on where along the beamline they are computed. The study demonstrates that care is required when measuring the emittance of a beam with relatively large energy spread.

The discussion of simulation codes centered on additional effects that need to be included in simulations. Most of the codes used for photoinjectors do not include wakefields. Both transverse and longitudinal wakefields need to be included to explain some of the effects seen in measurements. Also, the simulations do include the Shottky enhanced emission off the cathode. This mean that the time dependent emission off the cathode due to the changing rf and space charge field is not modeled and neither is the saturation effect as the charge nears the space charge limit. Finally, the simulations should be using measured values for thermal emittance and more accurate rise times for the drive laser. Despite the effects that are not included in the simulations, the agreement between experiment and simulations has substantially improved in recent years.

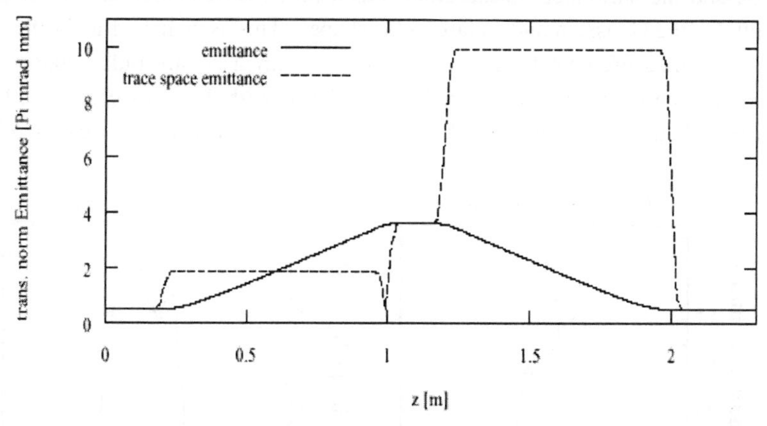

Figure 7. The normalized emittance and the normalized trace space emittance are plotted along a beamline. A beam with 5 mrad correlated beam divergence is launched at z = 0. A cavity at z = 0.2 m introduces a correlated energy spread of $2\ 10^{-3}$ which is compensated by another cavity at z = 1.0 m. A third cavity at z = 1.2 m introduces a correlated energy spread of $-2\ 10^{-3}$ which is again compensated by a fourth cavity at z = 2.0 m. The figure was provided by K. Floettmann.

5. New Experimental Techniques and Results

With the large number of photoinjectors in operation world wide, it was no surprise that a large number of presentations concentrated on experimental results. Section 5.1 describes several of the experiments most relevant to high brightness beams. In addition a surprising amount of time was spent discussing data analysis techniques. The importance of proper data analysis for comparison with simulation values was pointed out by C. Limborg in the course of a study comparing experimental results with PARMELA simulations from several different photoinjectors. Errors associated with ignoring space charge in the data analysis were also pointed out. Section 5.2 describes the data analysis issues and recommendations.

5.1. *Techniques and Results*

M. Uesaka presented results from Sumitomo Heavy Industries with emittance measurements as a function of laser pulse length and shape [1]. Nearly square or flat-top pulse shapes were generated using a spectral phase mask consisting of a liquid crystal modulator in the mid plane of a grating compressor. The pulse shapes were measured with a streak camera. Figure 8 shows both the laser pulse shape and the emittance measurements as a function of charge and laser pulse length for the Gaussian and square pulse shape. This is believed to be the first emittance measurement from a photoinjector with a square pulse shape drive laser. The emittance was measured to be 1.2 μm with 1 nC of charge with a 9 ps FWHM square laser pulse. The emittance is estimated to be a factor of two less

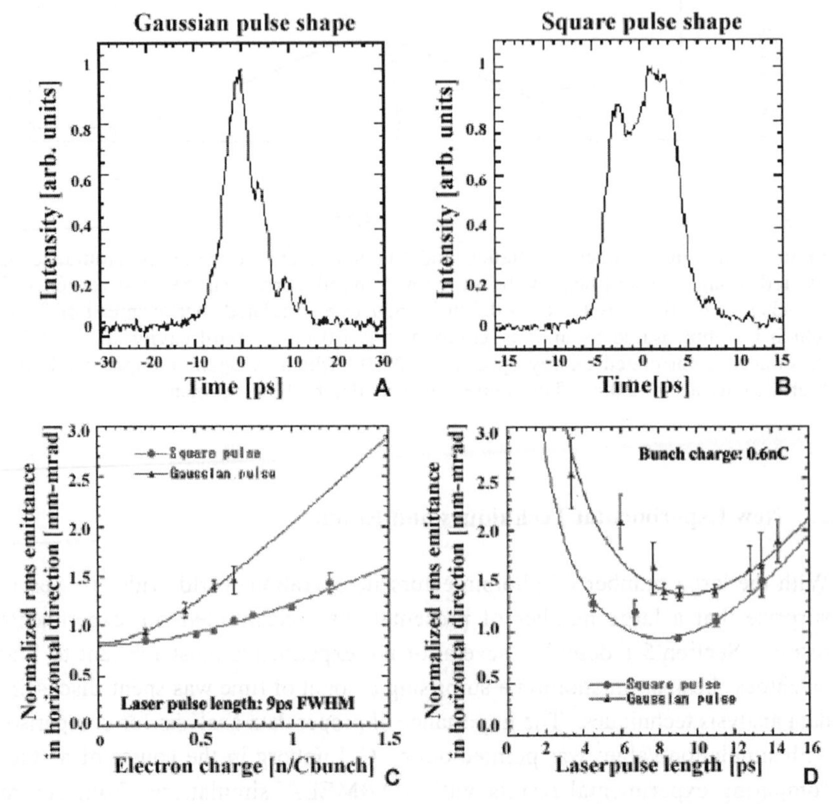

Figure 8. The measured laser pulse shape in (a) and (b) for the Gaussian and square pulse respectively with pulse length 10 ps FWHM. The emittance is plotted in (c) and (d) as a function of charge and laser pulse length respectively for both Gaussian and square pulse shapes. The figure was provided by J. Yang.

than the beam produced with an identical length Gaussian pulse. The thermal emittance can be extracted from the emittance as a function of charge curve and is estimated to be approximately 0.6 μm with an approximately 1 mm diameter Cu cathode.

J. Schmerge presented measured slice emittance results. Typical emittance measurements report only the projected emittance although simulations consistently show the emittance to be a function of longitudinal position. By introducing a time-energy correlation into the beam by appropriately setting the phase of the linac downstream of the gun, the emittance was measured as a function of time in a dispersive section using a quadrupole scan technique. The slice emittance measurements yield significantly more information than projected emittance alone since they reveal the presence of any longitudinal correlations and provide the means for removing them. Measurements were performed with 300 pc of charge with a 2 ps FWHM laser pulse as measured with a streak camera. The gun field was 110 MV/m with 30° extraction phase and the linac phase was set at +5° to appropriately chirp the beam exiting the linac at 30 MeV. Figure 9 shows the emittance as a function of slice for different values of the solenoid field. The projected emittance is plotted as slice 0 and the time axis is estimated at 200 fs/slice. The measurements show slice emittances as low as 2 μm with a Gaussian pulse shape and peak current of 150 A.

In addition to the transverse phase space measurements, a measurement of the longitudinal phase space was also performed. The longitudinal phase space was determined by measuring the energy spread at the exit of the linac as a function of linac phase. The technique is analogous to a quadrupole scan in transverse space. However, a quadratic and cubic term were added to the phase space ellipse description to account for the sinusoidal rf acceleration term and longitudinal space charge force respectively. The distribution at the linac entrance is mapped through the linac and is then fit to the measured energy spread exiting the linac. A large correlated energy spread is observed at the gun exit and is also shown in Figure 9. The energy spread is 8% FWHM and the longitudinal emittance is 4.6 keV ps.

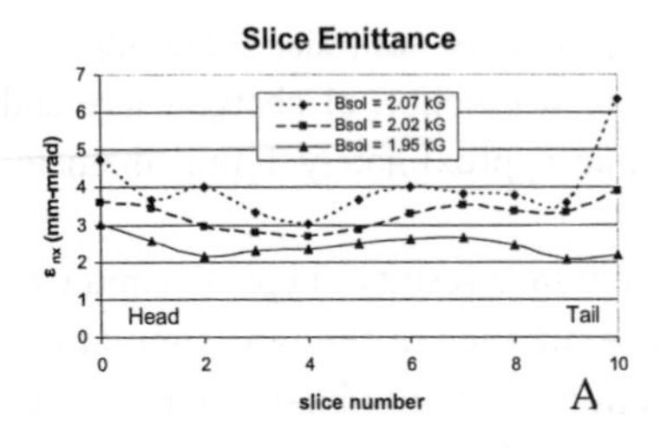

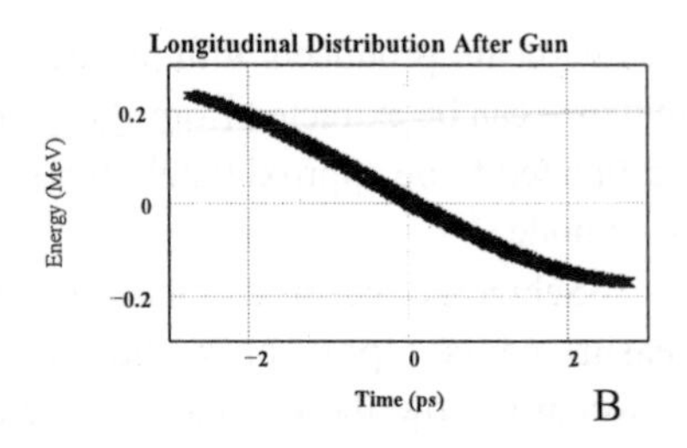

Figure 9. The emittance as a function of temporal slice is plotted for three different solenoid fields in (A). Slice 0 is the projection with slice 1 the head and slice 10 the tail of the beam. Each slice is approximately 200 fs long. The measured longitudinal distribution between the gun and linac is shown in (B). The beam has an approximately 8% correlated energy spread. The figure was provided by J. Schmerge and D. Dowell.

Additional longitudinal phase space measurements are under way at DESY Zeuthen on a 1.5 cell L-band gun with coaxial rf coupler. D. Lipka described the proposed experiment to use a Cherenkov radiator in a dispersive region with a streak camera to measure the longitudinal phase space. The experiment will allow the simultaneous measurement of energy spread, bunch length and correlations with better than 1 ps resolution. Figure 10 shows a schematic of the proposed experiment. Transverse emittance measurements directly exiting the gun are also planned using a pepper pot technique.

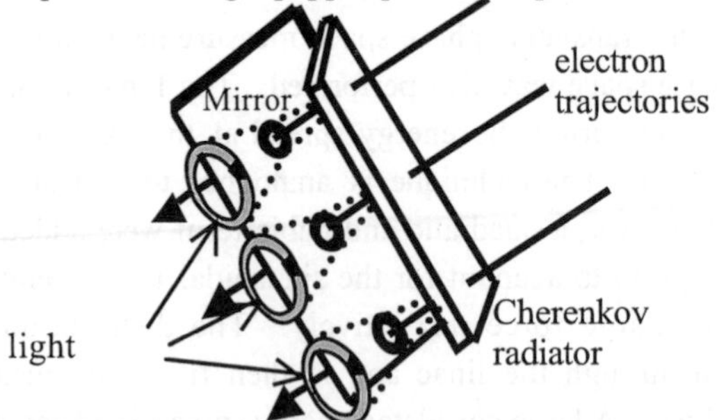

Figure 10. A schematic of the proposed Cherenkov radiator for measuring the longitudinal phase space. The light produced by the Cherenkov radiatior reflects of the mirror and is then incident into a streak camera. The figure was provided by D. Lipka.

Although the laser temporal and spatial profiles were identified as limiting factors in the achieved brightness from most existing rf photoinjectors, surprisingly little work was presented on improving the performance of the laser. H. Dewa did show a measured 33% decrease in the emittance by using a microlens array homogenizer to flatten the laser spatial profile. To date only one group has measured the emittance with a temporal flat-top laser pulse and no measurement has yet been performed with both temporal and spatial flat-top profiles. D. Palmer presented timing jitter measurement on a Ti:sapphire laser oscillator. The measured rms jitter between the laser and rf source was less than 500 fs.

5.2. *Data Analysis*

One point of concern for the experimentalist in the group was data analysis. The algorithm used to convert beam images into spot sizes and the subsequent fitting routine can have a large impact on the measured emittance. The difference between rms and Gaussian fits to the beam spot size can be larger than 20% leading to almost 50% difference in the reported emittance from identical data. As shown in section 4 the amount of truncation when attempting an rms beam size measurement can lead to significant errors. Also, the beam size calculated from a single line out of an image versus a projection can be substantially different depending on the beam shape and orientation. The method used for emittance measurements such as quadrupole scan, pepper pot, two screen, multiple screen and single shot techniques all have various errors and different assumptions built into the analysis which can lead to discrepancies between measurements on identical beams. Finally, almost all of the emittance measurements reported in the literature assume that space charge is negligible over the length of the measurement. This is not always true and can have a large impact on the result. Thus comparing results between different machines and techniques is often misleading.

J. Rosenzweig demonstrated the effect of space charge on the emittance fit in a quadrupole scan. Typical quadrupole scans assume that the space charge term is small compared to the emittance term in the envelope equation. However, at relatively low energies, high charges or large beam sizes the space charge term can be significant and may have a large effect on the measured emittance. In a quadrupole scan and with no space charge, the square of the beam size is a parabolic function of the quadrupole focal length in the thin lens approximation. J. Rosenzweig showed that the beam size is also parabolic with focal length when the emittance term is neglected and only the space charge term is considered. The beam size as a function of focal length is no longer symmetric about the minimum spot size when both terms are significant. The asymmetric shape of the curve is a clear sign of the space charge effect. It should be pointed out that space charge actually dominates when the beam size is large since the space charge term is inversely proportional to the beam size and the emittance term is inversely proportional to the beam size cubed. Thus space charge actually dominates when the beam is large with all other parameters being constant. Figure 11 shows the beam size as a function of quadrupole focal length with space charge only and with both the space charge and emittance terms included. As can be seen the curve with both terms is clearly asymmetric. However, despite the asymmetry the curve can still be fit without the space charge term although the fit will result in an error in the emittance.

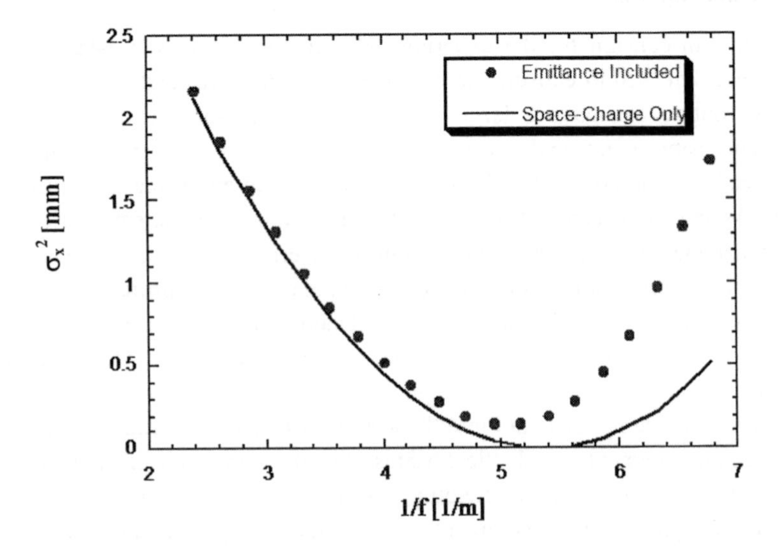

Figure 11. A plot of beam size versus quadrupole focal length is shown. The beam size is computed from the envelope equation. One curve uses only the space charge term and neglects the emittance term while the second curve includes both terms. The figure was provided by J. Rosenzweig.

In a typical quadrupole scan the emittance from a fit including space charge will be lower than the emittance obtained by neglecting space charge. This is due to the fact that the space charge force always acts to increase the beam size along the drift and the result is a larger beam at the screen than would be expected without space charge. In order to keep the beam size at the measured value when including space charge, the emittance used in the fit must be reduced. However, in multiple screen measurements with focusing elements between screens this is no longer necessarily true. Thus it is even more important to be sure space charge is negligible in multiple screen emittance measurements. Regardless of technique, space charge needs to be considered when analyzing emittance measurements.

The group agreed that in order to accurately compare results between groups and simulation, certain details of the measurement technique and analysis need to be reported. The general consensus was that full projected images and rms spot sizes should be used since the envelope equation is derived for rms quantities. The truncation level used in the rms calculation of beam size should be as low as possible and also reported in the literature. Finally the effects of space charge should be included in the analysis.

Acknowledgments

The author would like to acknowledge all the participants in working group A for their valuable and timely presentations as well as the figures they provided that were used in this paper. The participants also posed numerous interesting questions, comments, and suggestions during the course of the workshop. In addition the author appreciated the lively debate during the discussion phase and the helpful suggestions in the summary preparation period. Although everyone assisted with the group summary, I would specifically like to acknowledge the assistance of J. Lewellen and C. Limborg for their aid preparing and editing the working group A summary. Finally I must acknowledge the help of A. Mueller with the graphics.

References

1. J. Yang, F. Sakai, T. Yanagida, M. Yorozu, Y. Okada, K. Takasago, A. Endo, A. Yada and M. Washio, *Jrn. App. Phys. Rev.* **92**, 1608 (2002).
2. J.-G. Marmouget, A. Binet, Ph. Guimbal and J.-L. Coacolo, "Present performance of the low-emittance, high-bunch charge ELSA photo-injected linac" *Submitted to EPAC 2002*, Paris, June (2002).
3. See the LCLS home page at http://www-ssrl.slac.stanford.edu/lcls/
4. G.R. Neil, *Nucl. Instr. Meth. A* **483**, 14 (2002).
5. W.S. Graves, L.F. DiMauro, R. Heese, E.D. Johnson, J. Rose, J. Rudati, T. Shaftan and B. Sheehy, *Proceed. PAC 2001* 2227 (2001).
6. K. Batchelor, J.P. Farrell, G. Dudnikova, I.Ben-Zvi, T. Srinivasan-Rao, J. Smedley and V. Yakimenko, *Proceed. EPAC 1998* 791 (1998).
7. S.M. Gierman, in *The Physics of High Brightness Beams*, World Scientific, 511 (2000).
8. W.J. Brown, K.E. Kreischer, M.A. Shapiro, R.L. Temkin and X. Wan, in *The Physics of High Brightness Beams*, World Scientific, 454 (2000).
9. A.E. Vlieks, G. Caryotakis, R. Loewen, D. Martin, A. Menegat, E. Landahl, C. DeStefano, B. Pelletier, N.C. Luhmann, Jr., *SLAC-PUB-9354,* (2002).
10. W.Graves et al. "Experimental study of sub-ps slice electron beam parameters in a photoinjector" *Submitted to Phys. Rev. ST-AB*, (2002).

WHAT BRIGHTNESS MEANS

C. A. BRAU

Vanderbilt University, Nashville, TN 37235

Abstract: The relation between the emittance and the brightness of a charged-particle beam is reviewed, and several examples are used to illustrate the relationship for typical beams.

1. Introduction

Besides being part of the name of the conference in which we are participating, the concept we call the brightness of an electron beam gives us a useful way of summarizing the complex properties of the electron beams we create in the laboratory. We can use the measured brightness to compare our beams with each other and to estimate the usefulness of our beams for a variety of applications. Unfortunately, while the brightness of an electron beam is a well-defined quantity, experimental values are frequently reported erroneously, with an error that often exceeds two orders of magnitude.[1,2,3,4,5] It is particularly unfortunate since the error generally leads to an overestimate, and many of us will find our quoted brightness reduced by two orders of magnitude when the error is corrected. It is, however, important that our accounting and reporting practices not make the high-brightness electron-beam community the ENRON of the accelerator world, especially since we ourselves will have to use the guns we design to build the lasers, colliders, and light sources we have promised.

It is intuitively clear that the total current I is an important parameter of an electron beam, but it is also important to be able to collimate and focus the beam. The focus of an electron beam, which resembles that of an optical beam, is illustrated in Figure 1. We know from experience that as we try to focus the beam to a smaller area A, we must increase the convergence (and divergence) solid angle Ω, and that as we do so, the product $A\Omega$ remains essentially constant (until aberrations become important). These considerations led von Borries and Rhushka in 1939 to introduce the so-called brightness of the beam, which they defined as [1]

$$B = \frac{I}{A\Omega}. \tag{1.1}$$

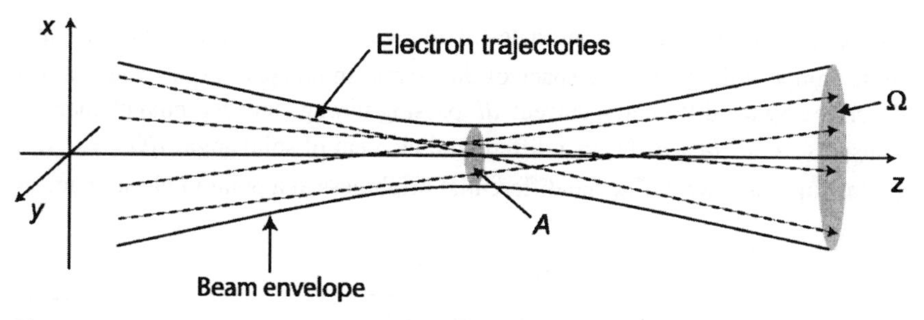

Figure 1. Geometry of an electron beam at the focus.

We also know from experience that if we use an aperture to filter out the worst part of the beam, we can focus it to a smaller area A, or with less divergence Ω, but only at the expense of the total current I. Typically, as the filtering is applied, the brightness (1.1) approaches a constant, which if the filtering is done correctly is the peak brightness $\hat{B}$ of the beam. In addition, we can focus the beam separately in each transverse direction, and in this case the product xx' remains constant, where x is the transverse dimension of the beam and $x' = dx/dz$ the divergence of the beam in the axial (z) direction. This leads to the concept of the transverse emittance ε_x (or ε_y in the other transverse direction) of the beam. As the beam is filtered, the emittance becomes smaller without limit (ignoring the quantum-mechanical uncertainty limit).

Because it summarizes so many of the detailed properties of an electron beam, the brightness has become a useful figure of merit for characterizing the quality of an electron beam. For example, it can be shown that under a variety of experimentally important circumstances, the spectral brilliance of a laser-synchrotron x-ray source is linearly proportional to the brightness of the electron beam. In addition, it can be shown that when the effective energy spread of the electron beam due to transverse electron motions is accounted for, the gain of a high-gain free-electron laser is ultimately limited by the brightness. Other parameters, such as the Budker parameter and the Pierce parameter, are also useful in certain circumstances. Nevertheless, brightness is the name of this conference, and we focus the remainder of this discussion on that parameter. The reason this discussion is necessary is that we generally do not measure the brightness directly, but measure instead the transverse emittance and then try to infer the brightness. It is actually the relation between emittance and brightness that requires some discussion.

2. Mathematical Definition of Brightness

In the following we ignore the longitudinal properties (energy and time, or phase) of the beam, and restrict our attention to the transverse properties of the beam. The transverse coordinates of an electron are x, x', y, and y', and these compose the 4-D phase space of the beam. In this phase space the beam is described by the differential current dI passing through a differential element of transverse area $dA = dxdy$, in a differential element of solid angle $d\Omega = dx'dy'$. Following Lenz, we define the differential brightness at a point in phase space as [2]

$$B = \frac{d^2 I}{dAd\Omega} = \frac{d^4 I}{dxdx'dydy'}. \tag{2.1}$$

That is, the differential brightness is just the current density in the 4-D transverse phase space. It is worth noting that the definitions (1.1) and (2.1) are universal throughout accelerator physics, electron microscope theory, and wherever electron beams are used. [4,5,6,7,8] As defined in (2.1), the brightness is not an overall property of the beam, although we can (and sometimes do) define the average brightness over some portion of the beam, as in (1.1). When we filter the beam to improve its overall brightness, the limit approached is just the peak brightness of the beam, which corresponds to the highest density in the 4-D phase space.

The emittance, on the other hand, is an overall property of the beam. It can be improved (reduced) by filtering the beam, but does not approach a limit (other than the quantum mechanical limit) as the filtering is increased. Specifically, the emittance is a measure of the size of the projection of the 4-D distribution on the $x - x'$ or $y - y'$ planes. This is indicated in Figure 2, where

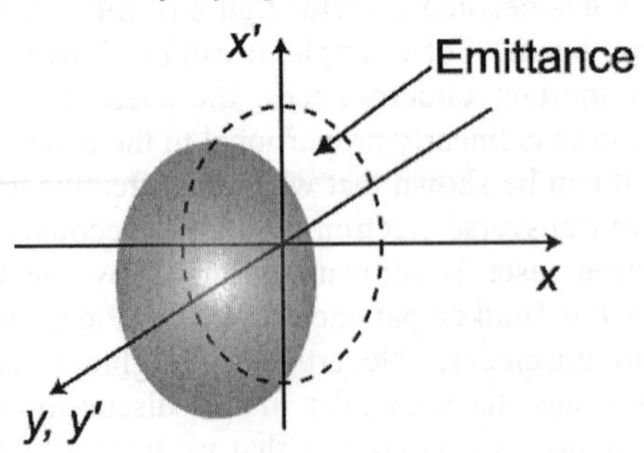

Figure 2. Emittance as the projection of 4-D phase space onto 2-D phase space.

we have taken some liberties by illustrating the projection of a 3-D distribution on the $x - x'$ phase plane. This is the best we can do, living in a 3-D world. For a hard-edged beam, the emittance is defined as the area of the projection, indicated by the dotted line in Figure 2, divided by π. For an upright elliptical distribution typical of a beam at its focus, the emittance is

$$\varepsilon_x = \frac{A_{xx'}}{\pi} = XX', \tag{2.2}$$

where X and X' are the semiaxes of the ellipse in the $x - x'$ plane, as shown in Figure 2. This is the classical definition of the emittance, and we call it the edge emittance ε_{xedge}.[1] For an irregular distribution in phase, the area $A_{xx'}$ is generally taken to be that of the smallest ellipse that encloses the current.[5] Historically, the emittance was often defined as the projected area itself, which is larger than (2.2) by the factor π. To avoid uncertainty in the definition, it became customary to report the emittance with the factor π shown explicitly, as in 4π mm-mradians.

In reality, of course, we work with beams whose edges are not so sharply defined, and as we have learned to deal experimentally with more detailed knowledge of our beams it has become customary to use the rms emittance of the beam,

$$\varepsilon_{xrms} = \sqrt{\langle x^2 \rangle \langle x'^2 \rangle - \langle xx' \rangle^2}, \tag{2.3}$$

where the brackets $\langle \ \rangle$ indicate an average over the beam, originally introduced by Sacherer.[4] As introduced at the same time by Lapostolle, the definition also included a factor of four, for reasons that will become clear shortly, and this is usually called the effective emittance.[4,6,9] A further factor of π can be included to make the definition congruent with the actual phase space area.

It is generally found that the collimation of the beam can be improved by accelerating the beam to higher energy. To remove this somewhat artificial change in the beam properties from the definitions, we introduce the) normalized) emittance

$$\varepsilon_{nx} = \beta\gamma\varepsilon_x, \tag{2.4}$$

and normalized brightness

$$B_n = \frac{B}{\beta^2\gamma^2}, \tag{2.5}$$

where γ and β are the usual relativistic parameters of the electrons in the) beam. If the acceleration is done carefully, the normalized emittance and normalized brightness are preserved.

3. Examples

Since the brightness is a detailed description of the electron beam in transverse phase space, whereas the emittance is an average over the distribution, the relationship between the emittance and the peak or average brightness depends on the details of the distribution of current in 4-D phase space. In this section we consider three representative models. All three models have ellipsoidal symmetry in 4-D phase space. However, to simplify the mathematics we assume in the following that the distributions are spherically symmetric. Since the transverse coordinates x, x', y, and y' can always be scaled to make the distribution spherical and the scaling restored at the end, this does not restrict the generality of the results.

K-V distribution: Before computers became common, the simplest 4-D phase space distribution (from the analytic standpoint) was that introduced by Kapchinskij and Vladimirskij.[10] It consists of a 4-D ellipsoidal shell of current, hollow inside. It may be shown that the projection of this distribution on the $x-x'$ or $y-y'$ plane produces a uniformly filled ellipse in the 2-D phase plane.[4] For a spherical distribution of radius R, a few tedious integrals show that the volume of the 4-D sphere is

$$V_4 = \frac{\pi^2}{2} R^4 . \tag{3.1}$$

Since the distribution is singular in 4-D phase space, it is not meaningful to speak of the brightness of the beam. However, we can define the average brightness as the total beam current divided by the volume of the sphere,

$$\bar{B} = \frac{I}{V_4} = \frac{2I}{\pi^2 R^4} . \tag{3.2}$$

Since the projection of the distribution on the $x-x'$ plane is a circle, the edge emittance is

$$\varepsilon_{xedge} = R^2 , \tag{3.3}$$

with a similar result for ε_{yedge}. From (3.2) and (3.3) we see that the average brightness of a beam with a K-V distribution is

$$\bar{B}_{K-V} = \frac{2I}{\pi^2 \varepsilon_{xedge} \varepsilon_{yedge}} = \frac{2I}{\varepsilon_{\pi xedge} \varepsilon_{\pi yedge}} , \tag{3.4}$$

where $\varepsilon_{\pi xedge} = \pi \varepsilon_{xedge}$ is the edge emittance defined as the phase space area, with the factor π included. Note that this famous, oft-quoted formula is valid only for the edge emittance, and only when the factor π is included. Similar remarks apply to the normalized emittance and brightness.

Since the projection of the K-V distribution on the $x-x'$ is a uniformly filled circle, the rms emittance is, by symmetry,

$$\varepsilon_{xrms} = \sqrt{\langle x^2\rangle\langle x'^2\rangle - \langle xx'\rangle^2} = \langle x^2\rangle = \frac{\int_0^R r^3 dr \int_0^{2\pi} \cos^2\theta d\theta}{\int_0^R r dr \int_0^{2\pi} d\theta} = \frac{R^2}{4}, \tag{3.5}$$

with a similar result for ε_{yrms}. In fact, it was this factor of ¼ that led Lapostolle to introduce a factor of 4 into his definition of the rms emittance.[9] From (3.2) and (3.5) we see that in terms of the rms emittance, the average brightness of a K-V distribution is

$$\overline{B}_{K-V} = \frac{1}{16\pi^2}\frac{2I}{\varepsilon_{xrms}\varepsilon_{yrms}}. \tag{3.6}$$

This differs from (3.4) by a factor of 16 (or $16\pi^2$, depending on which form of the equation we use).

Water-bag distribution: The next simplest (and more realistic) model for the distribution of an electron beam in 4-D phase space is the so-called water-bag model. In this model, the distribution consists of a uniformly filled ellipsoid. The peak and average brightness are the same, so in the spherical case

$$B_{W-B} = \frac{I}{V_4} = \frac{2I}{\pi^2 R^4}. \tag{3.7}$$

The projection of the water bag on the $x-x'$ plane is still a circle, but the circle is not uniformly filled. Instead, the density is greatest at the center, with a parabolic profile that vanishes at the edges.[4] The rms emittance is, by symmetry,

$$\varepsilon_{xrms} = \sqrt{\langle x^2\rangle\langle x'^2\rangle - \langle xx'\rangle^2} = \langle x^2\rangle = \frac{\int x^2 dx \int dx' \int dy \int dy'}{\int dx \int dx' \int dy \int dy'} = \frac{R^2}{6}, \tag{3.8}$$

where the (somewhat tedious) integrals are over the volume

$$x^2 + x'^2 + y^2 + y'^2 < R^2, \tag{3.9}$$

with a similar formula for ε_{yrms}. From (3.7) and (3.8) we see that the brightness of a beam with a water-bag distribution is

$$B_{W-B}=\frac{1}{36\pi^2}\frac{2I}{\varepsilon_{xrms}\varepsilon_{yrms}}. \tag{3.10}$$

Gaussian distribution: The next simplest (and more realistic) model for an electron beam is a Gaussian distribution in 4-D phase space. In the spherically symmetric case this may be expressed

$$B=\hat{B}\exp\left(-\frac{x^2+x'^2+y^2+y'^2}{R^2}\right), \tag{3.11}$$

where $\hat{B}$ is the peak brightness at the center of the beam, and R is the characteristic radius. The total current is

$$I=\int dx\int dx'\int dy\int dy'B=\pi^2R^4\hat{B}, \tag{3.12}$$

where the integrals are over all phase space. The projection of this distribution on the $x-x'$ plane is still Gaussian, and the rms emittance is, by symmetry,

$$\varepsilon_{xrms}=\sqrt{\langle x^2\rangle\langle x'^2\rangle-\langle xx'\rangle^2}=\langle x^2\rangle=\frac{\int x^2dx\int dx'\int dy\int dy'B}{\int dx\int dx'\int dy\int dy'B}=\frac{R^2}{2}. \tag{3.13}$$

Combining these results, we see that in terms of the rms emittance the peak brightness of a Gaussian beam is

$$\hat{B}_G=\frac{1}{8\pi^2}\frac{2I}{\varepsilon_{xrms}\varepsilon_{yrms}}. \tag{3.14}$$

This is the peak brightness (current per unit area per unit solid angle) that is found experimentally when the beam is passed through apertures that transmit only the most intense part of the beam.

In conclusion, it would seem that the Gaussian distribution is the best approximation to the 4-D current distribution of a real electron beam, and that (3.14) is the best formula for estimating the brightness from the current and the rms emittance. However, with modern instrumentation and analysis it should be possible in the future to determine the brightness directly from the experimental measurements.

References

1. B. von Borries and E. Rushka, Z. Tech. Phys. **20**, 225 (1939).
2. F. Lenz, *Theoretische Undersuchungen über die Ausbreitung von Elektronenstrahbündeln in rotationssymmetrischen elektrischen und magnetischen Feldern*, Habilitationsschrift, Aachen (1957).
3. F. Sacherer, IEEE Trans. Nucl. Sci. **NS-18**, 1105 (1971).

4. J. D. Lawson, *The Physics of Charged-Particle Beams*, 2nd edition, Clarendon Press, Oxford, 1988.
5. S. Humphries, *Charged Particle Beams*, John Wiley & Sons, New York, 1990.
6. M. Reiser, *Theory and Design of Charged Particle Beams*, John Wiley & Sons, New York (1994).
7. P. W. Hawkes and E. Kasper, *Principles of Electron Optics*, Volume 2, Academic Press, London (1989).
8. J. Orloff, *Handbook of Charged Particle Optics*, CRC Press, New York (1997).
9. P. M. Lapostolle, IEEE Trans. Nucl. Sci. **NS-18**, 1101 (1971).
10. I. M. Kapchinskij and V. V. Vladimirskij, "*Limitations of proton beam current in a strong focusing linear accelerator associated with the beam space* charge," Proc. Int. Conf, High Energy Accelerators, p.274, CERN, Geneva (1959).

BEAM QUALITY AND STABILITY IMPROVEMENTS FOR A SINGLE-CELL PHOTOCATHODE RF GUN

H. DEWA, A. MIZUNO, T. TANIUCHI, H. TOMIZAWA, T. ASAKA,
T. KOBAYASHI, S. SUZUKI, K. YANAGIDA, H. HANAKI

Japan Synchrotron Radiation Research Institute (SPring-8)
1-1-1 Kouto,
Mikazukicho, Sayogun, Hyogo 679-5198, Japan
E-mail: dewa@spring8.or.jp

M. UESAKA

University of Tokyo
2-22 Shirakatashirane,
Tokaimura, Nakagun, Ibaraki 319-1188, Japan
E-mail: uesaka@tokai.t.u-tokyo.ac.jp

Laser beam with high stability and quality to illuminate the photocathode is of great importance to generate an electron beam of high brightness from the photocathode RF gun system. In order to stabilize the electron beam, the laser system and the synchronization between the laser pulse and RF signal were improved. The synchronization with the time jitter of 1.2 ps was realized with a frequency feedback of the Ti:Sapphire oscillator and the RF generation from the laser pulses. Considering the emittance growth caused by the space charge effect, the ideal laser profile should be cylindrical with uniform density. The disturbed profile of the laser pulse coming from the third harmonics generator of Ti:Sapphire laser was homogenized as quasi-flattop with a microlens array. The horizontal normalized rms emittance of the electron beam was reduced to 2π mm-mrad for a charge per bunch of 0.1nC due to the improvement of spatial profile. The laser injection angle of 66 degrees causes a laser spot image deformation to an ellipsoidal image on the cathode. This leads to a lag of arriving time on the cathode as well as the expansion in horizontal diameter of the laser beam. Deformation of laser profile to an ellipsoidal image was also carried out with a pair of cylindrical lenses to get a round image on the cathode. The transverse beam emittance was improved to 5.4 πmm-mrad at 0.2 nC with the deformation optics, while it was 6.6 πmm-mrad at 0.1 nC without them.

1. Introduction

Future light sources using electron accelerators such as spontaneous self-amplified free electron laser or laser Compton scattering require electron guns that can generate a high brightness electron beam, namely, an electron beam with high charge and low emittance. So far thermal electron guns have been used as conventional electron injectors of linac. However, owing to a low energy injection of electrons, the space charge effect is so strong that the emittance of the electron beam increases before the electron beam arrives at the entrance of the linac. To reduce the space charge effect, the electron beam needs to be accelerated over MeV in a short distance. As a promising candidate of such a high-energy electron gun, a photocathode RF gun has been widely developed for generating a high brightness beam[1-4]. The photocathode RF gun can accelerate electrons emitted from the photocathode, which is illuminated with a UV-laser, to an energy as high as 4 MeV or more in an acceleration length of 2-3 cm. As the electron beam is accelerated rapidly to the order of MeV, the space charge effect is much smaller than that of the thermal electron gun. Since the pulse length of illuminating laser basically determines that of the electron beam, the electron beam generation with the bunch length of several picoseconds is possible without any other RF bunching system.

The quality of the electron beam strongly depends on the laser quality. The ideal laser profile is like a cylindrical distribution with a uniform density because the electron beam with a cylindrical distribution of an infinite length does not increase the transverse beam emittance. It is well known that the transverse electric field of the electrons is linear in such a distribution. The spatial and temporal distributions of the laser are, however, usually different from the ideal conditions. Therefore, the spatial or temporal profile shaping is a key technology to generate a low-emittance electron beam. In addition, the time jitter of the laser pulse should be minimized as small as possible because the laser pulse should be synchronized with the RF field of 2856 MHz fed into the RF gun cavity. The electron beam can be accelerated properly in the RF cavity only in a narrow range of the RF phase around an optimum phase. Furthermore the laser system should be very stable for generating the stable electron beam. Pulse-to-pulse laser stability determines the electron beam quality such as the beam emittance. Long-term stability for more than a month is necessary without any manual adjustment of laser if it is used for light source services for many users. The light source facilities such as SPring-8 are usually operated for 24 hours and longer than 2 or 3 weeks.

A single-cell photocathode RF gun has been developed in SPring-8 since 1996 [5-13]. For fundamental study of RF guns and the high gradient acceleration, we did not adopt a multi-cells cavity but a single-cell cavity [5,6]. The cathode is made of copper, which is a part of the cavity wall, and generates electrons by illuminating laser pulse with a wave-length of 263 nm, which is the third harmonics of Ti:Sapphire laser. The commercial laser system, however, was insufficient for our technical requirements such as the long stable operation, energy stability, laser profile, synchronization with an RF signal. For these reasons, the improvements of the laser and synchronization system have been carried out. These improvements increased the quality and stability of the electron beam, which were confirmed by the beam emittance measurement. The improvements of the laser beam profiles and the results of the emittance measurements are described in detail in this paper.

2. SPring-8 RF gun system

2.1. *RF gun cavity*

The Spring-8 RF gun is a single cell pillbox-type cavity with two symmetric RF ports. The schematic view and the parameters of the cavity are shown in Fig.1 and Table 1, respectively. One port is for RF input and the other for output to a dummy load. The dummy load was added to damp the loaded Q value of the cavity, which shorten the RF filling time of the cavity. It is expected to increase the limit of the maximum field strength and decrease the dark current. Such a symmetric design has an advantage to generate symmetric electric field in the RF cavity, therefore the electrons emitted from the cathode are accelerated straight forward without being bent with an asymmetric electric field at the center of the cathode. As there is only one cell, it is not necessary to consider the RF phase matching between cells. The UV-laser is incident through an input window at an incident angle of 66 degrees to the cathode surface. Considering the lower beam emittance, non-zero incident angle is not appropriate. The laser injection optics with the zero incident angle is also being preparing.

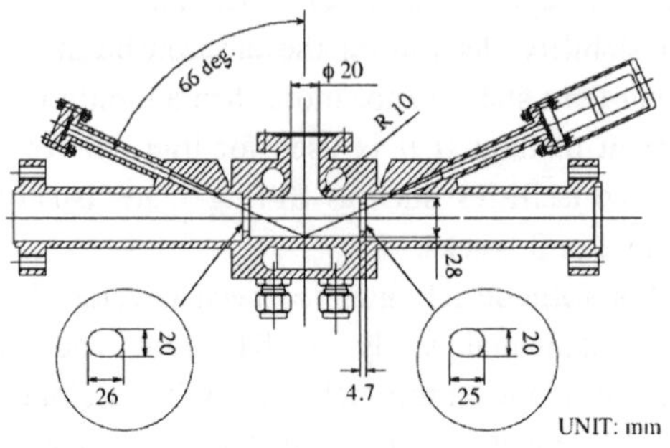

Figure 1. Schematic view of RF gun cavity.

Higher acceleration field can accelerate the electron beam with lower emittance especially in the high charge density region, which was confirmed by the three-dimensional beam simulation [13]. For this reason, a 35MW klystron (Mitsubishi PV3035) was replaced by the present 80 MW klystron (Toshiba E3712) in Dec 2001. The RF pulse length is varied from 500 ns to 1 μs according to the field gradient, because a long-pulse RF cause the RF breakdown and the more dark current. After an RF conditioning, the maximum field gradient in the center of the cathode of 135 MV/m has been achieved for an RF pulse length of 1 μs, and 175 MV/m for 500 ns. The field gradient of 135 MV/m and 175 MV/m correspond to the acceleration energy of 3.1 MeV and 4.1 MeV, respectively. The dark current at the field gradient of 175 MV/m is 0.15 nC/pulse.

Table 1. Parameters of the RF gun cavity

Frequency	2856 MHz
Intrinsic Q value	12400
External Q value for output port	3684
External Q value for input port	2786
Coupling coefficient for input port	1.01
Filling time	0.16 μs
Shunt impedance (for β=1)	1.16 MΩ

2.2. *Laser System*

The laser system adopted for our experiment is a titanium-sapphire femto-second laser followed by a third harmonics generator and a pulse length stretcher (Thales Laser, Alpha-10A/US). The detailed configuration of the laser system is shown in Fig. 2. The oscillator (Femto Laser Produktions GmbH, Femto source 20) generates infrared ultra-short pulses with a pulse duration of 20 fs and a frequency of 89.25 MHz. Chirped multi-layer dielectric mirrors are used to realize such a short pulse. Laser pulse trains are generated via self-mode-locking by means of Karr lens effect. The center wavelength of the laser is around 790 nm, the FWHM of the wavelength is more than 45 nm, and the average power is 250-600 mW. To synchronize the laser pulses with the RF signal of 2856 MHz, the cavity length of the laser oscillator is precisely controlled with a piezo-actuated mirror. The pump laser for the oscillator (Coherent, Inc., Verdi V-5) is a solid-state diode-pumped, frequency-doubled Nd:YVO_4 laser with a wavelength of 532 nm and a CW power of 5 W.

The output pulse from the oscillator is stretched to around 300 ps for the chirped pulse amplification. The pulses are selectively amplified at 10 Hz with a regenerative amplifier and a 4-pass amplifier up to about 38-40 mJ. To pump both amplifiers, the output laser pulse from a SHG of a Nd:YAG laser (Thales Laser, 5000 COMP FEMTO) with a pulse energy of 220 mJ is divided into 40 mJ for the regenerative amplifier and 180 mJ for the 4-pass amplifier. The optics and the pulse timing system for these amplifiers should be carefully adjusted so that the rms energy stability can be maintained less than 2%. The laser system is located in a clean room where the temperature drift is controlled within about one degree.

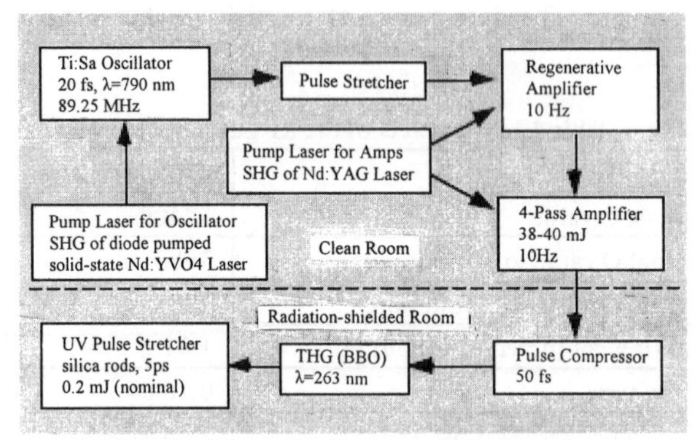

Figure 2. Schematic diagram of the laser system.

The laser pulse after these amplifications is transported to a radiation-shielded room for the RF gun. The laser transport from the 4-pass amplifier to the pulse compressor is approximately 7.5 m long. The laser pulse with a pulse duration around 300 ps is compressed to about 50 fs with a compressor that consists of a pair of gratings. The compressed infrared pulse is converted to its third harmonics (the wavelength is 263 nm). First BBO crystal is used to generate the second harmonics and the second BBO crystal is used to sum the fundamental and the second harmonics to generate the third harmonics. To generate larger energy at the third harmonics generator (THG), the spatial and temporal superposition of the fundamental and the second harmonics on the second BBO crystal should be accurately controlled with an optical transport line and optical delay line. The UV-laser pulse after THG is too short to generate a low emittance beam. According to the numerical results with the three-dimensional beam simulation, the optimum pulse length of the UV-laser is 20 ps to generate the electron beam with the lowest emittance [9]. Thus the UV-pulse should be stretched after the THG. Two silica rods of 45 cm in length were used to stretch the UV-pulse. Of course, the pulse stretching depends on the length of the rods and the peak power of the UV-laser. The temporal distribution of the stretched pulse was measured with a streak camera (Hamamatsu Photonics K.K., C6138 FESCA-200) with the minimum resolution of 200 fs. In figure 3, single-shot (right bottom) and 100-shot averaged (right top) pulse shapes are simultaneously shown. The measured pulse length of the UV-laser was approximately 5 ps (FWHM), which is shorter than the evaluated optimum length of 20 ps.

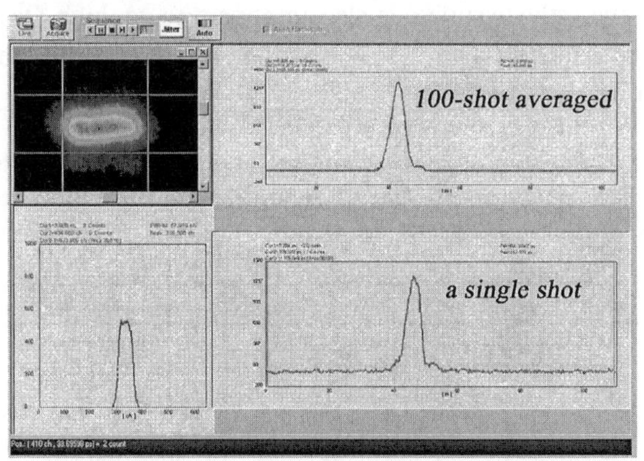

Figure 3. Streak image of UV-laser pulse

The stability of the laser system was insufficient when it was installed. Since the oscillator lost its mode-locking every several hours, the optics in the oscillator was re-adjusted and fixed tightly. The temperature of the recirculating chiller for the Ti:Sappphire crystal and the base of the pump laser was also optimized. The chiller (NESLAB, RTE-7) was replaced to NESLAB M-25 with a closed water circulation, because there were some problems due to the water dry-up every 2-3 weeks in the RTE-7. Because the strong air-flow from the air-conditioner in the clean room made the regeneration amplifier unstable, the strength of the flow were reduced by putting a borad under the air-flow exit of the air-conditionner. The environmental temperature of the BBO crystals in the THG is not stabilized. The UV-pulse has an energy fluctuation of 2-10 % and an energy drift of the long term operation. It will be improved when all of the laser system are re-installed in a new large clean room keeping a uniform temperature and the proper humidity (50~ 60 %) to get rid of static electricity on the laser optics, which is scheduled in the spring 2003.

As a material of the cathode, copper was chosen for the RF gun in consideration of easy fabrication, long life and laser availability. Table 2 shows the specifications of some materials for the photocathodes. Diamond has high quantum efficiency (Qe) and high durability. However, the laser wavelength should be be shorter than 227 nm. It can be obtained with a fourth harmonics generator (FHG) of Ti:Sapphire laser, but needs further developments of the FHG. Although Mg and Cs_2Te have more Qe than Cu, Mg is easily oxidized in the air and needs a surface treatment with a laser cleaning after an installation in vacuum and maintenance of high vacuum in the cavity. As for Cs_2Te, the life is short and need frequent exchanges of the cathode. In addition, for all cathode materials, Qe largely depends on the surface condition of the cathode. Therefore the control of the cathode surface is also important to get high Qe as well as the selection of the cathode material. It is necessary to study photocathode material. We are preparing to study for this subject in the test facility.

Table 2 Property of some candidate cathode materials for the RF gun. Qe is quantum efficiency. The wavelength corresponds to the work function of the materials.

Material	Qe	Wavelength	Life
Cu	$\sim 10^{-4}$	277 nm	durable
Mg	$\sim 10^{-3}$	339 nm	easily oxidized
Cs_2Te	$\sim 10^{-2}$	350 nm	short (< 1 month)
Diamond	$\sim 10^{-1}$	227 nm	most durable

2.3. *Synchronization*

For the generation of a low-emittance electron beam, the laser pulse should be synchronized with the RF signal at a possibly smaller time jitter. The required time jitter should be less than 1 ps, because the phase fluctuation in the RF system should be maintained less than about 1 degree. Therefore the stability of the time jitter is important for generating a low emittance beam. The total jitter δt between the laser and RF signal is given by

$$\delta t = \sqrt{\delta t_{laser}^2 + \delta t_{sync}^2} \tag{1}$$

where δt_{laser} is the time jitter of laser oscillator itself, and δt_{sync} is the jitter of the synchronization. The jitter of the laser oscillator is evaluated from the time jitter of the pulse period of the laser oscillator that occurs due to the expansion and contraction of the cavity length. The stability of the cavity length depends on the temperature fluctuation in the environment or the small mechanical vibration of the optics. The time jitter δt_{laser} is given by

$$\delta t_{laser} = \frac{2\delta L}{c} \tag{2}$$

where δL is the fluctuaton of the cavity length, c is the speed of light. The cavity length should be continuously controlled according to the variation of the frequency to keep the same frequency. In our laser oscillator, it is controlled with a piezo actuator (PI-Polytec, P-830.10) where a cavity mirror of 5 mm in diameter is mounted. The travel length of the actuator is 15 μm and the length resolution is 0.15 nm. The length of the piezo actuator is controlled with an external voltage from 0 to 100 V applied by a piezo amplifier (PI-Polytec, E-505).

To keep the exactly same frequency, we tested two kinds of feedback of the cavity length. One is PLL feedback to the external reference RF signal of the second harmonics of 89.25 MHz. The other is a frequency feedback by counting the frequency of laser pulses and then move the piezo acutuator to get exact 89.25 MHz.

Concerning the PLL feedback, a synchro-lock system (Thales Laser, Maestro) was tested to synchronize the laser pulse with an external RF signal of 178.5 MHz (1/16 of 2856 MHz). The circuit diagram of the synchronization system is shown in Fig. 4.

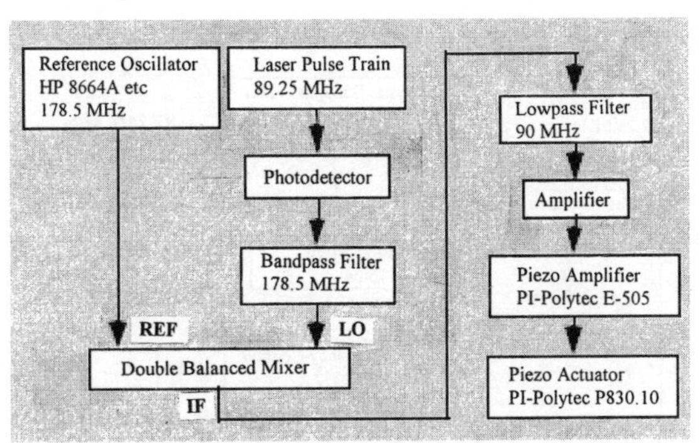

Figure 4. Schematic diagram of synchro-lock system

The jitter was measured with a sampling oscilloscope (HP, 54120B). As the synchronization largely depends on the characteristics of the reference RF signals, we compared three reference RF oscillators: HP 8664A, HP 83623B and a crystal oscillator (Lightwave, series 131). The jitter for the HP 8664A was the largest of 22 ps in a low noise mode and 11.4 ps in a normal mode. The difference was due to frequency noises at the side bands located 60 Hz apart from the oscillation frequency, confirmed by spectrum measurements. The jitters for the other two RF oscillators were about 8 ps, which were still insufficient for our requirements. The causes of the large jitter have not been explained yet. However, as one of the reasons, the time jitter depends on the frequency of the external reference RF signal. The time jitter is expected to be reduced by increasing the reference frequency to higher harmonics and improving the phase comparator.

The laser frequency feedback using a frequency counter was also tested. The feedback system configuration is shown in Fig.5. The laser pulses were detected with a fast InGaAs/Shottky photodetector (New focus, 1454). The frequency of the oscillator was counted with a precision pulse counter (HP, 53232A) and it is then compared with 89.25 MHz. A personal computer gets the frequency data from the counter and send a command to increase the voltage of a DC power supply (HP, E3632A) for the piezo-actuator if the frequency is smaller than 89.25 MHz, or to decrease the voltage in an opposite case. The rate of the feedback is very slow, aproximately 7 times in a second. Thus the jitter largely depends on the mechanical vibrations of the optics, and the feedback only prevents the drift of the frequency. The output voltage of the E3632A is amplified 10 times with the E-505. The frequency stability of the oscillator was measured for 10 minutes, and the standard deviation of the oscillation frequency was evaluated as 0.078 Hz.

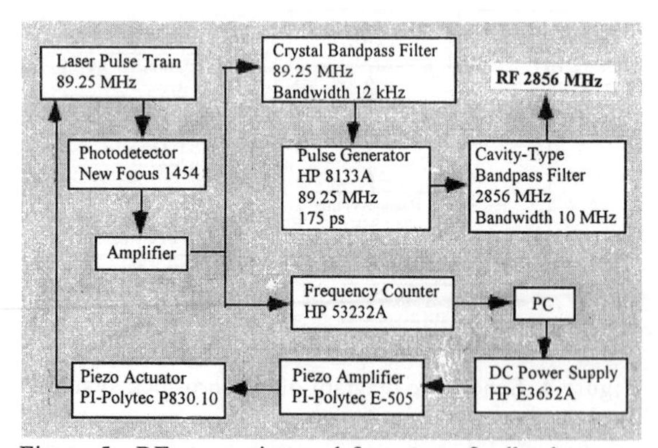

Figure 5. RF generation and frequency feedback system

In this feedback system, there is not any reference RF signal, therefore the RF of 2856 MHz was generated from laser pulses as shown in Fig. 5. A sinusoidal RF signal of 89.25 MHz was obtained with a narrow band-pass filter (BPF) of 89.25 MHz. The BPF is a crystal filter with a narrow bandwidth of 12 kHz. The filtered signal triggers a fast pulse generator (HP, 8133A). It generates pulses with an approximate pulse width of 175 ps. The pulse signal is synchronized with the RF signal of 89.25 MHz with a time jitter of about 1 ps and has higher harmonics components, especially at 2856 MHz. Therefore an

RF signal of 2856 MHz can be obtained by filtering the pulse signal with a BPF of 2856 MHz. The BPF is a cavity-type and the bandwidth is 10 MHz. This laser-driven RF generation system is inherently synchronized with the laser pulses without any feedback.

We measured the time jitter of the time delay between the laser pulse and the RF signal of 2856 MHz generated with the laser-driven RF generation system. The time delay between the laser pulse and the RF signal was measured with a digital oscilloscope (Lecroy, Wave Master 8500). The standerd deviation of the time delay was 1.4 ps (see Fig. 6). Considering the internal jitter of the oscilloscope, measured as 0.7 ps, the real system jitter was estimated as 1.2 ps. The time jitter is thought to occur mainly in the HP 8133A.

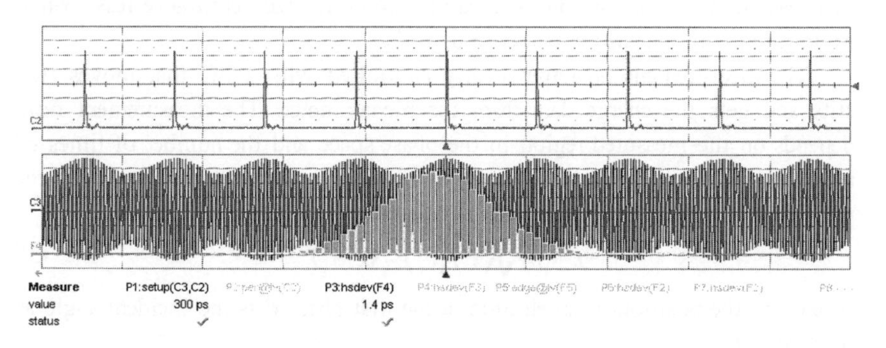

Figure 6. Measurement of the time jitter between the laser pulse and the RF signal. Top signal is laser pulses from the laser oscillator. Bottom signal is the RF of 2856 MHz. The distribution in the center bottom is the distribution of the time delay between the laser pulse and the RF signal.

2.4. *Solenoid magnets*

Two same solenoid magnets are installed for the emittance compensation [14]. The first magnet is located at a distance of 185 mm from the cathode surface and the second magnet is located at 385 mm from the cathode surface. The length of the both magnets is 90 mm. The inner and outer diameters are 80 mm and 300 mm, respectively. The optimized magnetic fields of the first solenoid magnet and the second one are 0.156 and 0.078 T, respectively, when the maxmum field gradient on the cathode in the RF cavity is 135 MV/m.

2.5. *Emittance measurement apparatuses*

The emittance of the electron beam is measured with a combination of double slits and a Faraday cup. The electrons that passed two slits are damped at the

Faraday cup. The charge of the electrons that stopped at the Faraday cup is measured with a highly sensitive picoammeter (Keithley 487). To reduce the pulse-to-pulse fluctuation, the charge of the electrons is averaged over ten or twenty measurements. The first slit for the vertical emittance measurement is located at a distance 650 mm from the cathode, and that for the horizontal measurement locates at 660 mm from the cathode. The width of the both slits is 0.3 mm. The slit is made of copper and the thickness is 8 mm. The second slits have the same dimensions as the first one. The distance between the first and second slits is 460 mm. The resolution of the unnormalized emittance measurement is estimated as 0.063 πmm-mrad, which corresponds to the normalized emittance of 0.42 πmm-mrad for the kinetic energy of 3.0 MeV. For this reason, it is not possible to measure the normalized emittance less than 1 πmm-mrad with a measurement error of 10 %.

The measurement is automatically controlled with a PC that moves two slits and gets the charge data from the picoammeter. The measurement time depends on the measured region in the phase space and the number of times of averaging. It usually took fifteen to thirty minutes for an emittance measurement run. The normalized rms emittance is evaluated by the definition formula

$$\varepsilon = \beta\gamma\sqrt{\langle x^2\rangle\langle x'^2\rangle - \langle x\cdot x'\rangle^2}, \tag{3}$$

where x is the position of an electron at the first slit, x' is the incident angle at the first slit.

Because the measured charge includes the dark current generated by the field emission, the charge component of the dark current should be subtracted from the total charge to decide the normalized rms emittance of photoelectron beam. The dark current component was assumed as uniform distribution in the measured region of the phase space, because the dark current has a beam emittance much larger than that of the photoemission. Gaussian fitting was used to determine the constant component in the distribution. Three matching parameters I_0, σ, I_{dark} were obtained by fitting the charge distribution function

$$I = I_0 \exp\left(-\frac{x'^2}{2\sigma^2}\right) + I_{dark} \tag{4}$$

to the measured charge distribution as a function of x' for the same x, which correspond to a set of data measured at the same position of the first slit but a different position of the second slit. These fittings were applied for all charge distributions measured with different x and then the obtained I_{dark} were subtracted from the original data. Finally the rms emittance was calculated from the corrected data. This correction is important, because the small dark current is distributed uniformly from the center to to outskirt of distribution. This induces a lot of error in the measurement of the rms emittance. Because there were a lot of procedures in the fitting process, it was not possible to calculate the rms emittance just after the measurement. Therefore the analysis was not suitable for the parameter search to find the minimum emittance condition.

For this reason, as a second way of determining the constant component of the dark current, we used so-called direct calculation: an analysis that a constant fraction of the maximum value in the measured two-dimensional distribution is subtracted as the constant component from the measured data. The comparison between two kinds of analyses is shown in Fig 7. Here the constant fraction was set as 5 %, which was decided so that the most of the dark current components in the distribution would disappear. The rms emittance estimated with this analysis decreased 18 % on average. The emittances evaluated with two analyses are different but roughly proportional each other except some data points. Faster analysis gave us an advantage to get the best parameters quickly in the experiments. Of course, all data were finally analyzed using the Gaussian-fitting to get the correct rms emittance.

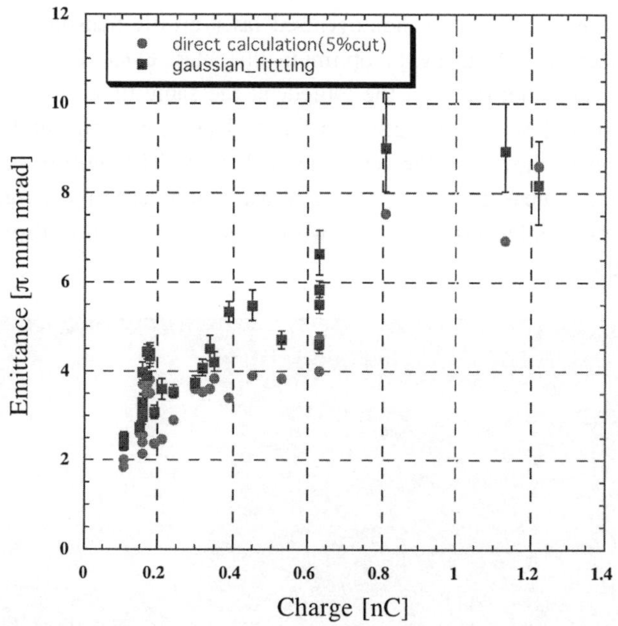

Figure 7. Comparison of rms emittance with two different analysis methods.

3. Improvement of Laser Profile and It's Results

3.1. *Homogenizer*

3.1.1 Optics of the homogenizer

The spatial profile of the UV-laser decides the quality of the electron beam of the photocathode RF gun. The final UV-laser profile is determined by the profile after the regeneration amplifier, the overlap of each path of the laser at the Ti:Sapphire crystal in the 4-pass amplifier, the laser overlap of the fundamental and the second harmonics at the second BBO crystal in the THG, and the profile deformation in the UV pulse stretcher. Because of the incompleteness of the optics, the emitted UV-laser has an inhomogeneous spatial profile. For this reason, we used a microlens array as a homogenizer. This microlens array is a collection of small hexagonal convex microlenses with a pitch of 250 μm. The transmission of this optical array is about 80 % in a region of ultraviolet. It is possible to homogenize any shape laser profile into a Silk-hat (cylindrical flattop) with combinations of an additional convex lens. The main difficulty to utilize this optics is how the homogenized laser profile transports toward the cathode surface with focusing to optimum diameter. Even if the whole wave front of laser does not reach on the cathode at the same time, the laser spot on the surface should be in the depth of a focus. The laser spatial profile without homogenizing is shown on the left hand in Fig. 8. The profile was spatially shaped by a microlens array as a quasi-Silk-hat profile (see on the right hand in Fig. 8). These profiles were measured with a laser beam profiler (Spiricon Inc., LBA300-PC).

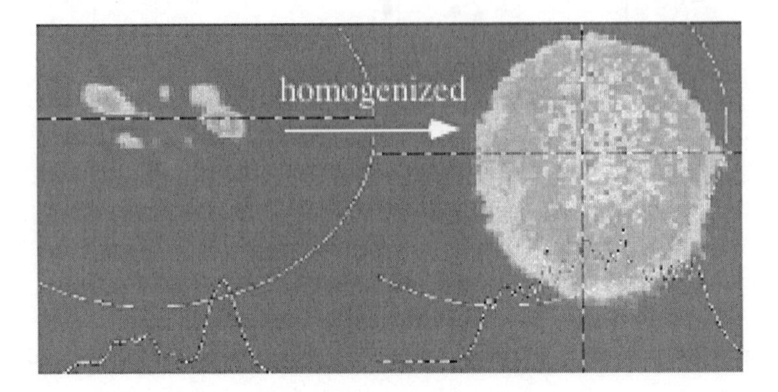

Figure 8. Profile improvement of the laser spatial profile with microlens array.

3.1.2 Emittance improvement with homogenizer

The emittance of the electron beam was measured with the homogenized laser profile. In the series of the measurements, the maximum electric field on the cathode was set at 135 MV/m. The charge of the accelerated electrons was varied by changing the attenuation ratio of the UV-laser attenuator located after the UV-pulse stretcher. The charge dependence of the horizontal normalized rms emittance is shown in Fig. 9. The emittance was measured three times and consistent data were obtained. The minimum emittance of 2.3 πmm-mrad was observed when the charge of the electrons at 0.1 nC/bunch.

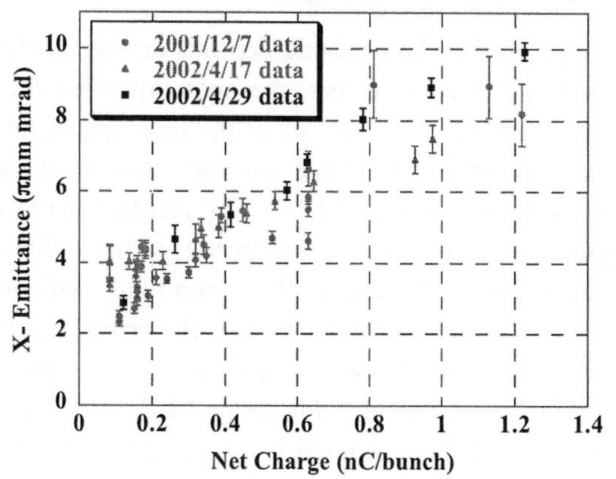

Figure 9. The charge dependence of the horizontal normalized rms emittance

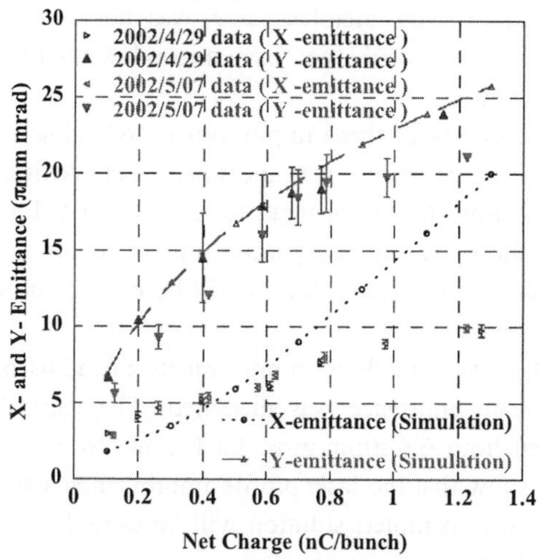

Figure 10. The comparison of the horizontal and vertical emittances. Dotted curves with circle or triangle indicate the results of the three-dimensional simulation.

The vertical emittance was also measured. The comparison of horizontal and vertical emittances is shown in Fig. 10. Because the laser incident angle is 66 degrees, the vertical emittance is much larger than horizontal emittance. The effect of the non-zero angle injection is described in the section 3.2. The reason the vertical emittance is larger is that the beam exchanges the horizontal coordinate with the vertical coordinates because of the beam rotation in the solenoid magnets. The three-dimensional beam simulation roughly agrees with the experimental results except the high charge region of the horizontal direction.

3.2. *Non zero-degree incident angle effect*

The laser spot image on the photocathode depends on the incident angle. When the laser incident angle is zero degrees, the spot image on the cathode is the same as that of the longitudinal projection of the laser spatial profile. But if the incident angle is not zero, the image on the cathode is deformed to an ellipsoidal shape as shown in Fig. 11. As the incident angle of the RF gun cavity is 66 degrees, the horizontal laser diameter on the cathode is 2.46 times wider than the laser diameter. Of course, this deformation will increase the beam emittance. In addition to the horizontal spread, the arriving time lag Δt at the cathode also occurs on the horizontal direction. The time lag causes an asymmetric distribution of the electron beam, and also increases the beam emittance. To decrease the effect of incident angle, the laser was deformed to an ellipse to get a round image on the cathode with two cylindrical lenses. The UV-laser profile deformed to an ellipse was measured with the beam profiler (see Fig. 12). In the optics, the microlens array was not used because the optics with the cylindrical lenses could not have a good focus on the photocathode with the microlens array.

To confirm the effects of these improvements of radius ratio of ellipsoidal spot image, the beam emittance was measured. The vertical normalized rms emittance decreased from 6.6 πmm-mrad for 0.1 nC to 5.4 πmm-mrad for 0.2 nC. These results show that the laser profile improvements are effective to get a low emittance beam. A complete solution will be carried out with a quasi zero-angle injection system.

To confirm the effects of these improvements of radius ratio of ellipsoidal spot image, the beam emittance was measured. The vertical normalized rms emittance decreased from 6.6 πmm-mrad for 0.1 nC to 5.4 πmm-mrad for 0.2 nC. These results show that the laser profile improvements are effective to get a low emittance beam. A complete solution will be carried out with a quasi zero-angle injection system.

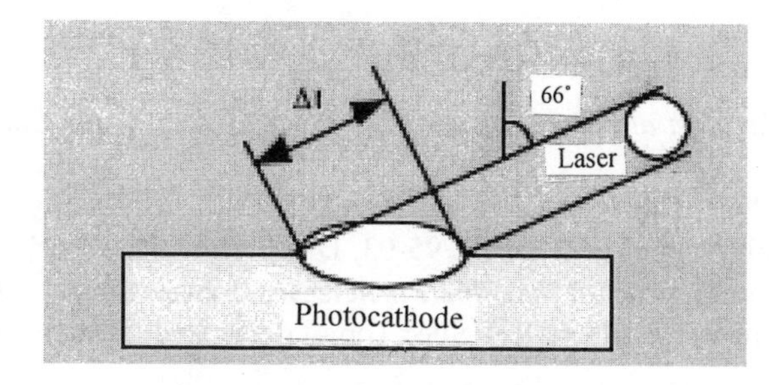

Figure 11. Non-zero incident angle laser injection on the photocathode.

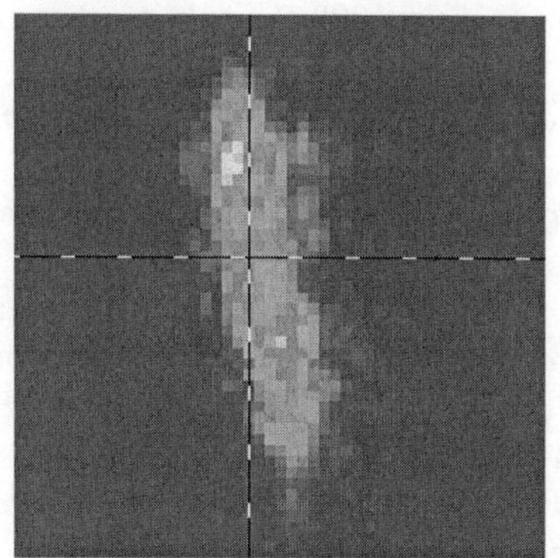

Figure 12. Laser spot image deformed to an ellipsoidal shape
(Δx: 0.60 mm, Δy:1.86 mm)

4. Summary

The electron beam quality and stability was improved by developing a low jitter synchronization system, stabilizing the laser system and shaping the laser profile. A low jitter synchronization system was realized with the laser frequency feedback system. The time jitter measured was about 1.2 ps, which is very close to our requirements. The environment around the laser system was also improved and it had a lot of contribution to realize long-time beam experiments and save the laser alignments. The uniform profile homogenized with the microlens array realized the normalized rms emittance of about 2 πmm-mrad at 0.1 nC. The effect of the non-zero angle injection was confirmed by simulations and experiments. In a separate experiment, deformation optics to get a circular profile on the cathode was realized with a pair of cylindrical lenses.

References

1. D. T. Palmer *et al.*, *Proc. of the 1995 Particle Accelerator Conf.* , (1995) 982
2. A VUV Free Electron Laser at the TESLA Test Facility at DESY-Conceptual Design TESRA-FEL 95-03, DESY Print June 1995.
3. S. Telfer *et al.*, *Proc. of the 2002 Particle Accelerator Conf.* , (2001) 2263
4. D.J.Gibson et. al, Phys.Rev.Spec.Top.- Acce. and Beams, **4**, 090101 (2001)
5. T. Taniuchi *et al.*, *Proc. of the 18th Int. FEL conf.* ,Rome (1996) II-137
6. T. Taniuchi *et al.*, *Proc. of the 19th Int. FEL conf.* ,Beijin (1997) II-97
7. T. Taniuchi *et al.*, *Proc. of the 1999 Particle Accelerator Conf.*, New York (1999) 2015.
8. A. Mizuno *et al.*, *Proc. of the 1999 Particle Accelerator Conf.*, New York (1999) 2749.
9. A. Mizuno *et al.*, *Proc. of the 2000 European Particle Accelerator Conf.*, Vienna (2000) 839.
10. H. Dewa *et al.*, *Proc. of the 23rd Int. FEL conf.* ,Darmstadt (2001) II-41
11. T. Taniuchi *et al.*, *Proc. of the 23rd Int. FEL conf.* ,Darmstadt (2001) II-43
12. H. Tomizawa *et al.*, *Proc. of the 2002 European Particle Accelerator Conf.*, Paris (2002), 1819.
13. T. Taniuchi *et al.*, *Proc. of the 21st Linear Accelerator Conf.*, Kyongju (2002), to be published.
14. B. E. Carlsten, Nucl. Instr. and Meth., **A285**, 313 (1985)

RECENT ADVANCES AND NOVEL IDEAS FOR HIGH BRIGHTNESS ELECTRON BEAM PRODUCTION BASED ON PHOTO-INJECTORS

M. FERRARIO, M. BOSCOLO, V. FUSCO, C. VACCAREZZA
Istituto Nazionale di Fisica Nucleare
Laboratori Nazionali di Frascati
Via E. Fermi 40, 00044 Frascati (Roma), Italy

C. RONSIVALLE
Ente Nazionale Energie Alternative
Via Enrico Fermi 45, 00044 Frascati (Roma), Italy

J. B. ROSENZWEIG
University of California Los Angeles
Department of Physics and Astronomy
Los Angeles, CA 90095

L. SERAFINI
Istituto Nazionale di Fisica Nucleare
Sezione di Milano
Via Celoria 16, 20133 Milano, Italy

Photo-injectors beam physics remains a fruitful and exciting field of research. New ideas have been recently proposed to achieve ultra-high brightness beams, as particularly needed in SASE-FEL experiments, and to produce flat beams as required in linear colliders. An overview of recent advancements in photo-injector beam physics is reported in this paper.

1 Introduction

The research and development of high brightness (high current, low emittance) beam production by photo-injectors has been driven in the last decade mainly by self amplified, spontaneous emission, free-electron laser (SASE FEL) applications. Beams with normalized emittances lower than 1 μm, with peak current of some kA, are required for example for the new x-ray SASE FEL projects [1,2]. A revival of longitudinal focusing techniques with a deeper understanding of emittance compensation theory [3] has opened up a new possibility of compressing the beam inside an RF structure or in a downstream drift, with a proper beam control employed through solenoid focusing to avoid the emittance degradation. This option may avoid the serious phase space degradations observed in magnetic chicans caused by coherent synchrotron radiation (CSR) emitted in the bends [4]. Kilo-ampere beams with low emittance have been predicted by simulations for the so-called velocity bunching configuration [5]. At Neptune (UCLA) [6], DUVFEL (BNL) [7] and PLEIADS (LLNL) [8] preliminary experimental results (in non-optimized beam lines) have verified the usefulness of this idea for strongly compressing photoinjector-derived beams, despite the space charge induced emittance degradation observed.

In a different context, a new technique has been recently proposed [9] and tested [10] by a DESY/FNAL collaboration, the so-called flat beam production, which is an important goal for linear colliders. It consists in a simple transformation of a magnetized round beam, with equal emittances in both transverse planes, produced by a photo-cathode embedded in a solenoid field, which is then followed by a skew quadrupoles triplet. With proper matching, a flat beam with high transverse emittance ratio (300) as required by linear colliders, may in principle be obtained. Experimental results have achieved so far an emittance ratio of 50.

An overview of these recent advancements in photo-injector beam physics is given in this paper.

Transverse normalized beam brightness is a quality factor of the beam defined as [11]:

$$B_{\perp} = \frac{2I}{\varepsilon_{n,x}\varepsilon_{n,y}} \tag{1}$$

where I is the bunch peak current and ε_n is the rms normalised emittance. The meaning of brightness can be understood by expressing the peak current by the transverse current density $I=J\sigma^2$ and the emittance at waist as $\varepsilon_n=\gamma\sigma\sigma'$ so that taking $\varepsilon_{n,x}=\varepsilon_{n,y}$ one obtains:

$$B_{\perp} = \frac{2J}{(\gamma\sigma')^2} \tag{2}$$

In this way, transverse brightness can be seen as the beam peak current density normalized to the rms beam divergence angle, i. e. a quality factor of a beam propagating with low divergence and high current density, as required for FEL interaction, or Thomson back-scattering applications.

By examining the SASE FEL scaling laws [12], one can understand the importance of a high brightness beam for achieving short radiation wavelength λ_r with short gain length L_g:

$$\lambda_r^{min} \propto \left(\frac{\delta\gamma}{\gamma}\right)\sqrt{\frac{\left(1+K^2/2\right)}{\gamma B_{\perp} K^2}} \tag{3}$$

$$L_g \propto \frac{\gamma^{3/2}}{K\sqrt{B_\perp\left(1+K^2/2\right)}} \tag{4}$$

where K is the undulator parameter.

Another way of viewing the requirements on the beam for FEL applications is through the emittance alone. The coherent part of the radiation from an undulator can be approximately described as an equivalent source at the undulator center [13], where the minimum photon beam phase space area is limited by diffraction and is given by

$$\varepsilon_{ph} = \sigma_{ph}\sigma'_{ph} = \frac{\lambda_r}{4\pi}, \tag{5}$$

where σ_{ph} and σ'_{ph} are the source radius and divergence respectively. In this picture the optimum condition during SASE interaction requires phase space matching of electron and photon beams in both transverse planes;

$$\varepsilon_{n,x} = \varepsilon_{n,y} \leq \gamma\varepsilon_{ph}, \tag{6}$$

a condition that allows to operate in the round beam configuration with $\varepsilon_{n,x}=\varepsilon_{n,y}$.

We now discuss the factors affecting the propagation and manipulation of the electron beams needed for advanced applications. As shown in Fig. 1, a high brightness electron beam is subject to transverse defocusing forces originated by space charge field, by emittance pressure, and by an equivalent emittance terms $\varepsilon_B = p_{\vartheta,o}/mc$ accounting for the centrifugal potential when the canonical momentum p_θ is different from zero.

The defocusing forces on the beam can be counteracted by external fields like solenoids, and alternating gradient transverse component of RF accelerating fields. Additionally, transverse motion is naturally damped by acceleration. Similarly, along the longitudinal axis defocusing forces are generated by space charge and longitudinal emittance, while a proper phasing of the RF field can be used to produce longitudinal focusing. The beam may also rotate inside a solenoid field and, in a more complicated way, also in a quadrupole channel. Rotation can be balanced or properly tuned as a consequence of the conservation of canonical angular momemtum, when the beam is generated in a gun embedded in a solenoid field.

High brightness beams experience two distinct regimes along the accelerator, depending on the laminarity parameter (the ratio between the space charge term and the emittance term in the transverse envelope equation):

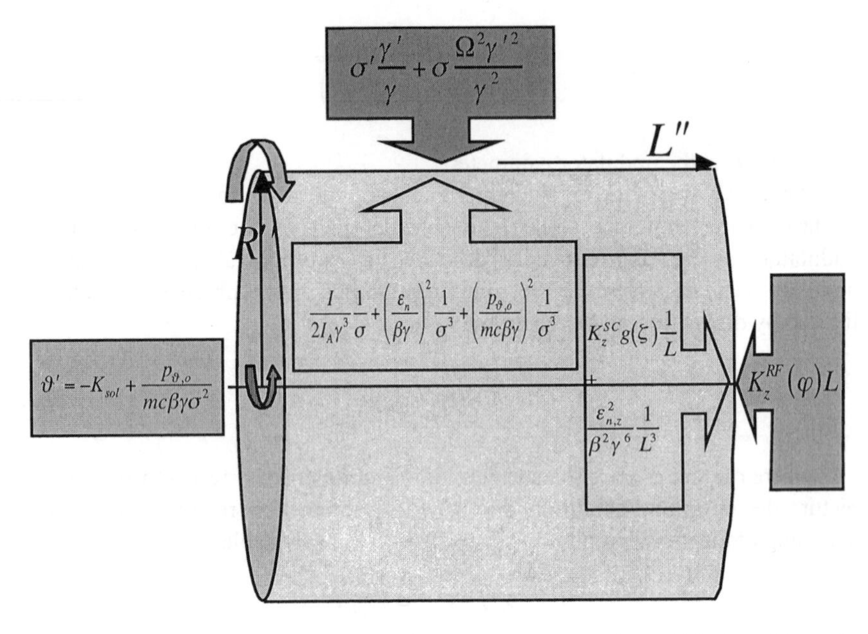

Figure. 1. A schematic representation of the envelope equations [14] describing the dynamics of a beam subject to its self field and external fields.

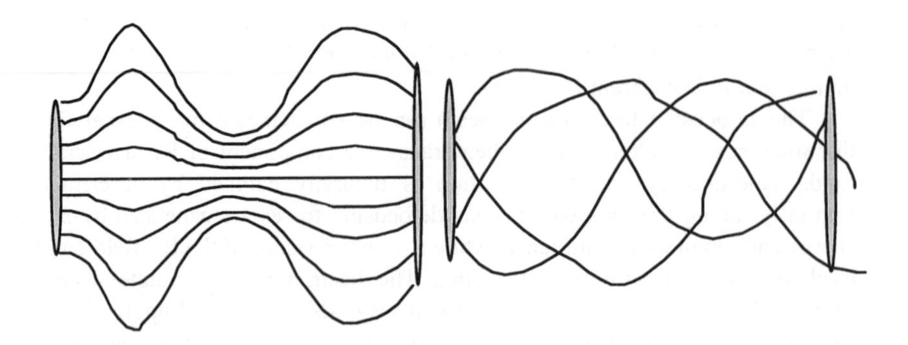

Figure. 2. Schematic representation of a quasi-laminar beam trajectories (left plot) and of an emittance dominated beam trajectories (right plot).

$$\rho = \frac{I\sigma^2}{2\gamma I_A\left(\varepsilon_{th} + \varepsilon_B\right)^2} \tag{7}$$

When $\rho >> 1$ the transverse beam dynamics are dominated by space charge effects, and the beam propagated in the quasi-laminar regime in which particle trajectories do not cross each other, as shown in Fig. 2 (left). Correlated emittance oscillations are observed in this regime [3], caused by the different local current along the bunch and by finite bunch length effects. By accelerating the beam, a transition occurs to the so-called emittance dominated regime, when $\rho << 1$, in this case the transverse beam dynamics are dominated by the emittance and trajectories are not parallel anymore, as shown in Fig. 2 (right).

In the following sections we describe how a beam can be manipulated in straightforward ways to obtain high brightness in FELs, and the possibility of high luminosity in linear colliders.

2 Emittance compensation: the concept of invariant envelope and the new working point

In a photo-injector, electrons are emitted by a photo-cathode located inside an RF cavity that is illuminated by a laser pulse, so that the bunch length and shape can be controlled on a picosecond time scale via the properties of the laser pulse [15]. The emitted electrons are rapidly accelerated to relativistic energies, thus partially mitigating the emittance growth due to space charge force effects. Nevertheless the phase dependent focusing forces that the electrons experience in the RF field [16] result in an RF induced emittance growth. In order to keep this effect small the transverse and longitudinal bunch dimensions have to be kept small. The increased particle densities lead, in turn, to increased space charge forces thus partially counteracting the beneficial effects of the high gradients available in RF cavities.

Since the early '80s it was clear that this large space charge-induced emittance growth in an RF gun is partially correlated and can be reduced by a simple focusing scheme first studied by B.Carlsten [17]. The space charge in fact acts to first order as a defocusing lens, the strength of which varies over the bunch length. The force is strongest in the middle slice of the bunch and decreases towards both ends. Therefore a fan-like structure appears in the phase space. After a focusing kick is applied by means of an external solenoid, the fan slices distribution tends to close in the following drift space until a minimum phase space area is reached, corresponding to a partial re-alignment of bunch slices. A residual emittance growth is nevertheless left in practical cases, caused by non-linear space charge fields within the bunch [18] and chromatic effect in the solenoid.

In order to avoid additional space charge emittance growth in the subsequent beam line, the emittance minimum has to be reached at high beam energy so that space charge forces are sufficiently damped. The beam has to be properly matched to the following accelerating sections in order to keep under control emittance oscillations and obtain the required emittance minimum at the injector exit.

A full theoretical description of the emittance compensation process [3] has demonstrated in fact that in the space charge-dominated regime, *i.e.* when the space charge collective force is largely dominant over the emittance pressure, mismatches between the space charge correlated forces and the external focusing gradient produce slice envelope oscillations that cause normalized emittance oscillations. It has been shown that to also damp these emittance oscillations secularly, the beam has to be propagated close to a the so-called invariant envelope, given by

$$\sigma_{INV} = \frac{1}{\gamma'}\sqrt{\frac{2I}{I_A\left(1+4\Omega^2\right)\gamma}} \tag{8}$$

where $\gamma = 1 + T/mc^2$ is the normalized beam kinetic energy while the normalized accelerating gradient is defined by $\gamma' = \frac{eE_{acc}}{m_e c^2} \approx 2E_{acc}$, E_{acc} is the accelerating field, I is the beam peak current in the bunch, and $I_A = 17$ kA is the Alfven current. The normalized focusing gradient is defined as:

$$\Omega^2 = \left(\frac{eB_{sol}}{mc\gamma'}\right)^2 + \left\{\begin{matrix} \approx 1/8 \;\; SW \\ \approx 0 \;\; TW \end{matrix}\right\} \tag{9}$$

for a superposition of magnetic field of solenoids and RF ponderomotive focusing by standing wave or traveling wave sections.

The invariant envelope σ_{INV} is an exact analytical solution of the rms envelope equation for laminar beams:

$$\sigma'' + \sigma'\frac{\gamma'}{\gamma} + \sigma\frac{\Omega^2\gamma'^2}{\gamma^2} - \frac{I}{2I_A\sigma\gamma^3} = \frac{\varepsilon_{th}^2}{\sigma^3\gamma^2} \approx 0 \tag{10}$$

where the thermal emittance term (r.h.s.) is considered negligible. This solution corresponds to a generalized Brillouin flow condition that assures emittance correction, *i.e.* control of emittance oscillations associated with the envelope oscillations such that the final emittance at the photoinjector exit is reduced to an absolute minimum.

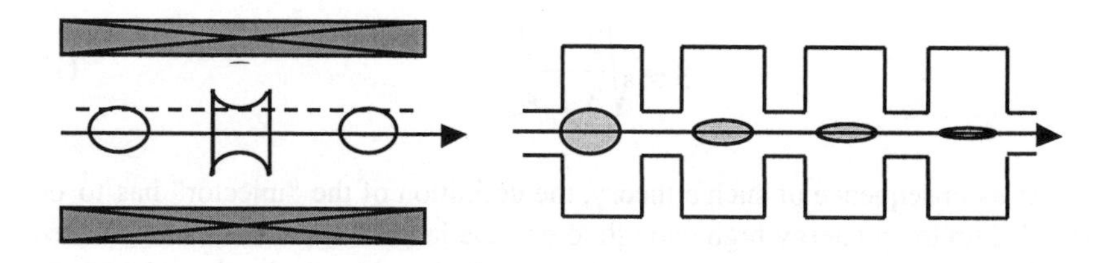

Figure. 3. Schematic representation of a Brillouin flow in a drift (left plot) and an invariant envelope in an accelerating structure (right plot).

In order to assure this condition, it is necessary to match two types of flow along the photoinjector (see Fig. 3): the invariant envelope inside accelerating sections and Brillouin flow, given by:

$$\sigma_{BRI} = \frac{mc}{eB_{sol}}\sqrt{\frac{I}{2I_A\gamma}} \tag{11}$$

in intermediate drift spaces.

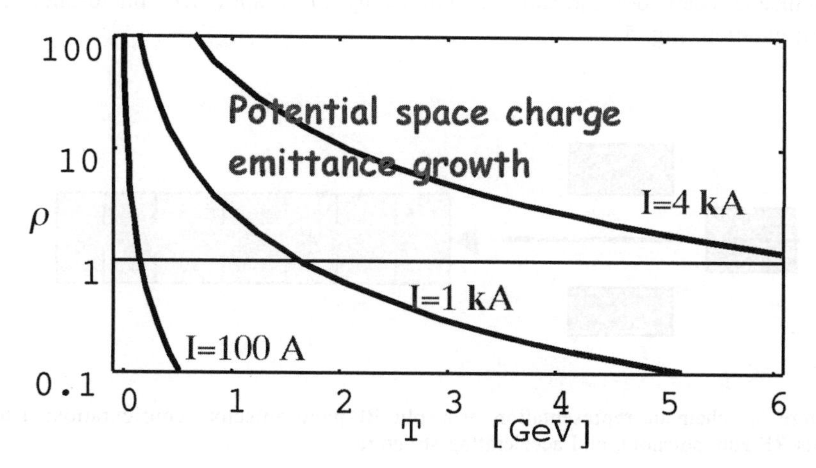

Figure. 4. Laminarity parameter ρ versus beam energy T for different beam currents.

By substituting the σ_{INV} expression (8) in laminarity parameter definition (7) one can see that the laminar regime extends up to an energy given by:

$$\gamma \approx \sqrt{\frac{2}{3}} \frac{I}{I_A \varepsilon_{th} \gamma'} \tag{12}$$

As a consequence of such a theory, the definition of the "injector" has to be extended up to an energy high enough to exit the laminar regime, as one can see from Fig. 4. The beam enters then in the so called emittance-dominated regime, where trajectories cross over dominates over space charge oscillations and the total normalized emittance remains constant in an ideal accelerator. For example, with the expected SPARC/FEL [19] injector parameters $I=100\ A$, $E_{acc}=25\ MV/m$ and an estimated [20] thermal emittance of 0.3 μm for a copper cathode with UV excitation, the transition occurs near 150 MeV.

Following the given matching condition, a new working point very suitable for damping emittance oscillations has been recently found [21] in the context of the LCLS FEL project [1]. It has been found that the correlated emittance can be damped to the level of 0.3 μm at 150 MeV for a uniform charged cylindrical bunch in this design. The main feature of such an effective working point are described here after.

Let us consider the emittance compensation regime of a space charge-dominated beam downstream the gun cavity of a split RF photo-injector configuration, Fig. 5.

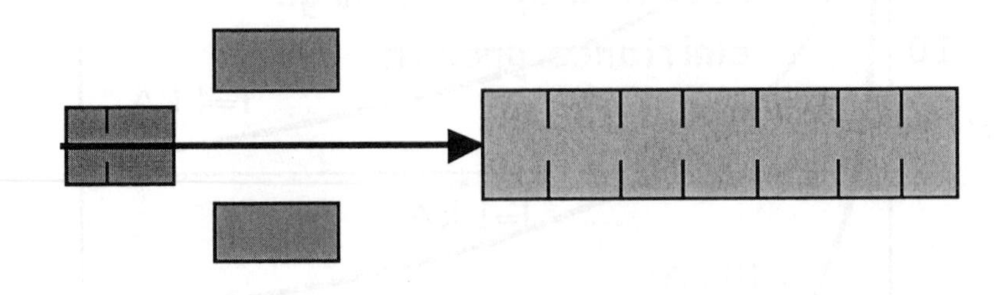

Figure. 5. Schematic representation of a split RF photo-injector configuration: 1.6 cells RF gun, solenoid, and accelerating structure.

In the drifting region downstream the gun the emittance evolution displays a double minimum, with the distance between the two minima decreasing with increasing of the solenoid strength.

It has been shown [21] that if the booster is located where the relative emittance maximum and the envelope waist occur, the second emittance minimum can be shifted to the booster exit and frozen at a very low level to the extent that the invariant envelope matching conditions are satisfied , see Fig. 6.

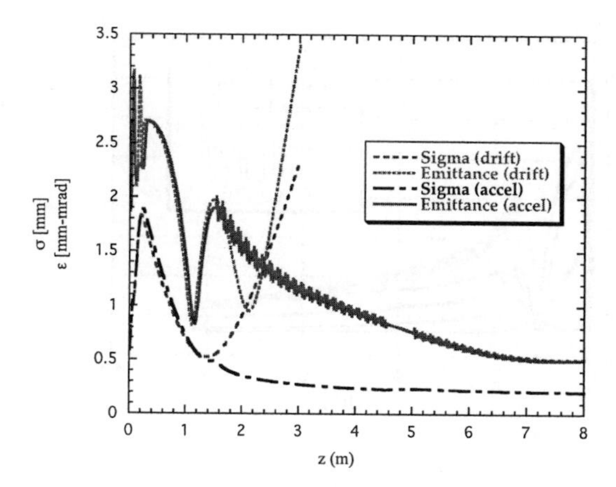

Figure 6. HOMDYN simulation of an optimized split photo-injector. Transverse rms beam size and normalized emittance, with and without booster.

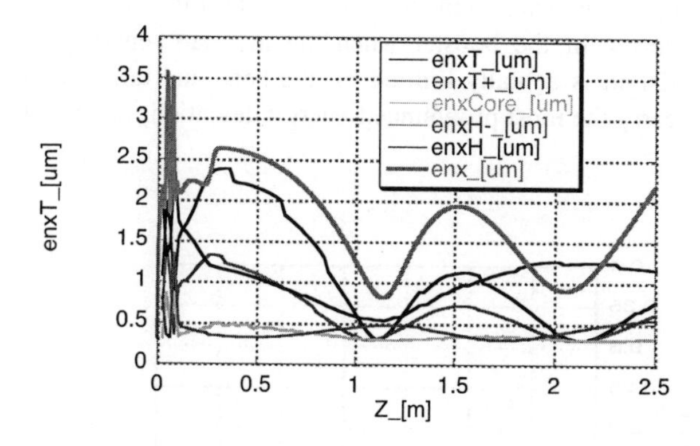

Figure 7. Total rms normalized emittance for an S-band gun with solenoid (red bold line) from the cathode along a 2.3 m long drift, together with the rms normalized emittance computed for five representative 400 μm long slices, located at the longitudinal extrema and in the core of the bunch.

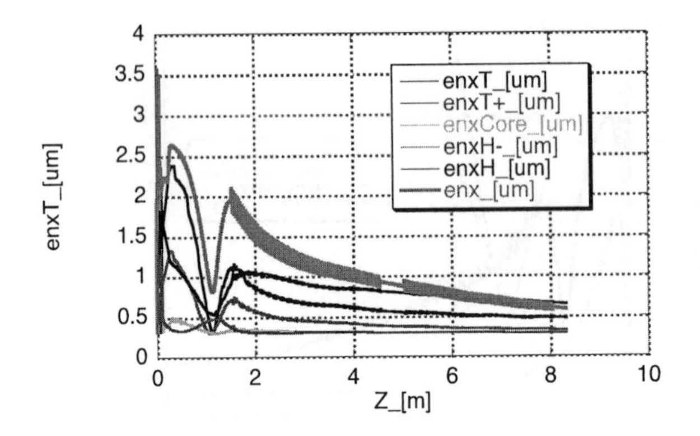

Figure 8. Same as Fig. 7 but with a booster located at z=1.5 m.

A slice analysis shows that the core of the bunch exhibits only minor emittances oscillation (it has very little effective variation of the defocusing space charge force) while the tails experience a a noticeble oscillation driving also the total emittance oscillation see Fig. 7 and Fig. 8. The emittance compensation occuring in the booster when the invariant envelope matching conditions are satisfied is actually limited by the head and tail slice behavior; these slices also carry the most pronounced energy deviation, as shown in Fig. 9.

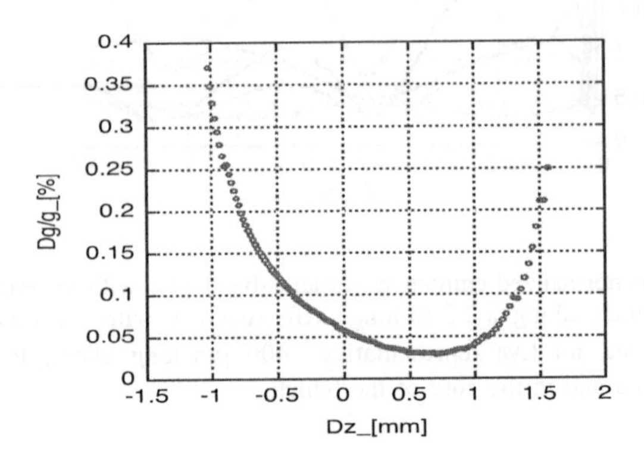

Figure 9. Energy deviation along the bunch at z=0.2 (solenoid centre) computed for a 400 μm long slice centered at Dz.

The origin of the double emittance minimum downstream of the solenoid location can be explained as mainly due to a chromatic effect occuring inside the solenoid. Following [3], a simple heuristic example helps to understand this process. Let consider a beam traveling in a long solenoid channel without acceleration, subject to its own space charge field and an external focusing force. Such a beam can be described by the following rms envelope equation, neglecting the thermal emittance term:

$$\sigma'' + k^2\sigma = \frac{Ig(\zeta)}{2I_A\sigma\gamma^3} \tag{13}$$

where $k = \dfrac{eB}{2mc\gamma}$ is the external solenoid parameter (depending on γ), ζ is the coordinate along the bunch and $g(\zeta)$ is the space charge field form factor, that for a uniform charged cylinder with radius R and length L is given by [21]

$$g(\zeta) = \frac{1-\zeta/L}{2\sqrt{(1-\zeta/L)^2 + A^2}} + \frac{\zeta/L}{2\sqrt{(\zeta/L)^2 + A^2}}, \tag{14}$$

where $A = \dfrac{R}{\gamma L}$ is the bunch aspect ratio.

An equilibrium solution of Eq. 13 can be easily found by setting $\sigma'' = 0$, leading to the already quoted Brillouin flow solution,

$$\sigma_{eq}(\zeta,\gamma) = \sqrt{\frac{Ig(\zeta)}{2I_A\gamma^3}}\,\frac{1}{k}. \tag{15}$$

Notice that each slice along the bunch, identified by its coordinate ζ, has a different equilibrium solution, not only because of the ζ dependence of the space charge field form factor, that from now on we will assume constant $g(\zeta) = 1$, but also because of the γ dependence of the solution. In case of a significant correlated energy spread along a cylindrical uniformly charged bunch with transverse size σ_c (assuming at the beginning each slice has the same σ_c) only the slice with the correct γ_o, if any, will be in equilibrium with a given external field B. Any other slice will oscillate around the initial size σ_c according to the solution of the linearized envelope equation about its equilibrium solution (assuming $\sigma_c' = 0$),

$$\sigma(z,\zeta,\gamma) = \sigma_{eq}(\zeta,\gamma) + \delta\sigma(\zeta,\gamma)\cos\left(\sqrt{2}k(\gamma)z\right), \tag{16}$$

where $\delta\sigma(\zeta,\gamma) = \sigma_c - \sigma_{eq}(\zeta,\gamma)$, as shown in Fig. 10.

Notice that in this case each slices has its own oscillation amplitude and its own oscillation frequency.

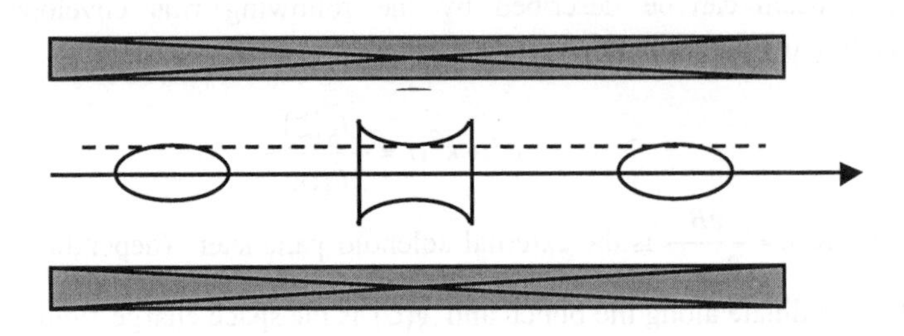

Figure 10. Schematic representation of a Brillouin flow. Notice that only the central slice is in equilibrium, the other slices are oscillating around theyr own equilibrium solution.

Let us compute now the emittance oscillations of a bunch with a linear energy spread correlation entering in a long solenoid at a waist. Taking into account only two representative slices with a slightly different energy: $\gamma_+ = \gamma_o(1+\delta_\gamma)$ and $\gamma_- = \gamma_o(1-\delta_\gamma)$, where $\delta_\gamma = \frac{\delta\gamma}{\gamma_o}$, a convenient way to calculate the emittance has been described in [22], the rms normalized emittance for a two slice bunch is given by

$$\varepsilon_n = \frac{\gamma}{2}\left|\sigma_+\sigma'_- - \sigma_-\sigma'_+\right|. \tag{17}$$

Substituting

$$\begin{matrix} \sigma_+ = \sigma_{eq+} + \delta\sigma_+ cos\left(\sqrt{2}k_+z\right) & & \sigma_- = \sigma_{eq-} + \delta\sigma_- cos\left(\sqrt{2}k_-z\right) \\ & \text{and} & \\ \sigma'_+ = -\sqrt{2}k_+\delta\sigma_+ sin\left(\sqrt{2}k_+z\right) & & \sigma'_- = -\sqrt{2}k_-\delta\sigma_- sin\left(\sqrt{2}k_-z\right) \end{matrix} \tag{18}$$

into to (17) one obtains, to first order in δ_γ,

$$\varepsilon_n = \frac{\gamma}{\sqrt{2}}k_o\left|\sigma_{eqo}\left(2\delta\sigma_o + \delta_\gamma\right) sin\left(\frac{\Delta k}{2}z\right)cos\left(\langle k\rangle z\right) + \delta\sigma_o^2\, sin\left(\Delta kz\right)\right| \tag{19}$$

where $\langle k\rangle = \frac{1}{\sqrt{2}}(k_+ + k_-) = \sqrt{2}k_o$ is the average beam plasma frequency, $\Delta k = \sqrt{2}(k_- - k_+) = 2\sqrt{2}k_o\delta_\gamma$ the modulating plasma frequency and $\delta\sigma_o = \sigma_c - \sigma_{eqo}$ is the envelope offset with respect to the nominal energy equilibrium solution. When $z << \frac{\pi}{2\Delta k}$ Eq. 19 can be approximated by the following expression,

$$\varepsilon_n = \frac{\gamma}{\sqrt{2}} k_o \delta\sigma_o \left| \left(\sigma_{eqo} \left(2\delta\sigma_o + \delta_\gamma \right) cos\left(\langle k \rangle z \right) + \delta\sigma_o \right) \Delta k z \right| . \tag{20}$$

In Fig. 11, the average envelope (Eq. 18) and emittance oscillations predicted by Eq. 19 are reported, and these qualitatively reproduce the double emittance minimum behavior around seen in the more complex photoinjector case discussed above.

When the initial envelope offset is not too far from the equilibrium solution (Eq. 15), the emittance oscillates with the average plasma frequency and periodically goes back to its initial value. By increasing the initial envelope offset the emittance evolution is dominated by the beating term and the original minimum is recovered only after a longer period, as shown in Fig. 12. This is a warning for beam operation in long solenoid devices in which the beam current varies, as will be discussed in the next section

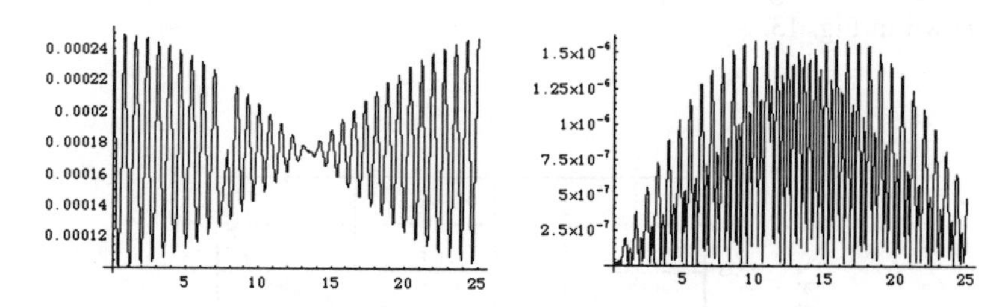

Figure 11. Envelope (left) and normalized emittance (right) evolution, as predicted by Eq. 19, reproducing the behavior of Fig. 7. Lower plots show a longer propagation length scales

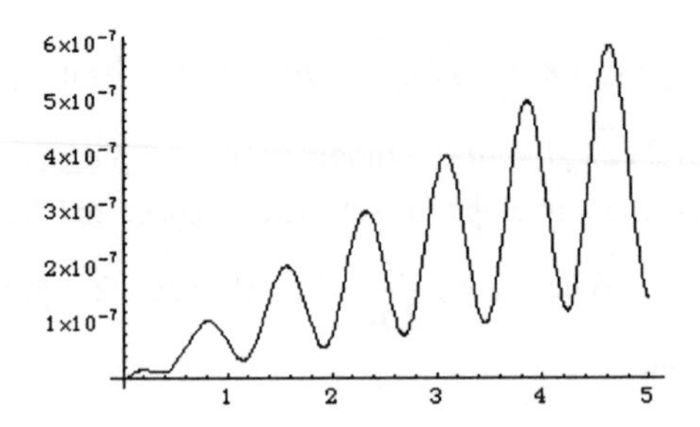

Figure 12. Normalized emittance as predicted by Eq. 19 with $\sigma_c=3\sigma_{eq0}$

This working point has been easily scaled to different gun RF frequency and bunch charge design through well-known techniques [23]. It might also be convenient to operate the injector at a different gradient with respect to the original design, while keeping the beam line unchanged. In this case the scaling has to preserve the main feature of the original bunch in terms of average and beating plasma frequencies. A reduced gun peak field results in a lower energy gain, hence a first scaling law can be deduced by the condition of keeping the average plasma frequency invariant,

$$B \propto \langle \gamma \rangle. \tag{21}$$

The beating plasma frequency is related to the relative energy spread along the bunch. The longitudinal phase space in the center of the solenoid (z=0.2 m) is shown in Fig. 13.

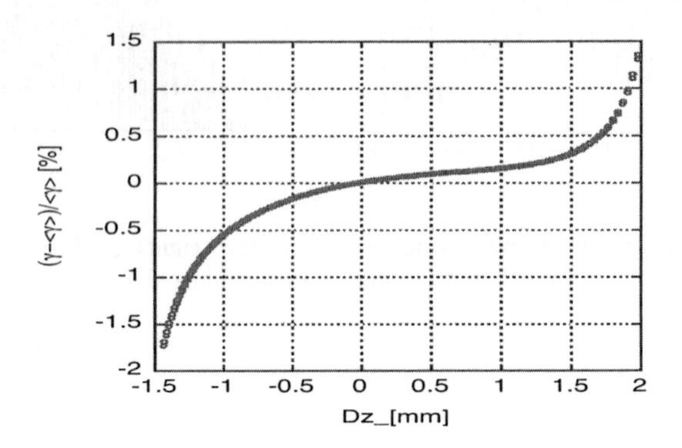

Figure 13. Longitudinal phase space in the center of the solenoid (z=0.2 m)

As one may easily note the energy correlation is dominated by a third order longitudinal correlation typical of the longitudinal space charge-induced energy spread. To keep such a correlation one has to keep unchanged the beam aspect ratio *A,* resulting in

$$\frac{\sigma_r}{\sigma_l} \propto \langle \gamma \rangle, \tag{22}$$

and the current density $J \propto Q/\sigma_l \sigma_r^2$ scales like [23]

$$J \propto \langle \gamma \rangle^3 . \tag{23}$$

For a constant charge scaling, Eqs. 22 and 23 taken together imply that $\sigma_l \propto \langle \gamma \rangle^{-5/3}$ and $\sigma_r \propto \langle \gamma \rangle^{-2/3}$.

For example scaling the original bunch parameters $R=1$ mm, $L=$ *10* ps and solenoid B=0.31 T, to operation with reduced gun peak field, from 140 MV/m to 120 MV/m, one obtains R=1.1 mm, L=12.7 ps and B=0.27 T. The original, unscaled (continuous line) and scaled (dashed line) envelopes and emittances are shown in Fig. 14. This comparison seems to indicate that our scaling considerations work quite well.

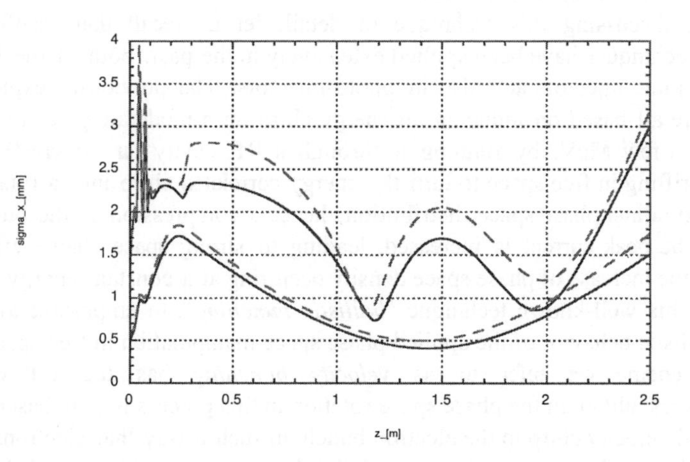

Figure 14. Original (continuous line) and scaled (dashed line) envelopes and emittances

The theory of the invariant envelope has been a very important tool to clarify the emittance compensation process. In the next section we discuss how the concepts introduced in this section can be extended to the case of a bunch under the effect of a longitudinal focusing.

3 High peak current: the concept of velocity bunching

Since the impact of magnetic compressors on the beam quality is a relevant and compelling topic, with the tendency to have serious emittance growth due to coherent synchrotron radiation effects in bends [4], a new method able to compress the bunch at moderate energies (tens of MeV), using rectilinear trajectories, and integrated in the emittance compensation process described in the previous section, has been recently proposed [5]. The so-called velocity bunching scheme is based on the weak synchrotron motion that the electron beam still undergoes at low energies in the RF wave of linear accelerating structures.

This combined bunching and emittance control technique has been extensively explored by numerical simulation studies of the entire photo-injector system that generates the ultra-short electron bunches, up to an energy 150 MeV. Additionally, recently the first experimental confirmations of the simulation and theoretical predictions have been observed at a number of laboratories [6,7,8]. Other experiments in the past showed some evidence of this process [24]. The promise is to attain 10-100 fs long bunches with charges ranging from 10 pC up to a few nC.

Before discussing this technique in detail, let us recall that rectilinear bunching techniques have been applied extensively in the past, both in the field of thermo-ionic injectors and also in photo-injectors. The previously explored schemes are all based on inducing an energy chirp on a moderately relativistic beam (say a few MeV, by running it through a RF cavity far off crest) and letting it drifting in free space to turn the energy correlated chirp into a rotation of the longitudinal phase space distribution, hence a compression of the bunch. Therefore the peak current is increased, leading to strong space charge effects caused by the increase of phase space density occurring at a constant energy. We will name this well-known technique *"ballistic bunching"*, to emphasize to the mainly ballistic behavior of the applied phase space manipulation in free space.

The scheme we refer to as *velocity bunching* has the following characteristics: although the phase space rotation in this process is still based on a correlated velocity chirp in the electron bunch, in such a way that electrons on the tail of the bunch are faster (more energetic) than electrons in the bunch head, this rotation does not happen in free space but inside the longitudinal potential of a traveling RF wave which accelerates the beam inside a long multi-cell traveling wave (TW) RF structure, applying at the same time an off crest energy chirp to the injected beam. This is possible if the injected beam is slightly slower than the phase velocity of the RF wave so that, when injected at the crossing field phase (no field applied), it will slip back to phases where the field is accelerating, but at the same time it will be chirped and compressed. The key point is that compression and acceleration take place at the same time within the same linac section, actually the first section following the gun, that typically accelerates the beam, under these conditions, from a few MeV (> 4) up to 25-35 MeV.

Therefore, the name *velocity bunching* refers to a correlation between particle velocity and position (phase) within the bunch, opposed to the magnetic compression technique which relies on a correlation between particle path length through the device and particle position (phase) within the bunch; since the chicane performing magnetic compression is often named *magnetic compressor*, the linac section performing *velocity bunching* is often named *RF (bunch) compressor* and velocity bunching itself is sometimes named *RF (bunch) compression.*

We show here how this method can preserve the low transverse emittance achieved at the exit of the photo-injector, reaching at the same time peak currents comparable with those produced by magnetic compressors.

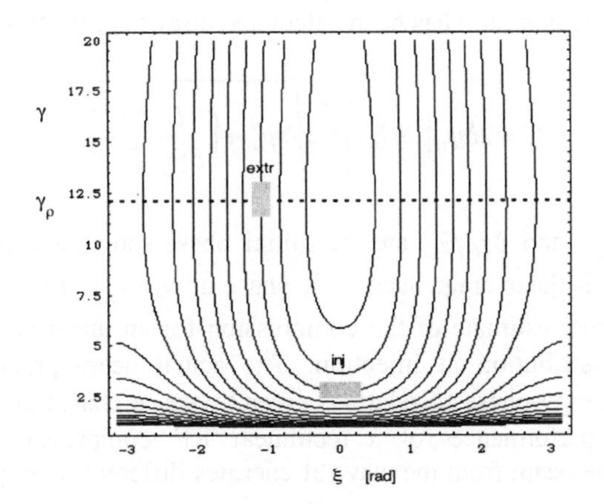

Figure.15 Phase space plots of a slow RF wave (γ_r finite) showing the basics of phase compression in a linac

The theoretical model explaining velocity bunching is based on a Hamiltonian treatment as follows. The interaction of a beam with a RF wave given by $E_z = -E_0 sin(\omega t - kz + \psi_0)$ is described by the Hamiltonian

$$H = \gamma - \beta_r \sqrt{\gamma^2 - 1} - \alpha cos\xi \quad , \tag{24}$$

where γ is the normalized electron energy, $\xi = kz - \omega t - \psi_0$ is its phase with respect to the wave, while $\alpha \equiv eE_0/mc^2k$ is the dimensionless vector potential amplitude and $\gamma_r = 1/\sqrt{1-\beta_r^2}$ is the resonant gamma of the wave (conventional traveling wave structures operate at $\beta_r = 1$, $k = \omega/\beta_r c$).

A wave whose phase velocity is slightly smaller than c, so that $k = \omega/c + \Delta k$, is characterized by $\beta_r = 1 - c\Delta k/\omega$, $\gamma_r = \sqrt{\omega/2c\Delta k}$ (assuming $c\Delta k/\omega << 1$).

The basic mechanism underlying the rectilinear compression effect is illustrated in Fig.15, where the contour lines of the Hamiltonian associated to a slow RF wave having $\gamma_r = 12$ (*i.e.* $\beta_r = 0.9965$ and $\Delta k = 0.0035\ \omega/c$) and $\alpha = 0.2$ are plotted. If the bunch is injected into the wave at zero phase (*i.e.* when the field of the wave is not accelerating), at an energy lower than the synchronous one (which corresponds to γ_r), it will slip back in phase and go up in energy, being accelerated by the wave). By extracting the beam from the wave at the time it reaches the resonant γ_r, *i.e.* when it becomes synchronous with the wave, we make the bunch undergo one quarter of synchrotron oscillation. In doing so, the beam is compressed in phase as depicted in Figure 15.

The compression ratio has been calculated elsewhere [5] to be:

$$C = 2\delta\psi_0 \left|\sin\bar{\xi}_{ex}\right| \Big/ \sqrt{\delta\psi_0^4 + \left(\frac{1}{\alpha\bar{\gamma}_0}\frac{\delta\gamma_0}{\bar{\gamma}_0}\right)^2} \tag{25}$$

where $\delta\psi_0$ and $\delta\gamma_0/\bar{\gamma}_0$ are the initial phase spread and energy spread of the bunch ($\bar{\xi}_{ex}$ is the average beam exit phase at $\gamma = \gamma_r$). This expression gives a good first order estimate of the compression for an uncorrelated longitudinal phase space distribution at injection. The actual beam produced by photo-injectors has energy-phase correlations given by space charge and RF effects, so that the best performances of a rectilinear RF compressor may occur for extraction of the beam from the wave at energies different than γ_r (as shown in the following), and can be enhanced with a longitudinal emittance compensation scheme, through use of a higher harmonic cavity to correct RF induced non linear energy spread.

In order to preserve the beam transverse emittance we have to integrate the longitudinal dynamics of the RF compressor into the process of emittance compensation, which is achieved for a beam at constant current by matching it on the invariant envelope. The analytical model is basically an extension of the invariant envelope theory [3], to the case of currents growing together with energy along the RF compressor. The analysis reported in the previous section is valid only for beams carrying constant peak current I , as usual in photoinjectors when no compression mechanism is applied (or space charge debunching is negligible). In order to extend the model to the case of RF compression (where I grows by a large factor) we have assumed that the current grows in the compressor at the same rate as the energy, i.e. $I = \frac{I_0 \gamma}{\gamma_0}$, where I_0 and γ_0 are the initial values for the current and the energy, respectively, at injection into the compressor. This assumption is derived by observations performed in several simulations of the RF compressor, indicating that best results in terms of final beam brightness are achieved under this condition, which indeed gives rise to a new beam equilibrium. It should be noted that this condition may be violated strongly near the end of the compression process, but as this occurs at high energy, the violation may not have serious consequences.

In fact, the rms envelope equation becomes in this case,

$$\sigma'' + \sigma' \frac{\gamma'}{\gamma} + \sigma \frac{\Omega^2 \gamma'^2}{\gamma^2} - \frac{I_0}{2 I_A \sigma \gamma_0 \gamma^2} = 0, \tag{26}$$

whose new exact analytical solution is

$$\sigma_{RFC} = \frac{1}{\Omega \gamma'} \sqrt{\frac{I_0}{2 I_A \gamma_0}}, \tag{27}$$

i.e. a beam flow at constant envelope (instead of $1/\sqrt{\gamma}$ as for the invariant envelope). This is dictated by a new equilibrium between the space charge defocusing term (decreasing now as $1/\gamma^2$) and the focusing and acceleration terms (imparting restoring forces to the beam): while for the invariant envelope equilibrium is achieved even in absence of external focusing, *i.e.* at $\Omega = 0$, in this case we need to provide external focusing.

Just for sake of comparison we notice that the solution for Brillouin flow (drifting beam at constant energy and constant current undergoing a rigid rotation in the solenoid field B_{sol}) becomes $\sigma_{BRI}^{BAC} = \frac{mc}{eB_0} \sqrt{\frac{I_0}{2 I_A \gamma_0}}$ in the case of current increasing linearly along the drift ($I = (\mu z) I_0$) for a corresponding growing solenoid field of the type $B_{sol} = \sqrt{\mu z} B_0$ (also in this case we obtain a constant envelope matched beam through the system, like for the case of RF compression). The beam size σ_{BRI}^{BAC} describes what typically is found in ballistic bunching to the beam envelope, which needs to be controlled by providing a ramped solenoid field to avoid envelope instability in this envelope dynamics.

What is relevant for the emittance correction process is the behavior of the envelope and associated emittance oscillations due to envelope mismatches at injection. Let us assume that the injecting envelope is mismatched with respect to the equilibriuum condition such that $\delta\sigma_{INV0} = \sigma_{INV} - \sigma_0$, or $\delta\sigma_{RFC0} = \sigma_{RFC} - \sigma_0$, or $\delta\sigma_{BRI0}^{BAC} = \sigma_{BRI}^{BAC} - \sigma_0$, depending on the type of equilibrium flow that the beam has to be matched on. A perturbative linear analysis of the rms envelope equations given above brings to these solutions for the evelope mismatches:

$$\delta\sigma_{INV} = \delta\sigma_{INVo} \cos\left[\sqrt{1/4 + 2\Omega^2}\ \ln\left(\frac{\gamma}{\gamma_o}\right) + \psi_o\right] \tag{28}$$

for the invariant envelope,

$$\delta\sigma_{RFC} = \delta\sigma_{RFC0} \cos\left[\Omega \ln\left(\frac{\gamma}{\gamma_0}\right) + \psi_0\right] \tag{29}$$

for its generalization in RF compressors, and

$$\delta\sigma_{BRI}^{BAC} = \delta\sigma_{BRI0}^{BAC} \cos\left[\left(\frac{eB_0\sqrt{\mu z}}{mc\gamma}\right)z + \psi_0\right] \tag{30}$$

for the ballistic bunching case.

These envelope mismatches produce emittance oscillations in laminar beams because of the spread in initial mismatches due to different slice currents [3]. The emittance behaviors for the three flow conditions turn out to be:

$$\varepsilon_n^{INV}(z) \approx \sqrt{\varepsilon_{off}^2 + \frac{I\left\langle\delta\sigma_{INV}^2\right\rangle}{\left(\frac{1}{4} + \Omega^2\right)\gamma'^2\gamma}}\ , \tag{31}$$

$$\varepsilon_n^{RFC}(z) \approx \sqrt{\varepsilon_{off}^2 + \frac{I_0\left\langle\delta\sigma_{RFC}^2\right\rangle}{\Omega^2\gamma'^2\gamma_0}}\ , \text{ and} \tag{32}$$

$$\varepsilon_n^{BAC}(z) \propto \sqrt{\varepsilon_{off}^2 + \frac{I_0\left\langle \delta\sigma_{BRI0}^{BAC\,2} \cos^2\left[\left(\frac{eB_0\sqrt{\mu z}}{mc\gamma}\right)z + \psi_0\right]\right\rangle}{B_0^2\gamma_0}}, \tag{33}$$

where the average $\langle \delta\sigma^2 \rangle$ is performed over the initial spread of mismatches in different bunch slices and ε_{off} accounts for the non linear and thermal contributions.

While the rms normalized emittance oscillates and adiabatically damps as $\gamma^{-1/2}$ in the invariant envelope case (ε_n^{INV}, constant current), it oscillates at constant amplitude along the RF compressor (ε_n^{RFC}), and with a frequency scaling similar to the invariant envelope case, *i.e.* $\frac{\Omega}{z}\ln\left(1+\frac{\gamma' z}{\gamma_o}\right)$, as compared with $\frac{\sqrt{1/4+2\Omega^2}}{z}\ln\left(1+\frac{\gamma' z}{\gamma_o}\right)$. In the case of ballistic bunching the emittance ε_n^{BAC} exhibits on the other hand a completely different scaling, with constant amplitude but an increasing frequency , as $\left(\frac{eB_0\sqrt{\mu z}}{mc\gamma}\right)$.

This is the basis how the transverse emittance can be corrected in the RF compressor; by connecting the two types of flow carefully (proper matching) we can make the emittance oscillate at constant amplitude in the RF compressor and smoothly connect these oscillations to a damped oscillatory behavior in the accelerating sections following the RF compressor, where the beam is propagated under standard invariant envelope conditions - this is possible because of the similar frequency behavior of the two flows.

A possible limitation can occur in all cases, which is due to the requirement of additional external focusing that, as discussed in the previous section, can drive a plasma frequency spread due to chromatic effects.

Table 1: RF compressor parameters

TW Section	**I**	**II**	**III**
Gradient (MV/m)	15	25	25
Phase (Deg)	-88.5	-64.3	0 (on crest)
Solenoid field (Gauss)	1120	1400	0

As an example the SPARC photoinjector [19] design assumes a 1.6-cell S-band RF gun of the BNL/SLAC.UCLA type equipped with an emittance compensating solenoid and followed by three standard SLAC 3-m TW sections, each embedded in a solenoid. The preliminary results of the first simulations show that with a proper setting of accelerating section phases and solenoid strengths it is possible, applying the compression method described above, to increase the peak current while preserving the beam transverse emittance. An optimized parameter set is shown in Table 1.

In order to obtain slow bunching of the beam (the current grows about at the same rate of the energy) and to increase the focusing magnetic field with the current during the compression process, we used the first two sections as compressor stages.

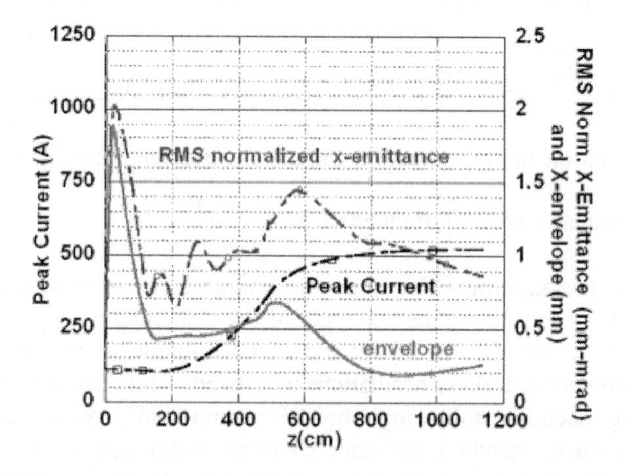

Figure 16: RMS normalized emittance, beam envelope and peak current vs the distance from the cathode.

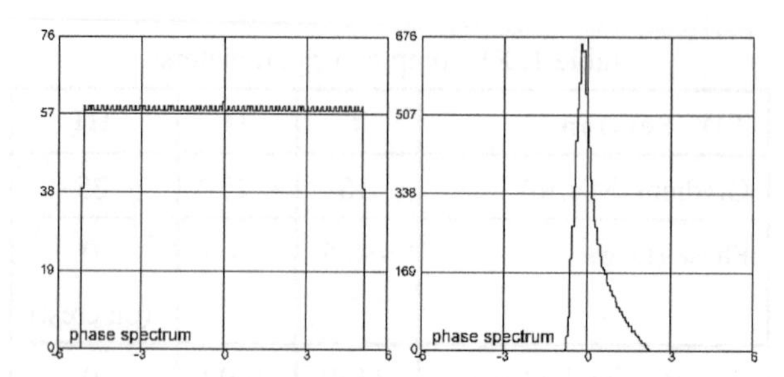

Figure 17: initial and final phase spectrum.

The plots in Fig. 16, of the peak current and the transverse normalized rms emittance (a thermal emittance of 0.3 mm mrad is included) as a function of the distance from the cathode computed by PARMELA for 10K particles, show that a peak current of 510 A can be reached with a transverse rms normalized emittance of 0.9 μm. The final beam energy is only 120 MeV, so additional care must still be taken after this point in order to properly damp residual emittance oscillation driven by space charge correlations.

The plots in Figures 17 and 18 show the evolution of the bunch during the compression as derived from PARMELA computations. One can see that the bunch temporal distribution that is uniform at the beginning tends to a triangular shape: so the value of the peak current in the plot of Fig. 16, that is simply scaled with the rms bunch length, in reality is an average current in the bunch corresponding to a larger value in the peak (almost double with respect to the average).

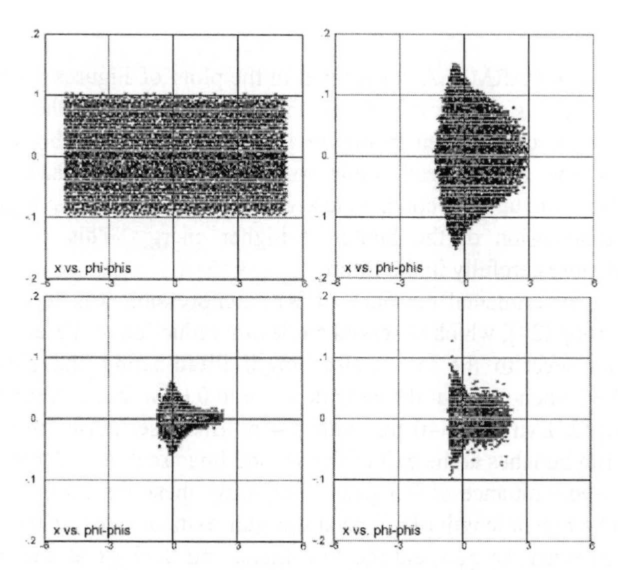

Figure 18: x-ϕ plots: Top, left plot: initial RF gun, top,right plot; output Section 1; bottom, right plot, output Section 2; bottom left plot: output Section 3.

From the point of view of the beam transverse dynamics, during the compression slices with different longitudinal position within the bunch experience different net focusing strengths. In particular the head of the bunch, which contains the maximum charge is defocused, while the tail tends to be focused or overfocused, as is shown in Figure 18, which displays the plot x-ϕ at different points of the compressor line.

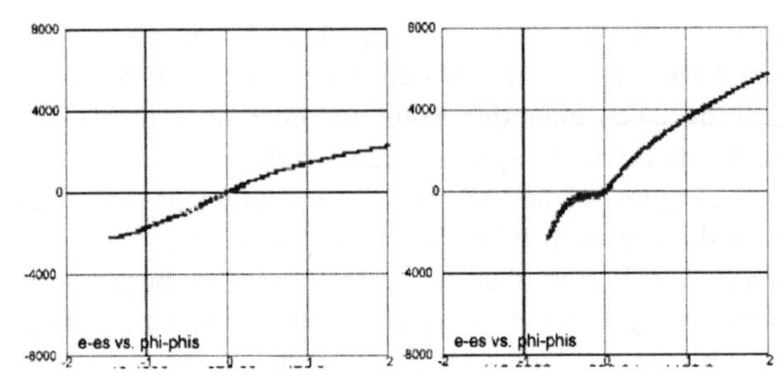

Figure 19: Energy-phase space: left: output Section 1 I=330 A, left: output Section 3 I=510 A.

According to PARMELA convention in the plots of Figures 18 and 19 the head of the bunch is on the left. From the point of view of the longitudinal phase space, as it can be seen in Figure 19, when the current becomes greater than 400 A the bunch head tends to loose the energy-phase correlation differently from the tail that contains less charge, which could be a problem for a further compression of the bunch at higher energy. This point will be investigated more carefully in the future.

The first experimental evidence of RF compression was reported by the BNL-ATF group [24], which observed a relevant reduction of the electron bunch length with respect to the laser pulse length illuminating the photocathode, whenever the launch phase at the cathode close to 0 (low, but increasing, electric field) was used. Extracting 40 pC with a 4 ps rms laser pulse, they measured 370 fs electron bunches at the exit of the second linac section (52 MeV) with an rms normalized emittance of 0.5 μm. Comparing these results to simulations, they inferred a bunch length of 1.6 ps at the gun exit, 0.63 ps at the end of the drift space between the gun and the first linac, and 0.37 ps at the exit of the entire system. Clearly the type of bunching process applied was mixed: velocity bunching in the gun, ballistic bunching in the drift and velocity bunching again in the linac. This last, anyway, was the smallest contribution and this explains the rather large emittance measured for the corresponding bunch charge.

It is interesting to note that the same authors recently proposed to repeat that experiment by decreasing the amount of bunching achieved in the drift and increasing the velocity bunching applied in the linac [25]. Simulations reported in this reference show that a 25 pC bunch can be extracted by launching a 4 ps laser pulse onto the cathode, to obtain a 1 ps electron bunch at the gun exit and a 15 fs bunch length at the exit of the second linac (almost no compression performed in the drift space). It is also interesting that the final value for the emittance is 1 μm in case of no additional focusing applied around the linac sections, while 0.5 μm can be achieved if a solenoid is located in the middle of the first linac section. This is the location where most of the velocity bunching is done, and the solenoid is used to control the beam envelope, thus confirming the theoretical predictions [25].

The first experiment dedicated to verify the velocity bunching concept was performed at UCLA PBPL [6], unfortunately in a photo-injector lay-out (Neptune Laboratory injector) that is clearly not optimized to perform RF compression. Indeed, the first linac section after the gun is a short (0.6 m) PWT linac, not long enough to perform a substantial acceleration together with bunching, *i.e.* to allow the bunch to slip back in phase with respect to the RF wave. However, they measured a 0.39 ps bunch length after the linac by using coherent transition radiation autocorrelator signal, and they reconstructed by simulations the beam dynamics to show that the bunching was still a mixed one, velocity bunching in the PWT linac and ballistic in the following drift.

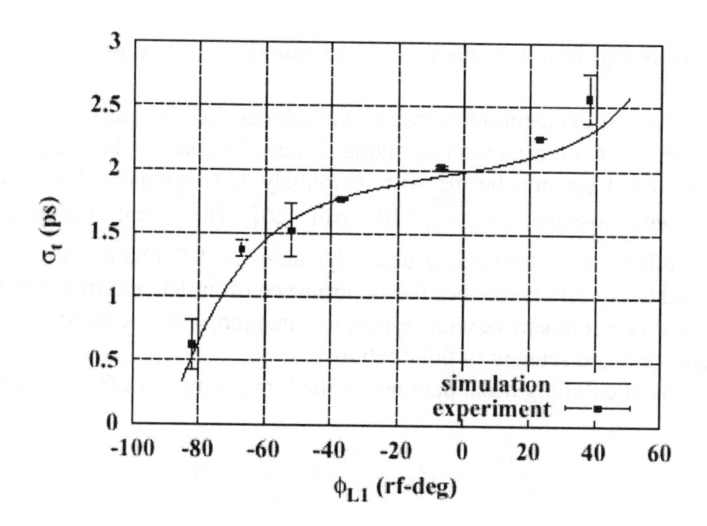

Figure 20. rms bunch length as a function of the phase of the first linac section.

An experiment performed recently with a photoinjector system operating long TW linac sections after the gun was done at the BNL-DUVFEL laboratory [7]. Here they excited the photocathode with a 1.15 ps laser pulse length extracting 200 pC of bunch charge, reaching a minimum 0.5 ps bunch length at the exit of the photoinjector system, at a beam energy of 55 MeV (no compression was performed in the gun neither in the drift space). No further compression could be applied to go below 0.5 ps because of lack of additional focusing: the beam envelope could not be controlled. The excellent agreement found in this experiment between simulations and experimental measurements, shown in Fig. 20, once again provides a nice confirmation that this technique is very promising in attaining fs electron bunches at high brightness. As a latest news, an ongoing experiment at the Livermore PLEIADES lab by a UCLA/LLNL team, is producing outstanding results both in terms of minimum bunch length (0.33 ps) and emittance achieved. Work is in progress, with quite good perspectives [8].

As a conclusion, a fully optimized dedicated photo-injector for application of the velocity bunching technique still does not exist: one of the missions of the SPARC project is indeed to design and commission such a system [19]. The full potential of such a technique has yet to be explored experimentally; high brightness, femtosecond-class bunches may be reached after a full exploitation this scheme.

4 Flat beams production: the concept of emittance exchange

In the context of electron-positron linear collider (LC) projects, the goals of electron sources are even more challenging. Linear colliders in fact require high charge, polarized electron beams with extremely low normalized emittances, with geometric average $\sqrt{\varepsilon_{nx} \cdot \varepsilon_{ny}} \approx 10^{-1}$ μm [26]. The recent analytical and numerical efforts in understanding beam dynamics in RF photo-injectors have again raised the question whether the performance of an RF electron gun based injector could be competitive with respect to a damping ring. A possible answer to this question is discussed in this section.

In terms of colliding beam parameters the luminosity in a LC is defined as

$$L = \frac{n_b N_e^2 f_{rep}}{4\pi\sigma_x^*\sigma_y^*} \times H_D = \frac{P_b}{E_{cm}} \times \frac{N_e}{4\pi\sigma_x^*\sigma_y^*} \times H_D \ , \qquad (34)$$

where $P_b = E_{cm} n_b N_e f_{rep}$ is the beam power, E_{cm} the center of mass energy, n_b the number of bunches per pulse, N_e the number of electron (positron) per bunch, f_{rep} the pulse repetition frequency, $\sigma^*_{x,y}$ the horizontal (vertical) beam size at the interaction point (IP), and H_D the beam-beam disruption enhancement factor [27].

A primary effect of the beam-beam interaction is an enhancement of the luminosity due to the pinch effect, *i.e.* the reduction of the of the cross section of both beams, occuring at the IP that is included in the luminosity definition through the factor $H_D \geq 1$. When electron and positron beams are intersecting, the defocusing electrostatic force diminishes by mutual space charge neutralization and only the focusing magnetic force plays a role, as a strong attractive transverse force between two opposite currents; both particle speicies are deflected towards the axis as in a focusing lens. The benefit of luminosity enhancement is reduced by the fact that particles emit synchrotron radiation in the strong electromagnetic fields of the opposite bunch, known as "beamstrahlung". The probability that a given particle will experience a significant energy loss before colliding with another particle in the opposite beam becomes high. The average fractional beam energy loss is approximately given by

$$\delta_E \propto \frac{r_e^3 N_e^2 \gamma}{\sigma_z \left(\sigma_x^* + \sigma_y^*\right)^2}. \tag{35}$$

Beamstrahlung leads to a large spread in the center of mass energies, reducing the accuracy during the measurement of a specific event, δ_E therefore has to be limited typically to a few percent. By choosing a large beam aspect ratio $\sigma_x^* >> \sigma_y^*$, δ_E becomes independent from the vertical beam size and luminosity can be increased by making $\sigma_y^* \sqrt{\varepsilon_{n,y} \beta_y^* / \gamma}$ as small as possible. Choosing $\beta_y^* = \sigma_z$ the luminosity can be expressed eliminating N_e as

$$L \propto \frac{P_b}{E_{cm}} \times \sqrt{\frac{\delta_E}{\varepsilon_{n,y}}} \times H_D. \tag{36}$$

An injector for a linear collider must provide a flat beam in order to reduce beamstrahlung effects at the interaction point (IP), thus implying $\varepsilon_{n,y} << \varepsilon_{n,x}$. But a production of a flat beam directly from the cathode surface would increase the difficulties for emittance compensation, easily achieved by means of a symmetric solenoid as discussed in the previous section. A flat beam is typically delivered by a damping ring. Nevertheless a transformation of a round beam derived from a photoinjector into a flat beam has been recently proposed [28] by means of simple linear beam optics adapter at the exit of a injector.

This transformation is possible with a magnetized beam, as produced by an rf gun with a cathode embedded in a solenoid field [9]. At the exit of the gun/solenoid system the beam has an angular momentum given by

$$p_\vartheta = \frac{1}{2} e B_{z,c} R_c^2 \quad , \tag{37}$$

where $B_{z,c}$ is the on cathode magnetic field, and R_c the laser spot radius. Both transverse planes are thus coupled by the beam rotation. Such rotation can be arrested by a suitable choice of a skew quadrupole triplet that, in addition, changes the emittance ratio according to the relation,

$$\frac{\varepsilon_x}{\varepsilon_y} = 1 + \frac{2\sigma_r^2}{\beta^2 \sigma_{r'}^2}\ , \tag{38}$$

for a beam with rms size σ_r and rms angular spread $\sigma_{r'}$. The final emittance ratio is thus simply variable by adjusting the free parameter $\beta = 2p_o/eB_{z,c}$, via the magnetic field on the cathode. Design studies based on this scheme [9] show that for a 0.8 nC charge one can obtain ε_{nx}=1.1x10^{-5} m and ε_{ny}=3x10^{-8} m, with an emittance ratio of about 370. Additional studies are under way for a better understanding of the space charge effects when the transformation is applied at low energy. A first successful demonstration of this method was recently achieved at the A0 experiment at FNAL [29].

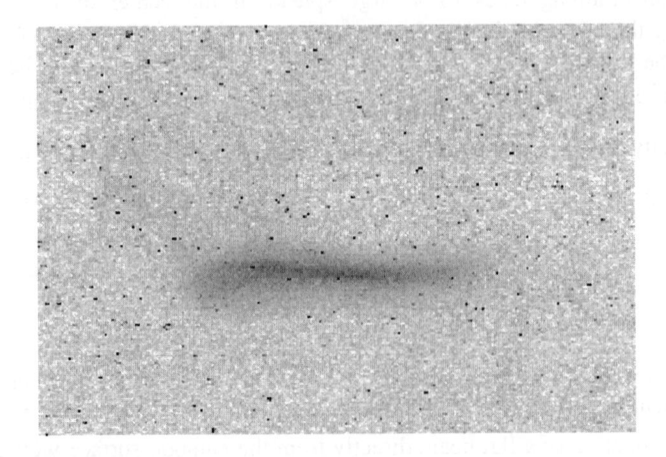

Figure 21 - Beam profile on OTR screen 1.2 m downstream of the third skew quadrupole at FNAL A0 flat beam experiment.

In Fig. 21 the beam image downstream the quadrupole triplet in the A0 experiment is shown, and it has been verified that the beam remains flat as it drifts farther downstream, an important experimental achievement that demonstrate the effectiveness of the linear beam optics adapter in the context of RF photoinjector. The measured ratio of emittances is about 50, with ε_{nx}=0.9 μm and ε_{ny} =45 μm for a 1 nC beam, which yields a geometric mean emittance that is still rather high at 6.5 μm. Additional experiments are foreseen in the near future to optimize the emittance compensation process.

Acknowledgements

We like to thank W. Decking, H. Edwards, D. Edwards, P. Emma, K. Floettmann, W. S. Graves, K. J. Kim, P. Musumeci, D. T. Palmer, Ph. Piot, V. Telnov, X. J. Wang, for the many with helpful discussions and suggestions.

References

1. Linac Coherent Light Source (LCLS) Conceptual Design Report, SLAC-R-593 (2002).
2. TESLA XFEL, Technical Design Report (Supplement), DESY 2002-167, TESLA-FEL 2002-09, (2002).
3. L. Serafini, J. B. Rosenzweig, "Envelope analysis of intense relativistic quasilaminar beams in rf photoinjectors: a theory of emittance compensation", Phys. Rev. E **55** (1997) 7565.
4. P. Emma, “Accelerator physics challenges of X-rays FEL SASE sources”, Proc. Of EPAC-02, Paris 2002.
5. L. Serafini, M. Ferrario, “Velocity Bunching in PhotoInjectors” , AIP CP 581, 2001, pag.87.
6. P.Musumeci, R.J.England, M.C.Thompson, R.Yoder, J.B.Rosenzweig, “Velocity Bunching Experiment at the Neptune Laboratory”, these proceedings.
7. P. Piot, L. Carr, W. S. Graves, and H. Loos, “Subpicosecond compression by velocity bunching in a photoinjector”, these proceedings.
8. G. Le Sage et al., “Ultrafast Materials Probing with the LLNL Thomson X-ray source”, these proceedings.
9. R. Brinkmann, Y. Derbenev, and K. Flöttmann “A low emittance, flat-beam electron source for linear colliders”, Phys. Rev. ST Accel. Beams 4, 053501 (2001)
10. K. Bishofberger et al., “Flat Electron Beam Production at FNAL - A Status Report”, Proc. Of LINAC 2002, Korea.
11. C. A. Brau, “What Brightness Means”, these proceedings.
12. J. Rossbach, E. L. Saldin, E. A. Schneidmiller, M. V. Yurkov, “Interdependence of Parameters of an X-ray FEL”, TESLA-FEL 95-06.
13. Linac Coherent Light Source (LCLS) Conceptual Design Report, SLAC-R-593 (2002)
14. M. Reiser, “Theory and Design of Charged Particle Beams”, J. Wiley & Sons, (1994).
15. PH. Piot, “Review of experimental results on high brightness electron beams sources”, these proceedings.
16. J. Rosenzweig and L. Serafini, “Transverse particle motion in radio-frequency linear accelerators”, Phys. Rev. E 49 (1994) 1599.
17. B. E. Carlsten, Nucl. Instrum. Methods A **285**, 313 (1989).

18. S. G. Anderson and J. B. Rosenzweig, “Non-equilibrium transverse motion and emittance growth in ultra-relativistic space-charge dominated beams, Phys. Rev. ST Accel. Beams 3(9), 094201 , (2000).
19. L. Serafini, “New perspectives and programs in Italy for advanced applications of high brightness beams”, these proceedings.
20. J. Clendenin et al., “Reduction of thermal emittance of rf guns”, SLAC-PUB-8284 (1999)
21. M. Ferrario et al., “HOMDYN Study For The LCLS RF Photo-Injector”, Proc. of the 2nd ICFA Adv. Acc. Workshop on “The Physics of High Brightness Beams”, UCLA, Nov., 1999, see also SLAC-PUB-8400.
22. J. Buon, “Beam phase space and emittance”, in CERN 94-01.
23. J. B. Rosenzweig and E. Colby, Charge and Wavelength Scaling of RF photoinjector design, TESLA-95-04.
24. X.J. Wang et al, Phys. Rev. E 54, R3121 (1996).
25. X.J. Wang and X.Y. Chang, *Femto-seconds Kilo-Ampere Electron Beam Generation*, in publ. in the Proc. of FEL 2002 Conf.
26. V. Telnov, Physics goals and parameters of photon colliders, Proc. of II Workshop on e-e- interaction at TeV Energies, To be published in Int. J. Mod. Phys. A.
27. K. Yokoya , *Beam-Beam Interaction in Linear Collider*, Proc. Joint USA/CERN/JP/Ru Acc. Sch. High Quality Beams, AIP 592 (2000).
28. Ya Derbenev, Adapting Optics for High Energy Electron Collider, UM-HE-98-04, Univ. Of Michigan, 1998.
29. D. Edwards et al., The flat beam experiment at the FNAL photoinjector, Proc. of LINAC 2000, Monterey.

SIMULATION CODES FOR HIGH BRIGHTNESS ELECTRON BEAM EXPERIMENTS

LUCA GIANNESSI

ENEA
Via Enrico Fermi 45,
00044 Frascati (Roma), ITALY
e-mail:giannessi@frascati.enea.it

The high brightness required by a short wavelength SASE FEL may be reached only with an accurate design of the electron beam dynamics from the generation in the rf-injector up to the undulator. The beam dynamics is affected by strong self consistent effects at injection, in the compression stage and during the FEL process. The support of numerical simulations is extensively used in the predictions of the beam behaviour in these non linear dynamical conditions. We present a review of available simulation tools, currently exploited in the design of a short wavelength free electron laser.

1. Introduction

We usually refer to a high quality e-beam as a beam with large brightness

$$B_n \equiv \frac{2I}{\varepsilon_{nx}\varepsilon_{ny}} \tag{1}$$

i.e. high peak current and small normalized emittances. The quality of the electron beam is the main limiting factor in reaching short wavelengths with a SASE FEL. In existing SASE FEL projects, low emittance beams are obtained within the present state of the art RF photocathode injectors and the peak current is increased by longitudinal compression. In these devices, as well as in the FEL amplification process itself, the interaction of the beam self fields with the beam, plays a significant role in the beam dynamics. A proper understanding of this dynamics is essential in the design of short wavelength SASE FEL devices, where the constraints on the beam "quality" are stringent. In this paper we review the most common techniques developed to study the beam behavior in strongly non linear conditions.

The three dynamical regimes that we will be analyzed are

- Injection
- Compression
- SASE FEL interaction

In all these regimes we have to predict the behavior of the same physical system, consisting of an ensemble of charged particles which interacts with electromagnetic fields. Despite of the common root of these problems, there is not a unique method for their solution and specialized techniques have been developed. The main difficulty, which is practically met in modeling the physics of such systems, is a common one, and consist in the large number of electrons contained in a single bunch of charge. The typical e-beam charge is of the order of 10^{-9} C, corresponding to $\sim 10^{10}$ electrons. A multi-body system of this size, cannot be efficiently simulated in all his complexity, even in a state of the art parallel computer. The widely used solution is that of simulating a reduced number of particles, with a scaled charge, each representing a large number of real electrons. The introduction of these macroparticles has some unpleasant consequences that need to be properly treated in a correct numerical implementation. This fact can be shown with a simple example. Let us consider a Gaussian bunch represented with a number N of macro-particles. The charge density, defined as

$$\rho(x,N)=\frac{1}{N}\sum_{j=1}^{N}\delta(x-x_j) \tag{2}$$

in Fourier space becomes

$$\rho(k,N)=\frac{1}{N}\sum_{j=1}^{N}e^{ikx_j} \tag{3}$$

that, in the limit of $N\rightarrow\infty$, reduces to the Fourier transform of the Gaussian distribution

$$\rho(k,N)=e^{-\frac{k^2\sigma^2}{2}} \tag{4}$$

In Fig. 1, it is shown the spectrum of the distribution obtained from Eq. (3), for N=100 and $N=10^4$. At low wave vector k, the behavior is that of a smooth distribution, correctly reproducing the Gaussian. At high k, the phase factors in the exponent of Eq.(3) are uncorrelated and the amplitude of the corresponding Fourier components scale as $1/\sqrt{N}$. It is evident the effect of a reduced number of numerical particles in the representation of a charge distribution. In a real beam the amplitude of this high frequency “noise”, commonly referred to as “shot noise”, is orders of magnitude lower than in the case of a “simulated” beam. A “noisy” charge distribution by itself should not be considered as a problem. It becomes a problem when the distribution enters as source for the electromagnetic fields that are driving the dynamics of the whole system. This feedback loop, charge distribution→fields→charge distribution, must be simulated only for frequency components where the distribution has the correct spectral behavior of the physical distribution. The other frequencies must be suppressed, otherwise the simulation will produce unphysical results.

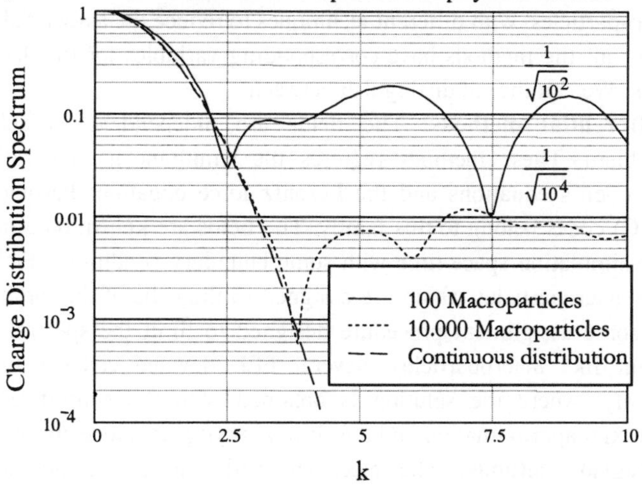

Fig.1 Spectrum of charge distribution for an analytical continuous distribution and for a distribution sampled with 10^2 and 10^4 macro-particles.

This frequency selection is one of the main distinguishing features of the three regimes of operation mentioned above. In an injector we have bunch lengths of few picoseconds, interacting at mm / sub-mm wavelengths. In a compressor the bunch length is reduced to few tens of microns, and the coupling wavelength can scale down to the micrometer range. These are broad band coupled systems where the high frequency components should be filtered for a correct numerical representation. In these broad-band conditions, filtering is generally obtained by giving a finite size to macro-particles [1,2,3]. The Fourier transform of a "Gaussian" particle of RMS extension σ in space is proportional to $\exp(-k^2\sigma^2/2)$. The effect of a spatial extension is that of a low band pass filter, but a similar result may be accomplished for radiation fields in the CSR case by spreading the particles in momentum space [4]. In a free electron laser, the coupling fields may have very high frequency components, but in a very narrow frequency bandwidth, which is proportional to the Pierce parameter ρ [4]. The input phase space is prepared to reduce the harmonic content in this spectral window with a "quite start" procedure (see sec. 4.1) and the narrowness of the spectrum is exploited in simplifying the equations for the fields and the particles dynamics.

The frequency response typical of the various regimes, are exploited in the numerical implementations to obtain correct results with a reduced number of macro-particles. While all the most widely known numerical models are based on macro-particles, with different techniques to reduce the associated numerical noise, not all the methods have corresponding variables for the description of the fields. We can indeed distinguish between

- **Differential methods:** The fields are independent variables of the problem. These methods requires the simultaneous integration of the Maxwell's equations and the Lorentz force equation. Particles In Cells (PIC) codes belong to this family. The fields are known on a mesh filling the simulation space-time with some boundary conditions. Fields sources are macroparticles which are assigned to the cells of the mesh according to some assignment procedure designed to limit the spectral content of point like macroparticles. Several FEL devoted codes belong to this family, where the solution is obtained with the simplification of the paraxial approximation and the slowly varying envelope approximation.
- **Integral methods:** The electromagnetic fields are not independent variables of the problem. The interaction between particles is calculated according to "forces" calculated from particles positions and velocities known at earlier times. Effects of plane walls can be included with the image charge method, radiative and retarded effects may also be included by the Lienard Wiechert retarded potential formalism [6].

The first class of "solvers" allows in principle the implementation of rigorous physical models, including almost all the aspects of the problem, such as the interaction with walls of any shape, any kind of retarded effects associated to the finite "time of flight" of signals, radiative effects, etc. There is a price to be paid for this physical accuracy, i.e. Maxwell's equations are partial differential equations and the solver must follow some important prescriptions. The fields are indeed known on a mesh, characterized by a given spacing $(\delta_x, \delta_y, \delta_z)$. According to the Nyquist theorem there is a limit in the highest wave vector that can be represented by the mesh,

$$k_{x,y,z} = \frac{1}{2\delta_{x,y,z}} \tag{5}$$

and consequently in the highest frequency that can be supported by the mesh,

$$\omega_{\max} = c \max(k_x, k_y, k_z) = \frac{c}{2\min(\delta_x, \delta_y, \delta_z)} \tag{6}$$

Higher frequencies will appear as low frequency components. For the same reasons the time step size is limited by this maximum frequency by the relation

$$\delta t < \frac{1}{\omega_{max}} \tag{7}$$

Longer time steps would cause instability in the solution [7]. In table I we have summarized typical numbers for the mesh and time step requirements in the three regimes of beam dynamics. The simulation volume is given by the product of the transverse mesh size (squared in a 3D simulation) by the typical longitudinal slippage length of the problem. The mesh size is the ratio between the simulation volume and the third power of the cutoff frequency. Assuming a limit of 10^6 mesh vertices for a numerical implementation, only the simulation of a 2D injector seems practicable. In the example of Table I we have assumed typical numbers for a "short" wavelength SASE FEL. The mesh size can be reduced in this case by taking advantage of the narrow bandwidth of the FEL gain, with the slow wave approximation used by many FEL devoted codes. There have been however successful attempts in the simulation of FELs with PIC codes developed for plasma dynamics, in the long wavelength regime and in conditions of reduced dimensions[8].

Table I. Typical mesh sizes for the simulation of the beam dynamics at the injection, compression and FEL stage

	INJECTOR	**COMPRESSOR**	**FEL**
Transverse mesh size	10 cm	1 cm	1 mm
Slippage length	3 cm	10 cm	1 μm
Cut off wavelength	100 μm	10 μm	1Å
# of Mesh vertices (3D)	10^8	10^{10}	10^{18}
# of Mesh vertices (2D)	10^5	10^7	10^{11}
Integration time	16 ps	60 ps	330 ps
Maximum step length	0.3 ps	0.03 ps	$3\ 10^{-19}$ s
# of time steps	$5\ 10^4$	$2\ 10^6$	10^{12}

This paper is organized as follows. In the next section we will review the methods used in the simulation of photo-injectors. Section 3 is devoted to an analysis of the techniques implemented for the simulation of compressors and of coherent synchrotron radiation effects and finally the last section is devoted to an overview of the simulation codes developed for short wavelength SASE FELs.

2. Injectors

We consider "injector" all the elements required to bring the beam from rest energy in the lab frame, up to an energy that is large enough to neglect effects of self interaction in uniform motion condition. A typical layout of a "split" system is shown in fig. 2.

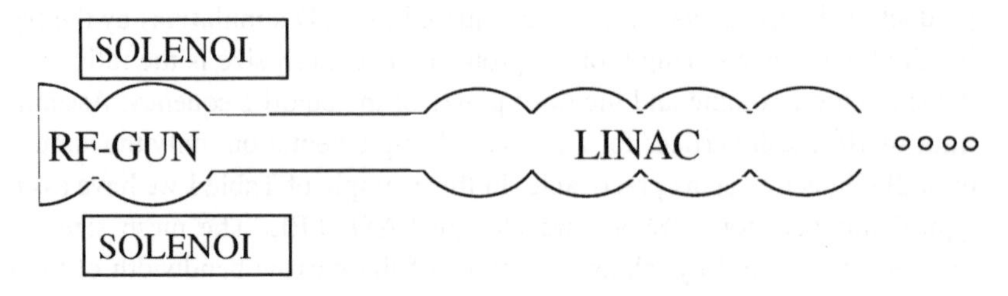

Figure 2. Typical layout of an "injector".

An injector is characterized by

1. A very high gradient, i.e. the beam becomes relativistic in the first half cell of the RF gun
2. The electron pulse length is much shorter than the radiofrequency period
3. An RF injector is a "quasi" axially symmetric device. Deviations from the axial symmetry may be due to in homogeneity of the cathode quantum efficiency, to dipolar terms in the accelerating fields, to asymmetry of the laser spot and to misalignments of the structure. Unless the study of one of these aspects is required, the assumption of axial symmetry greatly simplifies the problem by reducing the number of macroparticles and (or) the number of mesh points required.

The dynamics of such a system has been extensively described in [9]. The request of minimizing the induced emittance growth is accomplished by tuning the frequency of the first plasma oscillation that the beam executes according to the internal space charge fields and to the focusing due to rf forces and to the solenoid. When this frequency is correctly matched, the emittance has a minimum at an energy high enough that the contribution of the betatron motion associated to the thermal emittance overcomes that of the laminar motion. In ideal conditions, i.e. a beam which is flat transversally and longitudinally, this "emittance compensation" procedure brings the emittance at the end, almost to the same order of the initial thermal emittance. This "emittance compensation" concept [10], where magnetic focusing is used for compensating the space charge emittance growth, produced serious improvements in the beam brightness obtained at the injector exit [11], and, from the simulation point of view, substantially raised the requests in terms of numerical accuracy and resolution.

2.1. *PIC codes*

The injector is a good candidate for a particle in cell simulation in two dimensions. The restriction to a cylindrical symmetry provides an even larger extra advantage in the formulation of the problem, that goes beyond the simple reduction of particles/mesh points. In axial symmetry a closed sub-set of Maxwell equations, driven by the source $[\rho, J]$, can be fully described in terms of a scalar pseudo-potential

$$\Phi \equiv r \cdot H_{\varphi}(r,z,t) \tag{8}$$

obeying the scalar wave equation

$$\left[\frac{\partial^2}{\partial z^2} + \frac{\partial^2}{\partial r^2} - \frac{1}{r}\frac{\partial}{\partial r}\right]\Phi - \frac{1}{c^2}\frac{\partial^2 \Phi}{\partial t^2} = r\frac{\partial J_z}{\partial r} - r\frac{\partial J_r}{\partial z} \tag{9}$$

Electric and magnetic fields are recovered from the function Φ according to

$$\frac{dE_r}{cdt} = -\frac{1}{r}\frac{\partial \Phi}{\partial z} - J_r, \quad \frac{dE_z}{cdt} = -\frac{1}{r}\frac{\partial \Phi}{\partial r} - J_z, \quad \text{in } (\mathrm{r},\mathrm{z}) = (\mathrm{r}_m, \mathrm{z}_n) \tag{10}$$

The advantage in this formulation goes beyond the fact that the solution can be obtained solving a scalar wave equation in place of vector equations. The main advantage is that this formalism allows to get around the decoherence problems associated with Maxwell's equations in a conventional leap-frog approach. In Cartesian geometry, with the implementation of leap frog integration scheme, magnetic and electric components of the fields are not known at the same time and in the same position

$$\begin{aligned} &B_{k,l}(t+\delta t) = B_{k,l}(t-\delta t) + g\left[E_{k\pm 1/2, l\pm 1/2}(t)\right] \\ &E_{k+1/2,l+1/2}(t+\delta t/2) = E_{k+1/2,l+1/2}(t-\delta t/2) + f\left[B_{k\pm 1, l\pm 1}(t), J_{k\pm 1, l\pm 1}(t)\right] \end{aligned} \tag{11}$$

Where the functions f and g are the linear summation of the $\nabla\times\vec{B}+4\pi J$ and $\nabla\times\vec{E}$ terms respectively. This displacement between electric and magnetic components becomes critical because of the electric and magnetic field cancellation in the transverse force whose combined contribution scales with $1/\gamma^2$. This lack of cancellation brings longitudinally correlated forces inducing a nonphysical emittance growth [12]. For this reason 3D particles in cells codes developed for plasma physics applications are not reliable in the simulation of high brightness relativistic beams. In alternative 2D cylindrical codes as ITACA [13] or SPIFFE [14] have been developed and extensively used. The main limitations in their application are still related to the size of the mesh required in the simulation of short bunches. The mesh size is indeed grown with the introduction of the emittance compensation concept, both because of the larger spatial extension where the space charge dynamics must be followed and because of the increased numerical resolution required by the brightness improvement. As a final remark, we note that the frequency tuning, the balancing between cavities in a multi cell gun are all integrant part of the simulation and are affected by the beam parameters. The problems related to the determination of "external fields" and "space charge" fields are not fully separated as it is with integral methods, and from the practical point of view, the set up of a simulation may be in this case somewhat less handy.

2.2. *Integral methods*

The analysis of the beam dynamics developed in ref. [9] has pointed out that a beam with typical current of 100A may still be in space charge dominated condition even at quite high energies, exceeding 100 MeV. The request of such a high energy has extended longitudinally the "simulation volume" to several meters. At the same time the electron pulses of few picoseconds produced with photo-cathodes have increased the harmonic content of space charge fields in the bunch, reducing the required step size of the mesh for a PIC code. Simulation of long system with small mesh size tend to become time consuming and less practical. According to the example shown in Table I these consideration are even more important in an extension to the 3D domain. Codes based on integral methods do not suffer of this limitation. The evaluation of the fields from the phase space coordinates of the particles is done only in the positions instantaneously occupied by the particles. Some codes, as TREDI [15] or Atrap [16], take into account the effects due to the finite propagation velocity of signals calculating the fields according to the Lienard Wiechert formalism,

$$E = \frac{n \times \left[(n-\beta)\times\dot{\beta}\right]}{(1-\beta\cdot n)^3 |R|} + \frac{(n-\beta)\left[1-|\beta|^2\right]}{(1-\beta\cdot n)^3 |R|^2}\Bigg|_{ret}$$
$$B = n \times E\big|_{ret} \tag{12}$$

evaluated at the retarded time

$$t' = t - \frac{R(t')}{c} \tag{13}$$

This is accomplished in TREDI by storing in memory the histories of macro-particles trajectories, and by tracking back in time the source coordinates until the retarded condition (13) is satisfied. The effects of boundaries can be included only for flat walls by the image charge method. In practice only the cathode wall is considered. The above scheme still shows several drawbacks. The backward tracking procedure is time consuming and in a point to point interaction scheme the number of evaluations scales as the square of the number of macro-particles. An improvement of the method consists in evaluating the fields on a mesh surrounding the beam and by interpolating these fields to the particles positions. This allows to simulate a larger number of macro-particles with a defined mesh size, but still the major issue of the approach is constituted by the computation time. A simplification is obtained in the "static" approximation used e.g. in Parmela [17], in GPT[18,19], and in Astra[20]. This approximation consist in assuming that the beam relative energy spread is small, and that a reference frame where the beam may be considered at rest, exists. The effect is that of neglecting the finite velocity of signal propagation within the bunch. The energy spread $\Delta\gamma$ is proportional to the bunch length $\Delta\phi$

$$\frac{\Delta\gamma}{\gamma} \propto \frac{\gamma'\lambda_{RF}}{2\pi}(\beta\gamma)^2\Delta\phi << 1 \tag{14}$$

where $\gamma' = \frac{d\gamma}{dz}$. In these conditions (as e.g. in PARMELA "SCHEFF" and "SPCH3D" modes), the self fields are calculated by solving the Poisson equation for the electrostatic field in this moving frame. The fields are then transformed back to the laboratory frame where kicks to the particles are applied. In a photo-injector the electron bunch is short with respect to the RF period and when the beam becomes relativistic the approximation (14) becomes satisfied. The largest energy spread occurs in proximity of the cathode, where the beam is not yet relativistic and the quasi static approximation still works. The quasi static approximation may be critical in the simulation of long bunches as in the case of RF thermo-ionic guns, where the longitudinal phase spread leads to a large energy spread at the cathode.

A very efficient algorithm is obtained in HOMDYN [21], by considering a multi-envelope model based on the time dependent evolution of a uniform bunch [22],

$$\frac{d^2\sigma(z,\zeta_i)}{dz^2}+\frac{p'}{p}\sigma'(z,\zeta_i)+K\sigma(z,\zeta_i)=\frac{I(z,\zeta_i)g(z,\zeta_i)}{2I_0p^3\sigma(z,\zeta_i)}+\frac{\varepsilon_{n,th}^2}{p^2\sigma^3(z,\zeta_i)}\quad i=1,N \qquad (15)$$

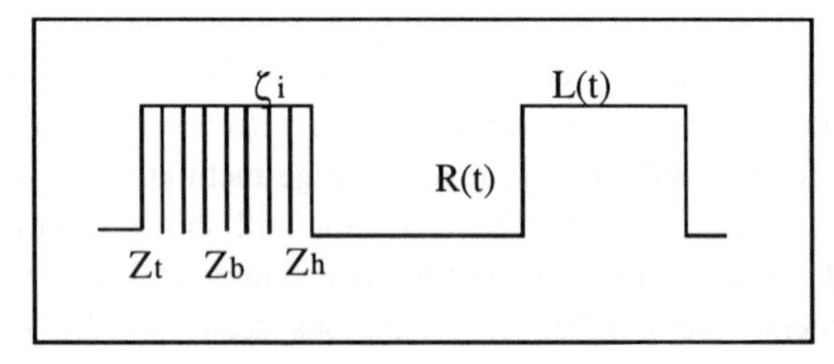

Fig. 3 Electron beam longitudinal profile and slicing procedure adopted in HOMDYN.

The electron bunch is sliced along the direction of propagation, as in Fig.(3).

The coordinate z in Eq.(15) represents the evolution along the beamline, and the coordinate $\zeta_i \equiv z_i - \beta_i c t_i$ indicates the position of a slice of the beam along the bunch. The function

$$g(\zeta_i)=\frac{1-\zeta_i/L}{\sqrt{(1-\zeta_i/L)^2+A_{r,s}^2}}+\frac{\zeta_i/L}{\sqrt{(\zeta_i/L)^2+A_{r,s}^2}}\quad A_{r,s}\equiv\frac{R}{\gamma L} \qquad (16)$$

represents the space charge interaction of a single slice with the whole beam, depending on the instantaneous bunch length and aspect ratio A. The above expression has been derived for a cylindrical beam of homogeneous density. The approximation in HOMDYN consist in considering the shape of the beam distribution unchanged along the beamline, except for the aspect ratio and the length. The beam emittances are calculated at each step as the projected emittances of all the slices of the beam,

$$\Delta\varepsilon_n^{cor}=\frac{1}{N}\sqrt{\sum_{i=1}^{N}\sigma(z,\zeta_i)^2\sum_{i=1}^{N}\sigma'(z,\zeta_i)^2-\left[\sum_{i=1}^{N}\sigma(z,\zeta_i)\sigma'(z,\zeta_i)\right]^2} \qquad (17)$$

Despite of these strong assumptions, HOMDYN allows an accurate determination of the working point which usually depends on a large number of input parameters as input phase, solenoid field, solenoid position, etc. The simulations results show only a weak dependence on the number of slices. A run with 40 slices of the first 10m of the SLAC injector [23] lasts less than 1' on a 2 GHz Intel P-IV [24].

2.3. *Comparison between codes*

In Fig. (4) we have shown a comparison of simulations of a split configuration similar to the one of Fig. (2), obtained with ITACA, PARMELA and HOMDYN. The dashed lines represent the radial emittance and the continuous line represents the beam envelope.

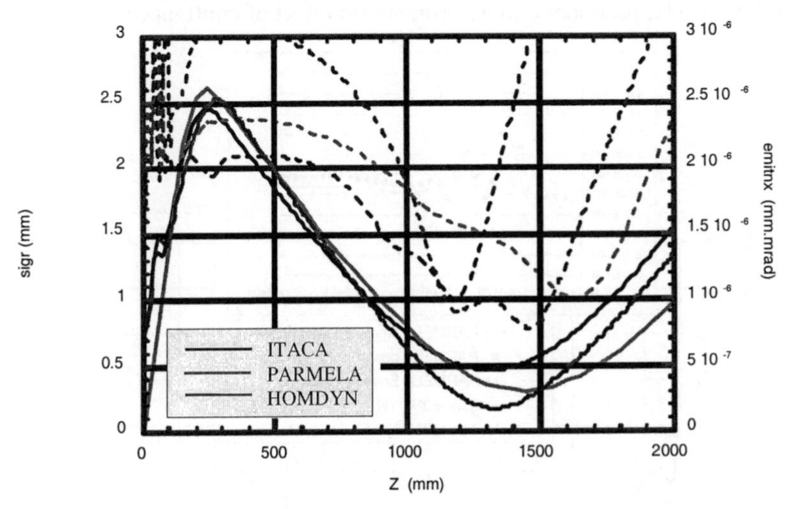

Fig. (4) Comparison between ITACA, PARMELA and HOMDYN obtained in a configuration based standard split injector configuration as the one shown in Fig. (2) *(Courtesy of L. Serafini)*

With the assumption of unchanged beam shape, HOMDYN preserves the linearity of transverse fields for each slice along the integration. For this reason reaches the lowest emittance. The behavior of the emittance relevant to ITACA was partially affected by the size of the mesh. The trend observed has shown an improvement of the agreement with a finer mesh [25]. The differences in the waist positions and emittances minima, suggest also that a slight difference in the definition of the solenoid field in the simulations may have also played a role (the sensitivity to the magnetic field integral is high). In Fig. (5) it is shown a comparison between PARMELA and HOMDYN for a split system including two TW linac sections, with the parameters of Tab. II.

Table II. Set of parameters for the simulation shown in Fig. (5).

Gun	BNL, 1.6 cells
Charge	1 nC
Spot radius	1 mm
Laser pulse (flat)	10 ps
Peak field	140 MeV/m
Linac	2 TW sections
Linac gradient	25 MeV/m

The approximation of uniform bunch preserved along the beamline on which HOMDYN rely, produces a more pronounced effect of emittance compensation.

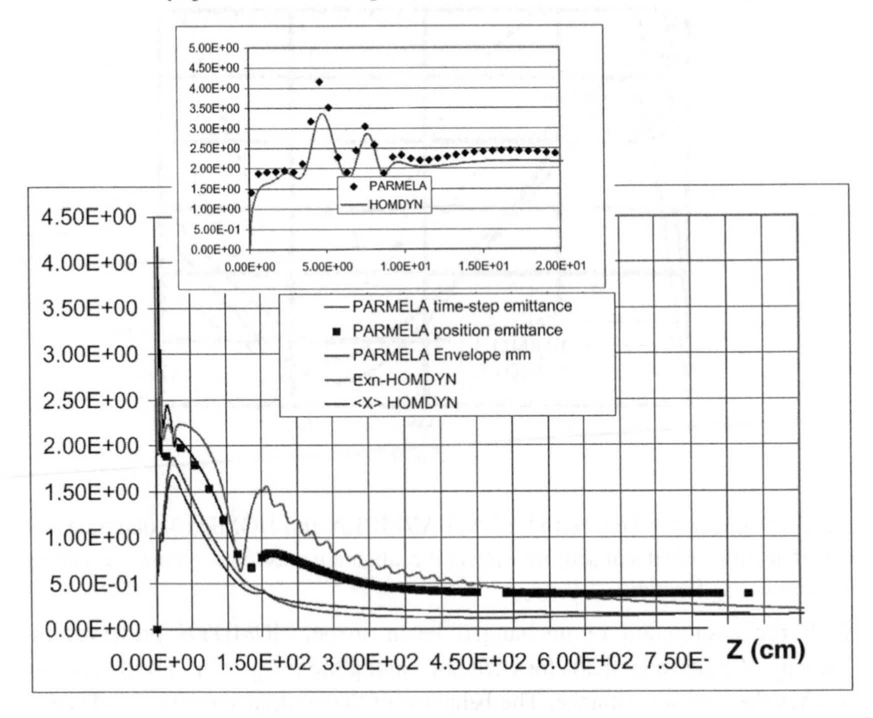

Fig. (5) Comparison between PARMELA and HOMDYN with the parameters listed in table II. The simulation with PARMELA has been obtained with 10^4 macro/particles, tim step of 0.1°, and a mesh of 20(r)x400(z) points (SCHEFF mode). The upper plot represent the first 20 cm of the simulation. *(Courtesy of M. Ferrario and C. Ronsivalle)*

The agreement is however reasonable and the position of the minimum of the emittance before the linac is reproduced. A comparison of HOMDYN with ASTRA [26], has shown a similar behavior. In Fig. (6) a comparison between PARMELA and TREDI with the same parameters is shown for the first 1.7m. Despite the differences in the models, the minimum projected emittance differs of about 10%. The working point of Tab.II which has been optimized with PARMELA and HOMDYN is not perfectly optimized in the TREDI simulation. Differences are probably be due to the different algorithmic representation of the fields and, in part, to the retarded effects which are included in TREDI. More details may be found in ref. [27].

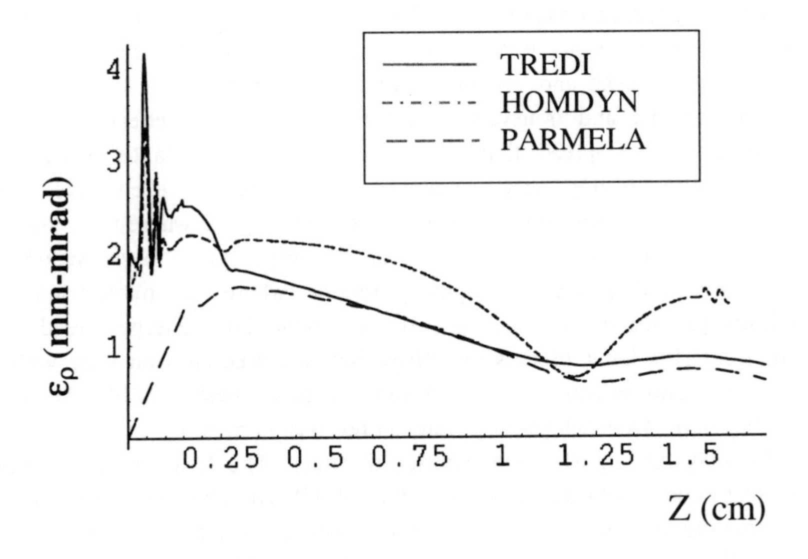

Fig. (6) Comparison between TREDI and HOMDYN and PARMELA with the parameters listed in table II. The radial emittance is plotted for the first 1.70 m along the beamline. *(Courtesy of M. Quattromini)*

One of the targets of the analysis of ref.[28], where PARMELA (LANL and UCLA) has been compared to other PIC codes, was the study of the limits of the quasi static approximation. The results have shown an overestimation of the PARMELA vs. PIC codes as MAGIC2D [29] emittances, of about 20%. The conclusions attributed part of this difference to the quasi static approximation adopted in PARMELA.

3. Compressor

The highest normalized beam brightness (1) obtained in a RF photo-injector is of the order of 10^{14} [30,31] while the beam brightness required for the design parameters of both the LCLS [32] and TTF [33] SASE FELs projects is approximately $4x10^{15}$. An improvement in the injector brightness, especially with the introduction of the RF compression [34,35] is expected, but the remaining gap should be filled with an increase of the normalized beam brightness obtained by magnetic beam compression. The scheme is the following: a correlated energy slope is induced by an off crest operation of the accelerating RF field. This correlation is exploited in a dispersive magnetic chicane to produce longitudinal focusing. One of the most challenging issues related to longitudinal compression is due to the coherent spontaneous radiation produced by the electrons in the tail of the bunch, that reaches the head of the bunch along the bend and interacts with the bunch itself. We have a broad band longitudinal and transverse interaction whose main effect is that of inducing an energy spread, that is transformed into an emittance increase by the longitudinal-to-transverse coupling of the chicane itself. We may distinguish two different regimes, a low frequency regime where the coherent spontaneous emission produces a correlated energy spread over the whole bunch [36], and a higher frequency regime where the micro-bunching instability [37-40] induces the growth of uncorrelated energy spread and emittances. This latter process resembles that of a free electron laser with a broad band gain associated to the chicane magnetic field [39,40]. The peak frequency of the feedback gain is located at few tens of microns.

We may distinguish two families of codes. The first consist of "first principles" codes with the typical structure of TREDI. The algorithm consists in evaluating the interaction between the macro-particles according to the Lienard Wiechert retarded potentials, in building a field representation over the particles positions and in advancing the particles of a time step. The second family consist in codes based on the "Line charge method", i.e. based on the expression of the longitudinal wake of a line charge distribution moving in a curved path [36].

3.1. *"First principles" codes*

While TREDI has been developed for the simulation of linacs and photoinjectrors dynamics, other codes as TRAFIC[4] [41,42] developed by M. Dohlus, T. Limberg and A. Kabel, and a "not named" R. Li code [43] (will be identified as "RL" in the following) have been specifically designed for the simulation of CSR effects. Both are based on first principles with noise suppression obtained by giving a finite size to macro-particles. The procedure adopted for the numerical noise suppression is the main difficulty and the main source of differences between the predictions of this class of codes. The relatively high frequencies involved would require a large number of macroparticles especially for the simulation of the microbunching instability, while the implementation of retarded effects and the preservation of causality in the evaluation of the integral over a finite dimension charge distribution is time consuming and severely limits the number of macroparticles that can be practically simulated. For this reason the RL code is limited to two spatial dimensions in the orbit plane. In TRAFIC[4] the beam has a twofold representation. It is represented as a continuous charge distribution when it is considered as a source of the electromagnetic fields, and an ensemble of point like particles when the effects of the fields are evaluated. The beam line is divided in slices and the "field generating" bunch is propagated along the beamline. The beam parameters along the beamline are calculated and stored. The field on "point like" particles following the beamline are then evaluated according to the retarded conditions and the information on the dynamics of the point like particles are then used to advance a new "field generating" bunch along the beamline. The iteration of this procedure leads to the self consistency of the method. TRAFIC[4] is capable of handling also boundary conditions of infinitely extended, perfectly conducting flat walls by image charge method. The computation time grows in this case and few thousand of particles require several days of CPU time, depending on the chicane complexity and length. A parallel implementation of the code is practically required in the simulation of these cases [44].

3.2. *"Line charge method"*

The second family of codes are based on the formulation of the wake potential due to a line charge distribution following an arc [36]

$$\frac{dE(s,\phi)}{d(ct)} = \frac{-2e^2}{\left(3R^2\right)^{\frac{1}{3}}}\left\{ s_l^{\frac{1}{3}}\left[\lambda(s-s_l)-\lambda(s-4s_l)\right] + \int_{s-s_l}^{s} \frac{dz}{(s-z)^{1/3}}\frac{d\lambda(z)}{dz}\right\} \quad (18)$$

where R is the radius of bend, λ the longitudinal charge distribution. This formulation has been extended by Stupakov and Emma to include the post-dipole region and is the basis for the CSR implementation in Elegant [45,46], in *CSR_CALC (P. Emma)* [47], *and in M. Dohlus line charge program* [48] .

The main assumptions are the following:

1. A one dimensional beam is assumed for CSR field calculations. The "real" beam is projected along the curvilinear coordinate s.
2. Transverse internal forces are neglected
3. Coulomb repulsion is neglected
4. Any change in the charge distribution at retarded time is neglected, the wake is calculated assuming that the charge distribution has not changed according to compression

The advantage is consistent in terms of computational time. In Fig. (7) it is shown a plot of the microbunching gain in the LCLS BC2 compressor [49] as a function of the initial modulation wavelength (prior to compression). The agreement with the theory (continuous lines) [38,39] is remarkable. It is even more remarkable the fact that the simulation, done with 2 10^6 particles, finished in less than 1 hour on a 1 GHz PC [50].

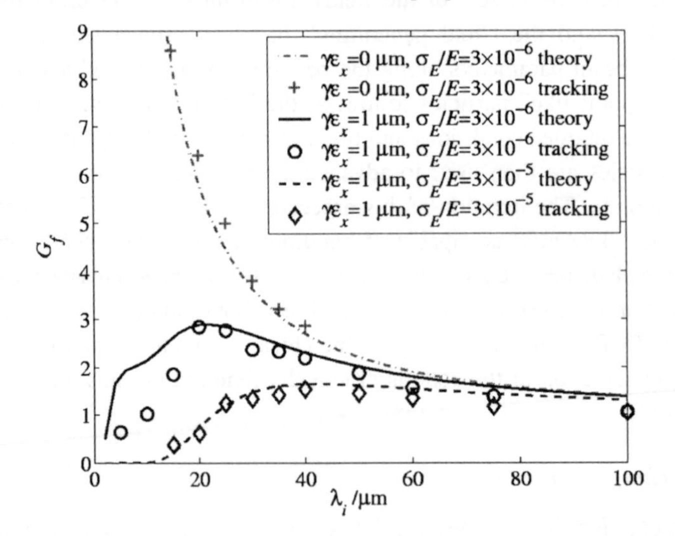

Fig.(7) : Gain spectrum of microbunching modulation in SLAC BC2 compressor at different input energy spread and emittances *(courtesy of P. Emma)*

The possibility of running such a large number of macroparticles allows the simulation of microbunching effects with a reasonably low numerical noise.

3.3. *Comparison between CSR devoted codes*

A comparison between the results predicted by the various mentioned programs has been obtained on a test case based on a four bends magnetic chicane in occasion of the ICFA Beam Dynamics mini workshop [51]. In fig.(8) it is shown the layout of the four bend chicane which served as a test case, with the parameters listed in Tab III.

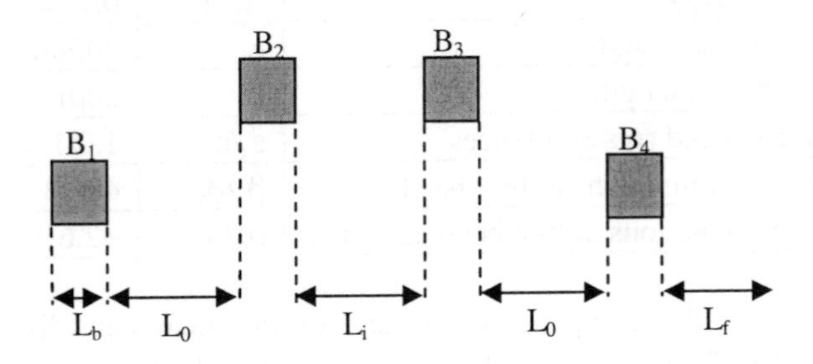

Fig. 8 Layout of the 4 bends test chicane considered for code benchmarking

Tab. III. List of the 4 bends test chicane parameters.

Chicane Parameters	Symbol	Value
Bend magnet length (projected)	L_b	0.5 m
Drift length B1->B2 and B3->B4 (projected)	L_0	5.0 m
Drift length B2->B3	L_i	1.0 m
Post chicane drift	L_f	2.0 m
Bend radius of each dipole magnet	R	10.35 m
Bending Angle	F	2.77 deg
Momentum compaction	R56	-25
2nd order momentum compaction	T566	+37.5 mm
Total projected length of chicane	L_{TOT}	13.0 m

The simulation were performed at electron beam energies of 0.5 GeV and 5 GeV with electron pulse shapes both Gaussian and Stepwise, but a complete comparison between all codes was available only at 5 GeV with Gaussian longitudinal current profile. The other electron beam parameters for the simulation are shown in Tab. IV.

Tab. IV e-beam parameters used in the benchmark

Electron beam Parameters	Symbol	Value
Nominal Energy	E	5 GeV
Bunch Charge	Q	1.0 nC
Incoherent Energy Spread	ΔE	10 keV
Linear energy – z correlation	A	36 m^{-1}
Initial rms energy spread	ΔE/E	0.72%
Initial rms bunch length	σ_i	200μm
Final rms bunch length	σ_f	20μm
Initial normalized rms emittances	$\varepsilon_x/\varepsilon_y$	1.0/1.0 mm-mrad
Initial betatron functions at first bend	β_x/β_y	40/13 m
Initial alpha functions at first bend entrance	α_x/α_y	+2.6/+1

The results of the comparison relevant to the Gaussian distribution are summarized in Tab V, where the energy change (δE) the energy spread (σ_E) and the normalized emittance in the orbit plane are given after the chicane. All the simulations except the one obtained with TREDI [4], have shown a good agreement with increments of normalized emittance between 30% and 50%. The simulation done with TREDI was obtained with only 10^3 macroparticles and was affected by a strong numerical noise. Part of the noise has been suppressed by discarding 15% of the "worse" macroparticles. TREDI simulations done after the workshop with up to 10^4 macroparticles have shown results closer (80% growth) to the one obtained with the other programs [52].

Tab. V - List of benchmarked codes and of the beam parameters at the end of the chicane

		δE	σ_E	ε
3D	TRAFIC4	-0.058	-0.002	1.4
"	TREDI	-0.041	0.017	2.3
2D	Program by R.LI	-0.056	-0.006	1.32
1D-Line Charge	Elegant	-0.045	-0.0043	1.55
"	CSR_CALC (P. Emma)	-0.043	-0.004	1.52
"	Program by M. Dohlus	-0.045	-0.011	1.62

4. Free Electron Laser

The simulation of SASE free electron lasers from first principles has been obtained both with PIC codes [8] and with Lienard Wiechert based algorithms [53] [54]. These methods, that have the merit of allowing a verification of FEL physics formulation from first principles, have been applied only to a limited number of specific cases. They indeed do not take any advantage of the fact that in a free electron laser the electron beam interacts with the periodic magnetic field of the undulator and with the optical field of a co-propagating e.m. wave with k-vector oriented in the z-direction. The FEL instability is characterized by a gain with a narrow bandwidth of the order of the Pierce parameter ρ [5,55],

$$\rho = \frac{1}{4\pi\gamma}\left[\frac{2\pi^2}{\Sigma_e}\frac{I_{peak}}{I_A}(Kf_B\lambda_u)^2\right]^{\frac{1}{3}} \tag{19}$$

where

$$f_B = J_0(\xi) - J_1(\xi), \quad \xi = \frac{1}{4}\frac{K^2}{1+\frac{K^2}{2}} \tag{20}$$

and where λ_u is the undulator period, γ the relativistic factor, $K = eB\lambda_u/2\pi mc$ the undulator strength , I/I_A is the ratio between the peak current and the Alfven current (17040A) and Σ_e the transverse electron beam cross section. Typically, in the case of short wavelength FELs, the Pierce parameter is smaller than 10^{-2}. For this reason almost all the codes devoted to FELs are based on the paraxial wave approximation (i.e. $k=k_z$) and on the Slow Varying Envelope Approximation (SVEA), which consist in assuming that the field may be written as the product of a term oscillating at the resonant frequency of the instability and propagating in the z-direction, and a slowly varying amplitude

$$E(z,r,t) = a(z,r,t)\exp(i(kz-\omega t)) \tag{21}$$

Where the function $a(z,r,t)$ satisfies the following inequalities,

$$\left|\frac{\partial^2 a}{\partial t^2}\right| << \omega^2 a \quad \text{and} \quad \left|\frac{\partial^2 a}{\partial z^2}\right| << k^2 a \tag{22}$$

This assumption leads to a significant simplification of the wave equation. The second derivatives of a in Maxwell's equations may be neglected and the wave equation for the field reduces to

$$\left[2ik_r\frac{\partial}{\partial z}+\nabla_\perp^2\right]a_r(r,z)e^{i\varphi_r}=-J(r,z)\left\langle\frac{Ke^{-i\vartheta}}{\gamma}\right\rangle \tag{23}$$

where the r.h.s. represents the source term depending on the beam current. The simplification consist in reducing the PDE for a fast oscillating wave in a the above equation for the slowly varying field that may be solved with a PDE solver. This is done assuming axial symmetry in TDA3D [56,57] and in GINGER [58,59] or in three Cartesian dimensions, in FELEX/N [60,61] and in GENESIS [62] were an Alternating Direction Implicit integration scheme is used [7]. In MEDUSA [63,64] the Source Dependent Expansion (SDE) technique has been implemented [65,66]. This technique consists in the expansion of the field a in terms of Gauss-Hermite functions with characteristic parameters depending on the source term $J(r,z)$. The expansion allows an efficient solution of the wave equation, representing the field with a reduced number of eigenfunctions.

The electron dynamics in MEDUSA is integrated by directly solving the Lorentz force equations with a fourth order Runge Kutta [7], with the fields given by the superposition of the laser and undulator fields respectively

$$\begin{aligned} m\frac{d(\gamma\mathbf{v})}{dt}&=-e\left(\mathbf{E}+\frac{\mathbf{v}}{c}\times\mathbf{B}\right)\\ \frac{d\gamma}{dt}&=-\frac{e}{mc^2}\mathbf{E}\cdot\mathbf{v}\end{aligned} \tag{24}$$

This "first principles" approach allows the possibility of accepting 3D field maps which includes the undulators and all additional field components (i.e., FODO lattices, etc.). The integration of the 3D Lorentz force equations facilitates the adaptation of MEDUSA to treat novel magnetic field configurations and beam distributions, as e.g. in the case of bi-harmonic undulators [67]. The drawback is that MEDUSA doesn't take advantage of the periodicity of the undulator magnetic fields, and several tens of Runge Kutta steps are required for the correct integration of the trajectory along a single undulator period. Most of the other FEL devoted codes (as e.g. GINGER, FELEXN, PROMETEO [58]) use the Kroll, Morton and Rosenbluth (KMR) approximation [5,59], consisting in assuming that the wiggling amplitude is small and that the frequency of any variation of the parameters entering in the FEL process is large with respect to the frequency of oscillation due to the undulator magnetic field, i.e.

1. Adiabatic change of undulator parameters, as e.g. magnetic errors, tapering

$$\frac{1}{\lambda_u}\frac{d\lambda_u}{dz} << 1, \quad \frac{1}{\lambda_u}\frac{dK}{dz} << 1 \tag{25}$$

2. The scale of variation of the e-beam parameters is large, i.e. the betatron period λ_β, is much larger than the undulator period, $\lambda_\beta >> \lambda_u$ and changes in current, energy etc., occurs on a scale of many periods.

Under these assumptions the particle trajectory is averaged over the undulator period and very few integration steps per period are required, with a consistent advantage from the CPU time point of view. When the effects of the laser field on the electron transverse motion can be neglected, the equations of motion can be furthermore simplified (as e.g. in GENESIS). Defining the variable $\zeta = (k + k_u)z - \omega t$ we have (in this example in circular symmetry, with $k_u = 2\pi/\lambda_u$, $k_0 = 2\pi/\lambda_0$)

$$\begin{aligned}
&\frac{d\gamma}{dz} = -k_0 \frac{a_L K}{\gamma\sqrt{2}}\sin(\zeta + \varphi_L), \\
&\frac{d\zeta}{dz} = k_u - k_0 \frac{1 + p_T^2 + \frac{K^2}{2} - \sqrt{2}a_L K\cos(\zeta + \varphi_{Lr})}{2\gamma^2} \\
&\frac{dp_\perp}{dz} = -\frac{1}{2\gamma}\frac{\partial K^2}{\partial r_\perp}, \\
&\frac{dr}{dz} = \frac{p_\perp}{\gamma}
\end{aligned} \tag{26}$$

The characteristic Colson's pendulum-like equation [5] is readily derived from Eq. (26) by neglecting the transverse dynamics terms and the term proportional to $k_u = 2\pi/\lambda_u$ in the r.h.s. of the second of Eq.(26). In the case of FELs operating with high beam energy and short wavelength $\lambda_u >> \lambda_0$, the assumption is well satisfied. The FEL pendulum equation in the conjugate variables,

$\zeta, \nu = d\zeta/\tau$ with $\tau = z / N\lambda_u$ is the equation of motion corresponding to the Hamiltonian

With the assumption of a steady state, uniform e-beam distribution, and

$$H = \frac{1}{2}\nu^2 - |a|\cos(\zeta + \varphi_L) \tag{27}$$

field *a* independent on *(r,z)*, both the coupled electrons equation of motion derived from the Hamiltonian Eq.(27) and the field equation (23), are periodic in ζ, with period 2π. The solution in this 0-dimensional case is obtained by imposing periodic boundary conditions to particles and fields [69]. The solution of this simple case may be used as the basic element for the extension to higher dimensional cases.

4.1. *Quiet start and modeling of the shot noise*

The approximations for fields (SVEA) and particles (KMR) can be considered as mutually consistent assumptions. With the KMR formulation we neglect wide bandwidth components in the field sources dynamics, while with the SVEA approximation we assume that the FEL gain is non zero in the same narrow frequency window centered around the lasing frequency $\omega_0 = 2\pi c/\lambda_0$. The width of this window is proportional to the parameter ρ defined in Eq. (19). The numerical representation of the electron beam spectral distribution must resemble that of the real distribution in this frequency interval. Randomly distributed macroparticles, unless their number is comparable to the real number of electrons in the bunch, provide a distribution with a much higher spectral content than the physical one. The correct noise level is introduced by exploiting the periodicity of the Hamiltonian (27). The beam is indeed represented as an ensemble of beamlets, each one with macroparticles distributed as

$$\zeta_i = \zeta_0 + i\frac{\pi}{n_p} + \delta_{i,k}, \qquad i = 0,1,...n_p - 1 \tag{28}$$

At $\delta_{i,k} = 0$ the harmonic content of such a beamlet is exactly zero at the fundamental wavelength and at the higher order harmonics $h = 2,...n_p - 1$ (see Fig. (9)).

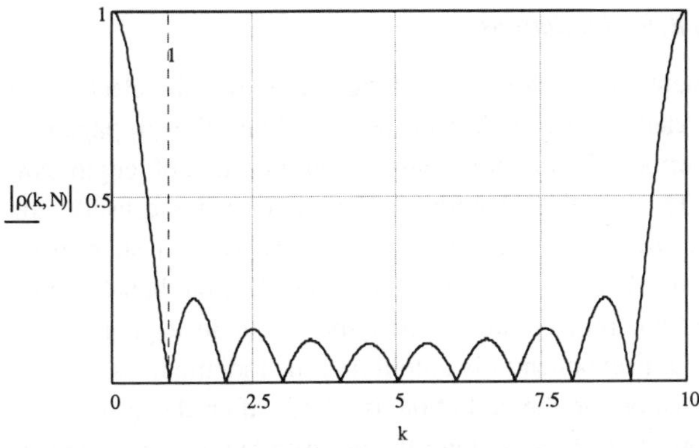

Fig. (9). Fourier transform of a beamlet composed by ten macroparticles. The spectral content of the distribution is zero at k=1..9

The number of macroparticles per beamlet should be larger than the corresponding number suggested by the simple application of the Nyquist theorem. With the definition (28), at the beginning the particles are equally spaced and 2**h* particles would be adequate for a noiseless quiet start of the *h* harmonic. When the dynamics is applied, this is not any more valid. First, the FEL couples bunching at a given harmonic *h* with the field growth at harmonics $h \pm 1$ [70]. Second, and this is even more critical, the particles position are shifted from equally spaced positions and a bunching contribution due to an insufficient original sampling frequency may appear. At saturation the macroparticles are bunched and phase space discontinuities in the lower density regions, strongly affect the bunching estimation, in particular at high order harmonics.

The effect of shot noise is introduced with the shifting factors $\delta_{k,i}$. A shot noise of amplitude proportional to the average macroparticles displacement is obtained. An algorithm to produce these displacements with the correct noise statistics in a 1D case is given by Penman and McNeil [71]. Particular care must be adopted in the generation of beamlets in multidimensional spaces to avoid undesired interferences that may arise from the dynamical effects (drifts, betatron motion etc.). An elegant description of the shot noise implementation in GINGER and the problems related to the generalization to multidimensional spaces is given in ref.[72].

4.2. *Time dependent simulations*

At the resonant wavelength the radiation slips over the electron bunch of one optical wavelength per undulator period. After the propagation in an undulator of N periods, the radiation passes over a portion of length $N\lambda_0$ of the electron bunch. The "slippage length" is the name of this causally connected region of length $N\lambda_0$. For typical numbers, in the case of short wavelength FELs, the slippage length is shorter than the electron bunch length. The radiation field is not propagating in free space, but in a gain medium (the electron beam) that is exponentially amplifying the radiation. As a consequence the phase information of the radiation is "lost" after the propagation over distance of the order of $\lambda_0/4\pi\rho$ along the electron bunch, which is commonly indicated as the "cooperation length". For the same reason the relative gain bandwidth is proportional to ρ. In the implementation of time dependent simulations in GINGER and in GENESIS a time window larger than a cooperation length, that is enclosing the whole electron beam, or part of it, is defined. This time window is then sampled and sliced as shown in fig.(9). The slices have a length of one optical wavelength (one of the "beamlet" defined in the previous section) and are separated by a distance Δ. In GINGER the simulation evolves at discrete time steps of length $\Delta\lambda_u/(\lambda_0 c)$ and at each time step the field relevant to one slice is shifted to the next slice. In GENESIS the simulation for the first slice is done up to the end of the undulator and then, the values of the fields are used to execute the simulation for the second slice. The procedure is repeated up to the end of all the slices. This procedure allows to keep in memory only one slice at the time, but from the conceptual point of view is essentially the same adopted in GINGER. The parameter Δ must satisfy some constraints. According to the Nyquist theorem the cutoff frequency is given by twice the sampling frequency, which is $1/\Delta$.

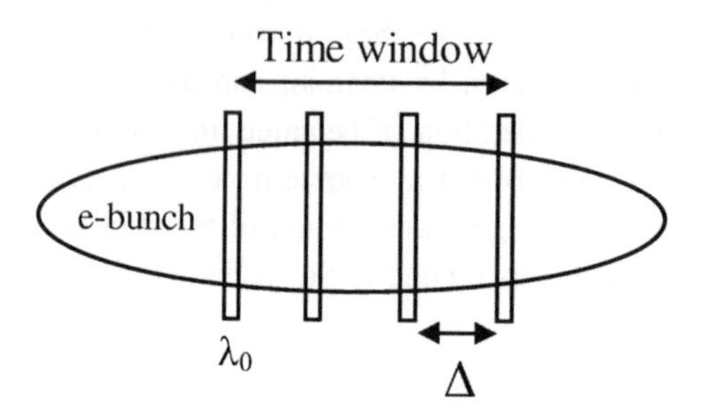

Fig. (9) Slicing of the e-bunch for time dependent simulations.

Imposing the condition that the band pass associated with the sampling procedure, must be larger than the FEL gain bandwidth, we find the condition

$$\Delta_{MAX} << \frac{\lambda_0}{4\pi\rho} \tag{29}$$

In general the simulation is done with $\Delta < \Delta_{MAX}/8$ to ensure that the whole gain bandwidth is contained in the numerical representation. The simulation time window should be larger than a slippage length. In GENESIS, when only a portion of the beam is simulated, the first slippage length may be ignored in the output, as it is used as the source of the field for the following slices. It is obviously retained in a simulation enclosing the whole beam. In GINGER the problem of simulating a portion of a long bunch is overcame with the possibility of setting a periodicity condition on the time window, and by assuming steady state parameters for the electron beam distribution along the bunch.

4.3. *Codes validation*

The output of these codes, provide many information on the saturation length and other characteristics of the FEL, as sensitivity to e-beam parameters, stability, spectrum, pulse duration, coherence, etc. There are a number of theoretical approaches that allows to obtain analytic estimation of some of these features [73-76]. These analytical formulations are useful in narrowing the range of the input parameters for the simulation, but specific effects as e.g. magnetic errors or alignment errors, or the effects of specific distributions as e.g. a "folding" of the input 6D electrons phase space, are conveniently analyzed with the numerical approach. For this reason the codes are widely used in the design of future sources and it is justified the ongoing interest in their validation against the available experimental data and the cross comparison of their results. The program MEDUSA has been extensively used in the prediction and the analysis of the LEUTL experiment [77,78]. In ref. [78] the most known multidimensional FEL codes have been compared on LEUTL parameters. The results have shown, in steady state mode, resonant wavelengths within 0.2%, and saturation lengths within 10%. The differences between the codes increased in the simulation of a multi segmented undulator with MEDUSA providing the largest saturation length. This agreement should be considered satisfactory, according to the differences in the models, in the algorithms and the large number of parameters entering in the simulation. The

VISA 800 nm FEL experiment has been reproduced in surprisingly good details with GENESIS [79,80].

4.4. *Harmonics*

When a SASE FEL approaches saturation, the modulation of the electron beam longitudinal distribution develops substantial Fourier components at higher harmonics of the fundamental and emission at the corresponding wavelengths occurs. There is widespread interest in this process, because emission on high order harmonics represents a significant resource to extend the wavelength tunability of a free electron laser [81-89]. We may distinguish two different cases:

- When the dynamics is driven by the fundamental and the harmonic emission is a byproduct of the phase space distortion
- When the electron dynamics is self consistently modified by the radiation field of the harmonics

In the first case the harmonics emission depends on the histories of particles along the undulator. The bunching coefficients and corresponding harmonic emission may be calculated in post-processing phase. The main required precaution, consist in a correct preparation of the beam quiet start at the desired harmonic. This method is used both GINGER and in GENESIS. FELEXN has the capability of operating at higher harmonics, but no coupling between the harmonics is included.

In MEDUSA, PROMETEO and in PERSEO (1D [69]) the harmonics are treated self consistently, by solving the particles dynamics with the fields given by the superposition of the fundamental and of the higher order harmonics. MEDUSA has been extensively used [90] in the prediction of the performances on higher order harmonics of the LEUTL experiment. Comparisons between the predictions of MEDUSA, PROMETEO and GINGER in conditions similar to those of the LCLS FEL and of LEUTL FEL are given in ref. [91] and [92] respectively. The sensitivity to the beam quality of the harmonic generation process has been analyzed. In both the tested cases the electrons longitudinal dynamics and the harmonic emission is mainly driven by the laser field on the fundamental harmonic. The results should not be affected by the lack of self consistency on harmonics of GINGER.

5. Conclusions

There are several other sources of intra-beam coupling and non linear collective effects that may play an important role in the design of short wavelength free electron lasers, as weak field effects, beam loading, etc. The three regimes analyzed in this review,

- Injection
- Compression
- FEL process

were characterized by quite well established families of investigation tools devoted to the simulation of the specific regimes and offered the opportunity to analyze, from the spectral point of view, the constraints posed by the numerical representation that is the main reason why the simulation of charged particles interacting with electromagnetic field in the wide conditions considered, cannot be accomplished by a single code.

In general we may distinguish between

- "First principles" codes and
- Semi-analytic codes

The first include most of the physical aspects of the specific problem. This completeness is usually paid in terms of complexity of the implementation and in terms of calculation time. In the second case the numerical implementation follows some "smart" theoretical approach which, according to some physical approximations, allows to point to some specific issue relevant to the problem with a fast relaxation of the usually large number of parameters involved. An example is provided by the "line charge method" in the calculation of the CSR microbunching. The number of particles simulated with this method is large enough to reduce at an acceptable level the numerical noise in the wide band spectrum of the CSR impedance. The approximations considered may be quite stringent, but the large number of macroparticles allows a much better representation of the beam statistics than with other "first principle codes". In general these two families of codes are both necessary and complementary.

The modelling aspects of the codes have been privileged, but several other features have great importance. An example is provided by the interface. The amount of data produced when 6D phase spaces with $10^4/10^6$ macroparticles are considered, is huge. The interface is meant not only in terms of a simple human interface for the analysis and visualization of the results. The necessity of different codes handling different regimes means that the exchange of phase spaces in a compatible format between different programs is a practical but essential problem. A significant effort in this direction has been done with the

SDDS [93] and HDF5 [94] data exchange format. Several examples of "start to end" simulation can be found in literature [95,96].

Last but not least aspect is the documentation and the sources availability. This is of great importance for the peer review possibility offered when the user is not faced by a "black box" but has the opportunity to analyse directly the sources and the equation he is dealing with. When documentation or/and sources were available, we have referenced the link to the web pages where this information may be retrieved from the network.

Acknowledgments

I am pleased to acknowledge the contribution of Marcello Quattromini for many useful discussions on various aspects of the fields regularization problem and for the patience shown in the development and execution of TREDI. I express my gratitude to Pino Dattoli for his support in preparing this manuscript and for the numerous discussions we had on various aspects of the FEL physics. I wish to acknowledge the contribution of Riccardo Bartolini, Luca Mezi, Massimo Ferrario, Jamie Rosenzweig, Luca Serafini and Titti Ronsivalle for many useful discussions, suggestions and for providing some of their results. I also gratefully acknowledge the contribution of Sandra Biedron, Michael Borland, Martin Dohlus, Paul Emma, Bart Faatz, William Fawley, Klaus Floettmann, Henry Freund, Andreas Kabel, John Lewellen, Torsen Limberg, Cecile Limborg, Steven Milton, Sven Reiche and Rui Li for their solicitude in providing informations, references and viewgraphs about their work.

References

1. J.M. Dawson, A. T. Lin, "Particle Simulation" in *Handbook of Plasma Physics*, Eds. M.N. Rosenbluth and R. Z. Sagdeev Vol. **2**, "Basic Plasma Physics, Elsevier Science Publisher (1984)
2. T. P. Wangler, "Introduction to Linear Accelerators", LA-UR-93-805, Los Alamos NM (1993)
3. M. Quattromini, in Proceedings of the *2nd Melfi School on Advanced Topics in Mathematics and Physics, Melfi, June 18-23 2000*, ed. by G. Dattoli, H. M. Srivastava and C. Cesarano, Aracne Editrice, Roma (2001)

4. L. Giannessi, *Il Nuovo Cimento* **A 112** (1999) 447. See also L. Giannessi, M. Quattromini, "An overview of TREDI and CSR test cases, at *www.desy.de/csr/csr_workshop_2002/csr_workshop_2002_index.html*
5. W. B. Colson, "Classical Free Electron Laser Theory" *Laser Handbook vol. VI*, ed. by W. B. Colson, C. Pellegrini, A. Renieri, North Holland, Amsterdam (1985)
6. J.D. Jackson, *Classical Electrodynamics*, ed. by John Wiley & Sons, 654 (1975)
7. W.H. Press, B. P. Flannery, S. A. Teukolsky, W. T. Vetterling, *Numerical Recipes*, Cambridge Univerity Press, New York, 615 (1988)
8. T. Lin, M. Dawson, H. Okuda, *The physics of Fluids* **17**, (1975) 1995
9. L. Serafini, J. B. Rosenzweig , *Phys. Rev.* **E 55**, 7565 (1997)
10. B. E. Carlsten, Nucl. Instrum & Meth. **A 285**, 313 (1989)
11. P. G. O'Shea, "RF Photoinjectors", in *Proceedings of the 2nd ICFA Advanced Accelerator Workshop*, ed. by J. B. Rosenzweig and L. Serafini, World Scientific (2000), p.17
12. L. Serafini, "Computational modeling of high brightness electron beam physics"in *Proceedings of the 2nd ICFA Advanced Accelerator Workshop*, ed. by J. B. Rosenzweig and L. Serafini, World Scientific (2000), p.17
13. L. Serafini and C. Pagani, Proceedings of 1st EPAC Conference, Rome, 866 (1988)
14. M. Borland, Summary of Equations and Methods Used in SPIFFE, APS/IN/LINAC/92-2, 29 (1992). The code SPIFFE is available in line at *http://www.aps.anl.gov/asd/oag/oagSoftware.shtml*
15. F. Ciocci, L. Giannessi, A. Marranca, L. Mezi, M. Quattromini, *Nucl. Instr. & Meth.* **A 393**, 434 (1997) (See also *http://www.afs.enea.it/gianness/tredi*. Will move soon to *http://www.tredi.enea.it*)
16. J.L. Coacolo et al. *Nucl. Instrum. & Meth.* **A 393**, 430 (1997)
17. See e.g. *http://laacg1.lanl.gov/laacg/services/parmela.html*
18. M.J. de Loos, S.B. van der Geer, "General Particle Tracer: A new 3D code for accelerator and beamline design"in *Proceedings of 5th European Particle Accelerator Conference*, Sitges, 1241 (1996)
19. "GPT User Manual", Pulsar Physics, Flamingostraat 24, 3582 SX Utrecht, The Netherland (*http://www.pulsar.nl*)
20. K. Floettmann, *Astra User Manual, see http://www.desy.de/~mpyflo/Astra_dokumentation/*
21. M. Ferrario et al. "Multi bunch energy spread induced by beam loading in a standing wave structure", *Particle Accelerators* **52**, 1 (1996)
22. I. M. Kapchinskj, V. V. Vladimirsky, in *Proc of II intern. Conference of High Energy Accelerators & Instrumentation, Geneva 14th/19th September* CERN, 274 (1959)

23. M. Ferrario, J. E. Clendenin, D. T. Palmer, J. B. Rosenzweig, L. Serafini, "Homdyn Study for the LCLS RF Photo-Injector", in *Proceedings of the 2nd ICFA Advanced Accelerator Workshop*, ed. by J. B. Rosenzweig and L. Serafini, World Scientific, 534 (2000)
24. M. Ferrario, private communication.
25. L. Serafini, private communication.
26. M. Ferrario, K. Floettmann, B. Grygorian, T. Limberg, Ph. Piot, "Conceptual design of the X-FEL Photoinjector" TESLA-FEL 2001-03 (2001)
27. L. Giannessi, M. Quattromini, "TREDI simulations for SPARC photoinjector", in these proceedings.
28. E. Colby, V. Ivanov, Z. Li, C. Limborg, "Simulation issues for RF Photoinjectors", in Proceedings of *International Computational Accelerator Physics Conference 2002* October 15-18, 2002 Michigan State University East Lansing, MI 48824
29. "User/Configurable MAGIC for electromagnetic PIC calculations", Computer Physics Communications 78 (1995) 54-86. Informations may be found at the link http://www.mrcwdc.com
30. P. Piot, "Review of experimental results on photo-emission electron sources", in these proceedings.
31. J. Yang et a., "Experimental Studies of Photocathode RF Gun with Laser Pulse Shaping", *Proceedings of the VIII European Particle Accelerator Conference*, 3-7/6/02 Paris (2002)
32. "LCLS CDR", SLAC-R-593 UC-414 (April 2002)
33. "TESLA TDR", TDR-2001-23, DESY (2001)
34. L. Serafini and M. Ferrario, Velocity Bunching in PhotoInjectors , *AIP CP* **581**, 87 (2001)
35. M. Ferrario et al. "Beam Dynamics Study of an RF Bunch Compressor for High Brightness Beam Injectors", *Proceedings of the VIII European Particle Accelerator Conference*, 3-7/6/02 Paris (2002)
36. E.L. Saldin, E.A. Schneidmiller, M.V. Yurkov, *Nucl. Instrum. & Meth.* **A 398,** 373 (1997)
37. E. L. Saldin et al., *Nucl. Instrum. & Meth.* **A 483**, 517 (2002)
38. S. Heifets et al., "CSR Instability in a bunch Compressor", SLAC-PUB-9165, March 2002
39. Z. Huang, K. J. Kim, *Phys. Rev. ST AB 5*, 74401 (2002)
40. S. Reiche, J. B. Rosenzweig, these proceedings
41. A. Kabel, M. Dohlus, and T. Limberg., *Nucl. Instrum. & Meth.* **A 455,** 185 (2000)
42. M. Dohlus, A. Kabel, and T. Limberg, *Nucl. Instrum. & Meth.* **A 445**, 338 (2000)
43. R. Li, *Nucl. Instrum. & Meth.* **A 429**, 310 (1998)

44. A. Kabel, “Coherent Synchrotron Radiation Calculations Using TraFiC4: Multi-Processor Simulations and Optics Scans”, SLAC-PUB-9352, August 2002
45. M. Borland, *Phys. Rev. ST AB* **4**, 070701 (2001)
46. M. Borland, “elegant: A Flexible SDDS-Compliant Code for Accelerator Simulation” LS-287, ANL, Argonne, IL 60439, USA. Elegant is available on line at *http://www.aps.anl.gov/asd/oag/oaghome.shtml*
47. P. Emma, private communication
48. M. Dohlus, private communication
49. P. Emma, “Accelerator Physics Challenges of X-Ray FEL SASE Sources”, in *Proceedings of the VIII European Particle Accelerator Conference*, 3-7/6/02 Paris (2002)
50. P. Emma, private communication
51. ICFA Beam Dynamics mini workshop on “Coherent Synchrotron Radiation and its impact on the dynamics of high brightness electron beams” held at DESY-Zeuthen in January 2002. (Available in line at the following address *www.desy.de/csr/csr_workshop_2002/csr_workshop_2002_index.html*).
52. L. Giannessi, M. Quattromini, “Simulation of CSR Effects on a 4 Bends Compressor With Tredi”, in these proceedings
53. L. R. Elias, I. Kimel, *Nucl. Instrum & Meth.* **A393**, 100 (1997)
54. L. Giannessi, P. Musumeci, M. Quattromini, *Nucl. Instrum. & Meth.* **A 436**, 443 (1999)
55. W. B. Colson, *Nucl. Instrum. & Meth.* **A393**, 82 (1997)
56. T. M. Tran and J. S. Wurtele, “TDA - A Three-Dimensional Axisymmetric Code For Free-Electron-Laser (Fel) Simulation”, *Computer Physics Comm.* **54**, 263 (1989). See also *http://www.desy.de/~tda3d*
57. B. Faatz, W. M. Fawley, P. Pierini, S. Reiche, G. Travish, D. Whittum, J. Wurtele, *Nucl. Instrum. & Meth.* **A393**, 277 (1997)
58. R. A. Jong, W. M. Fawley, and E. T. Scharlemann, “Modelling of induction-linac based free-electron laser amplifiers,” in *Modeling and Simulation of Laser Systems,* Proc. SPIE **1045**, pp. 18–27, 1989
59. Z. Huang, W. M. Fawley, “GINGER Simulations of Short Pulse Effects in the LEUTL FEL” *Proceedings of the 2001 Particle Accelerator Conference, Chicago* (2001) 2713
60. B. D. McVey, *Nucl. Instrum. & Meth.* **A250**, 449 (1985)
61. C. G. Parazzoli, “FELEXN, Boeing simulation code, version B08”, in *Proceedings of the X-Ray FEL Theory and Simulation Codes Workshop*, Stanford Linear Accelerator Center, Stanford University September 23 and 24, 1999, LCLS-TN-00-1, p. 161
62. S. Reiche, *Nucl. Instrum. & Meth.* **A429**, 243 (1999). See also *http://pbpl.physics.ucla.edu/genesis*

63. H. P. Freund, S. G. Biedron, and S. V. Milton, *IEEE J.Q.E.* **36**, 275 (2000)
64. S.G. Biedron, H.P. Freund, S.V. Milton, "3D FEL code for the simulation of a high-gain harmonic generation experiment, in: Harold E. Bennett, David H. Dowell (Eds.), *Free-Electron Laser Challenges II,* Proc. SPIE **3614**, (1999)
65. P. Sprangle, A. Ting, C. M. Tang, *Phys. Rev.* **A 36**, 2773 (1987)
66. P. Sprangle, A. Ting, C. M. Tang, *Phys. Rev. Lett.* **59**, 202 (1987)
67. G. Dattoli, L. Giannessi, P. L. Ottaviani, S. G. Biedron, H. P. Freund, S.V. Milton, "Two harmonics undulator and Harmonic generation in high gain Free Electron Lasers", to be published in *Optics Comm.*
68. G. Dattoli, M. Galli, P.L. Ottaviani, ENEA RT/INN/93/09 (1993) p. 5
69. An example of numerical integration of the FEL pendulum equation is given in *Perseo*, at the link *http://www.frascati.enea.it/gianness/perseo*
70. G. Dattoli, L. Giannessi, A. Torre, *J. Opt. Soc. Am.* **B 10,** 2136 (1993)
71. C. Penman and B.W.J. McNeil, *Opt. Commun.* **90**, 82 (1992)
72. W. M. Fawley, *Phys. Rev. ST AB* **5**, 070701 (2002)
73. G. Dattoli, A. Renieri, A. Torre, R. Caloi, *Il Nuovo Cimento* **D11**, 313 (1989)
74. M. Xie, *IEEE Proceedings for Pac95*, No. 95CH3584, **183**, 1996.
75. G. Dattoli, L. Giannessi, P.L. Ottaviani and M. Carpanese, *Nucl. Instrum. & Meth.* **A393**, 133 (1997)
76. G. Dattoli, P.L. Ottaviani,Optics Comm. **204**, 283 (2002)
77. J.W. Lewellen et al. *Nucl. Instrum. & Meth.* **A483**, 40 (2002)
78. S. G. Biedron et al., *Nucl. Instrum & Meth.* A445 (2000) 110
79. A. Tremaine, P. Frigola, A. Murokh, C. Pellegrini, S. Reiche, J. Rosenzweig et al., *Nucl. Instrum. and Meth.* **A 483**, 24 (2002)
80. S. Reiche, "Review of Application to SASE-FELs", in these proceedings.
81. R. Bonifacio et al. *Nucl. Instrum. & Meth.* **A 293**, 627 (1990)
82. R. Barbini et al. in Proceedings of *Prospects for a 1 Å FEL Laser Sag Harbor, New York, April 22-27, 1990* - BNL Rep-52273, J.Gallardo Editor.
83. R. Bonifacio et al., *Nucl. Instrum. & Meth.* **A296**, 787 (1990)
84. L.H. Yu, *Phys. Rev* **A44**, 5178 (1991)
85. F. Ciocci et al. *IEEE – JQE* **31**, 1242 (1995)
86. L. H. Yu, M. Babzien, I. Ben-Zvi, et. Al. *SCIENCE* **289**, 5481 (2000)
    S.G. Biedron, H. P. Freund, Z. Huang, K.-J. Kim, and S.V. Milton, "The Use of Harmonics to Achieve Coherent Short Wavelengths" in *Proceedings of the IEEE 2001 Particle Accelerator Conference*, p. 2704-2706
87. J. H. Wu, L. H. Yu, *Nucl. Instrum. & Meth.* **A475**, 104 (2001)
88. W. Brefeld et al. *Nucl. Instrum. & Meth.* **A483**, 80 (2002)
89. H.P. Freund, P.G. O'Shea, *Nucl. Instrum. & Meth.* **A483**, 449 (2002)

90. H. P. Freund, S. G. Biedron, S. V. Milton, *Nucl. Instrum.& Meth.* **A 445**, 53 (2000)
91. S. G. Biedron, et al., *Nucl. Instrum & Meth.* **A483**, 101 (2002)
92. S. G. Biedron et al. *Phys. Rev. ST AB* **5**, 030701 (2002)
93. M. Borland et al., *Nucl. Instrum & Meth.* **A483**, 268 (2002)
94. S. Reiche, C. Pellegrini, J. Rosenzweig, P. Emma, P. Krejcik, *Nucl. Instrum & Meth.* **A 483**,70 (2002)
95. Definitions and libraries for SDDS implementation may be found at the link http://www.aps.anl.gov/asd/oag/oagPackages.shtml
96. Definitions and libraries for HDF5 implementation may be found at the link http://hdf.ncsa.uiuc.edu/

BEYOND THE RF PHOTOGUN

O. J. LUITEN
Eindhoven University of Technology,
Center for Plasma Physics and Radiation Technology,
P.O. Box 513,
5600 MB Eindhoven, Netherlands
E-mail: o.j.luiten@tue.nl

Laser-triggered switching of MV DC voltages enables acceleration gradients an order of magnitude higher than in state-of-the-art RF photoguns. In this way ultra-short, high-brightness electron bunches may be generated without the use of magnetic compression. The evolution of the bunch during the critical initial part of the acceleration trajectory, the 'pancake' regime, where the space-charge induced deterioration is most severe, is investigated using a simple, but effective analytical model. We find an expression for the maximally achievable peak current that does not depend on the bunch charge. An expression for the normalized emittance is derived, which allows us to calculate the optimal beam radius. It is shown that both the peak current and the transverse emittance required for the most challenging applications can be attained *without* magnetic compression, if acceleration gradients of 1 GV/m can be realized. The results are confirmed by simulations with the GPT code, assuming a 1 GV/m acceleration field and a 50 fs laser pulse, generating 100 pC of charge. The model is complementary to simulations in the sense that it supplies useful scaling laws and improved understanding of the physics involved. Interestingly, we find that the highest brightness is achieved with the shortest photoemission laser pulses.

1. Extreme acceleration fields and ultra-short bunches

One of the major challenges facing accelerator physics is the development of an injector suitable for X-ray SASE FELs operating at sub-nm wavelengths [1,2]. Such an injector should deliver a beam with a brightness corresponding to, typically, a peak current of 1 to several kA in combination with an rms transverse normalized emittance less than $1\,\mu$m at an energy of at least a few times 10 MeV. The current approach is to use an RF photogun to generate electron bunches with a current of $\sim 100\,$A and an emittance of 1 to a few μm at an energy of a few MeV. After boosting the energy to a few times 10 MeV the bunches are compressed to the kA level in one or several magnetic chicanes. Unfortunately, however, magnetic compression gives rise to collective radiative effects, which spoil the emittance [3].

RF breakdown limits the maximally achievable acceleration field in S-band RF photoguns to $\sim 100\,\mathrm{MV/m}$. If substantially higher acceleration fields were available, high current, low emittance bunches could possibly be generated directly, without having to resort to magnetic compression. X-band RF photoguns presently under development hold the promise of $\sim 300\,\mathrm{MV/m}$ fields [4], but this is probably not yet sufficient. Recently it was realized, however, that by using laser-triggered spark gaps to switch DC MV voltages, fields up to 1 GV/m can be switched on and off again within 1 ns, which is too short for breakdown to occur [5,6,7]. This has ignited efforts to develop a 1 GV/m pulsed DC photogun that does not require magnetic compression [8].

In this paper we investigate theoretically the beam quality that can be achieved by photoemission of an electron bunch with an aspect ratio much larger than unity ('pancake' bunch) in a uniform 1 GV/m acceleration field. One of the interesting consequences of such extreme fields is that the bunch becomes relativistic before it has had a chance to expand significantly due to Coulomb repulsion: the *rest frame* bunch length remains much smaller than the bunch radius during the critical initial part of the acceleration trajectory. This so-called pancake regime allows a relatively simple, but effective analytical approach, which is the main subject of this paper. The analytical model is complementary to simulations presented in earlier work [8], in the sense that it supplies useful scaling laws and improved understanding of the physics involved.

The pancake regime has not received much attention up to now, mainly because in the fields encountered in RF photoguns the bunch expands so rapidly that a pancake description is hardly relevant. In fact, this rapid expansion, the so-called 'blow-out' regime, is considered to be beneficial by several authors [9], since it may lead to highly linear space charge fields. The general opinion is that extremely short bunches should be avoided during the initial stages of the acceleration process, on the intuitive ground that high space charge densities are always detrimental to the final beam quality. We hope to show in this paper that this is not necessarily true and that, *a fortiori*, shorter bunches may even lead to better beams.

The remainder of this paper is organized as follows: In Sec. 2 a model for space charge fields inside pancake bunches is presented, which allows an analytical treatment of the bunch evolution, while taking into account the full nonlinearity of radial space charge forces. In Sec. 3 the evolution of pancake bunches in longitudinal phase space is treated. We find an expression for the maximally achievable bunch current, which does not depend on the bunch charge. The evolution of the bunch according to the model is compared to results of simulations with the GPT code. In Sec. 4 we discuss the evolution of pancake bunches in transverse phase space. An expression for the normalized emittance is derived, which allows us to calculate the optimal beam radius. The model is compared to results of GPT simulations. In Sec. 5 the results of the two previous sections are combined in a discussion of the normalized brightness of pancake bunches. Interestingly, we find that the highest brightness is achieved with the shortest photoemission laser pulses. In Sec. 6 we end with conclusions.

2. Space charge fields inside pancake bunches

In this section an approximative description is presented of the rest frame space charge fields inside pancake bunches. Our bunch model is a uniformly charged circular cylinder (a "pill box") of length L, radius R and charge Q (see Fig. 1). A cylindrical coordinate system (r, ϕ, z) is defined in such a

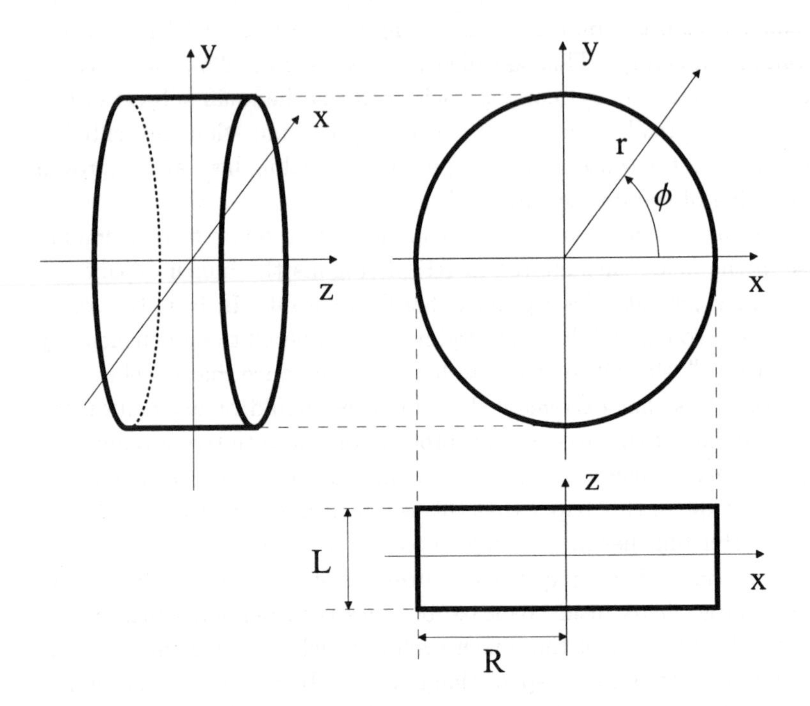

Figure 1. Pill box bunch geometry.

way that the z-axis is the symmetry axis of the bunch. The bunch extends in the z-direction from $z = -L/2$ to $z = L/2$. By definition pancake bunches have an aspect ratio $R/L \gg 1$: the rest frame bunch length is much smaller than the bunch radius. Due to the cylindrical symmetry of the bunch the azimuthal component E_ϕ of the electric field vector $\vec{E}$ is equal to zero. There are two nonzero components: the longitudinal component E_z and the radial component E_r.

2.1. *Longitudinal field*

For high aspect ratios the longitudinal field E_z on the symmetry axis inside a pill box bunch can be approximated by:

$$E_z(z) = \frac{\rho z}{\varepsilon_0} = \frac{\sigma z}{\varepsilon_0 L}, \tag{1}$$

where $\rho = Q/\pi R^2 L$ is the space charge density inside the bunch and $\sigma = Q/\pi R^2$ is the surface charge density associated with an extremely flat bunch. For our analysis we approximate the longitudinal field E_z everywhere inside the bunch by Eq. (1). In this way we slightly overestimate the magnitude of E_z and simplify our analysis by neglecting its dependence on r. In Fig. 2a the longitudinal field E_z is plotted as a function of position z. Outside the bunch ($|z| > L/2$) the field drops off again slowly, with a typical length scale $R \gg L$, which is not visible in Fig. 2a. The characteristic field strength of the problem is the surface charge field

$$E_s \equiv \sigma/2\varepsilon_0 = Q/2\pi\varepsilon_0 R^2. \tag{2}$$

Note that E_s is independent of the bunch length L. For typical bunch parameters $Q = 100\,\mathrm{pC}$ and $R = 0.5\,\mathrm{mm}$, the surface charge field has the value $E_s = 7\,\mathrm{MV/m}$. The front and the back of the bunch thus push each other apart with a constant force $2eE_s$, independent of the length L, as long as $L \ll R$.

2.2. *Radial field*

For high aspect ratios the radial field E_r for $r < R$ in the median ($z = 0$) plane of a pill box bunch can be approximated by [10]:

$$E_r(r) = \frac{\sigma}{\varepsilon_0}\, e_r(r/R), \tag{3}$$

with the dimensionless function e_r defined by:

$$e_r(u) \equiv \frac{1}{\pi u}\left(K(u^2) - E(u^2)\right). \tag{4}$$

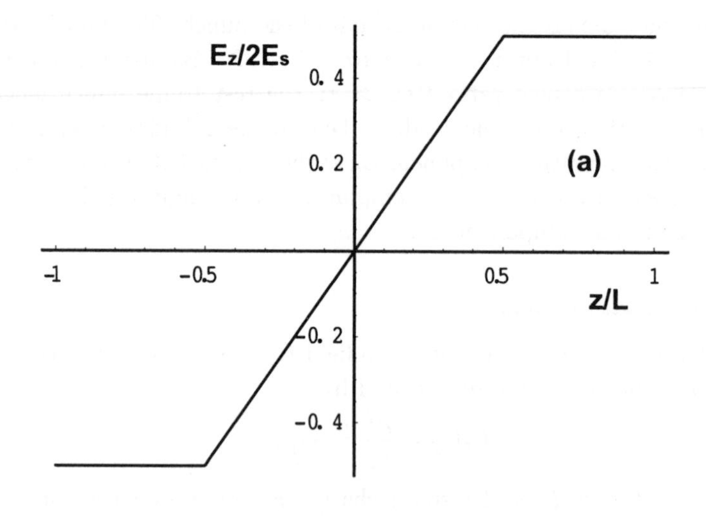

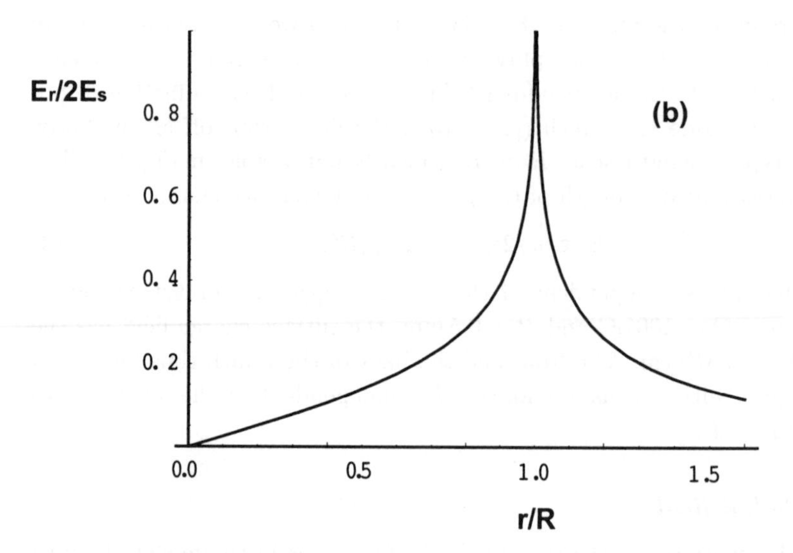

Figure 2. (a) Longitudinal field E_z of a bunch with length L as a function of the longitudinal position z. E_z is expressed in units $2E_s = \sigma/\varepsilon_0$, z in units L. (b) Radial field E_r of a bunch with radius R as a function of the radial position r. E_r is expressed in units $2E_s = \sigma/\varepsilon_0$, r in units R.

Here K and E are complete elliptic integrals of, respectively, the first and second kind [11].

In our analysis we use expression (3) to describe the radial field everywhere in the pancake bunch. In this way we slightly overestimate the magnitude of E_r and simplify the analysis by neglecting the small dependence on z. In Fig. 2b E_r is plotted as a function of radial position r. For small values of r the radial field is proportional to r and accurately described by $E_r = (\sigma/\varepsilon_0)r/4R$; for $r > R/2$ the dependence on r becomes increasingly nonlinear. For truly flat bunches ($L = 0$) the radial field diverges at $r = R$, which is of course not the case for finite values of R/L. This is not a problem, however, because the difference between the radial fields of a truly flat bunch and a pancake bunch with a high aspect ratio is only significant in a small region of radial size $\Delta r \sim L \ll R$ near the edge; the radial field in the largest part of the bunch is quite accurately described by Eq. (3). Note that the characteristic field strength of the problem is again the surface charge field E_s. The space charge forces pushing a pancake bunch apart are therefore of comparable strength in all directions.

2.3. *Comparison with spherical charge distribution*

It is instructive to compare the fields of a uniformly charged pancake with radius R and charge Q with the fields of a uniformly charged sphere with radius R and charge Q. The magnitude of the electric field pointing outward on the surface of the sphere is $Q/4\pi\varepsilon_0 R^2$, which is half the value of the surface charge field E_s, characteristic of the field strength on the surface of the pancake. It is quite remarkable that the space charge fields in a pancake bunch are only twice as large as in a uniformly charged spherical bunch of the same size and charge, even though the space charge density in the pancake bunch may be orders of magnitude higher.

3. Bunch evolution in longitudinal phase space

A laser pulse of duration τ_l excites an electron bunch, which is subsequently accelerated in the z-direction in a uniform field $\vec{E_0} = E_0\vec{e_z}$. By assumption the charge distribution is uniform from the outset and the photoemission process is prompt: the back of the bunch is created a time τ_l later than the front. For a laser pulse of duration $\tau_l = 50\,\text{fs}$ creating 100 pC of charge in an acceleration field $E_0 = 1\,\text{GV/m}$, the bunch length L immediately after creation is approximately 250 nm, *i.e.* $R/L = 2 \times 10^3$ for a typical bunch radius $R = 0.5\,\text{mm}$ and therefore truly pancake-like. The current at initiation is $I = Q/\tau_l = 2\,\text{kA}$.

3.1. *Beam current*

As the bunch is accelerated the time difference between the back and the front increases due to the longitudinal space charge forces and consequently the current goes down. Owing to the linear behavior of the longitudinal field E_z (Eq. (1)), the initially uniform charge distribution remains uniform during the expansion (as long as $R/L \gg 1$). The current during acceleration is therefore simply given by

$$I = \frac{Q}{\Delta t}, \tag{5}$$

with Δt the time difference between the back and the front of the bunch. In order to estimate the rate at which Δt grows, we start from the basic expression for the time t it takes an electron to travel from 0 to z in a field E_0, starting from zero velocity:

$$t(z) = \sqrt{\frac{2mz}{eE_0} + \frac{z^2}{c^2}}, \tag{6}$$

with e the elementary charge and m the electron mass. The duration Δt of the bunch at a position z is then given by $\Delta t(z) = \tau_l + t_{\text{back}}(z) - t_{\text{front}}(z)$. Realizing that the back of the bunch is effectively accelerated in a field lower than the front by an amount $2E_s = \sigma/\varepsilon_0$ (see Eq. (1)), and that generally $E_s \ll E_0$, we find to first order in E_s/E_0:

$$\Delta t(\gamma) = \tau_l + \frac{mc\sigma}{e\varepsilon_0 E_0^2}\sqrt{\frac{\gamma - 1}{\gamma + 1}}, \tag{7}$$

where we have expressed Δt in terms of $\gamma = 1 + eE_0 z/mc^2$, the Lorentz factor of an electron accelerated from 0 to z in a field E_0.

Note that this analysis only holds for the pancake regime, *i.e.* as long as $R \gg L = \gamma\beta c\Delta t$. Using Eq. (7) one can now show straightforwardly that for $E_0 = 1\,\text{GV/m}$, $R = 0.5\,\text{mm}$, and $Q = 100\,\text{pC}$ the pancake approximation is justified if $\gamma \ll 25$. Therefore it is probably a reasonably accurate description up to energies of a few MeV.

Equation (7) shows that the space-charge induced growth of the bunch duration saturates for high energies to a fixed value $mc\sigma/e\varepsilon_0 E^2$. For $E_0 = 1\,\text{GV/m}$, $R = 0.5\,\text{mm}$, $Q = 100\,\text{pC}$, and $\gamma = 10$ the increase in bunch duration Δt is only 22 fs, which is 10% less than the maximum value $mc\sigma/e\varepsilon_0 E^2$ and small compared to the laser pulse duration $\tau_l = 50\,\text{fs}$.

If we define the current parameter I_0 by

$$I_0 \equiv \frac{e\varepsilon_0 \pi R^2 E_0^2}{mc}\sqrt{\frac{\gamma + 1}{\gamma - 1}}. \tag{8}$$

then the following expression for the beam current can be obtained from Eqs. (5) and (7):

$$I = \frac{I_0}{1 + \frac{I_0 \tau_l}{Q}}. \tag{9}$$

From Eq. (9) it is clear that the current parameter I_0 is to be interpreted as the maximum beam current that can be obtained for given values of E_0, R, and γ. This is illustrated in Fig. 3, in which the beam current I, expressed in units I_0, is plotted as a function of the bunch charge Q, expressed in units $I_0\tau_l$. The maximum beam current is obtained if $Q \gg I_0\tau_l$, *i.e.* if

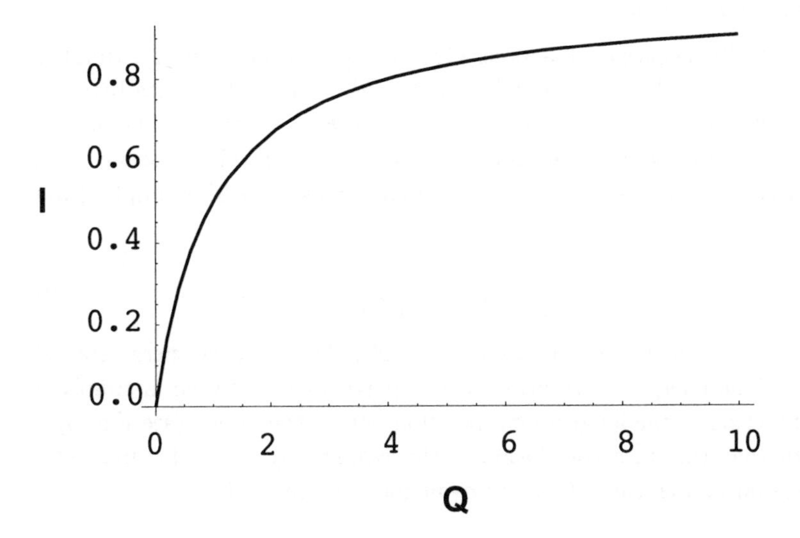

Figure 3. Pancake beam current I, expressed in units I_0, plotted as a function of the bunch charge Q, expressed in units $I_0\tau_l$.

the bunch duration is completely determined by space charge forces and independent of τ_l. Remarkably, the maximum current I_0 does *not* depend on the bunch charge Q. This can be understood as follows: If the charge Q is doubled, then the space charge forces (Eq. (1)) also become twice as large. For negligible τ_l the duration of the bunch Δt then also becomes twice as long (Eq. (7)), so the current $I = Q/\Delta t$ remains the same.

For $E_0 = 1\,\text{GV/m}$, $R = 0.5\,\text{mm}$, and $\gamma \gg 1$, the maximum current is 5 kA. Using $\tau_l = 50\,\text{fs}$ and $Q = 100\,\text{pC}$, we find $Q/I_0\tau_l = 0.5$, which means the beam current is a factor three lower than what is maximally achievable. This can be improved by using shorter laser pulses or higher bunch charges. Reducing the beam radius R is not an option because of the thermal emittance, as is discussed in the next section. Also the bunch charge cannot be increased indefinitely, because of the concomitant increase of the space charge induced emittance. The duration of the shortest laser pulse that can be realized, which is still capable of exciting $\sim 100\,\mathrm{pC}$, is about 5 fs [12]. The time scale for photoemission from copper, however, can be estimated to be about 10 fs [13], which is therefore in practice the absolute lower limit for the value of τ_l in our model. For $\tau_l = 10\,\mathrm{fs}$ and $Q = 100\,\mathrm{pC}$ the current is 70% of its maximum value I_0.

3.2. *Energy spread*

The longitudinal space charge forces give rise to an energy spread within the bunch. For a bunch accelerated from 0 to z the energy difference between the particles in the front and in the back is $2eE_s z = e\sigma z/\varepsilon_0$. The energy spread grows during acceleration proportional to the travelled distance, but the relative energy spread $\Delta\gamma/(\gamma-1)$ remains constant and is simply given by

$$\frac{\Delta\gamma}{\gamma-1} = \frac{2E_s}{E_0} = \frac{\sigma}{\varepsilon_0 E_0}. \tag{10}$$

For $E_0 = 1\,\mathrm{GV/m}$, $R = 0.5\,\mathrm{mm}$, and $Q = 100\,\mathrm{pC}$ the relative energy spread is 1.4%. The energy distribution of the particles in between is simply a linear function of the longitudinal position within the bunch (see Fig. 5).

Note that this nice correlation is the wrong way round for magnetic compression in a chicane. Only an α-magnet can be used.

3.3. *GPT simulations*

Prior to the above analytical approach, we performed simulations of acceleration of ultra-short bunches to 2 MeV in a 1 GV/m pulsed DC photocathode [8]. The pulsed DC photocathode consists of a flat copper cathode and an anode with a circular aperture, as is shown in Fig. 4. The cathode and anode are separated by 2 mm and the radius of the anode aperture is 0.7 mm. In the simulations the bunches were initiated with a 50 fs FWHM *gaussian* temporal distribution and a uniform radial distribution of 0.5 mm radius. The cylindrically symmetric 2-D space charge model of GPT was used.

In order to assess the validity of the model, simulations have also been performed in a hypothetical uniform 1 GV/m electrostatic acceleration field

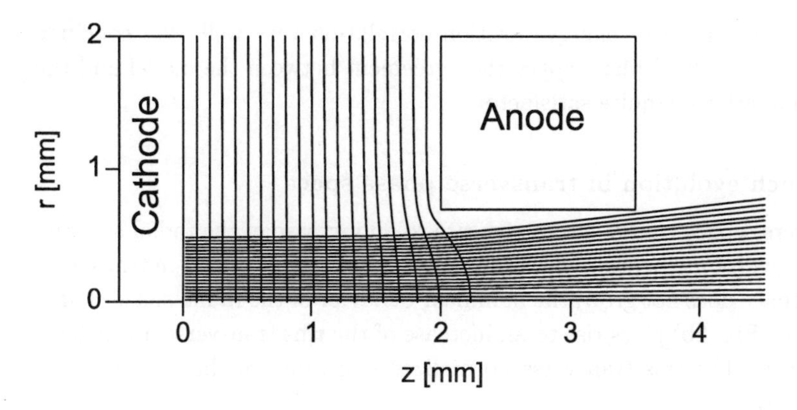

Figure 4. Diode with potential lines and sample particle trajectories.

without any radial components, with a 50 fs *top-hat* temporal distribution. In these simulations the bunch attains at 2 MeV an energy spread of about 26 keV , a bunch length of about 65 fs, and a peak current of about 1.5 kA, while the analytical model predicts 28 keV, 70 fs, and 1.4 kA, respectively. Fig. 5 shows the bunch distribution in longitudinal phase space. The crosses

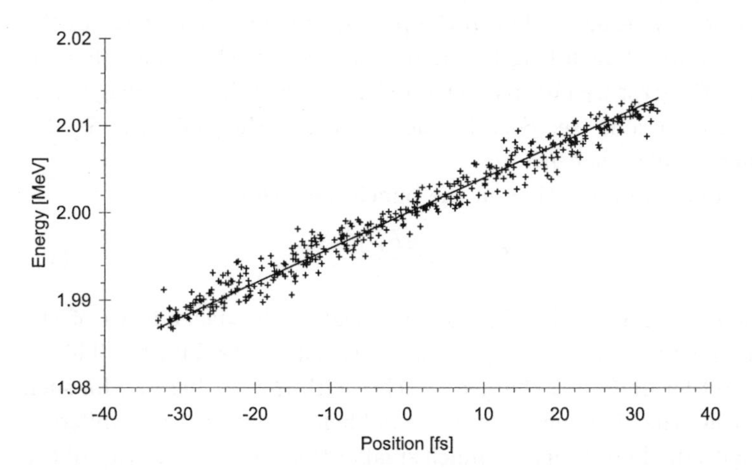

Figure 5. The bunch distribution in $z - E$ longitudinal phase space at a distance of $z = 2.0\,\text{mm}$ from the cathode surface; the solid line is the distribution according to the model.

are the result of the GPT simulations, calculated at a position $z = 2.0\,\text{mm}$; the solid line is the distribution according to the simple model. The model describes the positive energy–position correlation very well. We conclude that in longitudinal phase space the agreement between the model and the GPT simulations is quite satisfactory.

4. Bunch evolution in transverse phase space

Transverse space charge forces give rise to an increase of the radial size and the transverse momentum. Although the actual occupied area in transverse phase space does not grow, the nonlinear behavior of the radial space charge force (see Fig. 2b) gives rise to an increase of the rms transverse normalized emittance. The rms transverse normalized emittance in the x-direction is defined by:

$$\varepsilon_{n,x} = \frac{1}{mc}\sqrt{\langle x^2\rangle \langle p_x^2\rangle - \langle xp_x\rangle^2}, \tag{11}$$

where $\langle\rangle$ indicates averaging over the entire distribution and $p_x = mc\gamma\beta_x$ is the x-momentum, with $\beta_x \equiv v_x/c$ the transverse velocity normalized to the speed of light. In order to calculate space-charge induced emittance growth in the pancake regime, we make use of the following approximations:

(1) Radial displacement is negligible in the pancake regime; this can be justified by realizing that in the pancake regime the increase of the (rest frame) bunch length L is by definition much smaller than R. Since the (rest frame) space charge forces are of the same magnitude in all directions (see Sec. 3), the increase of the radial size is also much smaller than R.

(2) The radial forces acting on the bunch are given by

$$F_r(r) = \frac{eE_r(r)}{\gamma}, \tag{12}$$

where E_r is given by Eq. (3): we neglect the dependence of the radial forces on the longitudinal position in the bunch. This is correct for uniformly charged bunches with $R \gg L$, because for such bunches the difference between radial field profiles $E_r(r)$ at different longitudinal positions z is much smaller than the nonlinearity of the radial field profile at any particular position z. Consequently, the contribution of the z dependence of the radial fields to the growth of the effective area occupied in transverse phase space is much smaller than the contribution due to nonlinearity of $E_r(r)$. The z dependence can therefore safely be neglected in the calculation of the rms transverse normalized emittance. Also for this reason the well-established emittance compensation scheme based on the alignment of longitudinal bunch slices in transverse phase space [14,15] is not relevant in the pancake regime.

(3) The space-charged induced emittance growth is first calculated assuming zero thermal emittance. Subsequently, the finite value of the thermal emittance is taken into account by adding it in quadrature.

4.1. *Space-charge induced emittance growth*

The value of the transverse momentum p_x at a position $x = r\cos\varphi$ after acceleration during a time t is given by:

$$p_x = \int_0^t \frac{eE_r\cos\varphi}{\gamma}dt'. \tag{13}$$

Since we may neglect any radial displacement, the radial distribution and therefore the radial field E_r remain fixed. Using the basic relation $t(\gamma) = (mc/eE_0)\sqrt{\gamma^2-1}$, Eq. (13) may now be written as

$$\begin{aligned} p_x(r,\varphi) &= eE_r(r)\cos\varphi\int_0^t \frac{dt'}{\gamma} = \frac{mcE_r(r)\cos\varphi}{E_0}\int_1^\gamma \frac{d\gamma'}{\sqrt{\gamma'^2-1}} \\ &= \frac{mcE_r(r)\cos\varphi}{E_0}\log\left(\gamma+\sqrt{\gamma^2-1}\right) \end{aligned} \tag{14}$$

Note that the transverse momentum has the same functional dependence on transverse position as the radial field: for pancake bunches the shape of the distribution in transverse phase space directly reflects the behavior of the radial field. This is illustrated in Fig. 6, which shows the bunch distribution in $r - p_r$ phase space of a 50 fs, 100 pC, $R = 0.5\,\mathrm{mm}$ bunch accelerated to an energy of 2 MeV in a uniform 1 GV/m field. The crosses represent the results of GPT simulations. The solid curve is the phase space shape according to the model. The solid curve has exactly the same shape as the curve describing the radial electric field E_r inside the bunch, which is depicted in Fig. 2a.

Using definition (11) and Eq. (14) we find after some manipulation:

$$\varepsilon_{n,x} = C_1\frac{Q}{\pi\varepsilon_0 E_0 R}\log\left(\gamma+\sqrt{\gamma^2-1}\right), \tag{15}$$

with the constant C_1 given by

$$C_1 = \sqrt{\frac{1}{4}\int_0^1 u\left(e_r(u)\right)^2 du - \left[\int_0^1 u^2 e_r(u)du\right]^2} = 0.0375. \tag{16}$$

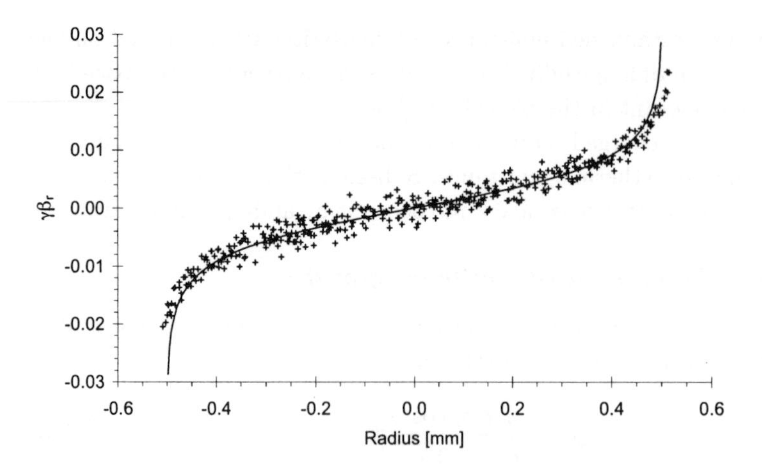

Figure 6. The bunch distribution in $r - p_r$ transverse phase space at a distance of $z = 2.0\,\text{mm}$ from the cathode surface, accelerated in a 1 GV/m field. The transverse momentum $p_r = mc\gamma\beta_r$ is expressed in units mc. The crosses represent the results of GPT simulations. The solid curve is the phase space shape according to the model.

The dimensionless function e_r is given by Eq. (4). Note that the constant C_1 contains all information on the radial field profile. Emittance growth of a pancake bunch with a different, non-uniform, radial charge distribution will also be described by Eq. (15), but with a different value of C_1.

4.2. *Thermal emittance*

We assume the photoemission proces creates electrons with an energy of 0.4 eV, which are ejected isotropically over $2\pi\,\text{sr}$ [8]. For a uniformly illuminated circular spot of radius R the thermal emittance can then be calculated straightforwardly using definition (11):

$$\varepsilon_{th} = C_2 R, \tag{17}$$

with $C_2 = 4.6 \times 10^{-4}$. A beam radius $R = 0.5\,\text{mm}$ thus implies $\varepsilon_{th} = 0.23\,\mu\text{m}$. Assuming the two contributions add up in quadrature, we arrive at the final expression for the rms transverse normalized emittance:

$$\varepsilon_{n,x} = \sqrt{\left[C_1 \frac{Q}{\pi\varepsilon_0 E_0 R} \log\left(\gamma + \sqrt{\gamma^2 - 1}\right)\right]^2 + C_2^2 R^2}. \tag{18}$$

Note that the emittance is independent of the laser pulse duration τ_l, because in the pancake regime the radial space charge forces are independent of the bunch length. The emittance increases with charge, because of radial space charge fields. For a given value of γ the emittance decreases with the strength of the applied acceleration field E_0, simply because for higher values of E_0, γ is reached in a shorter time t, and so the transverse momentum is less affected (see Eq. (13)). The dependence of the emittance on the beam radius R is more complicated: for small values of R the space charge contribution dominates and the emittance increases for *decreasing* R; for large values of R space charge forces become negligible, and the (thermal) emittance increases for *increasing* R. The explicit analytical form of Eq. (18) allows us to determine the optimal value R_{min} of the beam radius:

$$R_{\text{min}} = \sqrt{\frac{C_1}{C_2}\frac{Q}{\pi\varepsilon_0 E_0}\log\left(\gamma + \sqrt{\gamma^2-1}\right)}, \tag{19}$$

for which the emittance has its minimal value ε_{min}:

$$\varepsilon_{\text{min}} = C_2 R_{\text{min}}\sqrt{2} = \sqrt{2C_1C_2\frac{Q}{\pi\varepsilon_0 E_0}\log\left(\gamma + \sqrt{\gamma^2-1}\right)}. \tag{20}$$

Note that the minimal value of the emittance is $\sqrt{2}$ times the thermal emittance. For $E_0 = 1\,\text{GV/m}$, $Q = 100\,\text{pC}$, and $\gamma = 5$ we find $R_{\text{min}} = 0.8\,\text{mm}$ and $\varepsilon_{\text{min}} = 0.52\,\mu\text{m}$.

4.3. *GPT simulations*

In Fig. 7 the rms transverse normalized emittance of a 100 pC, 0.5 mm radius pancake bunch, accelerated in a uniform 1 GV/m field, is plotted as a function of z. The upper curve is calculated using Eq. (18), with $C_1 = 0.0375$; the lower curve is the result of GPT simulations. For small values of z ($z < 0.2\,\text{mm}$) the model and the simulations are in good agreement. For larger values of z the two curves deviate. At 2 MeV the simulations indicate an rms transverse normalized emittance of about $0.5\,\mu\text{m}$, while our model predicts $0.65\,\mu\text{m}$.

At first sight, this significant discrepancy may seem quite surprising, in view of the close agreement between the model and the simulations in describing the bunch in transverse phase space (see Fig. 6). A closer inspection of Fig. 6, however, reveals a discrepancy near the perimeter of the bunch: near $r = R$ the transverse momentum predicted by the model peaks more sharply than according to the simulations. This is due to the fact that for finite bunch lengths the radial field near the outer perimeter is slightly "softer" (less steep) than for the truly flat bunch, described by Eqs. (3) and

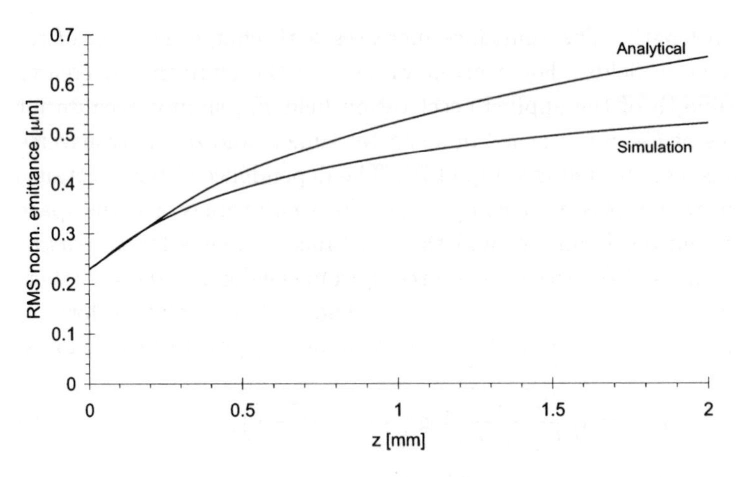

Figure 7. $\varepsilon_{n,x}$, in μm, plotted as a function of z, in mm. Upper curve: analytical model with $C_1 = 0.0375$; lower curve: GPT simulations.

(4). The analytical model thus overestimates p_r near the edge, but the overall transverse space description agrees quite well with the simulations, as long as the bunch is in the pancake regime. However, the divergent behavior of the model at the edge strongly affects the rms normalized emittance. The discrepancy may therefore be considered as an artefact associated with the particular definition of the transverse normalized emittance.

We thus conclude that also in transverse phase space the agreement between the model and the GPT simulations is quite satisfactory; one should however be careful in the evaluation of the transverse normalized emittance.

5. Normalized brightness

A comprehensive measure for beam quality is the normalized brightness, a Lorentz-invariant quantity defined by [1]:

$$B_n = \frac{N}{\varepsilon_{n,x}\varepsilon_{n,y}\sigma_z\sigma_\gamma}, \tag{21}$$

where N is the number of electrons in the bunch, $\varepsilon_{n,y}$ is the rms normalized transverse emittance in the y-direction, which is identical to $\varepsilon_{n,x}$ for the cylindrically symmetric beams under consideration, σ_z is the rms (lab frame) bunch length, and σ_γ is the rms energy spread, expressed in units mc^2.

However, since the energy spread is not a critical parameter, the beam quality is usually expressed in terms of the so-called transverse brightness

B_n^t, which is the 5-D phase space projection of the full 6-D normalized brightness B_n [1]. By convention the transverse brightness is defined as:

$$B_n^t = \frac{2I}{\varepsilon_{n,x}\varepsilon_{n,y}}. \quad (22)$$

Although, strictly speaking, the transverse brightness B_n^t is not a Lorentz invariant quantity, it is a useful measure of the beam quality for applications such as the X-ray SASE FEL. By substituting the values $I = 1\,\mathrm{kA}$ and $\varepsilon_{n,x} = 1\,\mu\mathrm{m}$, we find that X-ray SASE FEL operation at sub-nm wavelengths typically requires a transverse brightness of at least $B_n^t = 2 \times 10^{15}\,\mathrm{A\,m^{-2}sr^{-1}}$.

Using Eqs. (9) and (18), a general expression for the transverse brightness of pancake bunches can be written down straightforwardly. However, this somewhat lengthy and complicated expression does not give much additional insight. It is more interesting to directly investigate the *maximum* value of the transverse brightness that can be attained.

Using the results of Secs. 3 and 4 the optimal values of the beam current, $I = I_0$ (Eq. (8)), and of the emittance, $\varepsilon_{n,x} = \varepsilon_{\min}$ (Eq. (20)), can be substituted into Eq. (22), resulting in an expression for the maximum value of the transverse brightness $B_{n,\max}^t$ that can be attained:

$$B_{n,\max}^t = \frac{\pi e \varepsilon_0 E_0^2}{C_2^2 mc}\sqrt{\frac{\gamma+1}{\gamma-1}}. \quad (23)$$

Remarkably, $B_{n,\max}^t$ *only* depends on the acceleration field E_0 and γ, and is *independent* of the bunch charge Q. This can be understood as follows: the maximum current I_0 scales with $R_{\min}^2$ and the minimum emittance $\varepsilon_{\min}$ scales with $R_{\min}^{-1}$, so $B_{n,\max}^t$ is independent of $R_{\min}$. The only way I_0 and $\varepsilon_{\min}$ depend on the bunch charge Q is through $R_{\min}$ (see Eq. (19)), so $B_{n,\max}^t$ is independent of charge. In other words: either one chooses a high bunch charge with a large bunch radius, giving rise to a high beam current, but bad emittance; or one chooses a low bunch charge with a small radius, and thus a low beam current and very good emittance. The resulting transverse brightness, however, remains the same. For $E_0 = 1\,\mathrm{GV/m}$ and $\gamma \gg 1$ we find $B_{n,\max}^t = 7.7 \times 10^{16}\,\mathrm{A\,m^{-2}sr^{-1}}$, much higher than what is required for *e.g.* X-ray SASE FEL operation at sub-nm wavelengths.

The maximum value $B_{n,\max}^t$, however, can only be achieved for the maximum value of the beam current I_0, which requires that the laser pulse duration $\tau_l \ll Q/I_0$. Substituting again Eq. (19) into Eq. (8), this is equivalent to the condition:

$$\tau_l \ll \frac{C_2}{C_1}\frac{mc}{eE_0}\frac{\sqrt{(\gamma+1)/(\gamma-1)}}{\log\left[\gamma+\sqrt{\gamma^2-1}\right]}, \tag{24}$$

which can be approximated for $\gamma \gg 1$ by

$$\tau_l \ll \frac{C_2}{C_1}\frac{mc}{eE_0\log(2\gamma)}. \tag{25}$$

Interestingly, the condition for the optimum laser pulse duration τ_l is again independent of bunch charge Q. For a field $E_0 = 1\,\mathrm{GV/m}$ and $\gamma = 5$, we find that $\tau_l \ll 7.5\,\mathrm{fs}$, which cannot be realized in practice. To make a more realistic estimate of the achievable transverse brightness, let us assume $\tau_l = 10\,\mathrm{fs}$, $Q = 100\,\mathrm{pC}$, and $\gamma = 20$. Using Eqs. (19), (20), (9), and (8) we then find $R_{\mathrm{min}} = 1\,\mathrm{mm}$, with the corresponding value of the rms normalized transverse emittance $\varepsilon_{\mathrm{min}} = 0.65\,\mu\mathrm{m}$, and a beam current of $I = 6.3\,\mathrm{kA}$. For these parameters the aspect ratio $R/L \approx 5$, so the pancake description is not completely accurate anymore; the resulting numbers may still serve, however, as worst case estimates. The corresponding transverse brightness is $B^t_{n,\mathrm{max}} = 3\times10^{16}\,\mathrm{A\,m^{-2}sr^{-1}}$, still much higher than what is required for X-ray SASE FEL operation at sub-nm wavelengths.

6. Conclusions

An analytical model has been developed that accurately describes the evolution of high space charge density pancake electron bunches, accelerated in uniform fields. The model yields closed expressions for the current and the rms transverse normalized emittance. We find that, in the pancake regime, the maximally achievable peak current does not depend on the bunch charge, and the rms transverse normalized emittance does not depend on the bunch length. For the realization of high brightness pancake electron bunches the pulse duration of the photoemission laser should therefore be chosen as short as possible.

The model is in agreement with the results of GPT simulations of 100 pC bunches of 0.5 mm radius, created with a 50 fs laser pulse and accelerated to 2 MeV in a field of 1 GV/m. Under these conditions the pancake description can be used up to energies of a few MeV. For higher energies the rest frame aspect ratio of the bunch becomes too small for the pancake description to be valid, but the results of the model may still be useful then as worst case estimates. The energy range in which the pancake description is applicable may be increased by using either shorter laser pulses, smaller bunch charges, or larger bunch radii.

The model suggests that both the peak current and the transverse emittance required for X-ray SASE FEL operation at sub-nm wavelengths can be attained *without* magnetic compression, if acceleration gradients of 1 GV/m are realized and sustained up to energies of 10 MeV. The model may serve as a guiding tool for future investigations of more realistic geometries by means of numerical simulations.

Acknowledgments

The author would like to thank Marieke de Loos en Bas van der Geer of Pulsar Physics for the uniform-field GPT simulations. Useful discussions with Bas van der Geer, Marieke de Loos, Fred Kiewiet, and Marnix van der Wiel are gratefully acknowledged.

References

1. M. J. van der Wiel, in *The Physics of High-brightness Beams*, edited by J. Rosenzweig and L. Serafini, Conf. Proc. 2nd ICFA Workshop 1999 (World Scientific, Singapore, 2000).
2. J. Andruskov *et al.*, *Phys. Rev. Lett.* **85**, 3825 (2000).
3. B. Carlsten and T. Raubenheimer, *Phys. Rev. E* **51**, 1453 (1995).
4. D. J. Gibson, F. V. Hartemann, E. C. Landahl, A. L. Troha, N.C. Luhmann, Jr., G. P. Le Sage, and C. H. Ho, *Phys. Rev. ST AB* **4**, 090101 (2001).
5. T. Srinivasan-Rao and J. Smedley, in *Advanced Accelerator Concepts*, edited by S. Chattopadhyay, J. McCullough, and P. Dahl, AIP Conf. Proc. 398, (AIP, Woodbury, 1997), p. 730.
6. K. Batchelor *et al.*, in *Proceedings of the European Particle Conference, Stockholm, Sweden, 1998*, edited by S. Meyers *et al.* (IOP, London, 1998), p. 791.
7. T. Srinivasan-Rao, J. Schill, I. Ben-Zvi, K. Batchelor, J. P. Farrell, J. Smedley, X. E. Lin, and A. Odian, in *Proceedings of the IEEE Particle Conference, Stockholm, New York, 1999*, edited by A. Luccio and W. MacKay (IEEE, Piscataway, NJ, 1999), p. 75.
8. S. B. van der Geer, M. J. de Loos, J. I. M. Botman, O. J. Luiten, and M. J. van der Wiel, *Phys. Rev. E* **65**, 046501 (2002).
9. L. Serafini, in *Towards X-ray Free Electron Lasers*, edited by R. Bonifacio and W. A. Barletta, AIP Conf. Proc. 413 (AIP, Woodbury, NY, 1997), p. 321.
10. E. Durand, *Électrostatique I: Les Distributions*, masson et Cie (Paris, 1964).
11. *Handbook of Mathematical Functions*, edited by M. Abramowitz and I. A. Stegun, Dover Publications (New York, 1970).
12. S. Sartania, Z. Cheng, M. Lenzner, G. tempea, Ch. Spielmann, F. Krausz, and K. Ferencz, *Opt. Lett.* **22**, 1562 (1997).

13. R. Brogle, P. Muggli, and C. Joshi, in *Advanced Accelerator Concepts*, edited by S. Chattopadhyay, AIP Conf. Proc. 398 (AIP, New York, 1997), p. 747.
14. B. Carlsten , *Nucl. Instr. Methods Phys. Res. A* **285**, 313 (1989).
15. L. Serafini and J. B. Rosenzweig, *Phys. Rev. E* **55**, 7565 (1997).

REVIEW OF EXPERIMENTAL RESULTS ON HIGH-BRIGHTNESS PHOTO-EMISSION ELECTRON SOURCES

PH. PIOT

Deutsches Elektronen-Synchrotron (DESY), D-22603 Hamburg, Germany.
E-mail: Philippe.piot@desy.de

The generation of electron beam with high phase space density is one of the principal challenge for the next generation of ultra-short wavelength light sources and for the foreseen linear colliders projects. In this paper we review some of the recent experimental advances in the field of electron sources based on the photo-electric effect.

1 Introduction

1.1 Overview

In the recent years electron sources based on the photoelectric effect have supported operation of many accelerators ranging from colliders (e.g. SLC) to novel light sources (e.g. at ANL, BNL, DESY, JLAB). This kind of electron sources have been supporting advanced beam physics experiments such as plasma-driven accelerator (e.g. ANL, FNAL) and Thomson scattering (E.G. LLNL).

The photo-emission electron source (see earlier reviews from Sheffield [1] and O'Shea [2]) offers many advantages compared to the other popular technologies (e.g. thermionic emission). Firstly, there is a good handle on the initial conditions of the emitted electron bunch. Such initial conditions (transverse and longitudinal distribution) is controlled via the photocathode drive-laser parameters (pulse length, spot size on the cathode, intensity, etc...) and turn out to be an important feature for minimizing the transverse emittance growth. With the availability of ultra-short pulse lasers, the electron bunch length can be made small compared to the radio-frequency (rf) cavity wavelength thereby removing the need of sub-harmonic bunching schemes as in injectors based on thermionic sources. On the other hand, the charge density that can generally be reached with photocathodes is of the kA.cm^{-2} level, approximately two orders of magnitude above thermionic emitters. Finally the use of a polarized drive-laser enables the production of polarized photo-electrons in virtue of the angular momentum conservation.

Because of the charge density considered, the photo-electrons need to be quickly accelerated in order to significantly decrease the space charge forces within the bunch. Hence the photocathode is located on the back-plate of the half-cell of a radio-frequency cavity (rf-gun) or at an extremity of a high-voltage gap (dc-gun) as schematized in Figure 1.

An electron source is generally integrated within an "injector" consisting of accelerating section(s) and focusing element(s) configurated in such a way to take

Facility, Location	Gun Type f (GHz)	# of cell	Mission
IRDemo, JLab [6]	dc	–	high average power FEL
ATF, BNL [7]	rf, 2.86	1.6	multi-disciplinary facility
LEUTL, ANL [8]	rf, 2.86	1.6	UV SASE-FEL
SDL, BNL [9]	rf, 2.86	1.6	UV SASE-FEL
CTF, CERN [10]	rf, 3.0	1.5	CLIC test facility
AFEL, LANL [11]	rf, 1.3	10.5	IR SASE-FEL
AWA, ANL [12]	rf, 1.3	1.6	wakefield acceleration
GTF, SLAC [13]	rf, 2.86	1.6	LCLS test injector
TTF-1/2, DESY-HH [14]	rf, 1.3	1.6	TESLA test facility + VUV SASE-FEL
PITZ, DESY-Z [15]	rf, 1.3	1.6	injector test facility
FNPL, FNAL [16]	rf, 1.3	1.6	multi-disciplinary facility
MIT [17]	rf, 17.14	1.5	multi-disciplinary facility
Tokai [18]	rf, 2.86	1.6	multi-disciplinary facility
NEPTUNE, UCLA [19]	rf, 2.86	1.6	multi-disciplinary facility
PEGASSUS, UCLA [20]	rf, 2.86	10.5	multi-disciplinary facility
APLE, BOEING [21]	rf, 0.433	1.5	high average power FEL
ELSA, CEA [22]	rf, 0.144	1.5	high average power FEL + Thomson X-ray source
DROSSEL, FZR [23]	srf, 1.3	0.5	supercond. gun test facility

Table 1. List of some of the representative accelerator test facilities based on photo-electron sources.

advantage of the emittance compensation mechanism [3,4,5]. Some possible configurations are presented in Figure 2. For the rf-gun set-up, the two variants widely used are the so-called integrated and split injectors [24]. The former configuration (e.g. AFEL at LANL) uses long multi-cell rf-gun cavity with low peak electric field whereas the latter one (e.g. ATF at BNL) is based on a 1-1/2-cell rf-gun cavity operating with high peak electric field closely followed by a booster linac.

1.2 *Important parameters*

The figure-of-merit when considering the application of the electron source to drive free-electron lasers (FEL) or in the framework of linear colliders is the phase space density. Given the normalized transverse and longitudinal rms emittances $\tilde{\varepsilon}_{x,y}$ and $\tilde{\varepsilon}_z$ respectively and the bunch charge Q, we define the brightness as the ratio of the bunch charge to the hyper-volume it occupies in the 6-D phase space:

$$\mathcal{B} \doteq m_e c^2 \frac{Q}{\tilde{\varepsilon}_z \tilde{\varepsilon}_x \tilde{\varepsilon}_y}. \tag{1}$$

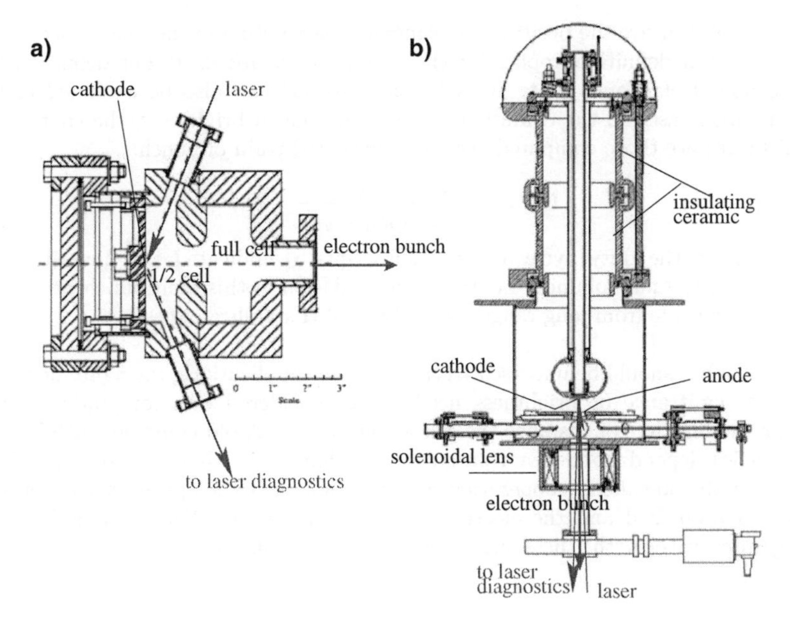

Figure 1. Schematic rendition of an rf-gun (**a**) and dc-gun (**b**) based electron sources.

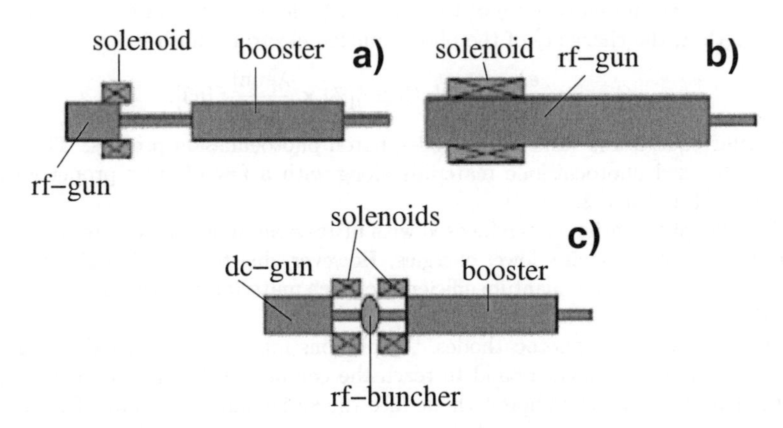

Figure 2. The various photo-emission source configuration used: rf-gun split (**a**) and integrated (**b**) design, and the dc-gun design (**c**).

Here $m_e c^2$ is the electron rest energy. In Eq. (1) the emittances are defined following:

$$\tilde{\varepsilon}_x = \frac{1}{\langle p_z \rangle} \sqrt{\langle (x - \langle x \rangle)^2 \rangle \langle (p_x - \langle p_x \rangle)^2 \rangle - \langle (x - \langle x \rangle)(p_x - \langle p_x \rangle) \rangle^2}, \qquad (2)$$

where x and p_x are the position and canonical momentum in the phase space. The same type of definition applies for the y and z emittances. In our definition the longitudinal emittance units are keV-mm. Eq. (1) can also be extended to the multibunch case. In the definition of such multibunch brightness, the charge and emittances are those computed over the considered train of bunch:

$$\mathcal{B}_M \doteq m_e c^2 \frac{Q_M}{\tilde{\varepsilon}_{z,M} \tilde{\varepsilon}_{x,M} \tilde{\varepsilon}_{y,M}} \times \mathcal{D}, \tag{3}$$

where $\mathcal{D}$ is the duty cycle and the subscripts $_M$ indicated the quantities are taken over the macrobunch. Sources of degradation of this multibunch brightness essentially come from long range wakefields and rf stability issues.

Though it should be underscored that for some applications, the figure-of-merits such as emittance or brightness need to be considered over longitudinal slices along the bunch. For instance the self-amplified stimulated emission (SASE) FEL properties depend upon beam parameters within slices whose lengths are of the order of the so-called "cooperation length", that is the slippage length between the radiation field and the electron for one gain length. For an X-ray FEL (1 Angstrom wavelength) these slices are typically ~ 1 μm long.

2 Photocathodes and drive-lasers

Given the quantum efficency η of the photocathode, the drive-laser energy $\mathcal{E}$ and wavelength λ, the charge Q of the photo-emitted electron bunch is:

$$Q = \eta \times \frac{e\lambda}{hc}\mathcal{E}, \text{or } Q[\text{nC}] \simeq \eta[\%] \times \frac{\lambda[\text{nm}]}{124}\mathcal{E}[\mu\text{J}]. \tag{4}$$

The latter equation is valid for a single-photon photoemission process. The most commonly used photocathode materials along with a few of their properties are summarized in Table 2.

Metallic photocathodes can be used with ultra-violet light as they are stable and "blind" to photons having lower energies. However, due to metallic reflection and internal absorption, the quantum efficiency of such materials is rather low (typically $\eta \sim 10^{-4}$).

In semiconductors photocathodes, the photons must impart enough energy to the electrons in the valence band to reach the conduction band and have enough energy left to escape the photocathode into the surrounding vacuum. The energy difference between the bottom of the valence band and the vacuum level is called the electron affinity E_a and the total energy required is $E_a + E_g$ where E_g is the energy gap between the top of the valence and the bottom of the conduction bands. Alkali Antimonide and Telluride (which are of the positive electron affinity ($E_a > 0$) type) have been used to produced unpolarized highly charged beams with a duty cycle similar to what is required for the TESLA linear collider.

Negative electron affinity (NEA) photoemitters are p-type semiconductors whose surface is treated to "bend" the band structure. In extreme cases, the vacuum level is lower than the bottom of the conduction band. The principal advantages

Material	typ. Q.E.	λ (nm)	lifetime (days)	vacuum req. (T)	in use at
Cu	2×10^{-4}	260	months	$\sim 10^{-7}$	ATF, SDL, Tokai,
Mg	2×10^{-3}	260	months	$\sim 10^{-7}$	ATF
Alkali Antimonide	5×10^{-2}	532	days	$\sim 10^{-10}$	BOEING
Alkali telluride	1×10^{-2}	260	months	$\sim 10^{-9}$	FNPL, TTF,
NEA GaAs (Cs)	5×10^{-2}	780	days	$\sim 10^{-11}$	IrDemo

Table 2. Commonly used cathode material and some of their typical characteristic.

of NEA photocathode include a longer wavelength operation and a higher quantum efficiency compared to the other photocathode types. An example of a NEA photoemitter is Gallium Arsenide (GaAs). It has been used as a photocathode in dc-guns to produce polarized beams with high duty cycle [25].

Although metallic photocathodes can stand relatively poor vacuum, their quantum efficency is at the 10^{-4} level. Hence the required laser energy for producing 1 nC of charge is of the order of 5 mJ. Driving a high average current accelerator, e.g. an energy-recovery linac-based light source, would require a ~100 W average power drive-laser if a metallic cathode would be used. Developing such a laser presents important technological challenges and instead, the so-called "ignited feedback regenerative amplifier" concept was proposed. In this latter scheme a part of the FEL radiation serves to drive the photocathode once a conventional drive-laser has been used for the start-up [27]. This concept was recently demonstrated at the TESLA test facility (TTF) where the light produced by the vacuum ultraviolet (VUV) SASE-FEL (wavelength 108 nm) was reflected on a downstream mirror back to the Cesium Telluride cathode to regenerate electron bunches [28].

The photocathode material along with the laser used to produce the electrons sets the initial "thermal" emittance that is the emittance generated because of the excess in kinetic energy of the photo-emitted electrons. The thermal emittance sets a lower limit on the beam emittance that can ideally be reached once the accelerator setup has been optimized. Given the excess kinetic energy K of the photo-emitted electrons, the thermal emittance $\tilde{\varepsilon}_{th}$ scales linearly with the drive-laser transverse spot size r on the cathode as [30]:

$$\tilde{\varepsilon}_{th} \simeq \frac{r}{2}\sqrt{\frac{2K}{3m_e c^2}}. \tag{5}$$

The latter equation assumes (1) the laser spot to be a uniform distribution and (2) the electrons to be emitted isotropically with the average kinetic energy K. To date, detailed measurements of thermal emittance have been reported for NEA GaAs and Copper cathodes only.

For GaAs cathodes, the transverse emittance measurements were performed at the Illinois/CEBAF polarized DC electron source [31]. The emittance was measured for different drive-laser wavelengths and spot sizes on the cathode. The deduced thermal transverse energy at $\lambda = 710$ nm was $K = 37$ meV. Recently, a technique to measure the energy distribution of electrons as a function of the transverse and longitudinal energies has been developed and applied to study the distribution of the photoemitted electron from a cooled ($T = 90$ K) GaAs cathode [32]. It was shown that the obtained population distribution has a sharp peak for transverse thermal energies of about 5 meV for $\lambda = 800$ nm. The comparison of the two aforementioned measurements supports that NEA photoemitters such as GaAs photoemit electrons that are in thermal equilibrium with the crystal lattice. Therefore by cooling this type of cathode, bunches with reduced transverse emittances may be obtained. However, the response time of such photoemitters is typically > 10 ps [29] and prohibits a precise tailoring of the initial time-distribution.

For Copper cathodes, a similar measurement was performed at the Source Development Laboratory (SDL) of BNL [33]. The emittance measurement was conducted using an S-band rf-gun. To avoid emittance dilution due to rf effects, the gun field and bunch charge were optimized to insure the emittance to be dominated by its thermal contribution. From this latter measurement, the thermal energy was estimated to be $K = 430$ meV for the wavelength $\lambda = 260$ nm, which is one factor of magnitude above the GaAs measurement at room temperature. In Reference [33], the large value of the measured thermal energy (theory predicts 260 meV) was attributed to the non perfect uniformity of the cathode surface: the roughness of the cathode surface can locally enhance the electric field [34] and thus impacts the bunch thermal emittance.

3 DC guns

Only one high-brightness photo-emission dc-gun was operated to drive an FEL, the Ir-Demo, at Jefferson Lab. The nominal dc voltage was 350 keV (corresponding to an electric field of ~ 6 MV/m on the cathode surface) and bunch charges up to 130 pC have been extracted from the GaAs cathode during nominal operation [35]. The gun was coupled to a so-called quarter CEBAF-cryomodule composed of two CEBAF-type 5-cell accelerating cavities with average accelerating gradient of about 10 MV/m. During the FEL operation [36], the electron beam emittance measured at the undulator location (at 45 MeV) was $\sim 8 \pm 1$ mm-mrad for 60 pC. Noteworthy, and this is the reason for this gun design, was the sustained lasing at 1.7 kW average power thanks to a bunch repetition rate of 75 MHz CW. At that time, the maximum gun voltage was limited by breakdown due to field emission. Present progress in treating the electrodes using ion implantation foresees the operation of similar dc guns with electric fields as high as 25 MV/m [37]. A team at JLab has also developed a laser, suited for GaAs cathodes, capable of operating with repetition frequency as high as 2.5 GHz [38] thereby providing the possibility to operate a linac downstream of the dc-gun in CW mode that is with a repetition rate equal to the rf frequency of the accelerator. Such a type of operation with a dc-gun is foreseen in the proposed energy recovering linac light source at Cornell [39].

	value	units
bunch charge	0.1	nC
laser pulse length (rms)	1	ps
peak cathode E-field	200	MV/m
beam energy	1.05	MeV
typical set of data:		
bunch charge	0.5	nC
beam energy	1.05	MeV
energy spread (rms)	1.5	%
$\tilde{\varepsilon}_{\perp}$ (rms)	1	mm-mrad
bunch length (rms)	1	ps

Table 3. Laser and rf parameters for the MIT rf-gun. A set of achieved beam parameters is also given.

4 Radio-frequency guns

Many groups operate test facilities that incorporate rf-gun photo-electron sources. The frequency of rf-guns presently under operation ranges from f =0.144 to 17 GHz. On the beam dynamics point-of-view if one could scales the rf cavity and photo-cathode drive-laser parameters such that the Vaslov equation is kept frequency-invariant [40,41], one would gain in operating at higher frequency since the brightness would scale as $\mathcal{B} \propto f^{-2}$. Indeed letting the Vaslov equation frequency-invariant requires the peak electric field to be scaled linearly with the frequency. In extreme cases this may result in problem such as rf-breakdown and field-induced heating. For instance simply scaling the presently achieved peak electric field (140 MV/m) for an S-band (f = 2.856 GHz) gun to an X-band (f = 17.136 GHz) would result in the unrealistic peak electric field value of 720 MV/m!

4.1 X-band

The Plasma Science and Fusion Center of MIT has worked toward the development of a high-frequency gun operating at 17.136 GHz. This work was motivated by the aforementioned favorable scaling of brightness with the frequency. The gun is a 1.5 cell accelerating structure and incorporates two side wall couplers (one on each cell). It also includes a solenoid for emittance compensation. The main parameters are gathered in Table 3. A multi-slit mask technique was used to measure the transverse emittance just downstream of the rf-gun. For 50 pC charge, the transverse emittance was estimated to be about 1 mm-mrad. To date the peak electric field achieved on the cathode is 200 MV/m and has been limited by rf-breakdown, resulting in a beam energy of 1 MeV approximately. A new gun (2.4 cell) will soon be built in order to boost the beam energy to 2 MeV. It will then be coupled to a 25 MV 17.136 GHz traveling wave linac.

4.2 S-band

The so-called BNL/SLAC/UCLA/ gun operating at 2.856 GHz is one of the most popular configurations. It is used at BNL (both at ATF and SDL), ANL (LEUTL), UCLA (Neptune), and in several Japanese Laboratories (e.g. at Tokai). Since its first version, the rf-gun has been gradually improved (symmetrization of the cavity, new solenoids, new laser port location,...).

At ATF, the gun has driven many experiments and has been use to demonstrate the saturation of the VISA SASE-FEL experiment. At this facility, the emittance has been measured to be 0.8 mm-mrad for a charge of 0.5 nC. Because of the low emittance and to avoid saturation of the beam viewers a special high-betatron function optics was set up and the emittance was inferred by a fit of the beam envelope observed on four viewers. The emittance has been measured at 70 MeV i.e. after acceleration through a 3 m SLAC-type traveling wave section and the measurement involved the minimization of transverse tails induced by dipole mode wake-fields along with beam-based alignment of the quadrupoles. Another study conducted at this facility aimed to investigate the transverse emittance dependence upon the transverse laser spot uniformity on the copper cathode. The laser uniformity is required to be very homogeneous transversely to insure within the bunch that the radial space charge field is linearly dependent on the radius (and thus does not yield a transverse emittance growth). Using neutral density matrix masks of different transmissions, the uniformity of the laser spot was varied and an increase of the emittance by a factor > 2 was observed as the uniformity was changed from a contrast of 0 (uniform spot) to 50 % [42].

At the source development laboratory (SDL) of BNL, an extended study on the DUV-FEL injector performance has been conducted. Two of the three accelerating sections that compose the accelerator have been operated at zero-crossing to impart a linear correlation between time and energy. The vertical emittance was then measured downstream of an horizontally bending spectrometer dipole. Therefore observing the vertical beam spot variation versus the strength of an upstream quadrupole for various energy (i.e. time) slice provides the slice emittance [44]. At 10 pC, slice emittances close the thermal emittance were measured (~ 0.7 mm-mrad over 200 fs time slices).
Bunch compression using velocity bunching has also been attempted, and peak currents similar to those obtained by magnetic compression were obtained by running the first accelerating section far off-crest [45].

At SLAC, the gun test facility (GTF) aims to optimize the LCLS injector and demonstrate the parameters required for the LCLS FEL project. Measurements of transverse and longitudinal emittance for various conditions have been conducted [46]. Noteworthy was the technique used to measure the longitudinal emittance: energy profiles downstream of the booster linac have been taken for various phase settings of the booster and a tomographic reconstruction technique in conjunction with particle tracking was used to infer the longitudinal phase space

	ATF	SDL	GTF	SHI	CLIC	units
bunch charge	1	1	1	1	100	nC
laser pulse length (rms)	3	2	1.3	4	4	ps
peak cathode E-field	120	100	120	100	105	MV/m
beam energy	70	200	30	14	40	MeV
typical set of data:						
bunch charge	0.5	0.2	0.2	1	100	nC
energy	70	70	30	14	40	MeV
energy spread (rms)	0.03(u)	0.05	0.025(u)	0.25	-	%
$\tilde{\varepsilon}_{\perp}$ (rms)	0.8	1.5	1.5	1.2	100	mm-mrad
bunch length (rms)	3	2	1.8	4	4.3	ps

Table 4. Typical photocathode drive-laser and rf parameters for various S-band guns. A set of achieved beam parameters is also given. All the gun presented in this Table operate at f=2.856 GHz apart from the CLIC gun for which f = 3.0 GHz. Note: "(u)" in the "energy spread" line indicates the quoted energy spread corresponding to the stochastic energy spread (correlations have been removed).

downstream of the rf-gun. The presently achieved parameters (see Table 4) are close to the required LCLS parameters for the low charge operation scenario.

In Japan, Sumitomo Heavy Industry (SHI) has performed beam measurements using a BNL/SLAC/UCLA type rf-gun coupled to an S-band linac [47]. Downstream of the linac, at energies of 14 MeV, the transverse emittance dependence on the bunch charge and drive-laser pulse shape and length have been investigated. For a uniform drive-laser time profile, the measured transverse emittance of 1.2 mm-mrad for a charge of Q = 1nC has been reported. The emittance has been measured using the quadrupole scan technique and Gaussian distributions were fitted to the profile in order to extract rms values of the beam size and to infer the transverse emittance.

At CERN, the CLIC linear collider test facility 2 (CTF-2) was constructed at CERN to study the generation of 30 GHz rf (required for the CLIC drive beam). It has been driven by a 3 GHz 1.5 cell rf-gun capable of producing the very high charge (100 nC) required for the drive beam. The electron beam requirements are essentially on the peak current, and up to 2.5 kA peak current were obtained after compression. The main parameters of the rf-gun are gathered in Table 4. The CTF-2 facility has also served as a test stand for life-time and quantum efficency studies of Alkali photocathodes [48].

4.3 L-band

An L-band gun operating at 1.3 GHz has been in use at LANL in the context of the Advanced FEL (AFEL) accelerator. The rf-gun consist of a 10-1/2 cell cavity. This gun alone has driven a far-infrared SASE FEL [49]. The slice emittance has been measured by performing streak camera measurements performed together with a quadrupole scan [50]. Thus the analysis of the beam size evolution versus the quadrupole strength for different time slices allows an indirect measurement of the slice emittance (measured slice emittance at 1 nC is about 1.6 mm-mrad). This rf-gun has demonstrated operation with 400 mA average current per macropulse. The gun design incorporates a Nitrogen vessel to support possible cryogenic operation thereby cutting ohmic losses in half.

The TESLA test facility (TTF) and the Fermilab/Nicadd photoinjector laboratory (FNPL) have both been operated with L-band (f = 1.3 GHz) rf-guns of the same design. This 1.6 cell rf-cavity was designed and manufactured at Fermilab [51]. The rf-gun was originally designed in the framework of the TESLA linear collider project to produce long macropulse trains and experimentally its operation with rf-macropulses of 800 μsec length was demonstrated. Typical transverse emittances measured in the 17 MeV section of the TTF injector are about 3.5 mm-mrad for a 1 nC bunch [52]. This rf-gun has driven a vacuum ultraviolet SASE-FEL in the saturation regime [53]. At FNPL, a series of measurements along with their comparison with numerical simulations have been performed on the gun [61]. Recently by immersing the photocathode in a magnetic field and using a linear transformation that decomposed the angular momentum into its transverse momentum components [62], flat beams (that is beams with $\tilde{\varepsilon}_y/\tilde{\varepsilon}_x \ll 1$) were produced [63]. As far as the TTF facility is concerned, a second generation gun that incorporate a co-axial input coupler is under commissioning at the injector facility of DESY-Zeuthen. The inclusion of a coaxial coupler with a doorknob transition renders the cavity axi-symmetric thereby avoiding the potential transverse emittance degradation due to the time-dependent kick induced by side coupler located at the full-cell as in the present design. This improved rf-gun is foreseen to be incorporated in the TTF-FEL user facility.

At Argonne, the AWA team uses a 1.3 GHz rf-gun to produce very high charge ($\sim$ 100 nC) bunches for beam-driven wakefield acceleration. Similarly to the case aforementioned of CTF-2 (previous section), the figure-of-merit for beam driven plasma acceleration is the peak current of the drive-beam rather than transverse emittances.

4.4 VHF and UHF band

In the 90's two low frequency guns have been operated at BOEING and CEA-Bruyère le chatel (France).

At Buyères-le-châtel, the ELSA facility incorporates a 144 MHz rf-gun followed by a linac section operating at 433 MHz. Because of the long bunch length (and thus very small linear charge-density) the achieved emittance at 1 nC is about 1 mm-mrad [55]. The ELSA accelerator has been operated with a macrobunch of 150 μsec at a frequency of 10 Hz, each macropulse consists of about 2000 bunches. The facility was designed to drive a high power infrared FEL and is about to be upgraded to an X-ray Thomson source [22].

	TTF	FNPL	AWA	AFEL	units
bunch charge	1-8	1-7	100	1-4	nC
laser pulse length (rms)	8	4-10		1.5	ps
peak cathode E-field	40	40	80	20	MV/m
beam energy	20	20		15-20	MeV
typical set of data:					
bunch charge	1	1	-	1	nC
energy	17	18	-	15	MeV
energy spread (rms)	0.03(u)	0.25	-	0.2	%
$\tilde{\varepsilon}_\perp$ (rms)	3.5	3.5	-	1.6	mm-mrad
bunch length (rms)	8	8	-	3.5	ps

Table 5. Typical photocathode drive-laser and rf parameters for various L-band guns. A set of achieved beam parameters is also given. All the gun presented in this Table operate at f=1.3 GHz. Note: "(u)" in the "energy spread" line indicates the quoted energy spread corresponding to the stochastic energy spread (correlations have been removed).

	ELSA	BOEING	units
bunch charge	1-10	1-7	nC
laser pulse length (rms)	26	10	ps
peak cathode E-field	25	25	MV/m
beam energy	19	5	MeV
typical set of data:			
bunch charge	1	3	nC
energy	19	5	MeV
energy spread (rms)	0.1(u)	2	%
$\tilde{\varepsilon}_\perp$ (rms)	1	5	mm-mrad
bunch length (rms)	25	11	ps

Table 6. Typical photocathode drive-laser and rf parameters for the two low frequency guns (ELSA and BOEING). A set of achieved beam parameters is also given. Note: "(u)" in the "energy spread" line indicates the quoted energy spread corresponding to the stochastic energy spread (correlations have been removed).

At BOEING defense and space center, a high average FEL including an energy recovery scheme was operated in the 90's. Noteworthy was the ability to produce electron bunches out of a 433 MHz rf-gun with the record duty cycle of 25% resulting in an electron beam average power of 160 kW [56].

These two low frequency designs operate with rather low electric peak field on the photocathode, about 25 MV/m, and with long (Gaussian) drive-laser pulse length of about 20 ps rms. Because of the latter fact, the charge density is reduced and so are the space charge forces.

4.5 Superconducting radio-frequency guns

The operation of a photocathode inside a superconducting resonator was proposed in Reference [58] and a first test was performed in the early 90's using a 2.8 GHz niobium cavity [59]. At that time, the cavity peak electric field was only about 1.5 MV/m. Recently, a major breakthrough was achieved by the DROSSEL collaboration: At Forschungzentrum Rossendorf, a superconducting rf-gun [57] was operated with a field as high as 15 MV/m. The proof-of-principle cavity consists of an half cell of the TESLA cavity geometry with its large diameter side terminated by a shallow cone (see Fig. 3) that provides some radial focusing to compensate for the absence of a solenoid [a] in the vicinity of the cavity [23]. Since the photocathode cannot be in thermal contact with the cavity, a special rf-filter (so-called "choke filter") was implemented to avoid the rf-power from flowing through the cathode coaxial manipulator system thereby preserving the high quality factor of the cavity. At a later stage, the DROSSEL collaboration envisiones to develop a multi-cell cavity superconducting rf-gun to serve as an injector for the ELBE [60] FEL.

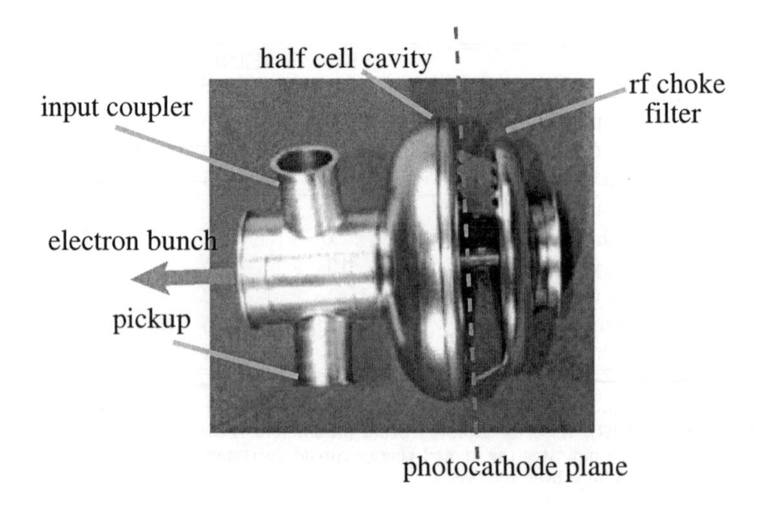

Figure 3. Picture of the Nb half cell cavity for the proof-of-principle srf-gun of Forschungzentrum Rossendorf (Courtesy of D. Janssen).

5 Conclusion & Outlook

Electron sources based on rf-guns have provided electron beams with the required high phase space density to drive short-wavelength single-pass FELs: recently

[a] a magnetic field above the critical field (for Niobium $B_c \simeq 200$ mT) would break the Cooper pairs responsible for the superconductivity.

two of these devices were operated in the saturation regime at ANL (LEUTL) and DESY (TTF). Rf-guns can sustain high peak field but cannot simultaneously provide long rf-macropulses: to date the highest duty cycle rf-gun was operated at BOEING with a peak electric field of only 25 MV/m, an electric field comparable to the foreseen achievable field in dc-guns. The latter type of gun having the advantage to be able to drive accelerators in CW mode as demonstrated in the operation of a kW-level average power FEL at JLab.

As far as the choice of the rf-gun frequency is concerned, Figure 4 compares the brightness of the various sources aforementioned. Given the peak current $\hat{I}$, the brightness defined as:

$$\mathcal{B}_{\perp} = \frac{\hat{I}}{\tilde{\varepsilon}_x \tilde{\varepsilon}_y} \tag{6}$$

has been compared since the longitudinal emittance (especially the uncorrelated energy spread) has not been measured in all the quoted experiments. In this figure there is a weak frequency dependence of the brightness. Though the gun with a significantly higher frequency (MIT gun) does not yet take advantage of the emittance compensation process. In the latter Figure, the brightness dependence on frequency is mainly dominated by the scaling of the peak current versus frequency[b] which is in any case increased by compressing the bunch.

The success at FNAL to generate flat beams directly out of the FNPL injector, if included in the design of a future linear collider, would relax the requirements on the electron damping ring. The round-to-flat beam transformation has also found applications in light sources [64,65]. If one would be able to generate polarized flat beams directly out of an injector, one might be able to remove the need for an electron damping ring. The topic of rf-gun for producing polarized electrons is under investigation [67,66], the main challenge being to operate GaAs photocathodes in an rf-gun (or finding another photo-emitter capable of producing polarized electrons). It was suggested that the plane wave transformer rf structure may be a good candidate for such an polarized rf-gun since it has a larger vacuum conductance (compared to standard gun cavities) and thereby offers a higher quality vacuum environment [68].

6 Acknowledgments

I wish to thanks the following individuals for providing materials: J.-P. Carneiro (DESY), D. Janssen (Forschungzentrum Rossendorf), V. Yakimenko F. Zhou (BNL/ATF), W. Graves (BNL/SDL), C. Sinclair (Cornell), Ph. Guimbal (CEA Buyères-le-châtel), R. Temkin (MIT), D. Dowell, J. Schmerge (SLAC/GTF), M. Useaka (Univ. Tokio), and J. Yang (Sumitomo Heavy Industry Ltd.) and K. Flöttmann for discussions. I am also indebted to J.-P. Carneiro and K. Flöttmann for reading and commenting the manuscript.

[b] typically we have $\hat{I} \propto \omega$ for the presently operated guns.

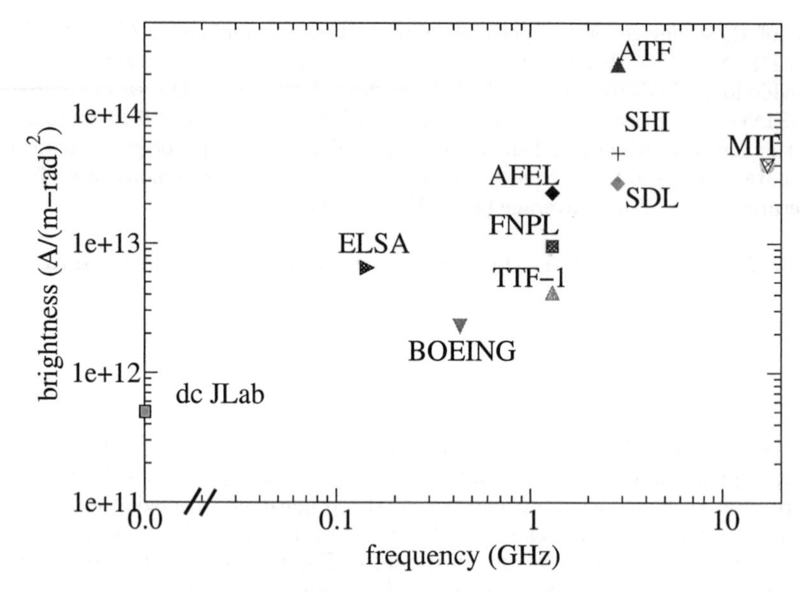

Figure 4. Comparison of the brightness of various sources quoted in the paper. In this plot the brightness has been computed using Eq.(6).

References

1. Sheffield R., proc. PAC1995 Dallas, 882-886 (1996)
2. O'Shea P., "Rf photoinjectors", proceedings of 1st ICFA workshop on "high brightness beam" UCLA, published by World Scientific (1999)
3. Carsten B., *Nucl. Instr. Meth.* **A285**:313-319 (1989)
4. Carsten B., *Part. Acc.* **49**:27-65 (1995)
5. Serafini L. and Rosenzweig J., *Phys. Rev.* **E55**:7565-7590 (1997)
6. Piot P. et al, proceedings EPAC1998 Stockholm, 1447-1449 (1998)
7. Wang X.J., et al. *Nucl. Instr. Meth.* **A375**:82-86 (1996)
8. Travish G., et al., proc. LINAC2000 Monterey (report SLAC-R-561), 899-901 (2000)
9. Yu. L.-H., et al., proc. PAC2001 Chicago, 2830-2832 (2001)
10. Braun H. H., et al., proc. PAC2001 Chicago, 720-722 (2001)
11. Sheffied R., et al., *Nucl. Instr. Meth* **A318**:313-319 (1989)
12. Conde M. et al., *Phys. Rev. ST Accel. Beams* **1**:041302 (1998)
13. Schmerge J. et al., *Nucl. Instr. Meth* **A483**:301-304 (2002)
14. Schreiber S., proc. EPAC2000 Wien, 86-88 (2000)
15. Bakker R., et al, proc. EPAC2002 Paris, 1813-1815 (2002)
16. Carneiro J.-P., et al, proc. PAC1999 New-York, 2027-2029 (1999)
17. Brown W. J., et al, *Phys. Rev. ST Accel. Beams* **4**:083501 (2001)
18. Uesaka M. et al., *IEEE trans. plasma science* **28**(4): 1133-1142 (2000)
19. Anderson S., et al., proc. PAC1999 New-York, 2006-2008 (1999)

20. Teifer S., et al., proc. PAC2001 Chicago, 2263-2265 (2001)
21. Dowell D., et al, proc. PAC1993 Washington, 2967-2969 (1993)
22. Guimbal Ph., et al., proc. EPAC2002 Paris, 1768-1770 (2002)
23. Janssen D.,and Volkov V., *Nucl. Instr. Meth.* **A452**:34-43 (2000)
24. Rosenzweig J.B., et al., proc. PAC1999 New-York, 2039-2041 (1999)
25. Sinclair C.K., proc. PAC1999 New-York, 65-69 (1999)
26. Sinclair C.K., ICFA beam dynamics newsletter **26**:31-34 (December 2001)
27. Kim K.J., Zholents A.A, Zolotorev M.S., Vinokurov N.A., *Nucl. Instr. Meth.* **A407**:380-384 (1998)
28. Faatz B. et al., "VUV-FEL driven rf-gun", proc. FEL2002 Argonne (in press)
29. Spicer W.E., Herrera-Cómez M.,"Modern theory and applications of photocathodes", report SLAC-PUB-63-06 and SLAC/SSRF-0042 (1993)
30. Flöttmann K., DESY-HH report TESLA-97-06 (1997)
31. Dunham B.M., et al, proc. PAC1995 Dallas, 1030-1032 (1995)
32. Orlov D.A., et al, *Appl. Phys. Lett.* **78**(18):2721-2723 (2001)
33. Graves W., et al, proc. PAC2001 Chicago, 1030-1032 (2001)
34. Wang J.W. and Loew G.A., SLAC-PUB-7684 Stanford Univ. (1997)
35. Engwall D., et al., proc. PAC1997 Vancouver, 909-911 (1998)
36. Neil G.R., et al, *Phy. Rev. Lett.* **84**(4):662-665 (2000)
37. Sinclair C.K., proc. PAC2001 Chicago, 610-612 (2001)
38. Hovater C., and Poekler M., *Nucl. Instr. Meth.* **A418**:280-286 (1998)
39. Gruner S., et al. "Study for a proposed phase I ERL synchrotron light source at Cornell university", CHESS TM-01-003 and JLAB-ACT-01-04 (2001)
40. Rosenzweig J. and Colby E., proc. PAC1995 Dallas, 951-953 (1996)
41. Lin L. C.-L., proc. PAC1995 Dallas, 957-959 (1996)
42. Zhou F., et al. *Phys. Rev. ST Accel. Beams* **5**:094203 (2002)
43. Yakimenko V., et al., , *Nucl. Instr. Meth* **A483**:277-281 (2002)
44. Graves, W. et al, proc. PAC2001 Chicago, 2224-2226 (2001)
45. Piot Ph. et al., "Sub-picosecond compression by velocity bunching in a photoinjector", this proceedings (also preprint TESLA-FEL-02-08)
46. Dowell D., et al., "Slice Emittance Measurements at the SLAC Gun Test Facility", submitted to proc. FEL2002 Argonne (in press)
47. Yang J., et al., proc. EPAC2002 Paris, 1828-1830 (2002)
48. Suberlucq G., "Développement et production de photocathodes pour la CLIC Test Facility", CLIC Note 299, CERN (1996)
49. Hogan M., et al, *Phy. Rev. Lett.* **81**(22):4867-4870 (1998)
50. Gierman S., "Enhanced quadrupole scan technique for slice emittance measurement", PhD thesis, University of California San Diego (defended in 1999)
51. Colby E. R., et al., proc. PAC1995 Dallas, 967-969 (1995)
52. Piot Ph., et al., proc. PAC2001 Chicago, 86-88 (2001)
53. Ayvazian V., et al., *Eur. Phys. Jour.* **D20**:149-155 (2002)
54. Schreiber S., et al., proc. EPAC2002 Paris, 1804-1806 (2002)
55. Marmouget J.-G., et al., proc. EPAC2002 Paris, 1795-1797 (2002)
56. Dowell D., et al., *Appl. Phys. Lett.* **63**(15):2035-2037 (2002)
57. Janssen D., private communication; H. Büttig et al., "First operation results of a superconducting rf-photoelectron gun", submitted to proc. FEL2002 Ar-

gonne (in press)
58. Chaloupka H. et al., *Nucl. Instr. Meth.* **A285**:327-332 (1989)
59. Michalke et al., proc. EPAC1992 Berlin, 1014-1116 (1992)
60. Büchner A., et al., proc. EPAC2000 Wien, 732-734 (2000)
61. Carneiro J.-P., "Etude expérimentale du photoinjecteur de Fermilab" thèse de l'Université Paris Sud XI (defended in 2001)
62. Brinkmann R., Derbenev Ya, and Flöttmann K., *Phys. Rev. ST Accel. Beams* **4**:053501 (2001)
63. Edwards D. et al., proc. PAC2001 Chicago, 73-75 (2001)
64. Lydia S., et al., this proceedings
65. Brinkmann R., proc. EPAC2002 Paris, 653-655 (2002)
66. Dikansky N.S., et al., proc. EPAC2000, 1645-1647 (2000)
67. Edwards D. (chairperson), ICFA mini-workshop on "Polarized rf-guns for linear colliders", FNAL April 18-20, 2001
68. Clendemin J., et al.,"A polarized electron RF photoinjector using the plane-wave-transformer design", submitted to *Phys. Rev. ST Accel. Beams* and preprint SLAC-PUB-8971 (2001)

TREDI* SIMULATIONS FOR HIGH-BRILLIANCE PHOTO-INJECTORS

L. GIANNESSI†AND M. QUATTROMINI‡

ENEA, Dipartimento Innovazione, Divisione Fisica Applicata, C.R. Frascati, Via E. Fermi 27, 00044 Frascati, Rome, Italy.

The TREDI Monte-Carlo program is briefly described, devoting some emphasis to the Lienard-Wiechert potentials approach followed to account for self-fields effects and the covariant technique devised to achieve regularization of electro-magnetic fields. Some guidelines to the choice of the correct parameters to be used in the simulation are also summarily sketched. The predictions obtained for the reference workpoint of the space-charge compensated SPARC project photo-injector are finally compared to those from other well-established simulation codes.

1. Introduction

The main issues in the development of coherent ultra-brilliant X-ray sources like the recently approved Sparc project[1] are the generation of ultra-high peak brilliance electron beams and the generation of higher order resonant harmonics in the SASE-FEL process, the former posing fairly strict requirements to the beam delivered at the undulator entrance. The typical scheme proposed for the accelerating system consists of a BNL/UCLA/SLAC type, 1.6 cell RF gun operated at 2.856 GHz with high peak field ($\geq$ 120MV/m) on the cathode (Cu or Mg). The gun, sorrounded by a focusing solenoid (peak field), delivers a beam of $\approx$ 6 MeV and is followed by a drift (up to 1.5 m from the cathode) and two TW linac sections operating in S-band boosting the beam to the final 150 MeV energy required to avoid further emittance growth due to space charge effects.

*joint icfa advanced accelerator and beam dynamics workshop. chia laguna, sardinia (italy), july 1-6, 2002.

†e-mail: giannessi@frascati.enea.it, tel: +39 06 9400 5180, fax: +39 06 9400 5334.

‡corresponding author. e-mail: quattromini@frascati.enea.it, tel: +39 06 94005718, fax: +39 06 9400 5334.

According to theoretical predictions[2] the working point for high brightness RF photo-injectors and the velocity bunching technique[3] were chosen, in order to achieve both longitudinal compression and emittance preservation. The relevant parameters of this scheme are resumed in table (1). In figure (1) are shown the profiles of RF gun (un-normalized) and solenoid fields.

Table 1. Photo injector parameters.

Peak accelerating field	120-140 MV/m
Frequency	2.856 GHz
Phase (beam head)	30
Charge	1.0 nC
Laser spot radius (homogeneous)	1.0 mm
Laser pulse length (flat-top)	10 ps (≈10 RF)
Solenoid peak field	3.09 kG

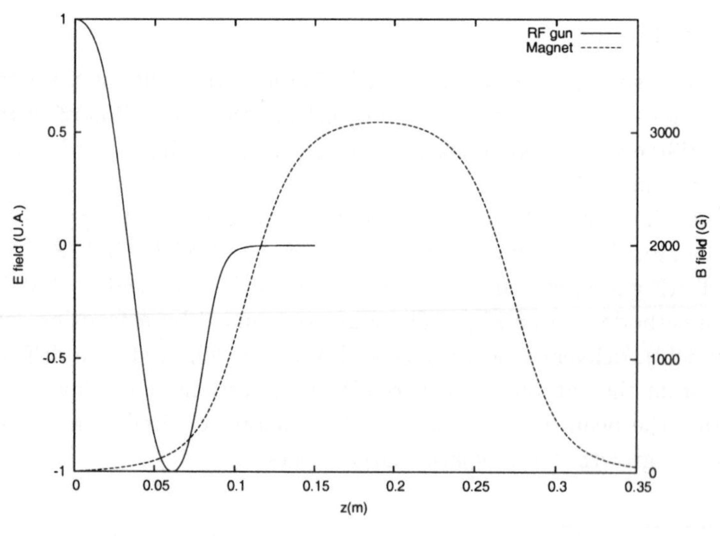

Figure 1. Electric (left scale, arbitrary units) and magnetic (right) longitudinal fields for the BNL/SLAC/UCLA RF Gun and focusing solenoid, respectively (Courtesy of D. Palmer and M. Ferrario).

For the above mentioned scheme theory and simulations (done typically with 2D axisymmetric codes assuming instantaneous propagation of space charge effects) nicely agree in predicting for transverse emittance both compensation and a graceful double-minimum, mainly a chromatic effect between the solenoid and the beam energy spread. It turns out[2] that the region close to the local maximum between the two minima is the optimal position for the first linac section, for this choice minimizes the beam emittance on a *slice* base, a concept deeply connected to that of co-operation length in FEL operation.

In this contribution we essentially discuss and compare the predictions mentioned above with those obtained with TREDI, a fully 3D Monte Carlo program devoted to simulations of charged beam dynamics by direct integration of particles trajectories, accounting for self-fields through Lienard-Wiechert retarded potentials[4].

2. Tredi in a nutshell

The development of TREDI was originally motivated[5],[6] by the necessity of simulating, e.g.: i) RF injectors in not-axisymmetric conditions (like those encountered in the high aspect ratio injectors proposed for future colliders); ii) the effects on emittance compensation schemes of axial symmetry breaking, possibly amplified by non-linearities of the system; iii) emittance growth in magnetic beam compressors due radiative/acceleration fields (discussed elsewhere in these proceedings [7]).

A detailed description of the simulation code and of its capabilities falls beyond the scope of this contribution and can be found elsewhere[8],[9]. Notwithstanding, it is worth remarking here that *principia prima* Monte Carlo's usually model the beam as a collection of mutually interacting objects ("macroparticles"), whose number - because of practical computer limitations - is necessarily bounded to a few thousands or a few millions at most. As a consequence - except perhaps for diluted systems where self fields can be neglected - suitable techniques must be devised to cancel the effects of many possible numerical artifacts, either leading to unreliable results or posing stability concerns[10]. For example, since in all practical cases each macroparticle mimicks a fairly large number ($> 10^4$) of "real life" electrons, one need to subtract the un-physical collisional contribution lurking in the model because of the huge electromagnetic fields that develop whenever macroparticles pass close to each other. While this poses a serious concern at low energies - where static (velocity) fields dominate - other kinematical conditions can be well a source of noise at fully relativistic regimes, making difficult any prediction about, e.g., Coherent Synchrotron Radiation (CSR) effects in magnetic compressors or chicanes.

The approach followed in TREDI is that of *dressing* macroparticles with a *form factor*, i.e. assuming the elementary dynamical objects to be extended rather than point-like distributions. While the working mechanism of this regularization technique is not new[11], much effort has been devoted to render the regularization procedure relativistically covariant[12] (at least for velocity fields) and reduce the artifacts introduced by the approximation of the model, e.g. to weaken the dependence on the number of macroparticles [13]. Quite expectedly, the size and the functional shape chosen for the form factors (typically gaussian) largely determine their filtering properties. A covariant approach to noise reduction in fields evaluation presents at least two valuable properties:

(1) in each reference frame, the smoothing procedure yields values of the fields that are connected through Lorentz transformations, as they should be. As a consequence, the reduction of harmonic content of fields spectra turns to be *naturally* independent from the dynamical regime of the beam;
(2) the calculation can be actually carried out wherever the dynamical variables of the particle are directly available, namely the accelerator ("lab") reference frame.

To do this, an "effective charge" can be defined in a covariant way, i.e. possessing the requested properties to perform the sought regularization as specified above. To this purpose, assume that a macroparticle is described in its own rest frame by a charge density

$$\rho\,(\vec{x}) = \frac{q}{\sqrt{\det \widehat{\sigma}}} \cdot \Theta\,[\Delta\vec{x}^{T} \cdot \widehat{\sigma}^{-1} \cdot \Delta\vec{x}] \tag{1}$$

where

$$(\Delta\vec{x} = \vec{x} - \vec{x}_{S})$$

and

$$\frac{1}{\sqrt{\det \widehat{\sigma}}} \int_{\Re^3} \Theta\,[\Delta\vec{x}^{T} \cdot \widehat{\sigma}^{-1} \cdot \Delta\vec{x}]\, d^3x = 1 \tag{2}$$

Here and in the following, the subscript $\mathcal{S}$ ("source") will refer quantities to the macroparticle generating the fields, so that $\Delta\vec{x} = \vec{x} - \vec{x}_{S}$ is the distance from the macroparticle's center. Quantity $\widehat{\sigma}$ (or $\widehat{\sigma}^{-1}$) in (1) is a matrix describing the macroparticles's *form factor* . As an example, for a

3D gaussian charge distribution

$$\Theta(z) = \frac{1}{\sqrt{(2\pi)^3}} \cdot \exp\left[-\frac{z}{2}\right] \tag{3}$$

$\widehat{\sigma}$ is the covariance matrix while for an upright "hard spheroid"

$$\Theta(z) = \begin{cases} \frac{3}{4\pi} \text{ if } z \leq 1 \\ 0 \quad \text{elsewhere} \end{cases} \tag{4}$$

is directly connected to ellipoid's semi-axes (note that $\sqrt{\det \widehat{\sigma}} = abc$):

$$\widehat{\sigma} = \begin{pmatrix} a^2 & & \\ & b^2 & \\ & & c^2 \end{pmatrix} \tag{5}$$

In the macroparticle's reference frame, where an observer at rest in $\vec{x}_O$ experiences a purely static electric field (provided, of course, the charge is not accelerating), an "effective charge" can be defined as the total charge included in the iso-density surface associated to the value of ρ at the observer point:

$$\mathcal{Q}_{\text{eff}}(\vec{x}_O) \equiv \int_{\mathcal{R}^2(\vec{x}) \leq \mathcal{R}_O^2} \rho(\vec{x})\, d^3x \tag{6}$$

where:

$$\begin{aligned} \mathcal{R}^2(\vec{x}) &\equiv \qquad \Delta\vec{x}^T \cdot \widehat{\sigma}^{-1} \cdot \Delta\vec{x} \\ \mathcal{R}_O^2 &\equiv \mathcal{R}^2(\vec{x}_O) = \Delta\vec{x}_O^T \cdot \widehat{\sigma}^{-1} \cdot \Delta\vec{x}_O \end{aligned} \tag{7}$$

Now, it is well known in electrodynamics that $\rho(\vec{x})$ is nothing but the time component of a charge density 4-vector $\mathcal{J}(x) = \mathcal{J}(t, \vec{x})$ which depends, in general, both on $\vec{x}$ and t. Moreover, signals propagate at the speed of light, i.e. the fields experienced by an "observer" in $(t_O, \vec{x}_O)$ have been generated by the infinitesimal "source" charge $\rho(\vec{x})\, dV$ centered at an event $(t, \vec{x})$ on the observer's (past) light cone:

$$c \cdot (t_O - t) = c \cdot \Delta t = \left|\overrightarrow{\Delta x}\right| = |\vec{x}_O - \vec{x}| \tag{8}$$

Retard condition (8) is fulfilled anyway, including the case of a still charge distribution, where - according to (6) and (7) - time delays are not relevant and what only matters is the geometrical shape as described by (1). On the other hand, consistency with special relativity suggests the quantity defined in (6) to be a Lorentz scalar (for it describes an electric charge), making desirable to cast it in a form which is manifestly covariant. This issue can be clarified observing that the very same result should be obtained in a reference frame where source particle moves at constant speed and time delays certainly play a rle . This task is accomplished assuming the elements of $\widehat{\sigma}$ to be the spatial components of a 4-D tensor $\widehat{\Sigma}$ reducing for a particle at rest to the special form:

$$\widehat{\Sigma}^{-1} = \begin{pmatrix} 0 & \vec{0}^T \\ \vec{0} & \widehat{\sigma}^{-1} \end{pmatrix} \tag{9}$$

It can be shown (see [12]) that for the generalized covariance matrix $\widehat{\Sigma}'$ the following relation holds in a reference frame (the "lab" frame) where the particles moves with constant speed $\vec{\beta}$:

$$\widehat{\Sigma}'^{-1} = \begin{pmatrix} \vec{\beta}^T \widehat{\sigma}'^{-1} \vec{\beta} & -\vec{\beta}^T \widehat{\sigma}'^{-1} \\ -\widehat{\sigma}'^{-1} \vec{\beta} & \widehat{\sigma}'^{-1} \end{pmatrix}$$

where

$$\widehat{\sigma}'^{-1} = \left[\widehat{1} + \frac{\gamma^2}{1+\gamma} \vec{\beta} \otimes \vec{\beta}^T \right] \cdot \widehat{\sigma}^{-1} \cdot \left[\widehat{1} + \frac{\gamma^2}{1+\gamma} \vec{\beta} \otimes \vec{\beta}^T \right]$$

is the (inverse) *form factor* in the laboratory frame while $\widehat{1}$ and $\vec{\beta} \otimes \vec{\beta}^T$ are shorthands respectively for

$$\widehat{1} = \begin{pmatrix} 1 & 0 & 0 \\ 0 & 1 & 0 \\ 0 & 0 & 1 \end{pmatrix} \tag{10}$$

and

$$\vec{\beta} \otimes \vec{\beta}^T = (\beta_1 \; \beta_2 \; \beta_3) \cdot \begin{pmatrix} \beta_1 \\ \beta_2 \\ \beta_3 \end{pmatrix}$$

$$= \begin{pmatrix} \beta_1^2 & \beta_1\beta_2 & \beta_1\beta_3 \\ \beta_2\beta_1 & \beta_2^2 & \beta_2\beta_3 \\ \beta_3\beta_1 & \beta_3\beta_2 & \beta_3^2 \end{pmatrix} \tag{11}$$

Note first that (7) generalizes to the following covariant form:

$$\begin{aligned} \mathcal{R}^2(\vec{x}) &\equiv \Delta x^T \cdot \widehat{\Sigma}^{-1} \cdot \Delta x \\ \mathcal{R}_O^2 &\equiv \mathcal{R}^2(\vec{x}_O) = \Delta x_O^T \cdot \widehat{\Sigma}^{-1} \cdot \Delta x_O \end{aligned} \tag{12}$$

and that the basic quadratic form is a Lorentz invariant by construction (primed quantities refer to lab frame):

$$\mathcal{R}_O^2 \equiv \Delta x_O^T \cdot \widehat{\Sigma}^{-1} \cdot \Delta x_O = \Delta x_O'^T \widehat{\Sigma}'^{-1} \Delta x_O' \equiv \mathcal{R}_O'^2$$

By standard manipulations the effective charge (7) can be reduced to a one-dimensional integral where the term $\sqrt{\det \widehat{\sigma}}$ simplifies (which is the rationale of explicitly factoring it out in the general expression of charge density (1) and normalization (2)):

$$\mathcal{Q}_{\text{eff}}(\mathcal{R}_O) = \int_{\mathcal{R}^2(\vec{x}) \le \mathcal{R}_O^2} \rho(\vec{x})\, d^3x = 4\pi \int_0^{\mathcal{R}_O} \Theta\left[\mathcal{R}^2\right] \mathcal{R}^2 d\mathcal{R} \tag{13}$$

so that for distributions (3) and (4) the effective charge (6) reads

$$\begin{aligned} \mathcal{Q}_{\text{eff}}(\mathcal{R}_O) &= q \frac{4\pi}{\sqrt{(2\pi)^3}} \int_0^{\mathcal{R}_O} \exp\left[-\frac{\mathcal{R}^2}{2}\right] \mathcal{R}^2 d\mathcal{R} \\ &= q \left[\operatorname{erf}\left(\frac{\mathcal{R}_O}{\sqrt{2}}\right) - \sqrt{\frac{2}{\pi}} \cdot \mathcal{R}_O \cdot \exp\left(-\frac{\mathcal{R}_O^2}{2}\right)\right] \end{aligned} \tag{14}$$

and

$$\begin{aligned} \mathcal{Q}_{\text{eff}}(\mathcal{R}_O) &= 3q \int_0^{\min(1, \mathcal{R}_O)} \mathcal{R}^2 d\mathcal{R} \\ &= q \cdot \begin{cases} \mathcal{R}_O^3 \text{ if } \mathcal{R}_O^2 \le 1 \\ 1 \quad \text{elsewhere} \end{cases} \end{aligned} \tag{15}$$

respectively.

The (6) can be used to regularize the electric field by the formula:

$$\vec{E}_{\text{eff}}(\vec{x}) = \frac{\mathcal{Q}_{\text{eff}}\left(\sqrt{\vec{x}^T \cdot \widehat{\sigma}^{-1} \cdot \vec{x}}\right)}{|\vec{x}_O|^3} \cdot \widehat{x}_O \tag{16}$$

Note that the electric field as defined in (16) is always radial, which is true only when the charge distribution is spherical (so one can make use

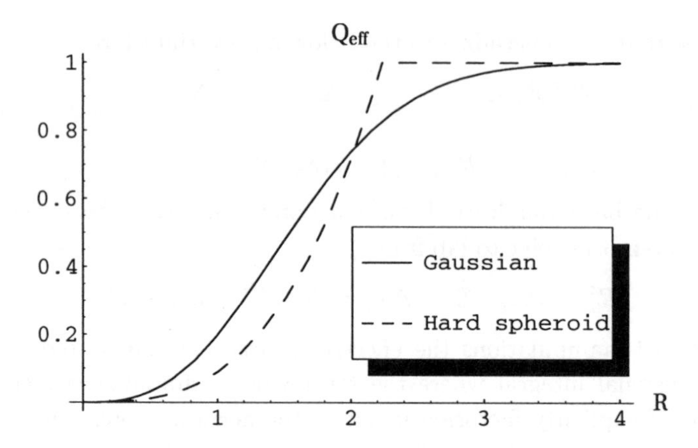

Figure 2. Fractional effective charge as a function of $\mathcal{R}$ for a macroparticle with gaussian (continuous line) and "hard spheroid" (dashed line) form factors.

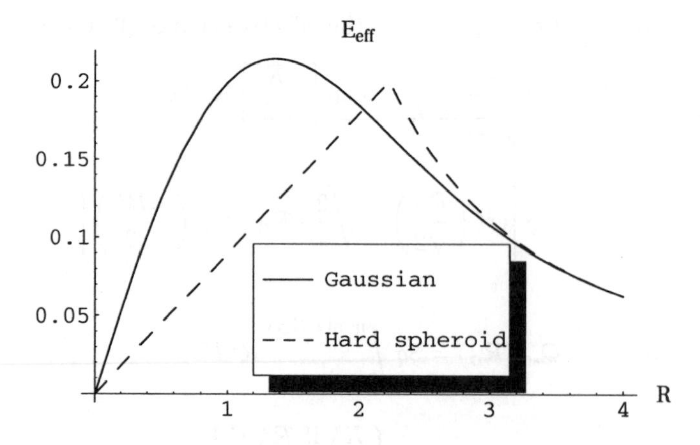

Figure 3. Relative effective field as a function of $\mathcal{R}$ for a macroparticle with gaussian (continuous line) and "hard spheroid" (dashed line) form factors.

of Gauss' theorem). Since the most sensible choice seems to assume the macroparticles' form factor to be a downscaled copy (i.e. same aspect ratio, reduced size) of the whole beam, in general this is not the case and the (16) is only an approximation. Notwithstanding the basic result (see, e.g. [14]) that for a charge distribution of the form (1) the charge on the outside of an iso-density surface does not contribute to the fields on the inside suggests that the main concept of "effective charge" as a key quantity to form the field remains valid. In a rigorous approach the "effective charge" $\mathcal{Q}_{\text{eff}}$ should be substituted by a tensor reflecting the f.f.'s anisotropy. Explicit evaluation of such a tensor, however, can be an awkward task for most but the simplest distributions (like 4). In figure (2) is shown the effective charge vs R as defined in formulæ(14) and (15). Figure (3) shows the effective electric field E_z vs $\mathcal{R}$ for a spherically symmetric macroparticle (i.e. for $\mathcal{R} \propto |\vec{x}_o|$) .

The size of macroparticles scales with their number as $\approx N^{\frac{1}{3}}$. The scale factor must be carefully "tuned" in order the superposition of the macroparticles (the "discrete approximation") to reproduce as much as possible the features of the continuous distribution representing the beam. This subtle topic will be described elsewhere[9].

3. Benchmark results.

In this section some results obtained on the Sparc benchmark case mentioned in the introduction will be shown in comparison with those obtained by other, well established numerical codes. Figure (4) plots the energy spread behaviour obtained with TREDI (solid line) and the Homdyn[15] simulation code (dotted line). The results are almost indistinguishable. In figure (5) the tranverse envelope (r.m.s) dimension are shown. In this case the peak dimension falls between the values predicted by Homdyn and Parmela[16] (dotted and dashed lines, respectively), the situation being reversed for the waist (for Parmela has the highest maximum and the lowest minimum and Homdyn the other way round). The overall agreement is fairly good, and it is worth remarking that both the maximum and the minimum of the envelope occur at the same positions for all the codes. The same holds for the longitudinal size and emittance (figures (6) and (8), respectively). The differences are much more marked for the radial emittance (figure (7)), for both TREDI and Parmela do not exhibit a double minimum effect as pronounced as Homdyn. It is worth remarking, however, that all the codes considered more or less agree in predicting the emittance compensation and the value of the minimum. The double minimum effect can be made to appear in TREDI as well, by slightly changing the parameter set. Figure (9) shows up Another simulation, done with all the parameters fixed except for the position of the focusing solenoid, which has been placed 1.0cm ahead.

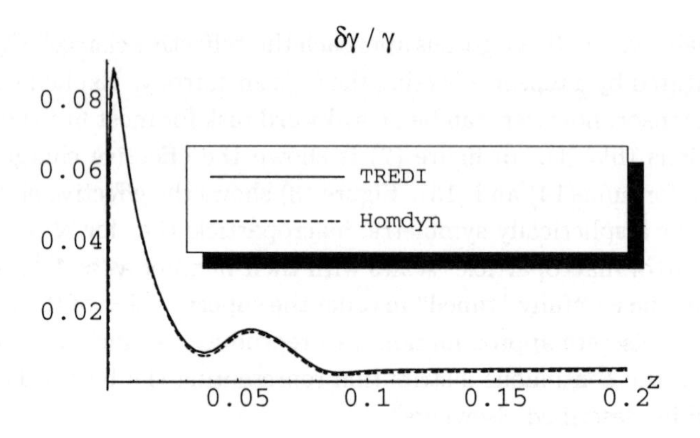

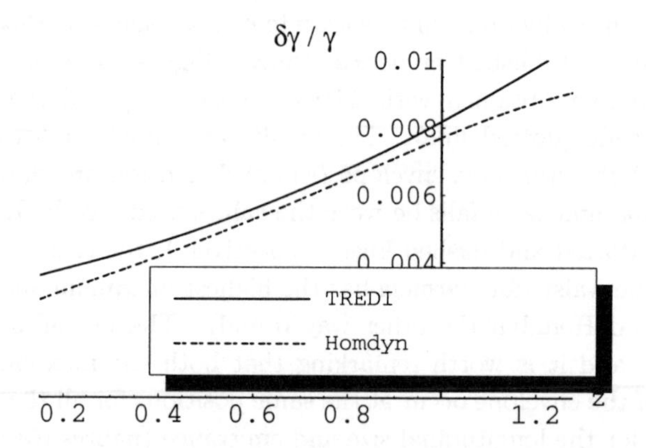

Figure 4. Relative energy spread in the RF gun (above) and ahead (below). Homdyn data (here and below) are courtesy of M. Ferrario.

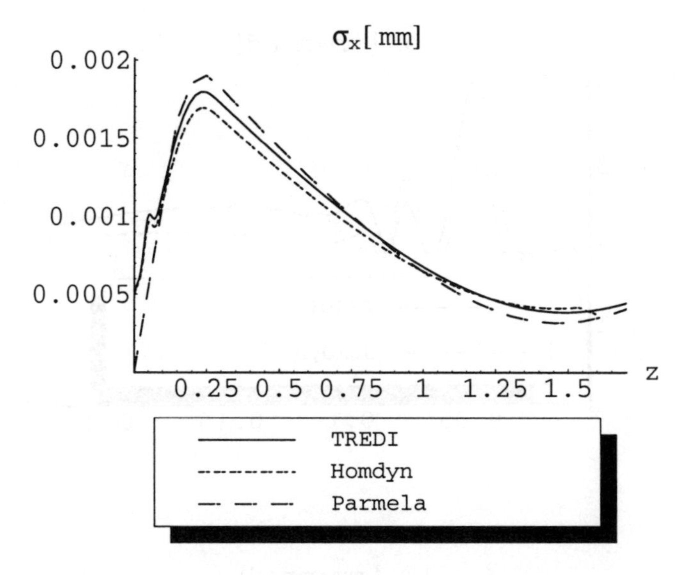

Figure 5. Radial envelope (mm). Parmela data (here and below) courtesy of C. Ronsivalle.

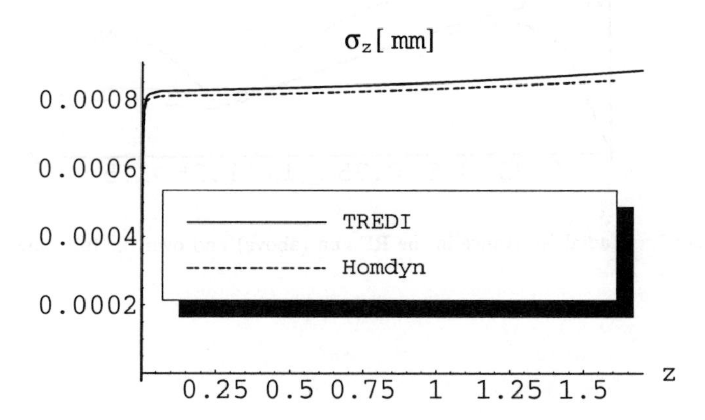

Figure 6. Longitudinal emittance (mm · keV).

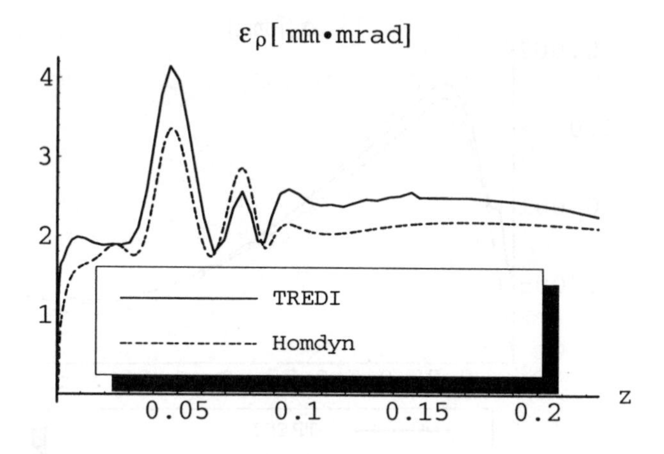

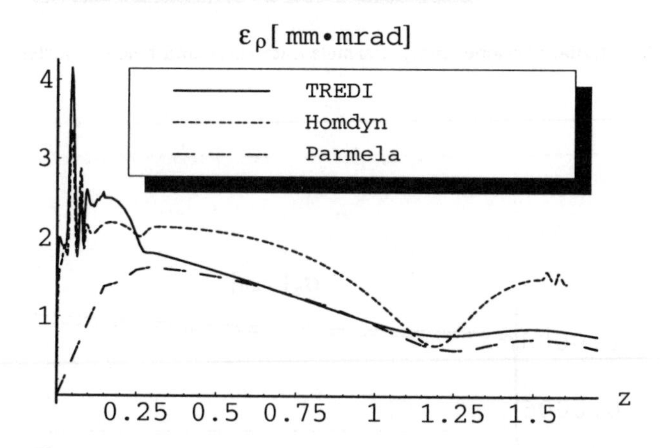

Figure 7. Radial emittance in the RF gun (above) and over the full range (below).

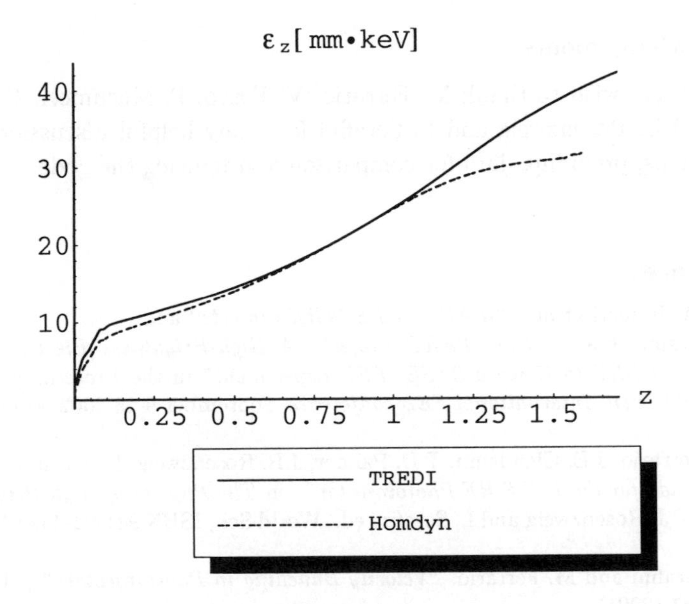

Figure 8. Longitudinal emittance (mm · keV).

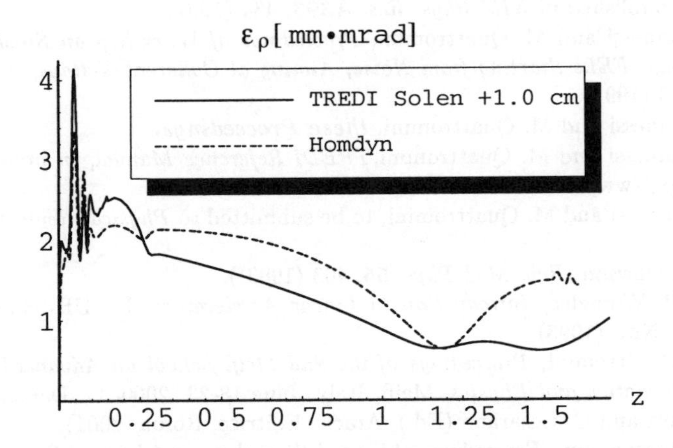

Figure 9. Radial emittance. The focusing solenoid has been placed 1.0cm ahead with respect to reference parameter set.

Acknowledgements

The authors wish to thank M. Ferrario, V. Fusco, P. Musumeci, C. Ronsivalle, J.B. Rosenzweig and L. Serafini for many helpful discussions and suggestions, providing data for comparison and revising the code.

References

1. see A. Renieri et al., *"SPARC and SPARX projects"* and D. Alesini et al., *"The SPARC Project: A High-Brightness Electron Beam Source at LNF to Drive a SASE -FEL Experiment"* in the *Proceedings of the FEL2002 24th International FEL conference*, September 9-13 2002 Argonne Il, USA.
2. M. Ferrario, J.E. Clendenin, T.D. Palmer, J.B. Rosenzweig, L. Serafini, *"Homdyn Study for the LCLS RF Photoinjector"* , in *The Physics of High Brightness Beams*, J. Rosenzweig and L. Serafini ed., World Sci., ISBN 981-02-4422-3, June 2000.
3. L. Serafini and M. Ferrario, *"Velocity Bunching in PhotoInjectors"* , *AIP CP* **581**, 87 (2001).
4. J.D. Jackson, *Classical Electrodynamics,* 2nd ed., p. 654ff, Wiley & Sons, New York (1975).
5. F. Ciocci, L. Giannessi, A. Marranca, L. Mezi and M. Quattromini, *Proceedings of 18th International Free-Electron-Laser Conference*, Rome, Italy, August 26-31, 1996, G. Dattoli and A. Renieri (Ed.), North Holland, Amsterdam (1997), published in *NIM Phys. Res.* **A393**, 434 (1997).
6. L. Giannessi and M. Quattromini, *Proceedings of Workshop on Single Pass, High Gain FELs Starting from Noise, Aiming at Coherent X-Rays, AIP C.P.* **413**, 313 (1997).
7. L. Giannessi and M. Quattromini, ***these Proceedings.***
8. L. Giannessi and M. Quattromini,*TREDI Reference Manual, in preparation.* See http://www.tredi.enea.it.
9. L. Giannessi and M. Quattromini, to be submitted to *Physical Review Special Topic.*
10. J. M. Dawson, *Rev. Mod Phys.* **55**, 403 (1983)).
11. Th. P. Wrangler, *Introduction to Linear Accelerators,* LA-UR-93-805, Los Alamos, NM, (1993).
12. M . Quattromini, *Proceedings of the 2nd Melfi School on Advanced Topics in Mathematics and Physics,* Melfi, Italy, June 18-23, 2000, G. Dattoli, H.M. Srivastava and C. Cesarano (Ed.), Aracne Editrice, Roma (2001).
13. M . Quattromini, *Proceedings of the 3rd Workshop on Advanced Special Functions and Related Topics in Differential Equations*, Melfi, June 24-29 2001.,to be published on Advanced Mathematics and Computation.
14. W. D. MacMillan, *Theoretical Mechanics. The theory of the potential,* Dover (1958).

15. M. Ferrario et al., "*Multi bunch energy spread induced by beam loading in a standing wave structure*", *Particle Accelerators* **52,** 1 (1996).
16. See e.g. http://laacg1.lanl.gov/laacg/services/parmela.html.
17. L. Giannessi, P. Musumeci and M. Quattromini, *Nucl. Inst. Meth.* ***A 436***, 443, (1999).

6D PHASE SPACE MEASUREMENTS AT THE SLAC GUN TEST FACILITY*

J.F. SCHMERGE, P.R. BOLTON, J.E. CLENDENIN, D.H. DOWELL,
S.M. GIERMAN, C.G. LIMBORG AND B.F. MURPHY

Stanford Linear Accelerator Center
2575 Sand Hill Rd,
Menlo Park, CA 94025, USA
E-mail: Schmerge@slac.stanford.edu

Proposed fourth generation light sources using SASE FELs to generate short pulse, coherent, X-rays require demonstration of high brightness electron sources. The Gun Test Facility (GTF) at SLAC was built to test high brightness sources for the proposed Linac Coherent Light Source at SLAC. The GTF is composed of an S-band photocathode rf gun with a Cu cathode, emittance compensating solenoid, single 3 m SLAC linac section and e-beam diagnostic section with a UV drive laser system. The longitudinal emittance exiting the gun has been determined by measuring the energy spectrum downstream of the linac as a function of the linac phase. The e-beam pulse width, correlated and uncorrelated energy spread at the linac entrance have been fit to the measured energy spectra using a least square error fitting routine. The fit yields a pulse width of 2.9 ps FWHM for a 4.3 ps FWHM laser pulsewidth and 2% rms correlated energy spread with 0.07% rms uncorrelated energy spread. The correlated energy spread is enhanced in the linac to allow slice emittance measurements by conducting a quadrupole scan in a dispersive section. The normalized slice emittance has been measured to be as low as 2 mm-mrad for beams with peak currents up to 150 A (300 pC with a laser pulse length of 1.8 ps) while the full projected emittance is 3 mm-mrad.

1. Introduction

High brightness electron beams are required to drive future 4th generation light sources and as injectors for linear colliders or future laser driven accelerators. A 4th generation light source called the Linac Coherent Light Source (LCLS) has been proposed at SLAC [1]. The LCLS is a SASE FEL at 1.5 Å wavelength which necessarily requires a small transverse and longitudinal emittance electron beam. To reach saturation in a 100 m long undulator, the current design of the LCLS requires an injector capable of producing a bunch with 1 nC of charge, transverse emittance of < 1 mm-mrad, energy spread of < 0.1 % and pulse length of 10 ps. The SLAC Gun Test Facility (GTF) was built to demonstrate the necessary emittance for the LCLS.

* SLAC is operated by Stanford University for the Department of Energy under contract number DE-AC03-76SF00515

The GTF consists of a Nd:glass drive laser, 1.6 cell S-band rf gun, emittance compensating solenoid, 3 m SLAC linac and diagnostic section. The diagnostics include a quadrupole doublet, multiple beam profile screens and a spectrometer magnet [2]. Figure 1 shows a schematic representation of the GTF also showing the phosphor screen installed downstream of the spectrometer magnet which is used to measure the energy spectra and beam sizes reported here.

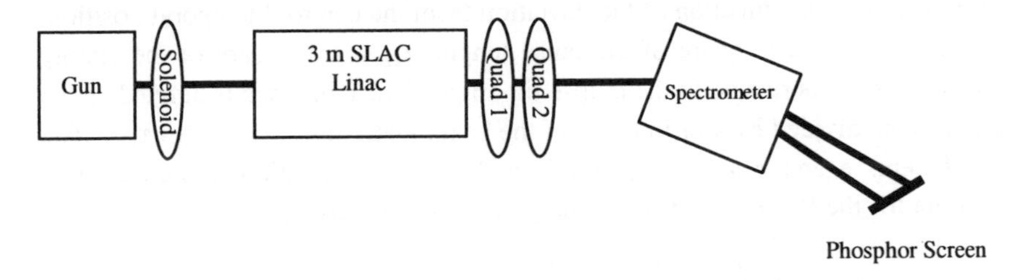

Figure 1. A simplified schematic of the GTF accelerator is shown including the gun, solenoid, linac, quadrupole doublet, spectrometer magnet and phosphor screen used to measure the beam sizes. Quad 1 focuses in the dispersive plane and Quad 2 in the non-dispersive plane.

Previously we have reported projected transverse emittance measurements of 1.5 mm-mrad with 200 pC of charge and a 2 ps FWHM laser pulse [3]. In this paper we describe the longitudinal emittance measurements and transverse slice emittance measurements technique using a linearly chirped electron bunch. The laser pulse length used in the experiments are Gaussian shaped with either 1.8 ps or 4.3 ps FWHM pulse length [3]. The UV pulse length was measured with a DRS Hadland FS300 streak camera with reported UV resolution of < 1 ps. The longitudinal emittance measurements are conducted by measuring the beam size on the spectrometer screen in a dispersive section as a function of the linac phase. The slice emittance measurements are obtained by performing a quadrupole scan on a linearly chirped electron beam using the same screen. Both techniques are described in greater detail in the next section.

2. Longitudinal Phase Space

2.1. *Measurement Technique*

The longitudinal emittance of the beam entering the linac is determined by measuring the energy spectrum downstream of the linac as a function of the linac phase. The longitudinal distribution can be described by three parameters -

pulse width, correlated energy spread and uncorrelated energy spread - analogous to the three independent beam matrix parameters in transverse phase space. However, space charge and the sinusoidal dependence of the applied RF field tend to distort the longitudinal emittance ellipses so that it is necessary to introduce two additional parameters to fully characterize the longitudinal emittance exiting the RF gun [4]. The energy spread of the beam, ΔE, is given by Equation 1 as a function of the deviation from the centroid temporal position, Δt. Here τ_{11} is the square of the pulse length, τ_{22} is the square of the energy spread, and τ_{12} is related to the ellipse tilt angle. The emittance is defined as the ellipse area divided by π or in terms of the beam matrix as $\varepsilon_\lambda^2 = \tau_{11}\,\tau_{22} - \tau_{12}^2$. The coefficients a and b are the quadratic and cubic perturbation terms added to account for the RF field and space charge forces respectively.

$$\Delta E = \pm\sqrt{\left(\frac{\tau_{12}}{\tau_{11}}\Delta t\right)^2 - \frac{\tau_{22}\Delta t^2 - \varepsilon_l^2}{\tau_{11}}} + \frac{\tau_{12}}{\tau_{11}}\Delta t + a\Delta t^2 + b\Delta t^3 \quad (1)$$

The initial longitudinal beam distribution given by equation 1 is mapped through the linac for each time slice using the linac energy gain equation shown in Equation 2 where E_t is the total energy after the linac, E_g is the energy of the particle exiting the gun, E_λ is the peak energy gain through the linac and θ_λ is the particle phase in the linac.

$$E_t = E_g + E_l\cos(\theta_l) \quad (2)$$

It is assumed that all particles travel at the speed of light, and no wakefields in the linac are included in this calculation. After populating an initial distribution with up to 200000 particles one can calculate the expected distribution at the linac exit according to the equation above and compare with the measured spectra. Using a least square error routine to minimize the error between the fit and measured value one can estimate the longitudinal emittance. Thus the measurement technique is very similar to the quad scan technique used to measure the transverse emittance except the data is fit to the entire spectrum because of the nonlinear phase space distortions instead of just the rms energy spread.

For the data reported here, energy spectra were recorded at ten different linac phases including the phase corresponding to the minimum energy spread and the maximum energy which is by definition 0° phase. The energy spectra were measured using a Pulnix TM7EX camera imaging the spectrometer phosphor screen, which is mounted at 45° with respect to the bend plane. The integrated spectrometer field along the curved beam trajectory was measured prior to installation for the absolute energy measurement. The spectrometer was calibrated to be 1.9 keV/pixel by varying the spectrometer current and measuring the resulting beam motion on the screen. The resolution of the spectrometer is estimated to be approximately 3-5 keV rms limited by the finite beam spot size at the input plane.

1.2. *Results*

Five spectra were captured at each phase setting along with a single background image with the laser off. The absolute phase was determined experimentally by varying the phase until the maximum energy was found and defining this phase as 0°. The minimum energy spread (0.1% FWHM) occurs at a phase of –13°. The energy spread was measured at 10 different phases in 5° steps on both sides of the minimum. The spectra were background subtracted and averaged over the five images.

The laser pulse length was set at 4.3 ps FWHM, with 38° injection phase with respect to the zero field and the gun was operated at a 110 MV/m gradient. The average energy of the particles exiting the gun is 5.8 MeV, the linac gradient was set at E_λ = 25 MeV and the bunch charge is approximately 150 pC. The solenoid field was operated at 2 kG to produce a waist near the spectrometer input plane so no other focusing elements were required. The five independent parameters in Equation 1 were varied to best fit the measured energy spectra with a least square error fitting routine and the results are reported in Table 1.

Table 1. The values of the fit parameters from Equation 1 using a least square error fitting routine. The parameters are fit at the linac entrance.

parameter	value	units
τ_{11}	1.5	ps^2
τ_{12}	-150	keV ps
τ_{22}	14000	keV^2
ε_l	4.6	keV ps
a	4.0	keV/ps^2
b	3.0	keV/ps^3

Figure 2 shows four of the measured spectra at the indicated linac phase along with the corresponding fits at the linac exit. The fit yields an e-beam pulse length of 1.2 ps rms (2.9 ps FWHM) with 2% rms energy spread (8% FWHM) at the linac entrance. The uncorrelated energy spread was computed to be 3.7 keV rms (.067%) by eliminating all correlations in the distribution and computing the remaining rms energy spread. The measured uncorrelated energy spread and longitudinal emittance is limited by the spectrometer's resolution since the resolution is estimated at 3-5 keV. The finite resolution of the spectrometer is not included in the analysis and accounts for the difference between the fit and the measurement for the lowest energy spread case shown in Figure 2.

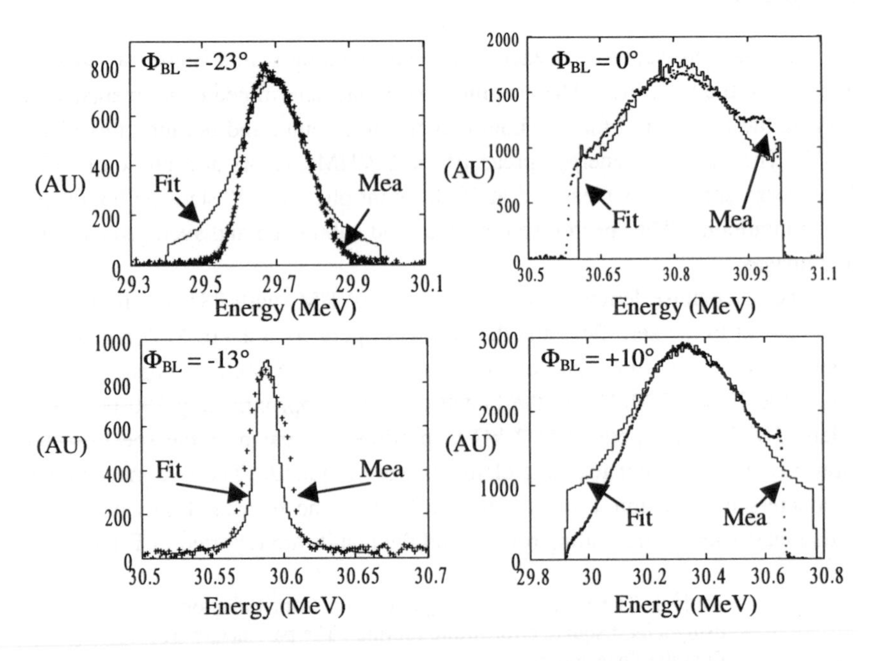

Figure 2. The measured energy spectra and fits in arbitrary units vs energy are shown for four different linac phases. The maximum energy occurs by definition at 0° and the minimum energy spread occurs at a linac phase of -13°.

Figure 3 shows the computed longitudinal distribution at the linac entrance as well as the linac exit for both the minimum energy spread and maximum energy cases. It can be seen that the energy spread at the linac entrance is dominated by the linear chirp while at the linac exit it is strongly dependent on the linac phase. The minimum energy spread occurs at the phase which nearly removes the linear term and only leaves higher order correlation terms plus the uncorrelated energy spread. At the linac entrance, the additional correlated energy due to the quadratic and cubic terms is 6.1 keV and 5.6 keV respectively for the rms pulse length of $\Delta t = 1.2$ ps. The strong presence of the cubic term in

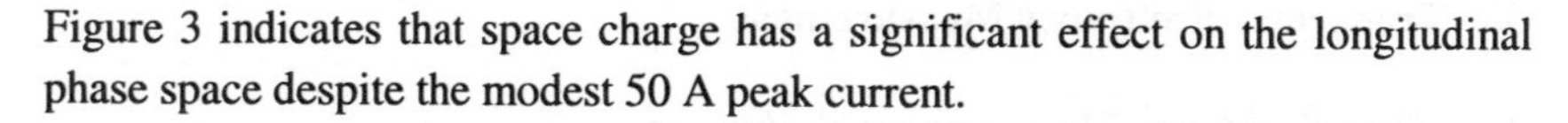

Figure 3 indicates that space charge has a significant effect on the longitudinal phase space despite the modest 50 A peak current.

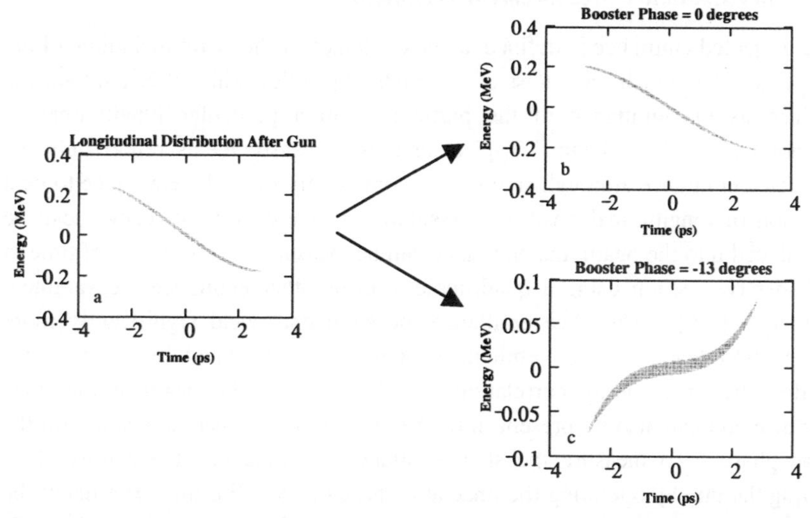

Figure 3. The computed longitudinal phase space at the linac entrance (a), the linac exit with a phase of 0° (b) and the linac exit with a phase of -13° (c). The linac phase can be adjusted to nearly remove the linear chirp resulting in the phase space shown in (c) with a dominant nonlinear term.

The measured values are consistent with PARMELA simulations except for the correlated energy spread. PARMELA only predicts an energy spread of approximately 1-2 % FWHM instead of the measured 8%. While this discrepancy is not yet completely understood, it may be partially due to the fact that longitudinal wakefields are not included in PARMELA. However, analytic calculations of the energy spread contribution due to the wakefield in the gun and linac predict energy spreads an order of magnitude less than the measurement. Alternatively, a large correlated energy spread from the gun can be produced if the ratio of the field in the two cells of the gun is not sufficiently close to unity. The gun fields are currently being measured using a bead drop technique and new calibrated probes installed so that the field can be measured at high power in both cells simultaneously to determine if a dynamic imbalance occurs due to rf heating and subsequent thermal distortions.

Finally phase slippage in the linac is not included in the analysis reported here. However, calculations of the energy spread due to slippage for parameters used in this experiment do not significantly alter the results since the bunch length is short and therefore all particles slip by nearly the same amount. Likewise almost no ballistic bunching or anti-bunching occurs in the linac.

3. Transverse Emittance Measurement

3.1. *Slice Emittance Measurement Technique*

The projected emittance is defined as the emittance of the beam including all the particles in the bunch regardless of longitudinal position while slice emittance is defined as the emittance of the particles from a particular longitudinal or temporal slice of the beam. Typical emittance measurements report only the projected emittance although simulations consistently show the emittance to be a function of longitudinal position. Assuming a time-energy correlation can be introduced into the beam, the emittance can be measured as a function of time in a dispersive section using a quadrupole scan or other emittance measurement technique [5-6]. The slice emittance measurements yield significantly more information than projected emittance alone. As mentioned in the previous section the time energy correlation already exists in the beam at the linac entrance and can also be present in the beam exiting the linac depending on the linac phase. To measure the slice emittance we enhanced the energy chirp exiting the gun by operating the linac at a phase of +5°. For this experiment the beam size is measured on the phosphor screen in the dispersive section, shown in Figure 1, as a function of quadrupole strength. The beam size in the non-dispersive plane is measured for 10 different temporal or energy slices instead of measuring the temporal or energy distribution as in the previous section. The complete chirped beam profile is acquired in a single shot and then sliced via software for analysis. In addition the software can also integrate in the time or energy axis to yield a projected emittance for comparison with the slice emittances. The longitudinal emittance measurement described in the previous section can be used to calibrate the temporal axis.

In addition to the quadrupole focusing elements in the GTF beamline, the focusing properties of the spectrometer magnet must also be included in the slice emittance analysis. Beam-based techniques were used to determine the spectrometer pole face rotations. This was necessary since the spectrometer was not fully characterized before installation. Subsequent magnetic measurements performed after the electron beam measurements were completed, resulted in almost no change in the computed emittances.

The number of slices that can be resolved is limited to the resolution of the energy or time measurement. Therefore the resolution depends on the minimum spot size that can be obtained at the spectrometer screen and the uncorrelated energy spread. For the GTF parameters, simulations show that the dispersive plane beta function can be kept approximately constant at 5 cm while the quadrupole scan varies the non-dispersive plane beta function from 1-10 m assuming identical Twiss parameters in the two planes at the entrance to the first quadrupole. Thus for a 30 MeV beam the beam size in the dispersive plane is less than 100 μm rms which is equivalent to approximately 2 pixels. For the uncorrelated energy spread reported in the previous section the rms width is also approximately 2 pixels. For the data reported below each of the ten slices is integrated over approximately 30 pixels. Thus the measurement is far from the resolution limit.

1.2. *Results*

The laser used for the slice emittance measurements was approximately 2 ps FWHM in the longitudinal dimension and nearly flat in the tranverse dimension due to a clipping aperature. The beam charge was 300 pC and the gun field was 110 MV/m with an injection phase of approximately 30°. The linac was operated at approximately 8 MV/m and +5° phase resulting in the measured total energy of 30 MeV.

Beam size measurements were made at three different solenoid settings and between 13 and 18 different quadrupole strengths for each solenoid setting. Five beam images are obtained at each quadrupole setting and an image with no laser beam incident on the cathode is used for background subtraction. The rms beam sizes are calculated after fully projecting each slice in the dispersive plane and then truncating the wings at 5% of the peak value to minimize the effect in the data analysis of non-beam-related pixels. The beam profile for each slice of a typical image is shown in Figure 4.

The average of the five beam widths are then used to fit the Twiss parameters. Figure 5 shows the normalized emittance, alpha and beta functions as a function of slice number for all three solenoid settings where slice 0 is defined as the full projected value of all ten slices. The phase space mismatch parameter is also plotted in Figure 5 which quantifies the relative misalignment between phase space ellipses. The mismatch parameter [7], ζ, is defined in equation 3 where α_n, β_n and γ_n are the nth slice Twiss parameters while α_0, β_0 and γ_0 are the projected Twiss parameters. A mismatch parameter of 1 means perfect alignment and should be less than 2 for good phase space matching.

$$\zeta_n = \frac{1}{2}(\gamma_n \beta_0 - 2\alpha_n \alpha_0 + \gamma_0 \beta_n) \leq 1 \tag{3}$$

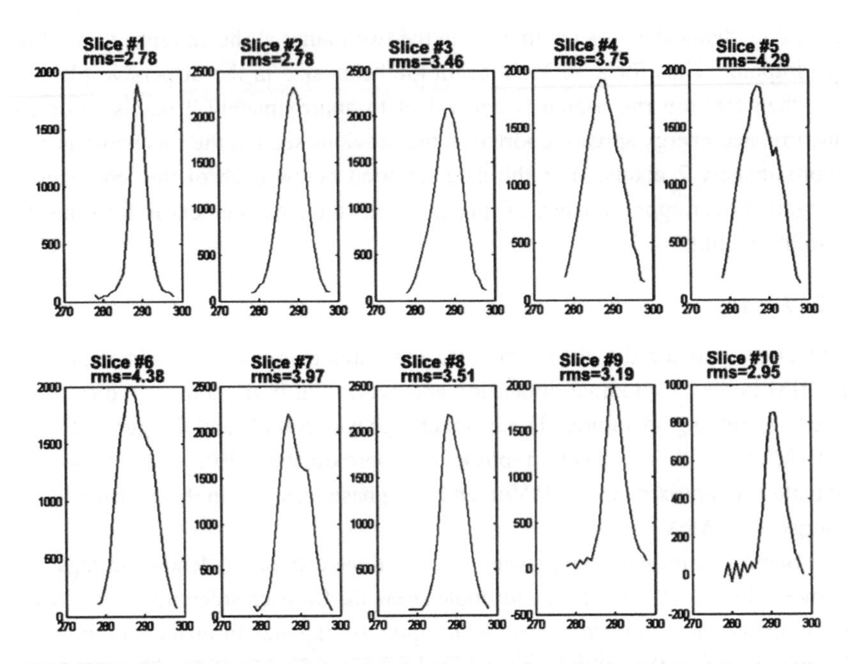

Figure 4. The slice beam projections are shown as a function of pixels. The calibration is 42 μm/pixel and the head of the bunch is slice # 1 and the tail is slice # 10. The rms width of each slice in pixels is reported above each projection.

Typically the head and tail data points should be ignored because the data is dominated by noise due to the small number of particles in these extreme slices. The minimum measured projected emittance is 3 mm-mrad while the minimum slice emittance is approximately 2 mm-mrad. However, the actual emittance minimum was probably not observed in this data set as the emittance is still decreasing as the solenoid decreases. Interestingly the slice emittances depend on the solenoid value. This may be due to centroid offsets between slices and will be investigated further.

The temporal axis for this data set was not calibrated although in principal it can be calibrated with the longitudinal emittance measurement. The calibration is estimated to be approximately 200 fs/slice since there are 10 slices and the laser pulse length was 2 ps FWHM. We have found this technique very powerful in discovering and removing longitudinal correlations in the transverse profile such as those due to wakefields and misalignments. The primary limitation of the technique is that the linac phase can not be varied as it must be properly set to chirp the electron beam.

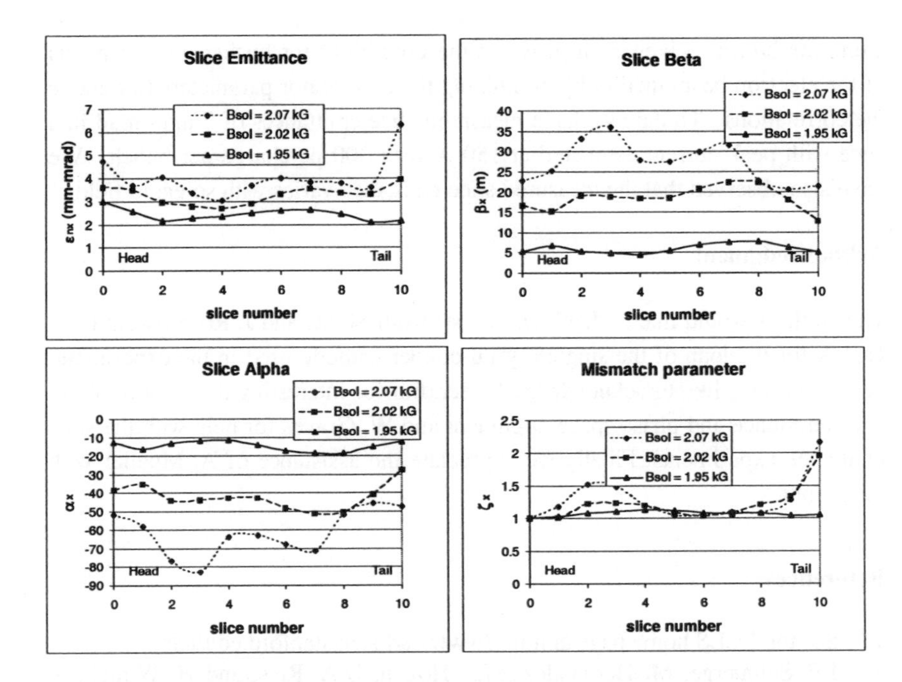

Figure 5. The Twiss parameters including normalized emittance and the mismatch parameter are plotted as a function of slice number for three different solenoid values. Slice 0 is the full projected value while slice 1 is the head and 10 is the tail.

4. Conclusions

We have shown results from a longitudinal emittance measurement using a phase scan technique which includes nonlinear terms in the fit. The results show a linear correlated energy spread of 8% FWHM between the gun and linac. The cause of this energy spread has not yet been determined. We are planning further experiments to discover the source of the correlated energy spread and eliminate it in order to improve the beam quality.

In addition we have demonstrated a slice emittance measurement technique by utilizing a quadrupole scan with a chirped electron beam in a dispersive section. This technique allows for quick measurement of the slice emittance since it requires no more time than a normal quadrupole scan measuring the projected emittance. The slice Twiss parameters yield much more information than a simple projected measurement since it reveals the presence of correlations along the bunch. Likewise it provides the possibility for further improvements of the electron beam quality by optimizing the accelerator parameters to remove the correlations. To date we have measured slice emittances of 2 mm-mrad in a slice with peak current estimated at 150 A with 300 pc charge per bunch. We have also observed that the measured slice emittance varies with solenoid field.

Acknowledgments

The Authors would like to thank D. Palmer from SLAC and J. Rosenzweig from UCLA for the loan of the single crystal copper cathode used in the experiment. We would also like to acknowledge L. Serafini for interesting discussions about slice emittance and phase space alignment and W. Graves for help with the slice emittance experiments. Finally we appreciate the assistance of A. Mueller with the graphics.

References

1. See the LCLS home page at http://www-ssrl.slac.stanford.edu/lcls/
2. J.F. Schmerge, M. Hernandez, M.J. Hogan, D.A. Reis, and H. Winick, *in SPIE* **3614**, 22 (1999).
3. J.F. Schmerge, P.R. Bolton, J.E. Clendenin, F.-J. Decker, D.H. Dowell, S.M. Gierman, C.G. Limborg, and B.F, Murphy, *Nucl. Instr. Meth. A* **483**, 301 (2002).
4. D.H. Dowell, S. Joly, A. Loulergue, J.P. deBrion and G. Haouat, *Physics of Plasmas* **4,** 3369 (1997).
5. C.Limborg, P.Bolton, J.Clendenin, D.Dowell, P.Emma, S.Gierman, B.Murphy, J.Schmerge, "PARMELA vs Measurements for GTF and DUVFEL" *Submitted to EPAC 2002*, Paris, June (2002).
6. W.Graves et al. "Experimental study of sub-ps slice electron beam parameters in a photoinjector" *Submitted to Phys. Rev. ST-AB*, (2002).
7. D.A. Edwards and M.J. Syphers., *An Introduction to the Physics of High Energy Accelerators* Wiley, p. 237, (1993).

PERFORMANCE OF THE TTF PHOTOINJECTOR FOR FEL OPERATION

S. SCHREIBER, J.-P. CARNEIRO, C. GERTH, K. HONKAVAARA,
M. HÜNING, J. MENZEL, P. PIOT, E. SCHNEIDMILLER, M. YURKOV

Deutsches Elektronen-Synchrotron
Notkestr. 85,
D-22603 Hamburg, Germany
E-mail: siegfried.schreiber@desy.de

The RF gun based photoinjector of the TESLA Test Facility Linac (TTFL) at DESY has been built to produce a beam close to TESLA specifications in order to test superconducting accelerating structures. With the installation of the TTF Free-Electron Laser (TTF-FEL) in the TTF linac, the injector has been gradually optimized to improve the gain of the SASE lasing process and to achieve saturation in the VUV wavelength region. The report describes the performance of the optimized injector in terms of longitudinal and transverse phase space.

1. Introduction

The TESLA Test Facility (TTF) operates an RF gun based photoinjector[1]. Among various experiments for the TESLA project, the photoinjector is used to drive the TTF-FEL free electron laser. To do this, excellent beam properties are essential. It requires a train of electron bunches, where each bunch has a high peak current in the kA range, a small transverse emittance in both planes in the order of a μm, and a small uncorrelated energy spread below 0.1 %[2]. Recently, the TTF-FEL achieved saturation in the VUV wavelength region (80 to 100 nm)[3]. One key issue for this success was the tuning of injector beam parameters. The optimized parameters differ from the design. In the following, the most relevant differences are described.

2. Overview and Design

A sketch of the the TTF Linac including the injector is shown in Fig. 1. The electron source is a laser-driven L-band 1 1/2-cell RF gun with a Cs_2Te cathode. The cathode is illuminated by a train of UV laser pulses generated in a mode-locked solid-state laser system synchronized with the

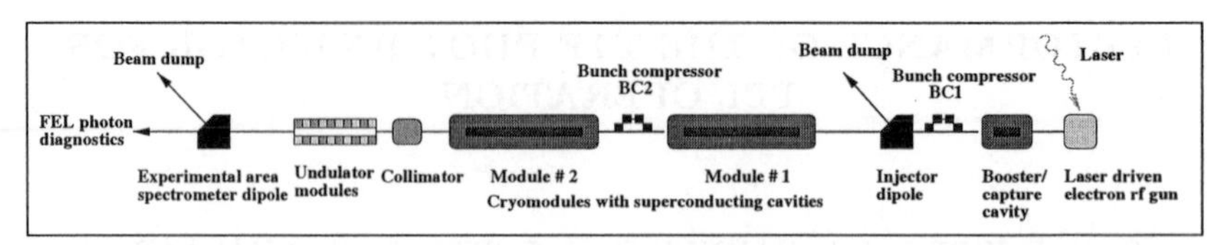

Figure 1. Schematic overview of the TTF-FEL linac phase 1 (not to scale). Beam direction is from right to left, the total length is 100 m.

RF (1.3 GHz)[4]. The RF gun section is followed by a booster, a standard TESLA 9-cell superconducting accelerating cavity operated at 11.5 MV/m. The beam energy measured at the energy spectrometer after the booster is 16.5 MeV. Further details of the injector can be found in [1]. The beam is accelerated by two 12 m long TESLA accelerating modules containing eight 9-cell superconducting accelerating structures each. After a collimation section, the beam is injected into the undulator modules with an energy of up to 300 MeV. Two bunch compressors are installed: BC1 is downstream of the booster cavity, BC2 between the accelerating modules. For more details refer to [3].

The requirements on the beam to drive a VUV free-electron laser demands a design of the injector which leads to high brilliance beams: high peak current, small transverse emittance in both planes and low energy spread – as listed in Table 1. To our knowledge, no injector has delivered a beam yet, which fulfilled all these required parameters at the same time. As an example, due to space charge effects it is not possible to produce the requested peak current of 500 A (phase 1) or 2.5 kA (phase 2) directly at the gun.

The design of the TTF-FEL[5] starts with a bunch length of 2 mm at the gun exit, a charge of 1 nC, a normalized emittance of 2 mm mrad, and an uncorrelated energy spread of 25 keV. In the design, the bunch is then compressed to 0.8 mm (BC1) prior to acceleration to 150 MeV. A second compression to 250 μm (BC2) leads to the required peak current. The pre-compression to 0.8 mm is necessary to keep the energy spread required for lasing below 0.1 %. The design parameters are listed in Table 1: (a) for the initial TESLA 500 design as defined in [6], (b) a revised set close to the TESLA parameters from [7], and (c) the parameters for the TTF-FEL[5].

3. Optimized beam parameters for the FEL runs

As a matter of fact, the TTF injector has to fulfill various demands on beam properties, which are in conflict with an optimized design especially

Table 1. Injector design parameters for TESLA related experiments and TTF-FEL phase 1 operation. TTFL(a) is the initial TESLA 500 design, TTFL(b) a set close to the revised design. Note: the actual performance may differ from this table. Refer to the text.

Parameter		TTFL		FEL
		(a)	(b)	
RF frequency	GHz		1.3	
Repetition rate	Hz		10	
Pulse train length	μs		800	
Pulse train current	mA	8	9	9
Bunch frequency	MHz	1	2.25	9
Bunch charge	nC	8	4	1
Bunch length (rms)	mm	1	1	0.8
Emitt. norm. (x,y)	μm	20	10	2
$\Delta E/E$ (rms)	%		0.1	
$\Delta E/E$ (bunch to bunch) (rms)	%		0.2	
Injection energy	MeV		20	
After 2nd compression				
Bunch length (rms)	mm			0.25
Bunch current	kA			0.5

for an FEL. As an example, for the measurements of higher order modes in TESLA cavities, a 54 MHz bunch train has been produced, where the charge along the train could be modulated with frequencies between 0.3 and 27 MHz[10]. These beam parameters are substantially different to those required for FEL operation and nevertheless have to be realized as well.

A difficult problem is the bunch compression at 20 MeV with the first bunch compressor (BC1). It has been designed for TESLA related experiments and meets the specifications in terms of bunch length and transverse emittance as in Table 1(a)[6].

Looking at the FEL case, a compression down to 0.8 mm is feasible however, with an expense in transverse emittance: simulation and measurements indicate an increase in transverse emittance mainly due to space charge effects. Simulations of the compressor suffer from the complicated space charge effects and do not give reliable predictions of the emittance. Measurements have been performed at the FNAL/NICADD Photo-Injector, a twin of the TTF injector. The results are reported in [8].

Initially, parameters of the RF gun have been adjusted to minimize the transverse emittance for a charge of 1 nC. The forward RF power is maximized 3 MW, a compromise between highest field on the cathode and reliable operation. This leads to a gradient on the cathode of 41 MV/m.

Other parameters have been optimized: The phase of the gun RF in respect to the laser pulses (the launch phase) has been chosen to be 40°, the laser spot flat top radius on the cathode is 1.5 mm. The field of the first and second solenoid is 0.105 T and 0.088 T respectively. For this parameters, we measure a transverse emittance of 3.0 ± 0.2 mm mrad[9]. The length of the photoinjector laser pulse is $\sigma_l = 7 \pm 1$ ps leading to an electron bunch length of $\sigma_z = 3.2 \pm 0.2$ mm[11]. The energy spread measured with the injector dipole is 22.1 ± 0.3 keV with a long tail induced by the acceleration process of the rather long bunch[1]. The energy spread measurement above is limited by the resolution of the optical system. An improved system is going to be set up soon.

The measured uncorrelated energy spread is well within the design, however, due to the long bunches from the RF gun, the RF induced correlation is large. Because of the predicted increase in transverse emittance by the low energy bunch compressor, we decided not to use the bunch compressor at low energy for the first tuning for FEL operation, only the second compressor at 150 MeV. With this compression scheme, we took advantage of the large correlated energy spread induced by the off crest acceleration. Off crest acceleration is required for magnetic chicane bunch compression. Because of the very small uncorrelated energy spread from the RF gun, its projection on the time/phase axis results in a high peak current spike with a long tail. This is illustrated in Fig. 2. It shows a simulation[12] of the longitudinal phase space performed with the beam parameters above and nominal compression settings for the second bunch compressor BC2, but with BC1 switched off. A bunch charge of 3 nC is used.

It is in fact this spike which carries the required peak current for lasing. The charge per bunch has been raised from 1 to 3 nC, but the laser spot size on the cathode is kept constant at 1.5 mm radius to keep the emittance small. A high acceleration gradient on the cathode of 41 MV/m reduces space charge effects and improves further the peak current. For 3 nC the total rms bunch length increases to 4 mm. A further increase of charge for the given laser spot size and gun gradient is not possible. Saturation of the charge extracted starts already at 3 nC, it is limited to about 4 nC. The launch phase has been slightly decreased from 40° closer to 30°.

With the operation mode described above, the SASE FEL reached saturation[3]. In the following, measurements of the transverse emittance and the longitudinal charge distribution for this operation mode are presented.

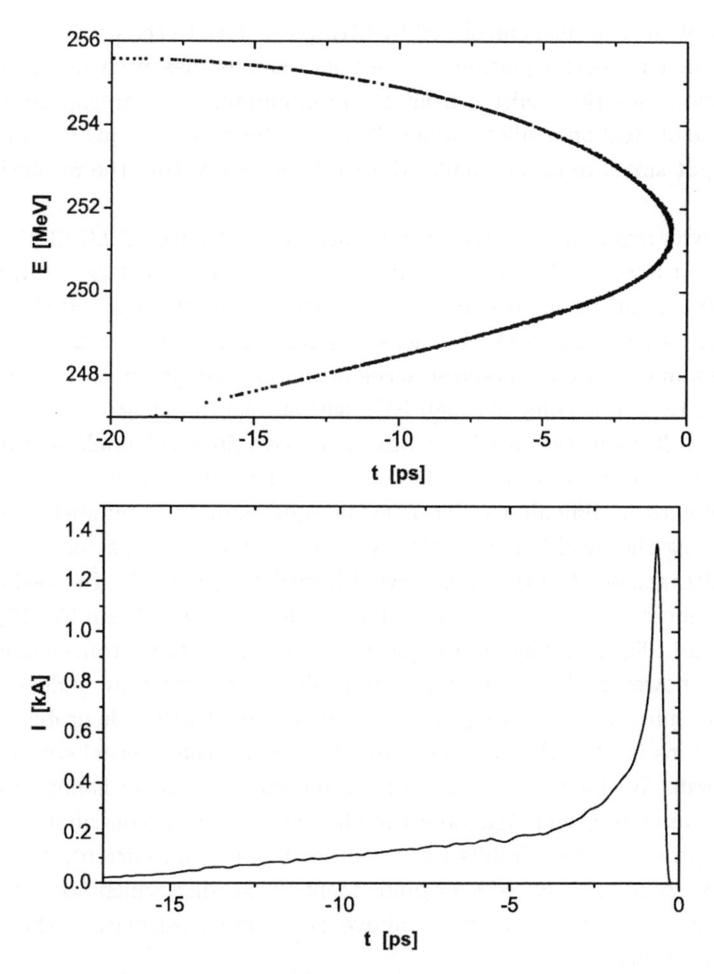

Figure 2. Simulation of the longitudinal phase space after bunch compression (upper) with its projection on the time axis (lower). Beam parameters of the improved operation mode used. Refer to the text for details.

4. Experimental set-up and results

4.1. *Transverse emittance*

The emittance has been measured with the quadrupole scan method at several places along the linac: downstream the booster cavity at 16.5 MeV, downstream the bunch compressor BC2 at 137 MeV, and at the entrance of the undulator with a beam energy of 246 MeV. In all cases, the beam size is measured as a function of the magnetic gradient of a quadrupole.

The emittance is then calculated by fitting the data to the prediction given by beam transport equations. The beam size is measured using optical transition radiation emitted from a thin aluminum layer on kapton and in the case of BC2 on a silicon wafer. The beta-function is adjusted in a way, that spot sizes are never smaller than 100 μm well within the resolution of 50 μm.

Downstream the booster we measure an emittance of 3.0 (3.2) $\pm$ 0.5 mm mrad horizontal (vertical). At 137 MeV we obtain a larger emittance of 8 (9) $\pm$ 2 mm mrad. It stays about constant along the linac: at the undulator entrance at 246 MeV the measured value is 11 $\pm$ 6 (7 $\pm$ 2) mm mrad. These numbers are for on-crest acceleration of a single bunch of 1 nC, bypassing BC1, but going through BC2 without compressing.

On full compression of the beam, for certain quadrupole settings, a break up of the transverse profile into two or three bunchlets is observed. This makes it difficult to give a meaningful emittance number. Simply projecting the total beam profile regardless of its internal structure yields an emittance of 14 (13) $\pm$ 2 mm mrad horizontal (vertical). Increasing the charge increases the measured emittance: for 2 nC we obtain 22 (19) $\pm$ 2 mm mrad. The result agrees roughly with the expectations from simulations (see [9]), however, the break up into bunchlets is not yet fully understood.

The emittance data are projected emittances. A principle problem with projections is, that they do not take into account the internal structure of the beam. We know, that after full compression the beam is not gaussian shaped anymore, and that only the slice of the bunch contributes to the lasing process, which fulfills the requirement for peak current, emittance and energy spread. Since we cannot identify the slice which lases, we can only use the emittance numbers above as an upper estimate of the lasing slice emittance.

However, from the measured properties of FEL radiation it is possible to deduce the value of the slice emittance. Using the measured gain length of 67 $\pm$ 5 cm[3] and a peak current in the range of 0.5 to 1 kA we get 4 to 6 mm mrad respectively (see Fig. 3). This estimate is in good agreement with the measurement in the injector. One would suggest, that after compression the beam keeps its injection emittance in its core, even if the projected full emittance is growing significantly.

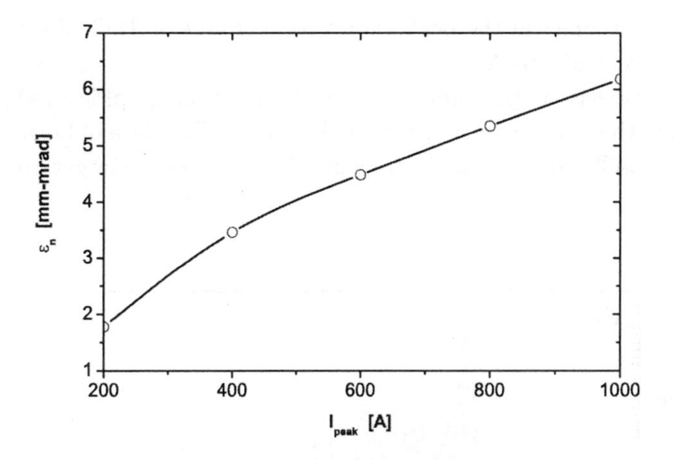

Figure 3. Estimated emittance of the lasing bunch slice for different peak currents of this slice, deduced from properties of the measured SASE radiation.

4.2. *Bunchlength*

For the bunch length measurements, we use synchrotron radiation emitted by the horizontally deflecting spectrometer dipole after the undulator (see Fig. 1). The optical part of the synchrotron radiation is guided by aluminum mirrors to a streak camera. It has an intrinsic resolution of 210 fs (FWHM)[13]. Details of the set-up are described in [11]. In order to reduce chromatic effects, a narrow-band wavelength filter ($\Delta\lambda = 5$ nm) has been used. The data presented here are obtained with the second fastest streak speed of 50 ps/10.29 mm, where the resolution is 200 fs sigma. The profiles have been taken when the beam was set-up to provide FEL laser radiation to experiments, close to saturation.

Figure 4 (A) shows an overlay of several measurements of the same longitudinal bunch profile. The average profile is shown in Fig. 4 (B). The profile has a clear leading peak and a long tail. The width of the leading peak is 650 ± 100 fs (sigma).

With the high resolution camera we have been able to see substructures in the longitudinal beam profile. This has not been possible in earlier measurements, where only rms numbers of the width could be given[11]. The rms width of the data presented here is consistent with the previous measurements.

From the measured profile, we can estimate, that about 30 % of the charge is contained in the peak. For a total charge of 3 nC, this results in a peak current of 0.6 kA.

For comparison, the profile obtained with tomographic methods[14] is overlaid to the streak camera profile in Fig. 4 (B). The data of both methods agree very well, except that the tomographic data show a larger but shorter tail.

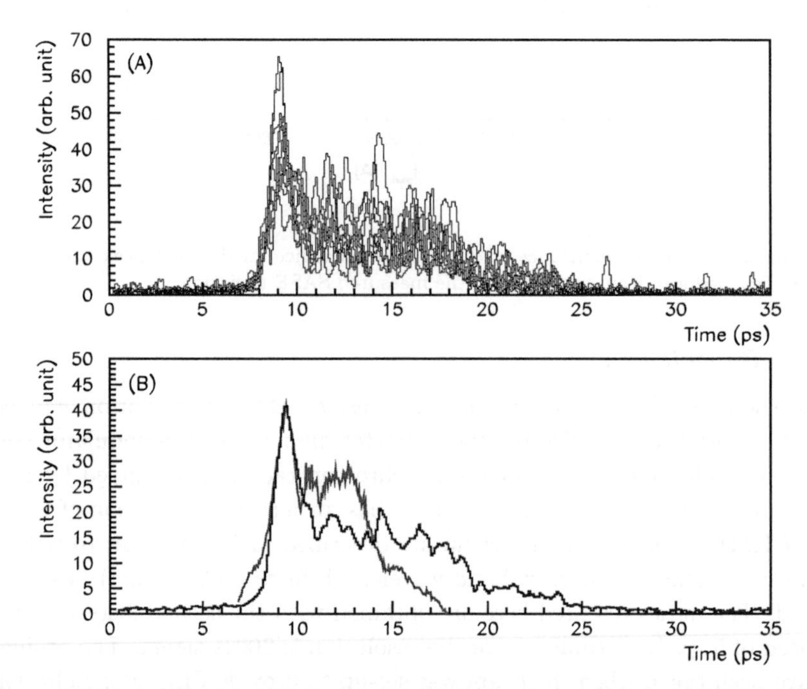

Figure 4. Several measurements of the same longitudinal beam profile obtained with a streak camera (A). The average over all profiles is plotted in (B), blue (long tail) curve. These data have been taken under beam conditions for lasing close to saturation. For comparison, a profile obtained with tomographic methods is overlaid in (B), red (short tail) curve.

Measurements have also been done with precompressed bunches using BC1. With a moderate compression down to 2 mm, the RF induced curvature in the longitudinal phase space is less pronounced and leads to a longer peak after the second compression stage closer to the design of 250 μm. The effect of lengthening or tailoring of the lasing bunch slice is clearly visible in the measured internal mode structure of the FEL radiation[15].

5. Discussion and Conclusion

To drive the TTF-FEL, the demands on both, the longitudinal and transverse phase space could not be fulfilled at the same time for the total bunch.

Full bunch compression in the injector at 20 MeV down to the design of 0.8 mm was not possible without spoiling the transverse emittance. The solution is to take advantage of the RF induced correlated energy spread when accelerating long bunches. After bunch compression, the profile exhibits a peak, which fulfills the requirement for the peak current while keeping the transverse emittance and energy spread small.

Acknowledgement

We like to thank our colleagues from INFN Frascati and Roma II for their support on the optical transition radiation devices.

References

1. S. Schreiber, "Performance Status of the RF-gun based injector of the TESLA Test Facility Linac", Proc. of the Europ. Particle Acc. Conf. EPAC2000, Vienna, Austria, p. 309.
2. J. Andruszkow et al., "First Observation of Self-Amplified Spontaneous Emission in Free-Electron Laser at 109 nm Wavelength", Phys. Rev. Lett. 85 (2000) 3825-3829.
3. V. Ayvazyan et al, "Generation of GW radiation pulses from a VUV Free-Electron Laser operating in the femtosecond regime", Phys. Rev. Lett., 88 (2002) 104802.
4. S. Schreiber et al., "Running Experience with the Laser System for the RF Gun based Injector at the TESLA Test Facility", Nucl. Instr. and Meth. A445 (2000) 427.
5. "A VUV Free Electron Laser at the TESLA Test Facility at DESY – Conceptual Design Report", DESY Print, June 1995, TESLA-FEL 95-03.
6. TESLA-Collaboration, ed. D.A. Edwards, "TESLA Test Facility Linac – Design Report", DESY Print March 1995, TESLA 95-01.
7. R. Brinkmann et al, "TESLA Technical Design Report, Part II The Accelerator", DESY 2001-011, http://tesla.desy.de/new_pages/TDR_CD/PartII/accel.html
8. J.-P. Carneiro et al., "Study of a Magnetic Chicane at the FNAL/NICADD Photo-Injector using Remote Operation from DESY", Proc. of the Europ. Particle Acc. Conf. EPAC2002, Paris, France, June 3-7, 2002, p. 1759.
9. Ph. Piot et al., "Emittance Measurements at the TTF Photoinjector", Proc. of the 2001 Partical Accelerator Conference, Chicago, Ill., USA, June 18-22, 2001, p. 81.
10. C. Magne et al., "Measurements with Beam of the Deflecting Higher Order

Modes in the TTF Superconducting Cavities", Proc. of the 2001 Partical Accelerator Conference, Chicago, Ill., USA, June 18-22, 2001, p. 3771.
11. K. Honkavaara, Ph. Piot, S. Schreiber, D. Sertore, "Bunch Length Measurements at the TESLA Test Facility using a Streak Camera", Proc. of the 2001 Partical Accelerator Conference, Chicago, Ill., USA, June 18-22, 2001, p. 2341.
12. Simulations with ASTRA; K. Flöttmann et al., `www.desy.de/~ mpyflo/ASTRA_dokumentation`.
13. Hamamatsu CS6138 FESCA-200.
14. M. Hüning, "Investigations of Longitudinal Charge Distribution in very short Electron-bunches", Proc. 5th Europ. Workshop on Diagnostics and Beam Instrumentation May 2001, Grenoble, France, p. 56.
15. V. Ayvazyan et al, "Study of the statistical properties of the radiation from a VUV SASE FEL operating in the femtosecond regime", FEL2002, Argonne, Ill., USA, Sept 9-13, 2002, to be published.

REVIEW OF RECENT DEVELOPMENT OF PHOTOINJECTORS IN JAPAN

M. UESAKA
Nuclear Engineering Research Laboratory, University of Tokyo
Tokai, Ibaraki, Japan
E-mail: uesaka@tokai.t.u-tokyo.ac.jp

H. IIJIMA AND K. DOBASHI
National Institute for Radiological Science
4-9-1, Anagawa, Inage, Chiba, Japan

J. YANG
Sumitomo Heavy Industries Co.
2-1-1, Tanido, Tanashi, Tokyo, Japan

S. MIYAMOTO
Himaji Institute of Technology
Sayo-Gun, Hyogo, Japan

Systematic developments of the photoinjectors for ultrashort and high quality electron beam works are under way in Japan. Sumitomo Heavy Industries succeeded in transformation of the Gaussian shape to the trapezoidal one in the temporal and transverse profiles of the drive laser and achieved 0.9 πmm.mrad with 1 nC/bunch. It is the best data in the transverse aspect. Himeji Inst. Tech. is operating very unique needle-shaped photocathode RF gun with the original Nd/Glass laser for IR-FEL and Compton scattering X-rays. U.Tokyo/JASRI(SPring8)/KEK/NIRS/BNL/etc. are developing and operating the S-band photoinjectors with Cu, Mg and Cs_2Te cathodes, and transmission-type one in near future. Further, U.Tokyo/KEK/NIRS are designing and constructing a new X-band RF-gun/linac/laser system to generate inverse Compton scattering hard X-rays(33-50keV) for intravenous angiography.

1. Mg Photoinjector by U.Tokyo/SPring8/KEK/etc

The Mg photoinjector (BNL Gun-IV type) had been constructed in cooperation with SPring-8, KEK, SHI, Waseda University and BNL. To decrease a dark current, the inner wall of the cavity was diamond-precise machined and the cathode plate was polished using diamond powders whose sizes were 3μm and 1μm in diameter. The gun was baked for 7 days at 150 °C and 48 hours at 120 °C before and after installation to an S-band linac. A vacuum condition is kept to be less than 5×10^{-10} Torr during gun operation. The aging of the gun was carried out carefully for 528 hours (22 days). Figure 1 shows the Fowler-Nordheim plot. Each point indicates the measured value after each aging term (triangle points indicate after 336 hours, squares after 413 hours and circles 528 hours). The enhancement factor β calculated by the fitting are 48, 36 and 20, respectively. The dark current was measured to be 600 pC/pulse for RF power of 6.6 MW, pulse width of 2 μs and repetition rate of 10 pps.

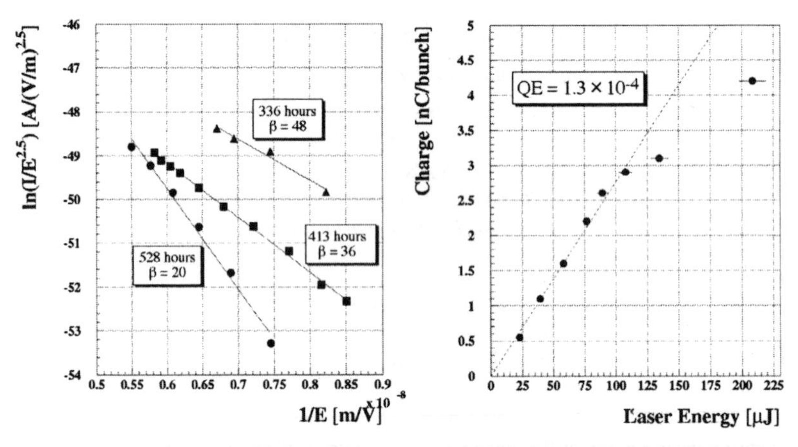

Figure 1. Fowler-Nordheim plot.

Figure 2. Electron bunch charge as a function of the laser energy.

The electron bunch charge as a function of the driven laser energy is shown in Fig. 2. During the measurement, the RF power is 6.6 MW, the RF pulse width 2 μs, the repetition rate 10 pps and a spot size of the laser larger than 3 mm in diameter. Magnitude of the solenoid coil, which is in the range of 1.0-1.8 kGauss, and laser injection phase for the cathode were optimized to charge-maximum. A dash line as shown in Fig. 2 is a fitting result of QE, where 5 points up to 100 μJ/pulse are used for fitting. The QE is calculated to be 1.3×10^{-4}. Generally the QE of the Mg without a treatment for the surface, such as the laser cleaning, is in the order of 10^{-5} due to the oxide-layer on the surface [1]. Our Mg cathode was kept in Helium gases immediately after diamond polishing, and the cathode was in the air for almost 2 days for installation. Therefore, we may consider that QE is higher owing to few oxide-layer. However, it is 1/10 as smaller as the expected value currently.

The electron bunch accelerated up to 22 MeV is compressed by the chicane-type magnetic compressor. Typical bunch duration is approximately 700 fs with the charge of 2 nC/bunch [2]. This ultrashort electron bunch is usually used as a pump beam for a pulse radiolysis method of chemical reaction of water. A probe laser light and the driven laser for the RF injector are supplied from the 0.3 TW laser. The 0.3 TW laser produces a laser light with the wavelength of 795 nm, the energy of 30 mJ/pulse, the pulse duration of 300 ps and the repetition rate of 10 pps. The laser light is guided by the optics in vacuum pipe and bellows, whose length is approximately 50 m in order to aviod fluctuation by air and realize precise synchronization between the pump beam and the probe laser. One laser light is compressed into the pulse width of 100 fs, and used as the probe laser. The other is guided to the third harmonic generator, which is provided the driven laser with the wavelength of 265 nm, the energy of a few hundred μJ/pulse and the pulse duration of a several picosecond

The synchronization test was performed to set the Xe chamber at the end of the linac. Cherenkov light emitted from the chamber is guided to the femtosecond streak camera with the probe laser. Figure 3 shows a typical streak image of the electron and laser timing profile.

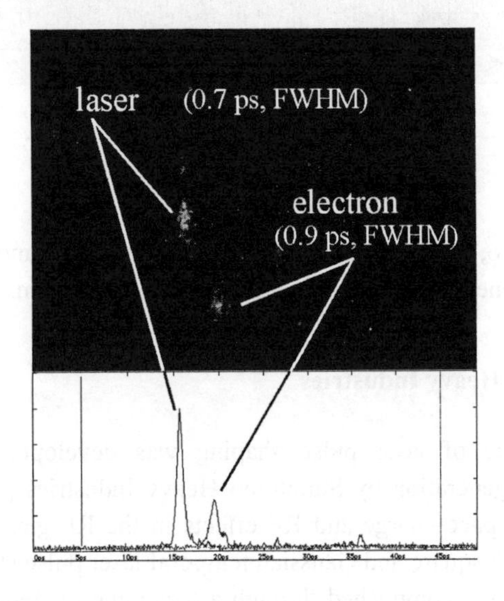

Figure 3. Streak image of compressed bunch

Before we had achieved the synchronization results of 330 fs for a few minutes and 1.9 ps for an hour [3]. However, the most important subject to be overcome is the suppress the long term drift of the time difference between the electron and laser. For the purpose, we renewed the water cooler of the resolution of 0.01 °C for the RF gun and tube, and the air conditioner of the resolution of 1 °C in the linac room. Even in much a temperature controlled environment, concrete walls and the cacuum pipe are deformed, which attribute the drift. We filled the pipe by N_2 gas in order to avoid its deformation of the pipe, especially the bellows parts, by air pressure. Finally, the timing jitter was measured to be 1.6 ps (rms) for 2 hours and 1.4 ps (rms) for 1 hour, as shown in Fig. 4.

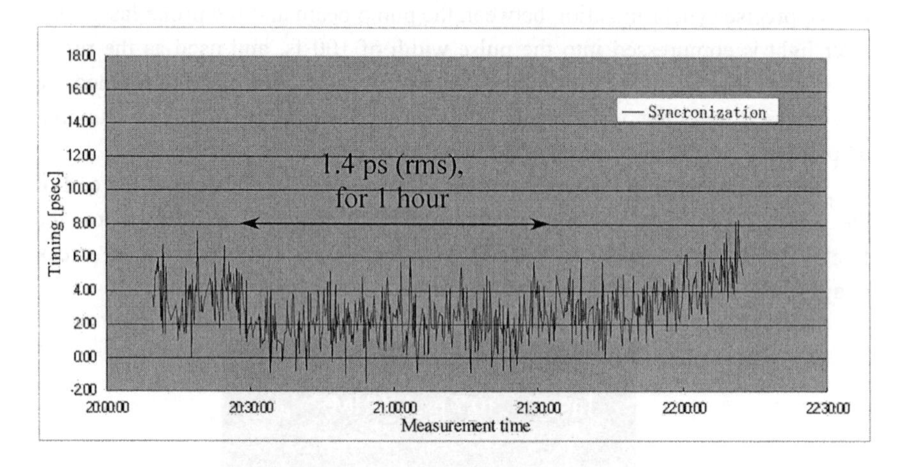

Figure 4. The synchronization result

At U. Tokyo, the transmission-type RF gun with a diamond cathode is being developed to generate high quality and dense electron beam.

2. Sumitomo Heavy Industries

A technique of laser pulse shaping was developed for low-emittance electron beam generation by Sumitomo Heavy Industries [4-6]. The emittance growth due to space charge and RF effects in the RF gun was experimentally investigated with square and Gaussian temporal laser pulse shapes. The temporal pulse shaper was accomplished through a technique of frequency-domain pulse shaping. The spectrum of the incident femtosecond laser pulse was dispersed in space between a pair of diffraction gratings separated by a pair of lenses (Fig. 5). A computer-addressable liquid-crystal spatial light modulator (LC-SLM) with 128 pixels was used as the phase mask. The resolution of the phase shift on LC-SLM was near 0.01 π. The pulse shaper was located between the oscillator and the pulse stretcher to reduce the possibility of damage on the optics.

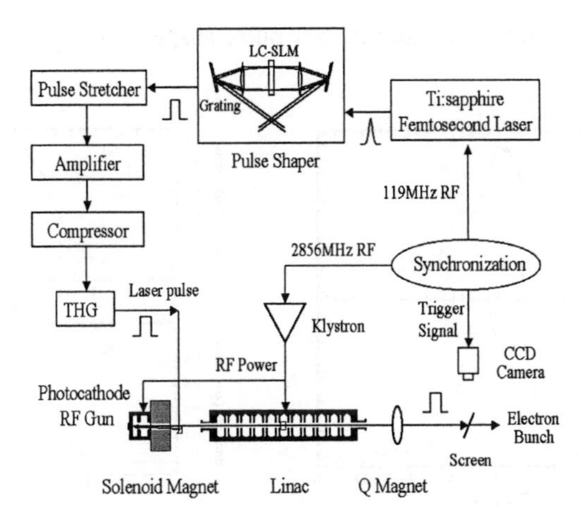

Figure 5. Experimental arrangement

The typical Gaussian and square-shaped temporal distributions of the UV laser pulses with a pulse length of 9 ps FWHM are shown in Fig. 6. The data was measured by an X-ray streak camera with a time resolution of 2 ps, resulting a rise time of 1.5 ps for the square pulse shape. The pulse-to-pulse fluctuation of the shaped pulse length was 7 %. The spatial profile of the laser beam on the cathode is shown in Fig. 7. The beam spot size was 1.2 mm and 0.4 mm FWHM in the horizontal and vertical directions, respectively. The normalized rms horizontal emittance measured as a function of the laser pulse length is shown in Fig. 8 for the Gaussian and square temporal pulse shapes. The electron bunch charge was fixed at 0.6 nC and the solenoid field was set to 1.5 kG which was optimal for compensating the space charge emittance at 0.6 nC. The data shows that the emittance increases at shorter and longer laser pulse length regions for both the Gaussian and square pulse shapes. This is behaved in emittance growth due to space charge and rf effects. The normalized rms horizontal emittance was also measured as a function of the bunch charge for the Gaussian and square temporal pulse shapes with a pulse length of 9 ps FWHM, as shown in Fig. 9. The measured data was fit as a function of

$$\varepsilon = \sqrt{(a'Q)^2 + b'^2}\ , \qquad (1)$$

where a' is a fitting parameter referred to space charge force, and b' in πmm-mrad is a zero charge emittance. It is found that the square pulse shape reduced the space charge force of about 50% comparing with the Gaussian pulse shape. Consequently, the optimal normalized rms emittance of 1.2 πmm-mrad at 1 nC was obtained by a square temporal laser pulse shape with a pulse length of 9 ps FWHM.

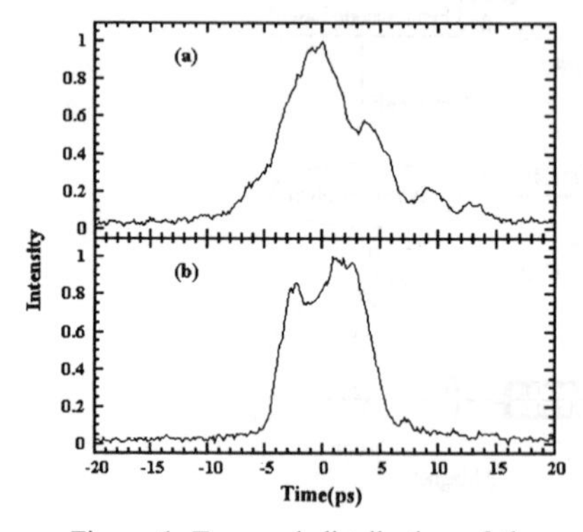

Figure 6. Temporal distribution of the Gaussian (a) and square (b) laser pulse.

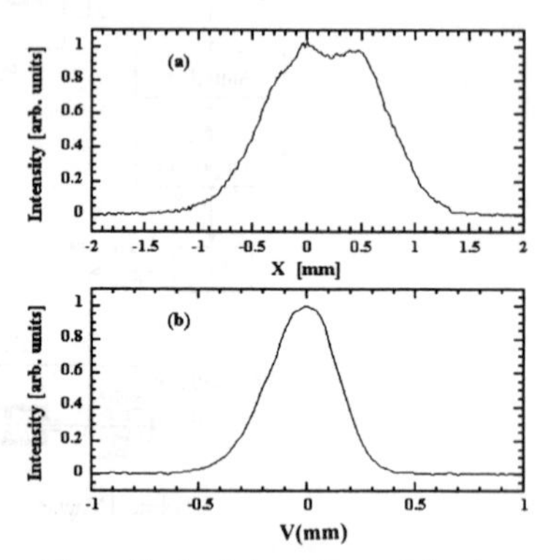

Figure 7. Spatial profiles of the laser beam in horizontal (a) and vertical (b) directions.

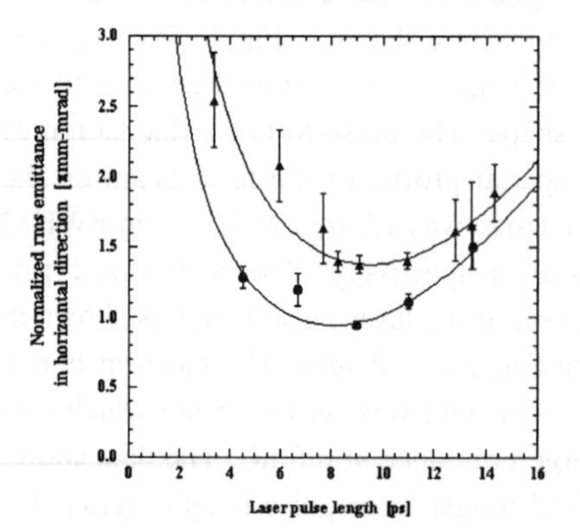

Figure 8. The emittance versus laser pulse length at 0.6 nC for Gaussian (triangle) and square (dot) pulse shapes.

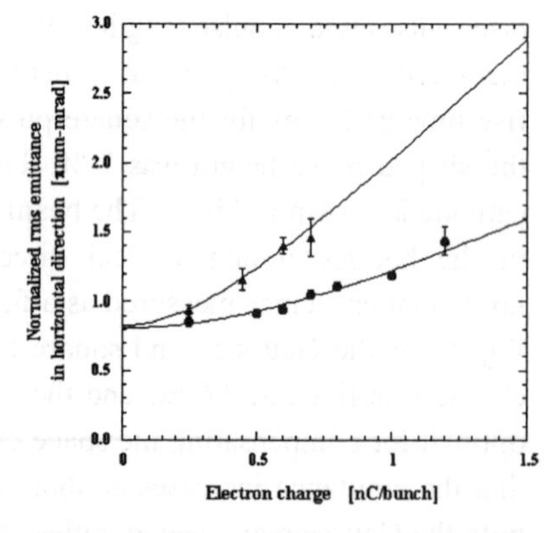

Figure 9. The emittance versus bunch charge for the Gaussian (triangle) and square (dot) pulse shapes at a pulse length of 9 ps FWHM.

3. Photo-Needle RF Gun

A laser-excited RF-gun using a tungsten needle photocathode (Fig.10a) was proposed. Design studies of gun performance with numerical calculations and a preliminary experiment of a needle-RF-gun were performed. A tip radius of a needle a low energy-spread and a relatively low field-emission current (Fig.10b). The results of design studies for our 15MeV linac (LEENA) RF-gun indicate that the optimum tip radius is 8.8 μm, and a peak photocurrent of 92 A with a pulse width of ~ 6 ps and an energy-spread of ~ 1 % can be obtained using a mode-locked Nd:YLF laser which has a wavelength of 351 nm, a micropulse duration of 10 ps, a peak intensity of 330 MW/cm^2, and an initial phase of 20 deg. In this design calculation, an enhanced quantum efficiency of photo emission due to the high electric field [7] is assumed.

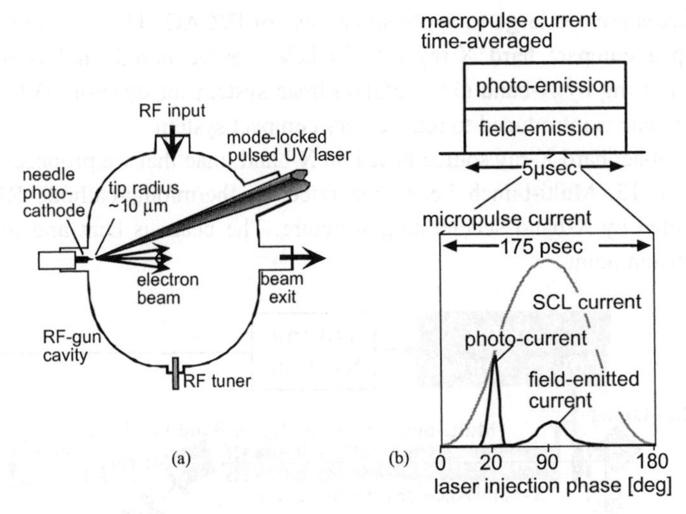

Figure. 10 Schematic view and scenario of photo-current emission
(a) Schematic of a laser-excited needle-RF-gun system.
(b) Field emission components in macropulse current and in micropulse current.

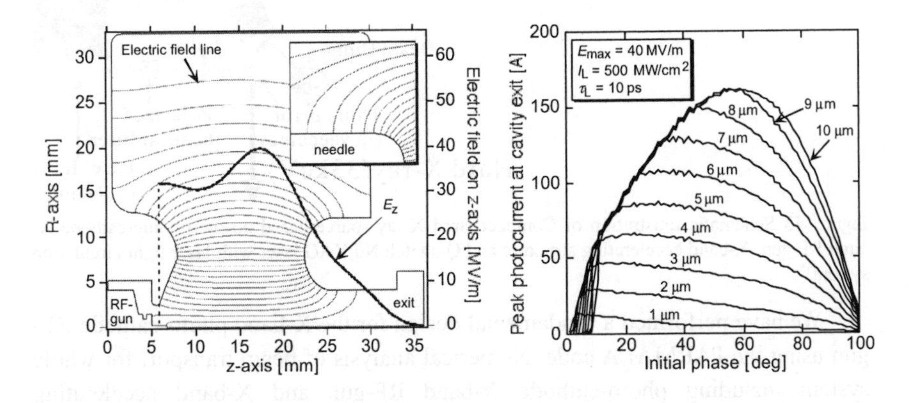

Figure 11. The calculated electric field distribution and the field-strength on the z-axis without a needle.

Figure 12. Calculated peak photo-current at the cavity exit.

4. RF GUNS FOR COMPACT HARD X-RAY SOURCE FOR ANGIOGRAPHY

Hard X-rays of 10~50keV are now very useful for medical science, biology, material science etc. For example, Dynamic Intravenous Coronary Arteriography (IVCAG) by a high quality monochromatic hard X-ray via Synchrotron Radiation (SR) is proposed and tested in some institutes. Most of SR sources are too large to apply spread use of IVCAG. Then, we are going to develop a compact hard X-ray (10~50 keV) source based on Laser-electron collision using by X-band (11.525GHz) linac system for dynamic IVCAG. The X-band linac is introduced to realize very compact system.

Compact hard X-ray source based on X-band linac that we propose is shown in Figure 13. Multi-bunch beam generated by thermionic-cathode RF-gun is accelerated by X-band accelerating structure. The beam is bent and focuses at the collision point.

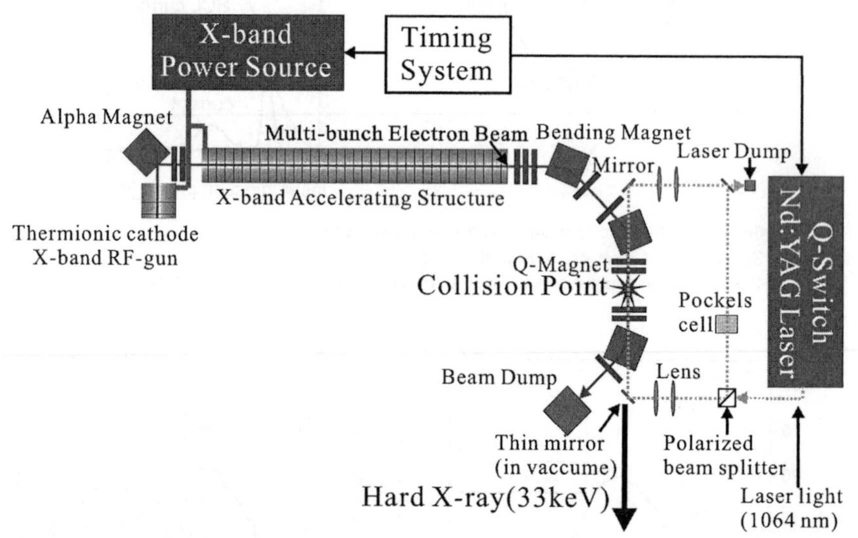

Figure. 13 Schematic illustration of Compact Hard X-ray source based on thermionic-cathode X-band RF-gun, X-band accelerating structure and Q-switch Nd:YAG laser with laser light circulation system.

We have performed a fundamental design for the X-band photo-cathode RF-gun using the PARMALA code. Numerical analysis of beam transport for whole system including photo-cathode X-band RF-gun and X-band accelerating structure is already presented[8]. Thermionic-cathode X-band RF-gun is under designing, and We assume beam parameter shown in Table 1.

Beam energy	56 MeV
Charge/bunch	20 pC
Bunch length	1.5ps(FWHM)
Beam Emittance	2.5, 2.5 πmm mrad

Table 1. Beam parameter.

To concentrate on R&D of the accelerator, we use existing laser system for laser-electron collision. To realize simple and compact system, we apply a Q-switch Nd:YAG laser with intensity 2J/pulse, repetition rate 10pps, pulse length 10ns(FWHM), and wavelength 1064nm, which is commercial product.

We choose a very simple system by focussing on only averaged X-ray flux. We construct the system with the thermionic-cathode RF-gun (20 pC/bunch, ~10^4 bunches/RF-pulse, 10paimm mrad) and Q-switch Nd:YAG laser (2J/pulse, pulse length 10ns in FWHM, repetition rate 10pps). We assume head-on collision for calculation of X-ray yield. Figure 14(left figure) indicates that optimal laser beam size is 82μm (rms) for electron beam size 100μm (rms) at the C.P. Each bunch collides to laser light with some time offset. X-ray yield of each bunch is shown in Figure 14(light figure). Thus, This system generates X-rays with $1.7x10^7$ photons/pulse ($1.7x10^8$ photons/s) that is sum of each bunch.

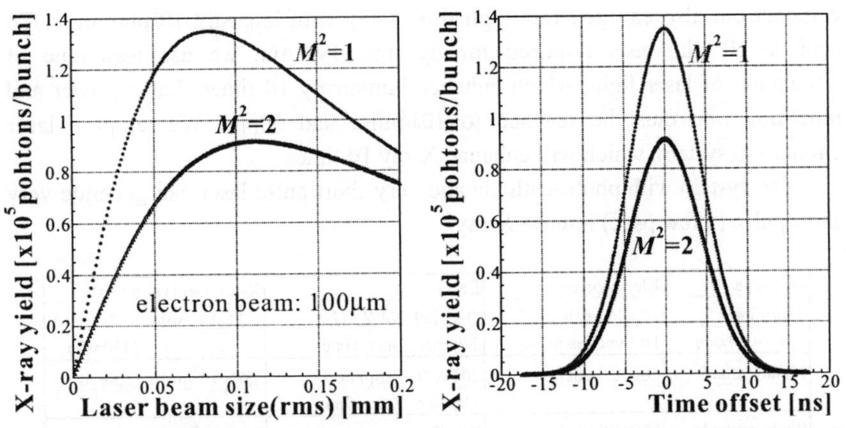

Figure 14. X-ray yield due to laser beam size at C.P.(left figure) and time offset between laser and electron beam (right figure).

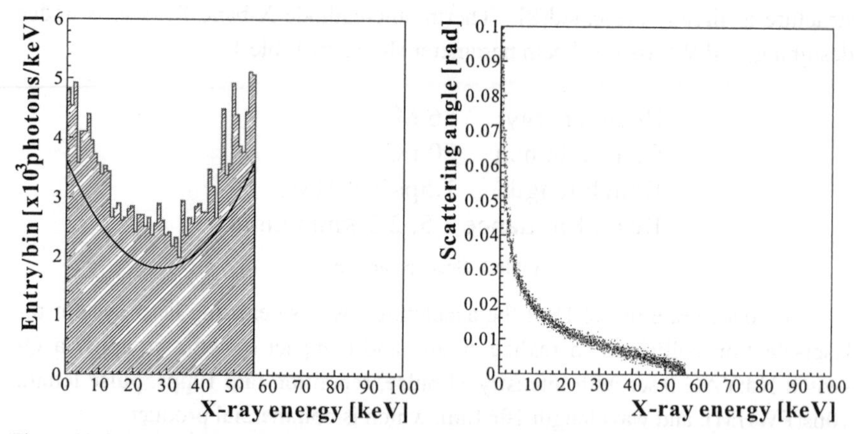

Figure 15. Energy spectrum and energy (left figure) vs. scattering angle (right figure) of X-ray in single bunch(20pC/bunch) collision with Q-switch Nd:YAG-laser(2J/pulse) . Scattering angle of 0 rad is direction of electron beam.

Energy distribution and angler distribution of generated X-ray is shown in Figure 15. Solid line indicates spectrum calculated by Klein-Nishina's formula[9] and Luminosity calculation, and Histogram shows the result of beam-beam interaction Monte-Carlo simulation code CAIN[10]. X-ray energy reached to 57keV at beam energy 56MeV.

X-ray yields of various laser system is summarized in Table 2. The system with Thermionic-cathode and Q-switch laser is not only very simple and compact but also can generate high flux X-ray with intensity 10^8photons/s. To achieve 10^{11}photons/s required for dynamic IVCAG, we use technique of circulation of laser light, which enhance luminosity 10 times. Laser power and repetition rate must be reached to 10J/pulse and 50pps. We adopt a laser circulation system, which will enhance X-ray 10 times.

The system with photo-cathode and very short pulse laser can generate very short pulse (a few psec) intense X-ray.

Gun type	Electroon beam	Laser	X-ray yield (photons)
Thermionic -cathode	20pC/bunch 10^4bunches/pulse	Q-switch Nd:YAG 2J/pulse, 10ns, 10pps	1.7×10^7/pulse (1.7×10^8/s) (430nW)
Photo-cathode	500pC/bunch	20TW Ti-Sapphire 1J/pulse, 50fs, 10pps	1.6×10^7/pulse (1.6×10^8/s)
Photo-cathode	500pC/bunch	Nd:Glass 10J/pulse, 10ps, <<1pps	2.1×10^8/pulse
Photo-cathode (Multi-bunch)	500pC/bunch 20bunches/pulse	15nJ/bunch+Super-cavity $(15\times N)$nJ/bunch, 7ps	$(6.3\times N)$/pulse

Table 2. X-ray yields of various laser system

Final target of this work is the integrated system for dynamic IVCAG shown in Figure 16. This system has X-band RF-source and moving arm including X-band linac, Q-switch Laser system with laser circulation system and X-ray detector. We can perform dynamic IVCAG easily and can get clear dynamic image of coronary artery with less distress for patients.

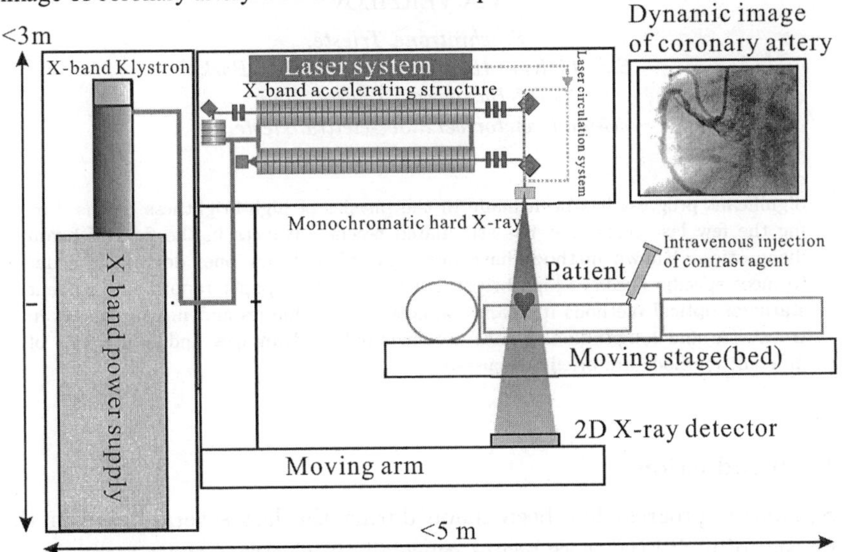

Figure 16. Final target of this work.

References

1. T. Srinivasan-Rao et al., *Rev. Sci. Instrum.* **69** 2292 (1998).
2. H. Iijima, et al., Proceedings of EPAC2002, 3-7 Jun. 2002, Paris,France.
3. M. Uesaka et al, Trans. Plasma Sci. 28 (2000) pp.1133-1142
4. J. Yang et al., *J. Appl. Phys.* **92** 1608 (2002).
5. J. Yang et al., Proceedings of EPAC2002, 3-7 Jun. 2002, Paris,France.
6. J. Yang et al., *Nucl. Instrum. Meth. A*, to be published.
7. T. Inoue, S.Miyamoto, S.Amano, M.Yatsuzuka, T.Mochizuki, submitted to Jpn. J. Appl. Phys. "Enhanced Quantumefficiency of Photocathode under High Electric Field".
8. A. Fukasawa, et al., Proceedings of 2nd Ashian Prticle Accelerator Conference, Sep. 17-21, Beijing, Chaina.
9. V. O. Klein and Y. Nishina, Z. Phys. 52.858,869(1928)
10. K. Yokoya, CAIN2.1e, contact Kaoru.Yokoya@kek.jp

OPTICAL DIAGNOSTICS OF HIGH-BRIGHTNESS ELECTRON BEAMS

V.A.VERZILOV

Sinchrotrone Trieste,
S.S. 14 Km. 163,5 -in Area Science Park
34012 Basovizza, Trieste, Italy
E-mail: victor.verzilov@elettra.trieste.it

Significant progress has been made in the physics of high-brightness beams during the few last years that has stimulated recent advances in the field of beam diagnostics. Known methods have been revised and new ones devised in order to meet specific needs of high-brightness beams. This paper reports the current status of optical methods in beam diagnostics. Techniques and measurements in transverse and longitudinal planes are reviewed. Advantages and limitations of different methods are briefly discussed.

1. Introduction

Significant progress has been made during the last several years in the physics of high-brightness beams especially in electron source technologies such as photocathode RF guns [1]. These new sources are capable of generating electron beams with extremely high particle densities in 6D phase space. Such beams are generally characterized by a relatively high charge per bunch combined with ultra-small transverse and longitudinal emittances that are achievable by control of the dynamics of the beam from the very first moment of photo-emission. To take full advantage of new electron sources in high-brightness beam applications the quality of the beam must be preserved whilst being accelerated, transported and manipulated (e.g. bunch compression).

This goal is impossible to reach without adequate beam diagnostics, which have to be sensitive to superior beam parameters and take into account specific properties associated with high-brightness beams, e.g., dense phase space populations lead to beam dynamics dominated by space charge forces up to energies of hundreds of MeV. Working experience accumulated up to date reveals that desired high-brightness beam diagnostic is required to satisfy a set of specific needs:

(i) resolution from several millimeters to a few micrometers in both transverse and longitudinal planes
(ii) large dynamic range both in terms of beam intensity and measuring interval
(iii) be non-invasive
(iv) be single shot capable
(v) be jitter-free and synchronized to the beam.

Known methods have been revised and new ones are devised in order to meet if not all at least some of these requirements. There exists a variety of different techniques, which are currently in use or considered promising, to control quite a long list of high-brightness beam parameters that are essential for commissioning and operation of an accelerator facility. This paper does not intend to give a comprehensive overview of a complete diagnostic system. Instead it reports the current status of optical diagnostics of high-brightness electron beams. Several reasons exist to concentrate on the optical methods since they are an essential part of the available tools. Furthermore this is the area where an enormous step forward has been made during the past few decades and has been done specifically in order to meet the needs of high-brightness beam diagnostics. Although there does not exist any accepted classification, one can define optical diagnostics as those based on analysis of photons generated by a beam in related processes or that make use of other optical methods and devices, such as lasers.

2. Transverse measurements

Having limited ourselves to optical methods, we obviously narrow a class of considered measurements to those were they are involved. The largest amount of information about a beam is extracted from its transverse and longitudinal profile measurements. These measurements are valuable in their own right or are an integral part of other techniques such as emittance or energy spread measurements. In the transverse plane a key issue is a resolution, since transverse dimensions can easily reach sub 100 μm levels depending on beam optics and energy. A series of theoretical and experimental studies have been performed recently on widely exploited devices like optical transition radiation (OTR) monitors and inorganic crystal scintillators.

The major advantage of OTR monitors is excellent time resolution $\sim$

λ/c. This allows one to resolve individual bunches in a bunch train and in principle go inside a single bunch which is not possible even with the best scintillators. Theoretical studies of the spatial resolution of OTR's were performed by several authors [2,3,4]. Similar to a classical optical system, the resolution of a TR monitor is determined by its angular acceptance θ and wavelength λ. However, if the intrinsic resolution of the optical system is determined by the so-called point spread function, the analogous function for TR vanishes at the center if a single particle is imaged. This makes the TR profile monitor fwhm resolution 3 times larger than the resolution of the optical system itself $\sim 0.51\lambda/\theta$. Nevertheless, with large acceptance optical systems, resolution as small as a few micrometers in the visible is realistic. If the fwhm resolution is of no concern, the rms one is, at

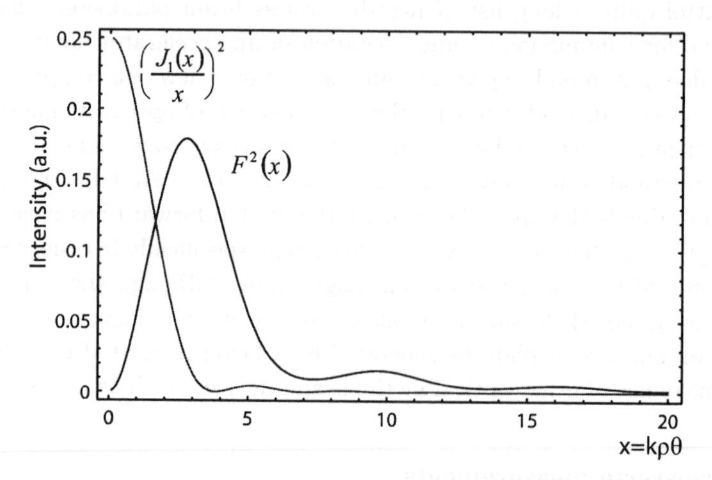

Figure 1. Point spread function $(J_1(x)/x)^2$ characterizes resolution of an ideal optical system. Function $F(x) = K_1(x)/\gamma\theta - J_0(x)/x$ describes the TR intensity distribution given by the image of a single particle. $\gamma\theta = 100$.

least at high energies. The long tails of the OTR intensity distribution in the image plane extend up to $\sim M\lambda\gamma$ (γ is the relativistic factor and M a optical magnification) and carry a significant part of the energy making rms measurements strongly dependent on the characteristics of the measuring device, i.e., dynamic range, threshold levels, etc.. Since tails are mainly created by photons emitted at small angles, it was suggested [3] to use an opaque mask to prevent them from passing through the optics. Excellent resolution of OTR was confirmed at high beam energies [5,6] and numerous low-energy experiments.

The major drawback of OTR is a relatively low photon yield that limits its application in some particular measurements like pepper-pot emittance measurements, where only a small fraction of the beam really interacts with a monitor. Providing much more light, new inorganic crystal scintillators, such as YAG:Ce, appeared to be a promising alternative to OTR monitors [7]. They remove the main constrains of traditional scintillators: grain structure, show good conformance to high vacuum and resistance to radiation damage. The response time can be as short as $\sim$ 100 ns. However, recent experiments demonstrated a big discrepancy between the beam sizes measured with OTR and wire scanning techniques, compared to those measured with scintillators. Differences in measuring beam size were found that ranged from 30-40% [8] to a factor of 2 [9]. This may indicate possible saturation effects for large charge densities [8] or evidence of collective effects exhibited by the beam itself [9] and limit the resolution of this type of monitor to a few tens of micrometers.

Both OTR monitors and scintillators are intercepting devices. For high-brightness beams of very small transverse dimensions, survival of equipment exposed to the beam becomes a serious problem. The temperature rise occurs nearly instantaneously and when localized in a small volume causes material heating leading to degradation of mechanical properties and even breakdown if thermal stress or melting limits are reached. The problem is more severe for bunch trains, although cooling due to thermal irradiation and diffusion may result in a temperature balance. The problem of monitor survival is a critical issue for the near future large-scale projects like the LCLS and, especially, the TESLA FEL. Thus, detailed EGS4 calculations [10] performed for several types of materials indicate that TR monitoring of the full TESLA FEL beam is likely to be impossible.

Synchrotron radiation, a well-known non-invasive technique, has for obvious reasons limited application on linear accelerators. Alternatively, the laser wire method is considered a very promising tool offering sub-micrometer scale resolution [11]. Application of laser wires for beam diagnostics is not limited to high energies only. At ATF 90° Tomphson scattering has been used for transverse beam profile characterization at a relatively low energy of 50 MeV [12]. In this experiment a terawatt level laser beam was scattered against the electron beam. Measurement of the transverse distribution was performed by scanning in steps of 10 μm the 30 μm laser beam across the tightly focused (66 μm FWHM) electron beam. The obtained results are in a good agreement with OTR measurements. Laser techniques lack single shot capability since they require many beam pulses to register a profile.

Another non-intercepting technique suggests making use of angular characteristics of diffraction radiation (DR) to extract beam size information [13]. Though this technique is very attractive due to its intrinsic simplicity, it is likely limited to high energy applications and still needs more studies to mature.

3. Longitudinal measurements

Small electron bunches are crucial for a number of high-brightness beam applications: FEL's, compton sources and perhaps to a lesser extent linear colliders. To avoid emittance blow-up due to space charge forces, electron bunches are usually generated by the gun with a length of several millimeters. After which they are pre-accelerated to an energy high enough to suppress space charge effects and shortened by a series of bunch compression devices to sub-millimeter lengths and sub-100 μm dimensions to meet design specifications. Measuring such bunches is always a challenge. Conventional streak cameras become very expensive and often do not operate at this time scale. Several new techniques have been specifically developed for ultra-short bunch diagnostics.

The very powerful method that has nearly unlimited potential towards ever shorter bunches is based on measuring coherent radiation spectra. Though different types of radiation have been considered in the framework of this technique, coherent far-infrared transition radiation is probably of most use on linacs and has been since the early 1990's [14,15,16,17], thanks to the ease of implementation and the flat radiation spectrum. The idea of the method lies in the fundamental principle of electrodynamics: an ensemble of charges emits radiation coherently at wavelengths larger than the dimensions of the radiating system.

When considering radiation from a bunch of N particles, phase correlations between different particles have to be taken into account. As a result, the total intensity

$$I_{tot}(\omega) = NI_{sp}(\omega) + N(N-1)F(\omega)I_{sp}(\omega)\,, \tag{1}$$

consists of the incoherent (first term) and coherent (second term) parts. Here $I_{sp}(\omega)$ is a single particle spectrum. It is not always stressed that Eq. 1 assumes all particles to be identical with no angular and energy spread. The form-factor F

$$F(\omega)=\frac{1}{N(N-1)}\sum_{k}^{N}\sum_{j\neq k}^{N}e^{i(\omega/c)\vec{n}(\vec{r}_k-\vec{r}_j)}\,, \tag{2}$$

is a term summing up all possible phase pair correlations. In Eq. 2 $\vec{n}$ is a unit vector in the direction of photon emission and $\vec{r}_k$ is the vector that points from the origin to the position of the k th electron in the bunch. The form-factor is a measure of the degree of coherence and its value is determined by how the wavelength relates to the bunch length. It approaches unity when the wavelength is comparable or greater than the bunch length because all particles emit more or less in phase. Otherwise, all particles emit independently and the form-factor and therefore the coherent component of radiation vanish.

Analysis of the whole region of the transition from incoherent to coherent emission allows one, in principle, retrieve information on the longitudinal bunch profile. The most complete analysis involves both real and imaginary parts of the Fourier transform of the charge distribution in a bunch [18]

$$\rho(\omega)=\int_0^\infty dz\rho(z)e^{i(\omega/c)z}=\sqrt{F(\omega)}e^{i\psi(\omega)}\,, \tag{3}$$

where the phase $\psi(\omega)$ can be calculated by means of the Kramers-Kronig relation, provided $F(\omega)$ is determined over the entire frequency interval

$$\psi_m(\omega)=-\frac{2\omega}{\pi}\int_0^\infty dx\frac{\ln\left[\sqrt{F(x)}/\sqrt{F(\omega)}\right]}{x^2-\omega^2}\,. \tag{4}$$

In this case the shape of the bunch is given by

$$\rho(z)=\frac{1}{\pi c}\int_0^\infty d\omega\sqrt{F(\omega)}\cos\left[\psi_m(\omega)-\frac{\omega z}{c}\right]\,. \tag{5}$$

If the phase term is neglected, data analysis is reduced to the cosine Fourier transform, which necessarily results in a symmetric bunch shape, an approximation that could be good or bad depending on the real bunch shape. Equations 3-5 assume that all particles emit coherently in the transverse plane. The latter is not true if the scale of longitudinal modulations of the charge density is much smaller than the transverse dimension of the bunch, as it takes place for microbunches formed as a result of developing instabilities in the FEL process [19,20,21].

For existing accelerating facilities, wavelengths of interest in radiation spectra lie in the mm and sub-mm range. Several experimental techniques have been used so far to measure radiation at these wavelengths. Fourier spectroscopy is the most wide-spread method, which consists in measuring in the time domain the autocorrelation function of the radiation pulse with an interferometer [22]. Time domain data is coupled to the frequency domain by means of the cosine Fourier transform. Because of this coupling data analysis can be performed in either of the domains [23].

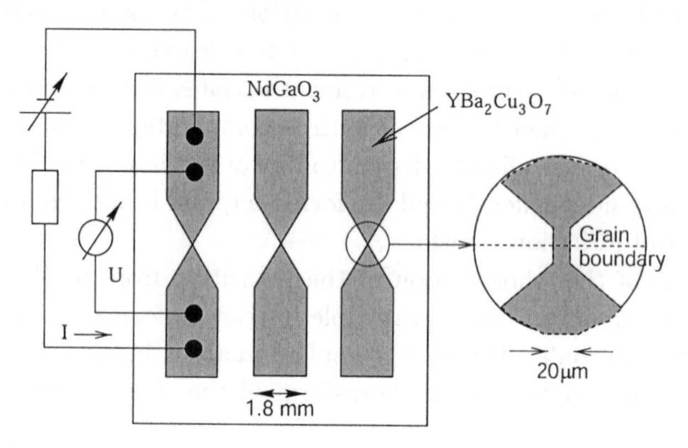

Figure 2. In the Josephson junction far-infrared spectrometer incident radiation changes the current created by cooper pairs penetrating through a thin insulating layer between two superconductors by the quantum mechanical effect.

An original spectrometer of radiation in the range 100-1000 GHz with a resolution of 1 GHz based on the Josephson junction effect was constructed and tested at TTF [24]. In this device, made of two superconductors separated by a thin insulating layer, cooper pairs penetrate from one superconductor to another by the quantum mechanical effect. Tunneled pairs create an electric current I through the junction that is changed by the incident radiation spectra $S(\omega)$ by a value

$$\Delta I = -\left(\frac{2e}{\hbar}\right)^2 \frac{R^2 I_c^2}{4I} \int_0^\infty \frac{S(\omega)d\omega}{\omega_s^2 - \omega_0^2} \,. \tag{6}$$

where $\omega_0 = 2eU/\hbar$ is the Josephson frequency, U the voltage across the junction and I_c the critical current through the junction. Measuring I and dI, $S(\omega)$ is retrieved using a Hilbert transform. A bunch length of 1.2 ps was measured in the experiment with this spectrometer.

Both considered techniques are indirect with at least one evident drawback: they need many pulses to measure a single spectrum. The most advanced device that can directly measure the radiation spectrum and, potentially, in a single shot is the polychromator, Fig. 3, developed at the university of Tokyo University [25]. In this device the incoming radiation is deflected by a grating to create a frequency-angle correlation. This generated correlation is resolved by the 10-channel-detector array. With an appropriate choice of grating, the device can be adjusted to measure the desired bunch length although in a very limited range. For the 1 mm pitch grating a bunch length of 0.9 to 1.6 ps was measured with a Gaussian fit to the data.

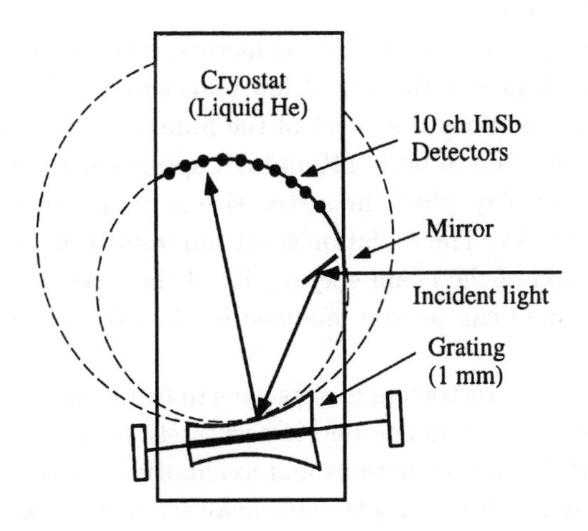

Figure 3. In the polychromator the radiation spectrum is measured by the 10-channel-detector array through the frequency-angle correlation generated by a grating.

There is also a non-intercepting analog of transition radiation: diffraction radiation, generated when a beam passes through an aperture made *ad hoc* in a special target, thus reducing interaction with the target material to a minimum. Radiation (a sort of wakefield) is produced as a result of diffraction on an aperture of virtual photons associated with the particle Coulomb field. Though the spectrum of DR is more complex than that of TR and data analysis complicated, this technique can be considered a good alternative for ultra-high power beams. DR was successfully applied for the short bunch diagnostic using hole [26] and rectangular [27] slit apertures. In the latter case, measurements of the bunch length were performed for a wide range (0-10 mm) of slit widths and showed very good agreement (within 2%) with TR data.

Main complication and uncertainty in the analysis of measured coherent spectra, independent of the nature of the measurement, is due to distortion. According to Eq. 1, the measured radiation spectrum is proportional to the product of the form-factor and the single particle spectrum, which is supposed to be well-defined by theory. In practice, unfortunately, there is a number of experimental factors which disturb measurements. In fact, all experimental data suffer to a different extent from the low-frequency cut-off. Possible reasons which cause the cut-off are the detector bandwidth, non-uniform spectral characteristics of optical elements in the radiation path, diffraction etc.

Apart from generic cut-off driving factors, there are more specific limitations for each kind of radiation. A few years ago it was anticipated [28] that under certain conditions the effect of the finite size of the emitting screen leads to a suppression of both TR and DR spectra at low frequencies. The effect comes into play when the screen size is comparable or smaller than the parameter $\sim \lambda\gamma$. The radiation spectrum from such a screen becomes a complex function of the beam energy, size of the screen and angle of emission. Spectral data can be also modified by the effect of the pre-wave zone [29].

Although all the distorting factors have to be taken into account in data analysis, they are often impossible to accurately compute or measure, thus provoking certain approximations and assumptions to be made on missing or corrupted data. For example, data analysis may consist in choosing an appropriate model for the bunch shape and introducing a filter function to account for the cut-off [23]. The model was originally applied to a Gaussian shape but can be easily extended to other bunch shapes as well [27].

As one can gather from the above discussion, the analysis of coherent spectra is, generally, a rather complex and time consuming procedure, so this kind of measurement can be hardly related to the "on-line" category. This limitation is eliminated in electro-optical sampling (EOS) . EOS is a new idea (at least in beam diagnostics) based on the Pockels effect. The effect is observable as a change of the polarization of light in a certain class of crystals when a strong external electric field is applied. Modulation of the polarization is proportional to the magnitude of the applied electric field E and the crystal thickness l:

$$\Delta\phi \propto (l/\lambda)E\,. \tag{7}$$

If an EOS crystal is placed near to the beam, the collective coulomb field of the particles is strong enough to cause an observable rotation of the polarization plane. There are two practical solutions to how EOS can be implemented. In the first scheme [30,31], a very short polarized laser pulse passes through an electro-optical crystal simultaneously with the bunch. The time dependence of the bunch electric field is sampled by varying the delay between the probe pulse and the transience time of the electron bunch and measuring the change in the polarization with an analyzer. Difficulties in this scheme lie in synchronization of the laser pulse and the beam with sub-picosecond accuracy.

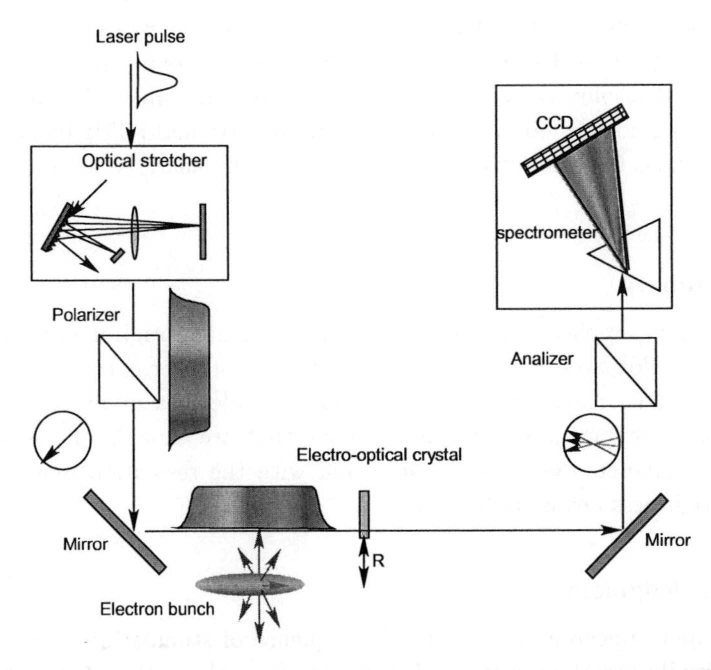

Figure 4. Single-shot EOS scheme. The bunch time profile is encoded by the EOS crystal into the polarization modulation of the stretched laser pulse with a linear frequency chirp.

The difficulty with synchronization is removed in a single-shot option where an initially short laser pulse is stretched using optical diffraction gratings to provide a long pulse with a linear frequency chirp. When the electric field of an electron bunch and the chirped optical pulse co-propagate in the electro-optic crystal, the polarization of various wavelength components of the chirped pulse are rotated by different amounts, corresponding to the magnitude of the local electric field. Thus, the time profile of the local electric field of the electron bunch field is linearly encoded onto the wavelength spectrum of the optical probe beam. An analyzer converts the modulation of the polarization of the chirped optical pulse into an amplitude modulation of its spectrum. The first very promising result using the single-shot option was obtained recently at the FELIX free electron laser facility in the Netherlands, where a piece of ZnTe crystal was placed at a distance of 1 mm from a electron beam and a 1.72 ps bunch length was measured [32].

Resolution in this method is determined by several factors but dominated by the chirp and width of the laser pulse. For the FELIX experiment it was as good as 300 fs. With the latest femtosecond laser technology a temporal resolution of $\sim$70 fs is realistically achievable. Though EOS has been demonstrated in various materials to have multi-THz bandwidth, contributions from non-linear effects still has to be analyzed at 100 fs time scale.

4. Summary

In summary, application of optical methods in high-brightness beam diagnostics allows one to obtain a large amount of information about the beam from its transverse and longitudinal profile measurements. A variety of techniques are currently available that are capable of measuring beam parameters over a wide range and with the resolution required by high-brightness beam applications.

Acknowledgments

This paper appeared as a result of a sequence of stimulating events that had initially been triggered by M. Ferrario whom the author is grateful for that.

References

1. X.J.Wang, Proc. of 2001 Particle accelerator conference, 81.
2. V.A.Lebedev, *Nucl. Instr. and Meth.* **A372**, 344 (1996).
3. X.Artru, R.Chehab, K.Honkavaara and A.Variola, *Nucl. Instr. and Meth.* **B145**, 160 (1998).
4. M.Castellano and V.A.Verzilov, *Phys.Rev. ST -AB* **1**, 062801 (1998).
5. J.-C. Denard et al., Proc. of 1997 Particle Accelerator Conference, 2198.
6. P.Catravas et al., Proc. of 1999 Particle Accelerator Conference, 2111.
7. W.S.Graves and E.D.Johnson, Proc. of 1997 Particle Accelerator Conference, 1993.

8. W.J.Berg, A.H.Lumpkin and B.X.Yang, Proc of XX International Linac Conference, 2000, 158.
9. A.Murokh, et al., Proc. of 2001 Particle Accelerator Conference, 1333.
10. N.Golubeva and V.Balandin, Private communication, 2001.
11. J.Frish, Web Proc. Nanobeam 2002, http://icfa-nanobeam.web.cern.ch.
12. W.P.Leemans et al., Proc. of 1997 Particle Accelerator Conference , 1984.
13. M.Castellano, *Nucl. Instr. and Meth.* **A394**, 275 (1997).
14. U.Happek, A.J.Sievers and E.B.Blum, *Phys.Rev.Lett.* **67**, 2962 (1991).
15. T.Takahashi et al., *Phys. Rev.* **E48**, 1479 (1993).
16. P.Kung, H.-chi Lihn and H. Wiedermannm, *Phys. Rev. Lett.*, **73**, 967 (1994).
17. W.Barry, Proc. of 1996 Beam Instrumentation, *AIP Conf. Proc.*, No. 390 (AIP, New York, 1997), p.173.
18. R.Lai and A.J. Sievers, *Phys. Rev.* **E50**, 3342 (1994).
19. J.Rosenzweig, G. Travish, A.Tremaine, *Nucl. Instr. and Meth.* **A365**, 255 (1995).
20. A.Tremaine et al., *Phys. Rev. Lett.* **81**, 5816 (1998).
21. A.H.Lumpkin et al., *Phys. Rev. Lett.* **86**, 79 (2001).
22. Hung-chi Lihn, , Proc. of 1996 Beam Instrumentation, *AIP Conf. Proc.*, No. 390 (AIP, New York, 1997), p.186
23. A.Murokh et al., *Nucl. Instr. Meth.* **A410**, 452 (1998).
24. M.Geitz., et al., Proc of 1998 European Particle Accelerator Conference, 1509
25. T.Watanabe et al., *Nucl. Instr. and Meth.* **A480**, 315 (2002).
26. Y.Shibata et al., *Phys. Rev.* **E52**, 6787 (1995).
27. M.Castellano et al., *Phys. Rev.* **E63**, 056501 (2001).
28. N.F.Shul'ga and S.N.Dobrovol'skii, *Pis'ma Zh. Eksp. Teor. Fiz.* **65**, 581 (1997) [*JETP Lett.* **65**, 611 (1997)]. N.F.Shul'ga, S.N.Dobrovol'skii and V.G.Syshchenko, *Nucl.Instr. and Meth.* **B145**, 180 (1998).
29. V.A.Verzilov *Phys. Lett.* **A273**, 135 (2000).
30. M.J.Fitch et al., *Phys. Rev. Lett.* **87**, 034801 (2001).
31. X.Yan et al., *Phys. Rev. Lett.* **85**, 3404 (2000.
32. I.Wilke et al., *Phys. Rev. Lett.* **88**, 124801 (2002).

PLANS FOR THE LASER SYSTEM OF THE SPARC TEST PHOTOINJECTOR

C. VICARIO, A. GHIGO AND F. TAZZIOLI
INFN-Frascati National Laboratory.
Via E. Fermi 40,
00044 Fascati, Rome, Italy
E-mail: carlo.vicario@lnf.infn.it

S. CIALDI AND I. BOSCOLO
INFN Dipartimento di Fisica Università di Milano
Via celoria 16, 20133 Milano Italy

The laser system proposed for SPARC photoinjector is described in this work. Lasers driving high brightness electron sources have to show specific performances: high pulse energy, uniform temporal and spatial distribution and low energy jitter from pulse to pulse. The emitted pulses have to be synchronized with a master oscillator within 1 ps time jitter to extract electron at exact phase of RF wave. Laser parameters tolerances have been studied with the aid of Parmela and Homdyn simulation code.

1. Introduction

The SPARC photoinjector[1] is required to produce a single bunch with 150 MeV energy, 1nC charge, 100 fs duration and with a normalized emittance of 1 mm-mrad. The laser system for SPARC electron source is required to deliver a 500 μJ pulse of UV photons to the photocathode at a repetition rate of 10 Hz. To meet the emittance requirements of the source, the laser pulse must have an adjustable pulse length, rectangular temporal shape and uniform transverse profile: nominally a hard edge at 1.0 mm radius. Finally, energy stability and pointing stability are important operational demands; the timing stability is also crucial to meet the energy stability requirements in the undulator. Simulations using Homdyn and Parmela have demonstrated that the emittance growth is still acceptable for 1 ps time jitter and 1 ps pulse rise-fall time. The pulse-to-pulse energy jitter tolerance is 5% that corresponds to 2% at the fundamental frequency.

Table 1. Laser minimum requirements.

Parameter	Requirement
Operating Wavelength	260 nm
Pulse energy on cathode	500μJ (Q.E.=10^{-5})
Energy jitter (in UV)	5 % rms
Temporal pulse shape	Uniform (10% ptp)
Transverse pulse shape	Uniform (10% ptp)
Pulse rise time (10-90%)	< 1 ps
Pulse length	2-10 ps FWHM
Repetition rate	10 Hz
Laser-RF jitter	< 1ps rms
Spot diameter on cathode	Circular 1 mm radius
Spot diameter jitter	1% rms
Pointing Stability	1% diameter rms

2. Laser system description

To obtain most of the performances summarized in the table, the third harmonic of commercially Ti:Sa can be used. Pulse shaping technique and energy stabilization need some dedicated R&D. The oscillator, pumped by CW frequency doubled Nd:YAG laser, is chosen to deliver 100 fs pulse, 780 nm wavelength, 79+1/3 MHz repetition rate that corresponds to the 36 sub-harmonic of S-band RF accelerating frequency. The oscillator pulse train has to be synchronized with an external master clock, using a feedback loop to limit the time drift within one RF period phase degree. The synchronization system has to control the laser oscillator repetition rate, changing the cavity length, and the phase between laser and accelerating field. To improve the time stability of the system it is useful to control the temperature of the environment. The energy produced by the oscillator is tens of nJ so it is necessary to make use of an appropriate number of amplifier stages. The energy delivered by the entire system has to be close to 20 mJ. In fact the frequency tripling, the spatial shaping, the diagnostics and the transport line to the photocathode have been estimated to introduce energy losses up to 97.5%.

To produce the required energy two amplifiers are needed. The better design that guarantees an easier alignment is a regenerative amplifier plus a multipass stage. Green pump lasers excite these two amplifiers. Pulse to pulse energy jitter produced by commercial high-energy multipass amplifier is larger than 3%.

In amplifiers for sub-picoseconds pulses, the peak power must be limited to avoid damage to optical components. Chirped pulse amplification is used to

reduce the peak power in the amplifier. The large bandwidth of the Ti:Sa oscillator permits the pulse to be stretched to hundreds of picoseconds using the dispersive region between a pair of gratings. After amplification, the process can be reversed to compress the pulse to the original or any greater width. A pinhole and a position-dependent attenuation filter are inserted after the compressor to obtain the uniform transverse profile. The photon frequency has to be tripled by a third harmonic generator. The efficiency of third harmonic conversion is very low for large bandwidth pulse at the most 10%. Initial energy jitter and ripples in time or space can grow during the conversion process, due to the nonlinearity. This means that the constraints on the spatial and temporal uniformity and energy stability before conversion will be more severe than those required at the photocathode.
Just before the gun, a grating and a cylindrical lens compensate the time skew and the elliptical spot. These effects are induced by the 72-degree incidence on cathode. The simulations show that 20% of ellipticity determines the doubling of emittance. With conventional lens it is possible to compensate completely this effect if laser beam position and beam spot remain stable.

3. Time Pulse Shaping

The Ti:Sa oscillator's large bandwidth allows the pulse to be shaped in time by manipulating its Fourier transform. In principle an arbitrary shape can be reconstructed with resolution equal to the input pulse length. To shape a given input pulse to an arbitrary temporal waveform both the spectral amplitude and the phase have to be modified. We propose to adopt the system in which the pulse shaper is placed before the amplifier to avoid the high-energy insertion losses and the optical damage of the modulation device itself. However in this layout only phase modulation can be adopted to preserve stretcher efficiency. In fact the amplitude modulation of the frequency spectrum to obtain a 10 ps flat top pulse cuts 95% of the original band. With phase-only modulation only the intensity profile of the output pulses can be specified, but it is the most important factor since the phase profile has no consequence in electron photoemission.
A liquid crystal mask in the Fourier plane of the 4-F non-dispersive optic arrangement can be used for phase manipulation[2]. Liquid crystals act as pixels that introduce a controlled phase delay on different portions of spectrum spatially dispersed by two gratings. For phase modulation the use of collinear acousto-optic TeO_2 modulator has been recently demonstrated[3]. The phase modulation is achieved by varying the path of different frequencies. A chirped

acoustic wave couples with different optical frequencies at different positions. The matching rotates the ordinary axis linear polarization toward the extraordinary axis. The refraction index along the extraordinary axis is different from that along the ordinary one and thus the phase delay is modulated. The acousto-optic modulator performs a continuous frequency modulation and can manage a bandwidth up to 200 nm, one order of magnitude larger than the capability of liquid crystals. The alignment of acousto-optic crystal is less critical than liquid crystal mask. Both liquid crystal mask and acousto-optic collinear device are computer addressable, so the phase function can be controlled by measuring the time profile after the frequency tripling to pre-compensate gain and conversion distortions.

4. Conclusion

Most of the laser requirements for high brightness electron sources are satisfied by commercial systems. The pulse shaping of high-energy pulse has to be investigated. The energy jitter that is achievable is higher than requested. Improvement in photocathode efficiency would simplify the problem.

Acknowledgments

We would like to thank Massimo Ferrario and Valeria Fusco for their simulations contribution to the present work.

References

1 L. Serafini and al, *An R&D Program For a High Brightness Electron Source at LNF* Proceeding of Epac 2002, Paris France.
2. A. M. Wiener, D. E. Leaird, J. S. Patel and J. R. Wullert *Opt. Lett.* **15**, 326 (1990).
3 F. Verluise and al, *Opt. Lett.* **25**, 572 (2000).

WORKING GROUP B

Bunch Compression

WORKING GROUP B SUMMARY : COLLECTIVE EFFECTS AND INSTABILITIES DURING ELECTRON BEAM PULSE COMPRESSION

S. ANDERSON
Lawrence Livermore National Laboratory, Livermore, CA, 94550, USA

T. LIMBERG
Deutsche Elektronen Synchrotron (DESY), 22607 Hamburg, Germany

In working group B, the implications that collective effects have on the process of pulse compression were discussed. Various physical effects, from space-charge to coherent synchrotron radiation were examined, in the context of diverse scenarios ranging from magnetic pulse compression to velocity bunching. A particular emphasis was placed on computational results, with some notable new results.

1 Session talks

In the first session, M. Gianessi reported on the status of CSR simulations with the 3-D tracking code TREDI. Improvements to the code removed problems with numerical noise. New results for the benchmark case of the ICFA workshop 2002 in Zeuthen were presented (Table 1):

Table 1: Results from Zeuthen benchmarking workshop with new numbers for TREDI.

		δE	σ_E	ε
3D	**TRAFIC4**	**-0.058**	**-0.002**	**1.4**
	TREDI*	**-0.001**	**0.001**	**1.85**
2D	**Program by R.Li**	**-0.056**	**-0.006**	**1.32**
1D	**Elegant**	**-0.045**	**-0.0043**	**1.55**
	CSR_CALC	**-0.043**	**-0.004**	**1.52**
	Program by M. Dohlus	**-0.045**	**-0.011**	**1.62**

The following conclusions were drawn:

- The agreement with other codes is improved with the new macro-particles model
- The CPU time is greatly reduced, this allows to run a larger number of macro-particles, but not yet sufficient to simulate microbunching.
- Fields regularization requires more work (still time consuming, not covariant)

P. Emma presented a “CSR Microbunching Study in the 'Benchmark' Chicane Using a 1-D Tracking Code”, where a flat-top bunch (beam parameters in Table 3) with an initial periodic intensity modulation is sent through a simple magnet chicane (parameters in Table 2) and the modulation amplification due to the CSR fields is calculated. The results in the form of gain curves for different emittance and energy spread values were shown. These results were compared to of analytic calculations using a 1-D model of P.Emma and E.Stupakov which is based on the original theory of Saldin et al..

F. Stulle also showed calculations on CSR instability gain, in this case done with the 3D-Code TraFiC4. He presented results are on total gain, and showed the development of the gain through the chicane. It is close to a two-step amplification process as assumed in the analytical model by Saldin et al.

Table 2: Parameters for simple magnet chicane

Chicane parameters	**symbol**	**value**	**unit**
Bend magnet length (not curved length)	L_B	0.5	m
Drift length (projected; B1-B2 & B3-B4)	ΔL	5	m
Drift length (B2-B3)	L_c	1.0	m
Post-chicane drift length (after B4)	L_f	2.0	m
Bend angle per dipole magnet	$\|\theta\|$	2.770	deg
Bend radius of each dipole magnet	$\|R\|$	10.35	m
Momentum compaction factor	R_{56}	–25.0	mm
2nd-order momentum compaction factor	T_{566}	+37.5	mm
Total projected-length of chicane	L_{tot}	13.0	m
Vertical half-gap of bend magnets	g	12.5	mm

Table 3: Beam parameters of modulation amplification calculations

Electron beam parameters	**symbol**	**value**	**unit**
Nominal energy	E_0	5.0	GeV
bunch charge	q	0.5 & 1.0	nC
Incoherent rms relative energy spread	$(\Delta E/E_0)_{u\text{-rms}}$	2.0	10^{-6}
Linear energy-z correlation	a	+36.0	m^{-1}
Total initial rms relative energy spread	$(\Delta E/E_0)_{rms}$	0.720	%
Initial rms bunch length	σ_{zi}	200	μm
Final rms bunch length	σ_{zf}	20	μm
Initial rms norm. emittances	$\gamma\varepsilon_{x,y}$	1.0, 1.0	μm
Initial beta-functions at 1st bend entrance	$\beta_{x0,y0}$	40, 13	m
Initial α-functions at 1st bend entrance	$\alpha_{x0,y0}$	+2.6, +1.0	

P. Piot followed with "Velocity bunching at a DUV-FEL", reporting on work carried out with L. Carr, W. Graves and H. Loos from Brookhaven national Laboratory. Good agreement between measurements and simulations are seen for achieved bunch shortening and longitudinal profiles.

G. Stupakov discussed "Wakefield due to roughness in a pipe with rectagular cross section". He gave the following motivation:

- In a recent paper, A. Mostacci et al. Calculated the wakefield for a rectangular waveguide with corrugated walls. The result – loss factor proportional to δ/a does not agree with what one expects from the round pipe model, earlier studied by K. Bane and A. Novokhatski (BN).
- The result of this paper was used to estimate the roughness impedance of the LCLS, with the conclusion that "the results differs by 2 orders of magnitude "from BN calculations.
- It was also used to estimate the impedance of the LHC beam screen.
- We do not discuss here if this is a good model for the real roughness.

The following conclusions were made:

- We found synchronous modes and calculated the loss factors for the waveguide of rectangular cross section with 2 corrugated walls. Our result for the loss factor is a factor of $\sim w/\delta$ larger than published by A. Mostacci et al.
- By order of magnitude, it agrees with the case of the round pipe (w,q _ pipe radius). It also agrees with the problem where the corrugation is imitated by a thin layer of dielectric coating.

The next topic was a paper of Geloni, Botman, Luiten, van de Wiel and Dohlus, Saldin, Schneidmiller and Yurkov: "Transverse self-fields within an electron bunch moving in an arc on a circle". The paper makes a statement on the cancellation between centrifugal and longitudinal forces acting on particles induced by coherent synchrotron radiation (CSR).

A slide of M. Dohlus' talk at the ICFA workshop in Zeuthen was shown, that displays, for the case of a 3-D treatment, that the cancellation cannot be perfect. Kwang-Je Kim and Zhirong Huang opened the next days session with presentations of their analytic work on CSR: "CSR Microbunching: Derivation" and "CSR Microbunching: Application". K.-J. Kim presented the theory of the model and concluded that it

- is a compact derivation of HKS (Heifets, Krinsky and Stupakov) paper
- treats the energy modulation spectrum
- agrees for high gain with the SSY (Schneidmiller, Saldin and Yurkov) theory

Z. Huang solves an integral that, given any initial condition (density and/or energy modulation) determines the final micro-bunching. The result is expressed as the gain=$b_{final}/b_{initial}$. He estimated for the LCLS case the initial bunch modulation from shot noise and pointed out that with a gain less than 100, this should be a small effect and reminded us to watch out for numerical noise in simulations.

G. Stupakov, in his talk "Beam microbunching in bunch compressor due to CSR wake" covered similar ground and concluded:

- A linear theory of CSR instability in bunch compressors has been developed that takes into account bunch compression, energy spread and the transverse emittance. The gain factor for a given bunch compressor is calculated by solving numerically the integral equation.
- Results show good agreement with elegant simulations for LCLS BC2, but not so good agreement for the benchmark BC.

M. Venturini: "CSR effects in storage rings" reviewed recent observations of CSR in electron storage rings in form of radiation bursts. He presented two case studies: A compact e-ring for a X-rays Compton Source and the Brookhaven NSLS VUV Storage Ring. He uses a parallel plate model of CSR impedance and calculates the beam dynamics with CSR in terms of 1D Vlasov and Vlasov-Fokker-Planck equations.

He works with linear theory on CSR-driven instability and involves numerical solutions of the VFP equation to take into account the effect of nonlinearities. He concludes that his model reproduces the main features of the observed CSR effect.

In the last session, S. Reiche presented "Comparison of the Coherent Radiation-induced Microbunching Instability in an FEL and a Magnetic Chicane", exploring similarities and differences in these two different regimes of CSR.

Applying FEL theory to a magnet chicane he can develop a 'low gain' model which shows a gain curve in agreement with the Heifets, Krinsky and Stupakov model for the first half of the chicane. However, it is difficult to incorporate incoherent energy spread into the model. In the high gain regime, he identified a dimensionless ρ parameter for CSR microbunching,that is analogous to the familiar FEL gain parameter, but is much larger than unity, indicating that the bunching occurs on a length scale smaller than a radius of curvature in the chicane magnets.

P. Emma reported finally about work carried out with M. Cornacchia: "Transverse to Longitudinal Emittance Exchange". The *concept* of the emittance exchange, uses a transverse deflecting cavity in a magnet chicane. Mr. Emma concluded:

- System potentially reduces transverse emittance
- Also increases longitudinal emittance, possibly damping the CSR instability
- Bunch length is compressed (all in last bend)
- Moves injector challenge to longitudinal emittance
- Must avoid CSR energy spread increase in 1st bends
- Scheme may have other uses not yet considered.

PHASE SPACE DISTORTIONS ARISING FROM MAGNETIC PULSE COMPRESSION OF HIGH BRIGHTNESS ELECTRON BEAMS

S. G. ANDERSON,* J. B. ROSENZWEIG, P. MUSUMECI, M. C. THOMPSON

UCLA Department of Physics and Astronomy
405 Hilgard Ave, Los Angeles CA 90095

We report detailed measurements of the transverse phase space distortions induced by magnetic chicane compression of a high brightness, relativistic electron beam to sub-ps length. A strong bifurcation in the phase space is observed when the beam is strongly compressed. This effect is analyzed using several computational models, and is correlated to the folding of longitudinal phase space. The impact of these results on current research in collective beam effects in bending systems, and implications for future short wavelength free-electron lasers and linear colliders are discussed.

1. Introduction

Future applications of very low emittance, high current — high brightness — relativistic electron beams in future high energy physics accelerators[1,2] or fourth-generation light sources[3] are dependent on compression of the bunch from many ps to sub-ps rms lengths. As the beams produced in the highest brightness electron sources, rf photoinjectors, are highly space-charge dominated[4], this limit on pulse length for a given charge, or equivalently, peak current, is due to collective, plasma-like behavior. Further, the beam's plasma frequency in the photoinjector is specified by geometric and external field considerations needed to obtain emittance compensation[5,6], and thus in practice one may always expect space-charge driven pulse lengthening. Compression of the beam after the injector is therefore necessary to obtain the currents needed for applications[2].

The most common method of pulse compression uses a magnetic chicane[7], which when combined with off-crest acceleration in the radio-frequency (rf) linear accelerators (linacs) used to boost the beam energy

*e-mail: anderson@stout.physics.ucla.edu

U, allows a rearrangement of the electrons' longitudinal positions. As the path lengths of highest momentum p electrons are shortest, a negative (z, p) correlation longitudinally focuses, or compresses the electrons. This process comes at a price, however, in that collective fields may severely distort the horizontal (x, bend plane) phase space. This distortion arises directly through transverse forces, or indirectly, when a longitudinal force changes the electron energy during bending, giving a betatron error through momentum dispersion mismatch. For ultra-relativistic ($\gamma = U/m_e c^2 \gg 1$, $v \simeq c$) electrons, the energy changes induced during the motion are expected to arise mainly from coherent synchrotron radiation (CSR). These energy changes may be so pronounced that a newly predicted microbunching instability develops[8].

Previous studies of collective effects during the chicane compression process have been carried out at CERN, where transverse emittance growth and changes in the momentum spectrum[9,10] were observed in the compressed beam. These studies were performed in the 40-60 MeV energy range, and the observed emittances and momentum spectra were compared to predictions from the simulation code TRAFIC4[11]. From this comparison, evidence for strong CSR emission was deduced, with the implication that a significant component of the collective effects in the experiments were due to acceleration fields. Other measurements of magnetic compression have been performed with higher energy beams at ANL, DESY and BNL[12], with CSR again playing a significant role in causing the observed emittance growth and longitudinal phase space distortions. It is notable that modeling of these experiments, which have increasingly concentrated on longitudinal phase space characteristics, has not reproduced some of the most striking aspects of the data. In particular, the severity of the beam's momentum spectrum modulation, and concomitant microbunching deduced from the emitted coherent radiation spectrum, has not been seen in simulation.

While these previous measurements have allowed tests of collective effects in high brightness beam compression, the distortion of the transverse phase space has been quantified only through examination of the rms normalized emittance, $\varepsilon_{n,x} = \beta\gamma\sqrt{\langle x^2\rangle\langle x'^2\rangle - \langle xx'\rangle^2}$, where $\beta\gamma = p/m_e c$ is the average normalized beam momentum. In the experiments reported here, the beam's horizontal phase space is directly sampled using a slit-based system, allowing an accurate, shot-by-shot phase space reconstruction. These reconstructions give an unprecedented view of the collective transverse dynamics of the beam as it undergoes compression, revealing a filamentary structure reminiscent of the strongly modulated longitudinal phase space in experiments mentioned above. Further, these measurements were performed with a lower electron beam energy (below 12 MeV), and thus explore a regime where velocity fields (space-charge) may play a dominant role. Such lower energy compression may be needed for experimental applications, such as Thomson scattering production of sub-ps x-ray pulses[13], which demand only moderate (20-50 MeV) final energies.

2. Experimental Methods and Results

The measurements presented here were performed using the rf photoinjector in the Neptune Advanced Accelerator Laboratory[14]. The photoinjector beam is created using a 1.6 cell rf photocathode gun[15]. The 0.25 nC, 4 MeV beam produced by the gun is then emittance compensated and accelerated further with a plane wave transformer (PWT) linac[16]. The beam exits the PWT with an energy near 12 MeV and the remainder of the photoinjector consists of the compressor and various beam diagnostics. The gun and PWT are independently phaseable, allowing us to use the PWT to impart the proper (z, p_z) correlation for compressing the beam.

The electron bunch is compressed using the compact chicane magnet system[14] illustrated in Fig 1. The compressor has $R_{56} = \partial(\delta z)/\partial(\delta p/p) =$

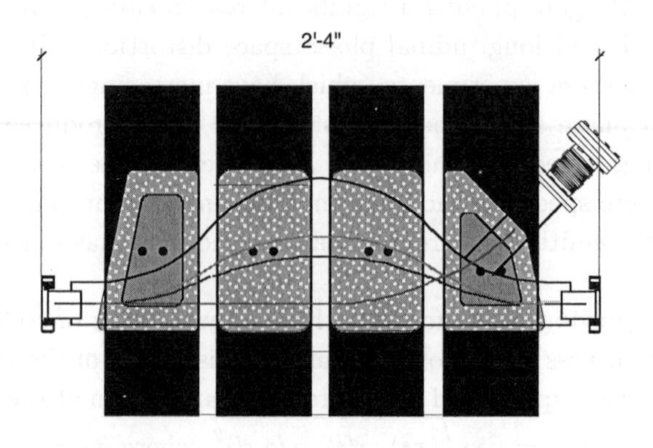

Figure 1. The Neptune chicane compressor.

3.5 cm (in TRANSPORT[17] notation) at the design 22.5° bend angle, which optimally compresses the beam when injected into the PWT at roughly $\theta = \tan^{-1} \lambda_{rf}/2\pi R_{56} = 25°$ ahead of the peak accelerating phase. Horizontally focusing edge angles are included at the entrance and exit of the chicane, to avoid excessive vertical focusing that would cause the beam to undergo a ballistic waist inside of the fourth magnet if the natural vertical focusing were not mitigated.

As the inner and outer sets of magnets in the chicane are identical pairs, the residual dispersion η_x, was minimized experimentally by using trim coils in the inner set to eliminate the beam horizontal (bend plane) centroid motion at a profile monitoring screen downstream of the chicane exit as a function of energy. The calculated normalized emittance growth attributable to the measured residual η_x is negligibly small, below 10^{-7} m-rad.

In order to study the beam's horizontal phase space (x, x') evolution for differing degrees of compression, we used used an array of diagnostics to measure the beam's bunch length and transverse phase space characteristics. The bunch length diagnostic we employed is a polarizing Michelson interferometer which analyzes coherent transition radiation (CTR)[18] emitted from the electron beam's impact on a metal foil[19]. The horizontal emittance of the beam is measured using a slit-based system[20,21,22] (1D pepper-pot), in which the beam is collimated by a slit-mask into narrow beamlets. This creation of beamlets both mitigates the strong space-charge forces and allows the reconstruction of the trace space and rms emittance of the original beam, as described in Refs. [21] and [22].

To obtain the second moments of the transverse phase space distribution, and thus calculate an emittance, we must first determine the true rms width of each beamlet. In the reconstruction of the phase space, however we wish to retain information about the structure of the beamlets. This point is especially relevant to the compressed beams produced in this experiment as the beamlets created by collimation of these beams have a great deal of structure. In fact, as we show below, for strongly compressed beams the beamlets can break into distinct parts and the intensity profile of a single beamlet may have separate peaks. In order to simultaneously account for the beamlet structure and compute rms beamlet sizes, we fit the beamlet intensity profile to a sum of Gaussians with positions x_j, amplitudes A_j, and widths σ_j. Then the mean square size of a beamlet is given by

$$\langle x_i^2 \rangle = \frac{\sum_j A_j \sigma_j \left[(x_j - \langle x \rangle)^2 + \sigma_j^2 \right]}{\sum_j A_j \sigma_j} \tag{1}$$

The phase space is then reconstructed as a set of x-axis positions defined by the slit separation and x'-axis curves given by the fit functions of the beamlets.

In order to evaluate the effects of compression on the horizontal phase space, we first determined the degree of pulse compression. Here the bunch length was measured at various PWT phases and chicane magnet settings. An example of these measurements is given in Fig 2, which shows the bunch length versus compressor magnetic field strength. The rms pulse length σ_z, was found to vary between 4 ps uncompressed and 0.6 ps at full compression, in agreement with both linear transport models and TREDI[23] simulations. The agreement between measured σ_z and simulation indicates that the compression system functioned as expected and provides a benchmark for parameters such as PWT linac phase and beam energy.

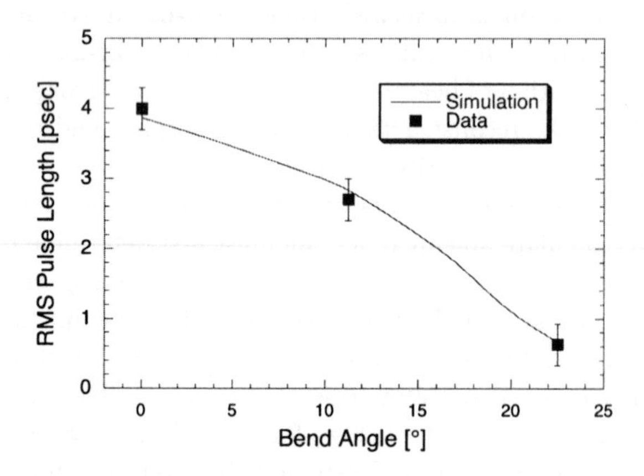

Figure 2. Measured rms pulse length versus compressor bend angle. The solid line shows the pulse length predicted by TREDI simulations.

The slit system was then employed to measure the phase space distribution and rms emittance of the beam versus bunch compression. This was accomplished by setting the bend angle to 22.5° and varying the PWT phase. Measurements holding the PWT phase constant and varying the chicane's R_{56} gave similar results.

Figure 3 shows an image of the beam passing through the slit-mask and the phase space reconstruction based on that image for a beam that does not compress in the chicane (when the PWT phase is set to minimize energy spread.) In Fig. 4 the PWT phase has been set to maximally compress the beam. Here the slit image shows that the beamlets clearly split into separate lines. This splitting indicates that the phase space of the compressed beam is bifurcated, as shown in the phase space reconstruction given in Fig. 4(b). In other words, distinct parts of a beamlet have the same position at the slit but differing mean angle, and drift to different positions at the detection screen. This splitting was observed to increase as the bunch length was shortened and had a dependence on vertical position within the beamlet; it was greatest in the vertical center of the beam, where the transverse space-charge forces are at a maximum.

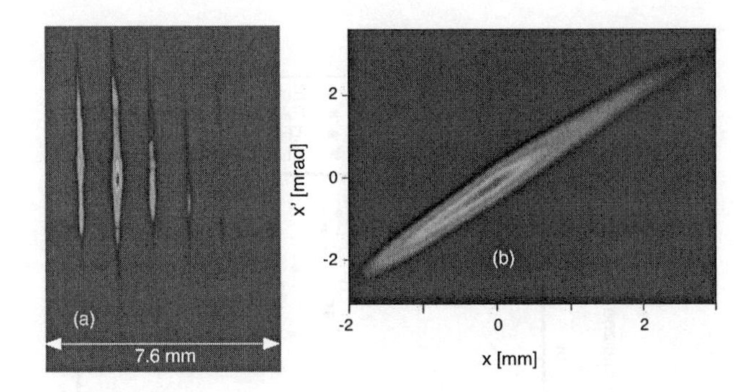

Figure 3. (a) Image of an uncompressed beam passing through the slit-mask. (b) Trace space derived from the image in (a).

Slit images like those in Figs. 3 and 4 were recorded for differing PWT phase and $\varepsilon_{n,x}$ was calculated from those images. The result is given in Fig. 5. The plot shows that the emittance increases from an uncompressed value of 6 mm mrad to about 20 mm mrad at full compression. We note also a sharp change in the emittance data at a PWT phase of 72°, the point in compression where the onset of phase space bifurcation is observed.

The vertical emittance $\varepsilon_{n,y}$, was found to be unaffected by the level of pulse compression. Quadrupole scan measurements produced values of $\varepsilon_{n,y} \approx 7$ mm-mrad at all linac phases and $\varepsilon_{n,x}$ in agreement with the slit-based measurements. These scans were deemed valid even in the presence of significant space charge forces, because $\varepsilon_{n,x}$ is quite large, and the horizontal

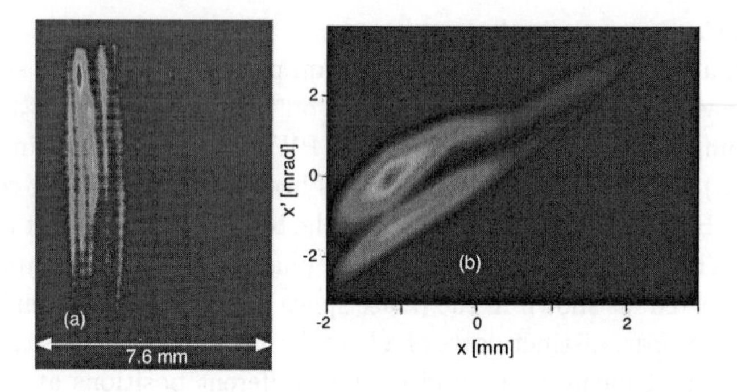

Figure 4. (a) Image of a compressed beam passing through the slit-mask. (b) Trace space derived from the image in (a).

beam size is large during the vertical scan[21].

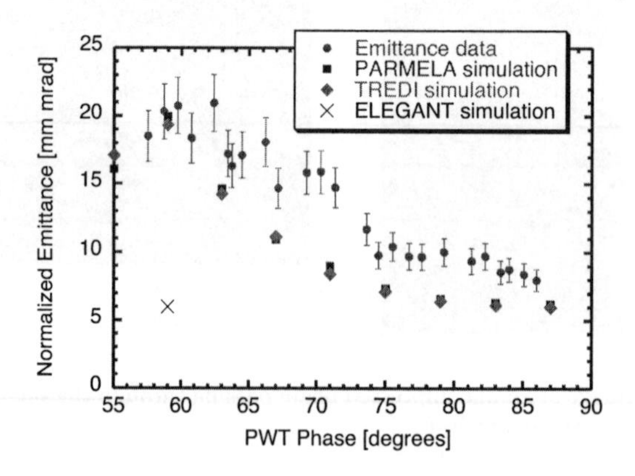

Figure 5. Normalized horizontal emittance as a function of PWT injection phase.

The bifurcation of the phase space shown in Fig. 4 was a consistent feature of the data collected. The severity of this effect, as measured by the emittance growth $\Delta\varepsilon_{n,x}$, is strongly dependent on the degree of compression (Fig. 5). In addition, it was found to be sensitive to the horizontal beam size at the entrance of the chicane $\sigma_{x,0}$, with $\Delta\varepsilon_{n,x}$ =4 mm-mrad for $\sigma_{x,0}$ =2.2 mm, while for $\sigma_{x,0}$ =1.1 mm, $\Delta\varepsilon_{n,x}$ = 24 mm-mrad was observed. For the 2.2 mm case, the phase space bifurcation was nearly eliminated. The behavior of these two cases is dramatically shown in the slit images of Fig. 6.

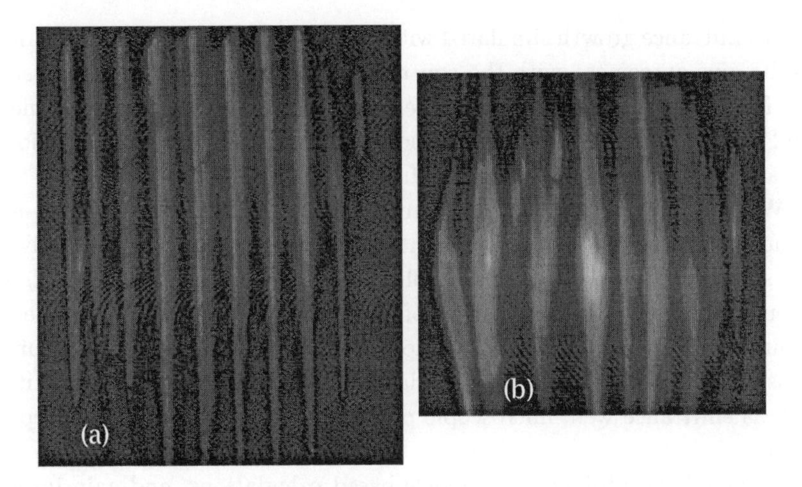

Figure 6. (a) Slit image used to compute the emittance growth for $\sigma_{x,0} = 2.2$mm. (b) Slit image used to compute the emittance growth for $\sigma_{x,0} = 1.1$mm.

3. Analysis

We employed several different computer codes, which model different physical processes, to simulate the evolution of the beam and gain information about the sources of the observed emittance growth in this system. The first code used was PARMELA[24], both to provide input phase space distributions to other codes, and to run through the chicane magnets using the 3D point-by-point space-charge calculation. This calculation uses a "quasi-static" approximation, which means that the effects of radiation fields and non-inertial space-charge[25] are *not* included in the physics model.

TREDI, a particle tracking code that keeps a full history of each particle and computes Lienard-Wiechart potentials to push them, was also used to simulated the compressor. As such, TREDI implicitly accounts for both the velocity and acceleration fields in the problem. Both TREDI and 3D PARMELA are computationally intensive, and thus 2500 simulation particles were used in both simulation approaches.

The results of these simulations are given in Fig. 5, and indicate that velocity field space-charge forces contribute most of the emittance growth observed, as the TREDI results closely agree with those from PARMELA, and both follow the experimental results. The PARMELA distributions were also used as input to ELEGANT[26], to calculate the emittance growth due to CSR alone. The ELEGANT results indicate that CSR effects were not important in the emittance growth simulated with TREDI. While the PARMELA/TREDI results give agreement with the emittance data at both weak compression and full compression, the threshold effect, of rapid emittance growth near the PWT phase of 72° is not reproduced by simulation. In the simulations, the associated (x, x') distribution bifurcation is also notably weaker.

We now note some relevant characteristics of the computational approaches used. The first is that PARMELA and TREDI employ relatively few simulation particles N_s to resolve effects which involve correlations throughout the full 6D phase space of the beam. Use of small N_s also leads to numerical Coulomb heating, and loss of coherent phase space structures. Thus, although we expect the calculations to approximate rms quantities such as emittance well, microscopic phase space details may not be reproduced.

To avoid the problems of particle-based calculations, and gain insight into the role of space-charge in this system, we employed a model simulation based on beam slices. In this model we break the bunch up into longitudinal slices (in z) and track the evolution of each slice's centroid and rms envelope. Each z-slice is modeled as a uniform ellipsoid of charge of some initial momentum offset, and zero momentum spread. Space-charge forces, in the quasi-static approximation, are calculated to give the slice-to-slice centroid force; the gradient of these forces is also used to calculate the focusing effects of the slices on each other. These focal forces, as well as the slices' self-forces, are implemented in rms envelope equations. The details of this calculation are provided in the appendix below.

In the absence of collective forces, the configuration space of centroids (x, z) shows the slices forming a "U" shape that folds down into a line in x as the beam traverses the final chicane dipole, where the relative slice positions in z do not change much. The impact space-charge has on this folding beam can be understood qualitatively; as the two halves of the beam are forced together, they repel, and at the exit of the compressor two components in the beam with different transverse (x) momenta are obtained. Space-charge repulsion during the folding has a maximum value roughly proportional to $\sigma_{x,0}^{-1}$, thus providing a simple explanation of the observed dependence of $\Delta\varepsilon_{n,x}$ on the beam size. Further, in our model, the bifurcation of the (x, x') space is enhanced by the gradient in the forces due to nearby slices; their repulsion falls with distance, and when the slice ellipsoids are disjoint, they serve to focus each other to produce a smaller σ_x.

Initial conditions for the slice simulations were obtained by use of rms information obtained from PARMELA. The results of the slice simulations

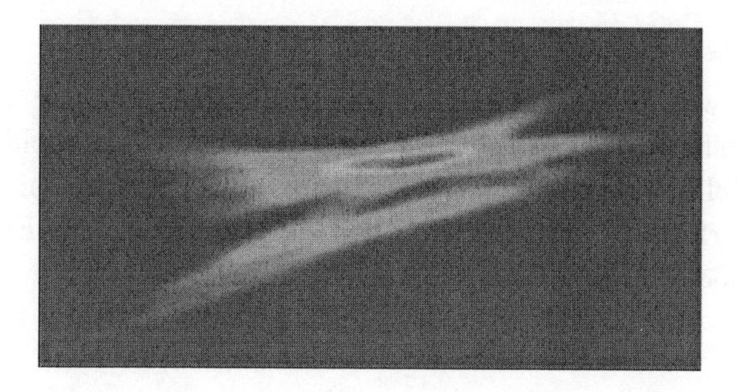

Figure 7. Contour plot of the horizontal trace space produced by the slice model simulation after the compressor.

produce much more notable phase space bifurcations (generated from the combination of the simulated horizontal centroid and ellipsoid dimensions) than those obtained from the particle codes. Even so, the phase space features, as shown in Fig. 7, are not as intricate as in the measurements. This is partially true because the focusing of each ellipsoid is only linear, whereas the repulsive forces for nonoverlapping ellipsoids is quite nonlinear. In addition, there are undoubtedly pre-existing non-ideal phase space structures in the beam which are not included in any of our models. These phase space artifacts may arise from temporal structure in the laser pulse on the cathode, and non-uniform emission from the photocathode. These effects serve to provide an initial filamentary structure to the phase space, which can be amplified by the mechanisms we have discussed.

The study of such non-ideal beam effects will be pursued in further computations and experiments. Two comments are relevant at this point: first, non-ideal beam effects may also help to resolve the differences between experiments and simulations in other compression experiments. Second, we note that in all codes we employed, as significant $\Delta\varepsilon_{n,x}$ was observed after the chicane, where the compressed beam distribution relaxes, converting nonlinear field energy (density nonuniformities) to emittance (phase space nonuniformities). The process of emittance compensation essentially images spatial nonuniformities from the cathode to the entrance of the chicane, thus maximizing the nonlinear field energy there. This conversion process is known to produce a maximum $\Delta\varepsilon_{n,x}$ and associated phase space distortion after one-quarter of a plasma oscillation[27,28] ($\lambda_p = \sqrt{\pi\gamma^3/r_e n_b}$), where n_b is the beam density. For the parameters in this experiment, the length between the chicane exit and the slits was in fact very close to $\lambda_p/4$, especially for cases where $\sigma_{x,0}$ was small.

Acknowledgments

The authors wish to acknowledge C. Joshi and C. Pellegrini for their support of Neptune efforts, and thank R. Agustsson, S. Boucher, T. Holden, M. Loh, H. Suk, S. Telfer, and A. Tremaine for their contributions to this work. This work was performed with support from U.S. Department of Energy, grant No. DE-FG03-92ER40693.

Appendix

The function of the code BENDER, and the heuristic model of the beam as a set of longitudinal slices that fold onto each other in the chicane are described above. Here, the calculation of the electric fields produced by a uniform density ellipsoid of charge are given, as these field are used in BENDER to derive the space-charge forces between slices as the beam is compressed.

The electrostatic potential both inside and outside the ellipsoid has been derived by Kellogg[29] using ellipsoidal coordinates. Using Kellogg's notation, the basic equation for the ellipsoid of charge is

$$\frac{x^2}{a^2} + \frac{y^2}{b^2} + \frac{z^2}{c^2} = 1, \qquad \text{with} \quad a^2 > b^2 > c^2, \tag{A.2}$$

where a, b, and c are the semi-major axes of the ellipsoid. If we define the functions $f(s)$ and $\varphi(s)$ by

$$f(s) = \frac{x^2}{a^2+s} + \frac{y^2}{b^2+s} + \frac{z^2}{c^2+s} - 1, \tag{A.3}$$

$$\varphi(s) = \left(a^2+s\right)\left(b^2+s\right)\left(c^2+s\right), \tag{A.4}$$

then the ellipsoidal coordinates are given by the roots of the cubic equation, $f(s)\varphi(s) = 0$. This equation has three real roots (as it must), λ, μ, and ν, obeying the relation

$$-a^2 \le \nu \le -b^2 \le \mu \le -c^2 \le \lambda. \tag{A.5}$$

The equation $\lambda =$ constant defines the surface of an ellipsoid, with $\lambda = 0$ giving the surface of the charge distribution.

The details of the mathematical formalism needed to solve the potential problem are given by Kellogg and others[30,31]. The result of these computation yields the external and internal potential of the ellipsoid,

$$U_{ext} = \pi abc\rho_0 \int\limits_{\lambda}^{\infty} \left[1 - \frac{x^2}{a^2+s} - \frac{y^2}{b^2+s} - \frac{z^2}{c^2+s}\right] \frac{\mathrm{d}s}{\sqrt{\varphi(s)}}, \tag{A.6}$$

$$= -\alpha(\lambda)\, x^2 - \beta(\lambda)\, y^2 - \gamma(\lambda)\, z^2 + \delta(\lambda), \tag{A.7}$$

and

$$U_{int} = \pi abc\rho_0 \int\limits_{0}^{\infty} \left[1 - \frac{x^2}{a^2+s} - \frac{y^2}{b^2+s} - \frac{z^2}{c^2+s}\right] \frac{\mathrm{d}s}{\sqrt{\varphi(s)}}, \tag{A.8}$$

$$= -\alpha_0 x^2 - \beta_0 y^2 - \gamma_0 z^2 + \delta_0, \tag{A.9}$$

with

$$\alpha(\lambda) = \pi abc\rho_0 \int\limits_{\lambda}^{\infty} \frac{\mathrm{d}s}{(a^2+s)\sqrt{\varphi(s)}}, \qquad \text{and} \quad \delta(\lambda) = \pi abc\rho_0 \int\limits_{\lambda}^{\infty} \frac{\mathrm{d}s}{\sqrt{\varphi(s)}}, \tag{A.10}$$

where $\beta(\lambda)$ and $\gamma(\lambda)$ are found from $\alpha(\lambda)$ by replacing a with b, and c, respectively.

The electric fields are, of course, found by differentiating the potential, and in the case of the internal fields the result is found immediately from Equation (A.9),

$$\begin{aligned} E_{x,int} &= 2\alpha_0 x, \\ E_{y,int} &= 2\beta_0 y, \\ E_{z,int} &= 2\gamma_0 z. \end{aligned} \tag{A.11}$$

The external fields will not in general be a linear function of only one variable, since U_{ext} is a function of λ. For example, the x-component of the electric field must be found through

$$E_{x,ext} = -\frac{dU_{ext}}{dx} = -\frac{\partial U_{ext}}{\partial x} - \frac{\partial U_{ext}}{\partial \lambda}\frac{\partial \lambda}{\partial x}. \tag{A.12}$$

The remarkable result here is that

$$\frac{\partial U_{ext}}{\partial \lambda} = \frac{\partial}{\partial \lambda}\left[\pi abc\rho_0 \int\limits_{\lambda}^{\infty} \frac{-f(s)}{\sqrt{\varphi(s)}}\mathrm{d}s\right] = \frac{-f(\lambda)}{\sqrt{\varphi(\lambda)}} = 0, \tag{A.13}$$

since $f(\lambda) = 0$ by definition. Therefore, the external fields are given by

$$\begin{aligned} E_{x,ext} &= 2\alpha(\lambda)\, x, \\ E_{y,ext} &= 2\beta(\lambda)\, y, \\ E_{z,ext} &= 2\gamma(\lambda)\, z. \end{aligned} \tag{A.14}$$

From the simulation point of view, one must first compute the ellipsoidal coordinate λ using the cartesian coordinates and the dimensions of the ellipsoid, then compute the functions α, β, and γ. The coordinate λ (as well as μ and ν) is found by solving the cubic equation, $w^3+Aw^2+Bw+C=0$, with

$$\begin{aligned} A &= a^2+b^2+c^2-x^2-y^2-z^2 \\ B &= c^2\left(a^2+b^2-x^2-y^2\right)+b^2\left(a^2-x^2-z^2\right)-a^2\left(y^2+z^2\right) \\ C &= c^2\left(a^2b^2-b^2x^2-a^2y^2\right)-a^2b^2z^2 \end{aligned} \tag{A.15}$$

In the present case, where $a < b < c$, the auxiliary quantities Q, R, and θ given by

$$\begin{aligned} Q &= \frac{A^2-3B}{9}, \\ R &= \frac{2A^3-9AB+27C}{54}, \\ \theta &= \cos^{-1}\left[\frac{R}{\sqrt{Q^3}}\right], \end{aligned} \tag{A.16}$$

are used to find the ellipsoidal coordinates,

$$\begin{aligned} \lambda &= -2\sqrt{Q}\cos\left(\frac{\theta+2\pi}{3}\right)-\frac{A}{3}, \\ \mu &= -2\sqrt{Q}\cos\left(\frac{\theta-2\pi}{3}\right)-\frac{A}{3}, \\ \nu &= -2\sqrt{Q}\cos\left(\frac{\theta}{3}\right)-\frac{A}{3}. \end{aligned} \tag{A.17}$$

All that remains at this point is to evaluate the integrals in Equation (A.10) to determine α, β, and γ. The form of these functions varies with the symmetry of the ellipsoid being considered, with spherical ($a=b=c$), prolate spheroidal ($a>b=c$), oblate spheroidal ($a=b>c$), and ellipsoidal ($a>b>c$) cases possible. In the case of spherical symmetry the integrals of Equation (A.10) are easily evaluated, and the expected result, $E_x = qx/r^3$, is obtained. The results in the other cases were tabulated in Ref. [32], and are given here.

For the general ellipsoid $(a > b > c)$, α, β, and γ are given by

$$\alpha(\lambda) = \frac{2}{\left(a^2 - c^2\right)^{3/2} k^2} \left[F(\phi, k) - E(\phi, k)\right],$$
$$\beta(\lambda) = \frac{2}{\left(a^2 - c^2\right)^{3/2} k^2 k'^2} \left[E(\phi, k) - k'^2 F(\phi, k) - \frac{k^2 \sin\phi \cos\phi}{\sqrt{1 - k^2 \sin^2\phi}}\right],$$
$$\gamma(\lambda) = \frac{2}{\left(a^2 - c^2\right)^{3/2} k'^2} \left[\frac{\sin\phi \sqrt{1 - k^2 \sin^2\phi}}{\cos\phi} - E(\phi, k)\right], \tag{A.18}$$

where we use the quantities

$$\phi = \sin^{-1}\left[\sqrt{\frac{a^2 - c^2}{a^2 + \lambda}}\right], \quad k = \sqrt{\frac{a^2 - b^2}{a^2 - c^2}}, \quad \text{and} \quad k' = \sqrt{1 - k^2}, \tag{A.19}$$

and the elliptic integrals

$$E(\phi, k) = \int_0^\phi \sqrt{1 - k^2 \sin^2\theta}\, \mathrm{d}\theta,$$
$$F(\phi, k) = \int_0^\phi \frac{\mathrm{d}\theta}{\sqrt{1 - k^2 \sin^2\theta}}. \tag{A.20}$$

For the prolate spheroid $(a > b = c)$, we obtain

$$\alpha(\lambda) = \frac{2}{\left(a^2 - c^2\right)^{3/2}} (u - \sin\phi),$$
$$\beta(\lambda) = \frac{2}{\left(a^2 - c^2\right)^{3/2}} \left[\frac{\sqrt{\left(a^2 - c^2\right)\left(a^2 + \lambda\right)}}{c^2 + \lambda} - u\right], \tag{A.21}$$
$$\gamma(\lambda) = \beta(\lambda),$$

with

$$u = \ln\left(\sqrt{\frac{1 + \sin\phi}{1 - \sin\phi}}\right). \tag{A.22}$$

Finally, for the oblate spheroid $(a = b > c)$, we have

$$\alpha(\lambda) = \frac{2}{\left(a^2 - c^2\right)^{3/2}} (\phi - \sin\phi \cos\phi),$$
$$\beta(\lambda) = \alpha(\lambda), \tag{A.23}$$
$$\gamma(\lambda) = \frac{2}{\left(a^2 - c^2\right)^{3/2}} (\tan\phi - \phi).$$

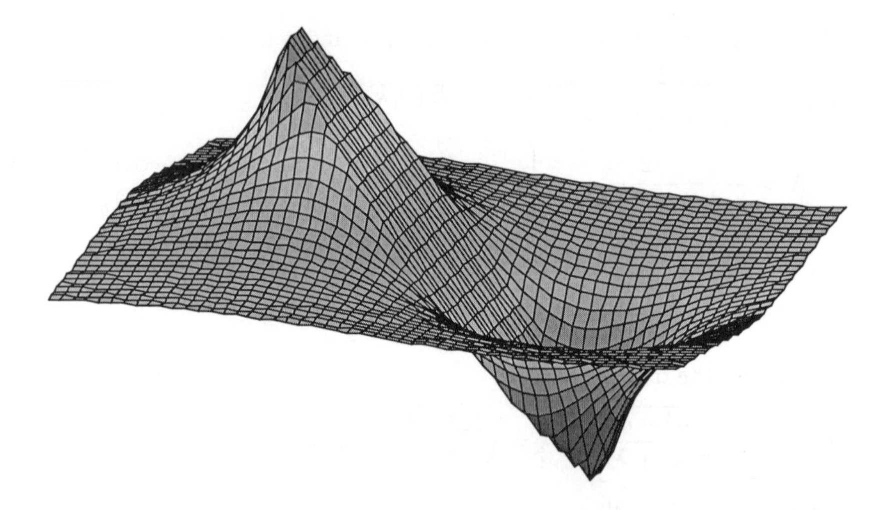

Figure 8. One component of the electric field (E_x) produced by a uniformly charged ellipsoid, and plotted for points in the $x - y$ plane. Inside the distribution the field is linear in x and independent of y and z. For this example the semi-major axes of ellipsoid are $(a, b, c) = (5, 1, 2)$

References

1. International Study Group. "International study group progress report on linear collider development." Technical Report SLAC-R-559, Stanford Linear Accelerator Center (2000).
2. J. B. Rosenzweig, N. Barov, and E. Colby. "Pulse compression in rf photoinjectors: Applications to advanced accelerators." *IEEE Trans. Plasma Sci.*, **24**, p. 409 (1996).
3. The LCLS Design Study Group. "Linac coherent light source (lcls) design study report." Technical Report SLAC-R-0521, Stanford Linear Accelerator Center (1998).
4. Martin Reiser. *Theory and Design of Charged Particle Beams* (Wiley, New York, 1994).
5. B. E. Carlsten. *Nucl. Instrum. Methods Phys. Res., Sect. A*, **285**, p. 313 (1989).
6. Luca Serafini and J. B. Rosenzweig. *Phys. Rev. E*, **55**, p. 7565 (1997).
7. M. James, J. Clendenin, S. Ecklund, R. Miller, J. Sheppard, *et al.* "Update on the high-current injector for the stanford linear collider." *IEEE Trans. Nucl. Sci.*, **NS-30**, p. 2992 (1983).
8. G. Stupakov and S. Heifets. "Beam instability and microbunching due to coherent synchrotron radiation." *Phys. Rev. ST Accel. Beams*, **5**, p. 054402 (2002).

9. H. Braun, F. Chautard, R. Corsini, T. O. Raubenheimer, and P. Tenenbaum. *Phys. Rev. Lett.*, **84**, p. 658 (2000).
10. H. Braun, R. Corsini, L. Groening, F. Zhou, A. Kabel, *et al. Phys. Rev. ST Accel. Beams*, **3**, p. 124402 (2000).
11. M. Dohlus, A. Kabel, and T. Limberg. In *Proceedings of the 1999 Particle Accelerator Conference*, p. 1650 (IEEE, 1999).
12. ICFA Beam Dynamics mini workshop: Coherent Synchrotron Radiation and its impact on the beam dynamics of high brightness electron beams. See `http://www.desy.de/csr/csr_workshop_2002/csr_workshop_2002_index.html` (2002).
13. G. P. LeSage, S. G. Anderson, T. E. Cowan, J. K. Crane, T. Ditmire, *et al.* In *Advanced Accelerator Concepts: Ninth Workshop*, edited by P. L. Colestock and S. Kelley, AIP Conf. Proc. No. 569 (AIP, New York, 2001).
14. J. B. Rosenzweig, S. G. Anderson, K. Bishofberger, X. Ding, A. Murokh, *et al. Nucl. Instrum. Methods Phys. Res., Sect. A*, **410**, p. 437 (1998).
15. D. T. Palmer. *The next generation photoinjector.* Ph.D. thesis, Stanford University (1998).
16. R. Zhang. In *1995 Particle Accelerator Conference*, p. 1102 (IEEE, 1995).
17. K. R. Crandall and L. Young. "Computer codes for particle accelerator design and analysis." Technical Report LA-UR-90-1766, Los Alamos National Laboratory (1990).
18. U. Happek, A. J. Sievers, and E. B. Blum. *Phys. Rev. Lett.*, **67**, p. 2962 (1991).
19. A. Murokh, J. B. Rosenzweig, M. Hogan, H. Suk, G. Travish, *et al. Nucl. Instrum. Methods Phys. Res., Sect. A*, **410**, p. 452 (1998).
20. Claude Lejeune and Jean Aubert. "Emittance and brightness: Definitions and measurements." In *Advances in Electronics and Electron Physics*, Supplement 13A, p. 159 (1980).
21. S. G. Anderson, J. B. Rosenzweig, G. P. LeSage, and J. K. Crane. *Phys. Rev. ST Accel. Beams*, **5**, p. 014201 (2002).
22. S. G. Anderson. *Creation, manipulation, and diagnosis of intense, relativistic picosecond photo-electron beams.* Ph.D. thesis, University of California, Los Angeles (2002).
23. F. Ciocci, L. Giannessi, A. Marranca, L. Mezi, and M. Quattromini. *Nucl. Instrum. Methods Phys. Res., Sect. A*, **393**, p. 434 (1997).
24. L. Young and J. Billen. Technical Report LA-UR-96-1835, Los Alamos National Laboratory (1996).
25. B. E. Carlsten and T. O. Raubenheimer. *Phys. Rev. E*, **51**, p. 1453 (1995).
26. M. Borland. Technical Report LS-287, Argonne National Laboratory Advanced Photon Source (2000).
27. O. A. Anderson. "Internal dynamics and emittance growth in space-charge-dominated beams." *Part. Accel.*, **21**, p. 197 (1987).
28. S. G. Anderson and J. B. Rosenzweig. "Non-equilibrium transverse motion and emittance growth in space-charge dominated beams." *Phys. Rev. ST Accel. Beams*, **3**, p. 094201 (2000).
29. O. D. Kellogg. *Foundations of Potential Theory* (F. Ungar Pub. Co., New

York, 1929).

30. W. E. Byerly. *An Elementary Treatise on Fourier's Series and Spherical, Cylindrical and Ellipsoidal Harmonics with Applications to Problems in Mathematical Physics* (Ginn & Company, Boston, 1893).
31. S. Chandrasekhar. *Ellipsoidal Figures of Equilibrium* (Yale University Press, New Haven, 1969).
32. L. B. Tuckerman. "Inertia factors of ellipsoids for use in airship design." Technical Report 210, National Advisory Committee for Aeronautics (1926).

VELOCITY BUNCHING EXPERIMENT AT THE NEPTUNE LABORATORY*

P.MUSUMECI, R.J.ENGLAND, M.C.THOMPSON, R.YODER, J.B.ROSENZWEIG
Department of Physics, University of California at Los Angeles, 405 Hilgard Ave, Los Angeles, CA 90095, USA

In this paper we describe the rectilinear compression experiment at the Neptune photoinjector at UCLA. The electron bunches have been shortened to sub-ps pulse length by chirping the beam energy spectrum in a short S-band high gradient standing wave RF cavity and then letting the electrons undergo velocity compression in the following rectilinear drift. Using a standard Martin Puplett interferometer to characterize coherent transition radiation from the beam, we measured bunch length as short as 0.4 ps with compression ratio in excess of 10 for an electron beam of 7 MeV energy and charge up to 300 pC. We also measured slice transverse emittance via quad scan technique after a 45 degrees dispersing dipole. Three-dimensional simulations agree with the observed emittance growth.

1. Introduction

In recent years electron beam users have increased their demands for high brightness beam in short sub-ps pulses [1-3]. These beams find applications in the advanced accelerator community for injection into short wavelength high gradient accelerators, or as plasma wake-field drivers, and in the light source community for short wavelength SASE Free Electron Laser and for Thompson-scattering generation of short X-ray pulses. Recent designs of systems capable of delivering high brightness very short electron beams include the use of conventional photoinjectors in conjunction with magnetic compressors [4]. While the magnetic compression scheme has been proved successful in increasing the beam current, the impact on the phase space has been shown to be quite relevant: performing the compression at low energy, the space charge forces are not strongly suppressed by the relativistic cancellation of electric and magnetic fields and their emittance-damaging effect becomes significant especially in bending trajectories [5]. In the case of compression at higher energy [6], one has to deal with the deleterious effects on the longitudinal as well as the transverse phase space of Coherent Synchrotron Radiation. Phase

* This work is supported by U.S. Department of Energy grant DE-FG03-92ER40693.

space filamentation and in general emittance growth jeopardize the goal of achieving the desired high brightness.

An alternative scheme that could preserve the phase space quality while still shortening the beam to sub-ps bunch length has been recently proposed as an injector for a X-ray Free Electron Laser, by Serafini and Ferrario [7]. This scheme, commonly known as "velocity bunching" is an elaboration of the old idea of RF rectilinear compression to the RF photoinjector system. In general, for every application in which compression at low energy is required, it seems that velocity bunching is an efficient alternative to magnetic compression The idea is based on the weak synchrotron motion that the beam undergoes at moderate energies in the RF wave of a linac accelerating structure. The compression happens in a rectilinear section so that the damage suffered by going through bending trajectories is avoided. . At ATF, in the recent past, the bunching effect of a 1.6 cell RF gun when operating at launching phases far off crest has been already observed [8]. A main ingredient of Serafini&Ferrario recipe to produce high brightness sub-ps electron beam is to integrate this compression section in the emittance compensation scheme, by keeping the transverse beam size under control through solenoidal magnetic field in the region where the bunch is compressing and the electron density is increasing.

A small variation inside this framework is the thin lens version of velocity bunching. Here the synchrotron motion of the electrons inside the RF structure is very limited. There is almost no phase advance inside the longitudinal lens and all the bunching happens in the drift following the linac.

In this paper we describe an experimental study of this configuration. At the Neptune laboratory at UCLA there is a 1.6 cell gun and a PWT standing wave linac that was used to test this idea. In the next section we draw the schematics of the experimental layout, and show the results. We measured the bunch length by using the Coherent Transition Radiation technique. We then varied the electric field in the linac and measured the best compressing phase for different gradients. The results agree with our simple model for ballistic bunching. In the last sections of the paper, we turn our attention to the transverse dynamics. The strong correlation between longitudinal position and energy allowed us to use a 45 degrees dispersing dipole to select a portion of the beam with a relatively small energy spread (few percent). In this way, we were able the emittance of the central slice of the electron beam via quad scan technique. As the phase of the RF accelerating wave in the PWT Linac, towards the compressing phase, we observed emittance growth. We present three-dimensional simulations that agree with the experimental results. It is important to note that the beamline at the Neptune photoinjector was not optimized for a rectilinear compression experiment, in the sense that no solenoidal magnetic fields are present to match the increasing space-charge forces and there is no post acceleration to remove the induced energy spread. We also studied a system optimized for ballistic bunching, the proposed injector for the Orion Research facility [9]. Here the solenoids wrapped around the accelerator cavities should keep the beam under control and the simulations show that the scheme can achieve high current preserving phase space quality.

2. Neptune experimental layout

The Neptune facility at UCLA currently operates as an injector for advanced accelerator experiments. At the same time the Neptune photoinjector is being used for high brightness beam dynamics studies like emittance growth in bends [5] and negative R56 compressors [10]. The accelerator can be tune up for ballistic compression.

A 266 nm 4 ps rms long laser pulse hits a single crystal copper cathode inside a 1.6 cell BNL-SLAC-UCLA RF gun. The photoelectrons generated are then accelerated by the RF fields and go through the emittance compensation solenoid. At this point the beam can be energy chirped inside a 6+2 ½ cell S-band PWT RF cavity. There is the capability of controlling independently and measuring the phases of the two accelerating structures allowing us to test the ballistic bunching scheme. Downstream of the linac a aluminum foil can be inserted and the transition radiation generated is collected by a parabolic mirror and reflected to a Martin Puplett autocorrelator for pulse length diagnostic. There are also 4 chicane dipoles along the beamline and two of them can be turned on in a 45 degrees dipole mode. On the 45 degrees beam line there is a quadrupole lens and a Yag screen for emittance measurement via quad scan.

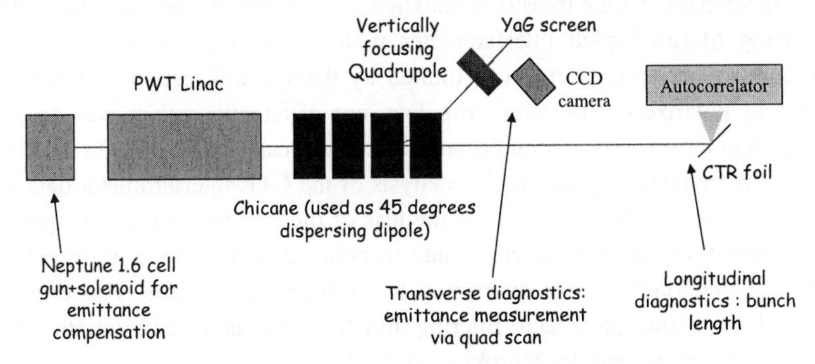

Figure 1. Layout of the Neptune experiment

3. Longitudinal dynamics

Ballistic bunching can be viewed just as the thin lens version of the more general velocity bunching. In this configuration, the phase advance of the electrons going through the longitudinal lens (the PWT linac) is few RF degrees and all the bunching happens in the following drift. The important difference with the long RF-structure slow compression version of the velocity bunching is that the beam is extracted still very close to the zero phase of the RF bucket and the RF non-linearities that usually dominate the final bunch length are greatly reduced.

One simple way to understand the ballistic bunching is to think to the time of arrival difference for particles having different velocities. When the time of arrival difference compensates the difference in the longitudinal position, the bunch length will be minimum and that is the maximum compression point. A first order approximation to describe the ballistic bunching can be written as:

$$\frac{\Delta p}{p^3}\cdot L = \frac{E_{linac}\cdot\cos(\phi)\cdot k\cdot\Delta z}{\left(E_{gun}+E_{linac}\cdot\sin(\phi)\right)^3}\cdot L = \Delta z \tag{1}$$

where L is the distance from the RF structure, E_{linac} is the energy given by the PWT linac and E_{gun} out of the gun, k is the RF wavenumber and ϕ is the Linac phase. This relationship is strictly valid at first order, neglecting space charge and phase advance inside the PWT Linac.

3.1. *Measurement*

We measured the pulse length in the frequency domain by Coherent Transition Radiation interferometry. The electrons hit an aluminum foil inserted in the beam path and the transition radiation is collected and reflected towards a polarizing Martin-Puplett interferometer with two Golay cell detectors. The resolution of the interferometer is limited by the spectral response of the two wire grid polarizers. The wire grid does not efficiently reflect wavelengths shorter than the wire separation distance that in our case is 100 μm. On the other side of the frequency spectrum, the analysis of the CTR interferometer data has to be done taking into consideration the loss of the low frequencies component in the transition radiation spectrum due to poor vacuum window transmission and more importantly to the diffraction of the long wavelengths. This effect is included by a one parameter filtering function that is fitted from the data following previous work by Murokh et al. [11].

Note that the measurement was performed on a beam only moderately relativistic, the divergence ($\sim 1/\gamma$) cone of the radiation was quite big. Because the Martin Puplett is a polarizing interferometer, and because the polarization of transition radiation depends on the direction of propagation, misalignment and losses in the collection of the divergent cone can result in systematic errors in the measurement.

For 250 pC of charge and 70 degrees off crest in the PWT cavity, scanning the moving arm of the interferometer, we obtain the interferogram shown in fig.1. The data analysis gives for the pulse length 0.39 ps.

It is worth noticing the compressed beam is shorter than what we were ever able to get with magnetic compression for comparable beam charge, confirming the fact that in this case a more linear part of the RF wave is sampled.

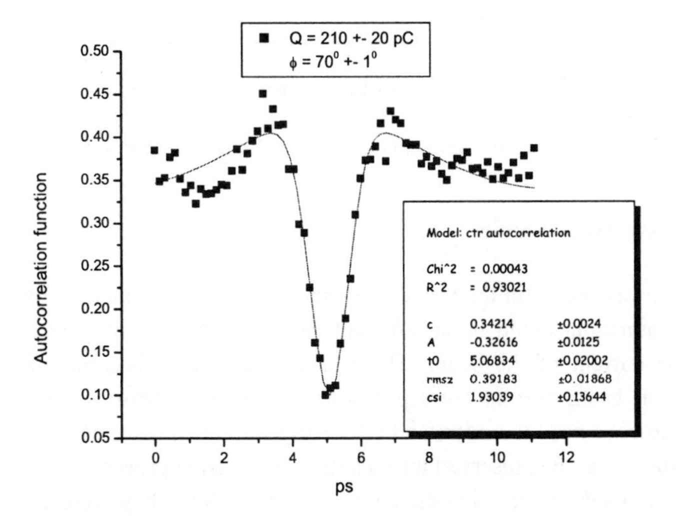

Figure 2. Autocorrelation results

The predictions from the first order approximation given in (1) have been experimentally verified by measuring the compression phase ϕ changing the energy gradient in the Linac. The RF-cavity phase can be measured with a very small error by mixing the RF fields inside the structure with a reference RF-clock, at the same time the phase for maximum compression is easily determined by maximizing the Coherence Transition Radiation energy on the bolometer detectors. The agreement with the analytical formula is very good. Note that there is an important cancellation effect. As we decrease the energy gradient in the Linac, the beam is getting less energetic and less rigid to a rotation in the longitudinal phase space so that the adjustment of the phase to maintain the compression condition is minimal.

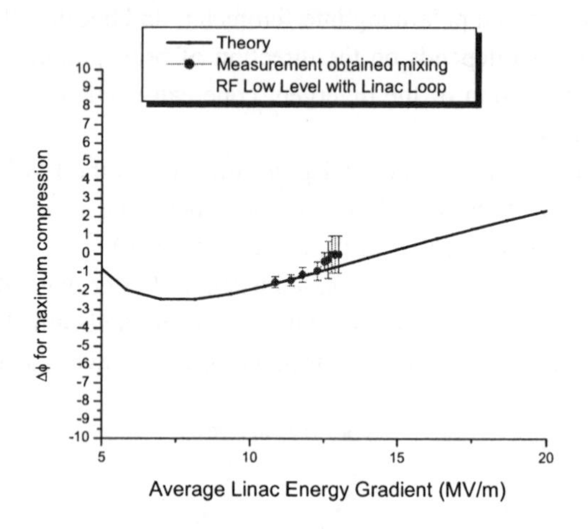

Figure 3. Phase of maximum bunching vs. Linac accelerating gradient

4. Transverse dynamics

As it was shown in the last section, the Neptune experiment confirmed that ballistic bunching could be an efficient and compact way of increasing many folds the current of the beam. The question to answer becomes then, if the increase in beam current corresponds to a relative increase in brightness, in other words we need to understand what happen to the transverse phase space, if the emittance can be preserved through the compression process.

For optimal compression the beam runs through the high gradient structure far from the crest of the RF wave so that the energy spread at the exit of the Linac is very big. For example for the case in which the focus of the longitudinal lens is 3 m downstream on the beamline, the RF phase was set 70 degrees off crest, resulting in a energy spectrum extending from 5 MeV to 9 MeV. This would not be a problem in a system where the beam energy can be boosted up by additional accelerating cavities to quickly remove the relative energy spread, but at the Neptune photoinjector there is no such capability. This is a limitation to the determination of the transverse projected emittance because the energy spread translating in an angle spread will appear to all measurement techniques (that are trace space measurements) as unphysical transverse emittance.

On the other hand, the energy is correlated with the longitudinal position of the beam and with a small window of acceptance in energy, a longitudinal slice of the beam can be selected. Experimentally, we used the 45 degrees dispersing bending dipole configuration to select a beam slice over which a vertical quad scan emittance measurement was performed. We studied the vertical phase space of the electron beam scanning the phase of the linac to understand the transverse dynamics of the beam as it was undergoing compression.

4.1. *Slicing the beam with the 45 degrees dispersing dipole*

Because changing the linac phase that is the main compression knob, has also the effect of changing the energy of the beam, it is important to ensure that always the same part of the beam hits on the small acceptance YAG screen (few degrees of bending angle that is few % of energy spread). We need to be able to set the dipole current to keep at the 45 degrees bending angle always the same portion of the beam as its energy changes. This is accomplished first by measuring the full spectrum of the beam as the linac phase is changed, then individuating one reference slice (in our case is the central slice or the maximum current slice for our gaussian beam distribution) and finally adjusting the dipole current to analyze always the reference slice. In the figure we can see the energy change of the central reference slice of the beam as the linac phase is varied. This curve incidentally allows us an independent determination of the RF cavities accelerating gradients. We found in good agreement with the RF measurements the energy gain in the 1.6 cell gun to be 4 MeV corresponding to a 80 MV/m gradient and in the PWT linac 8 MeV corresponding to a 40 MV/m accelerating gradient.

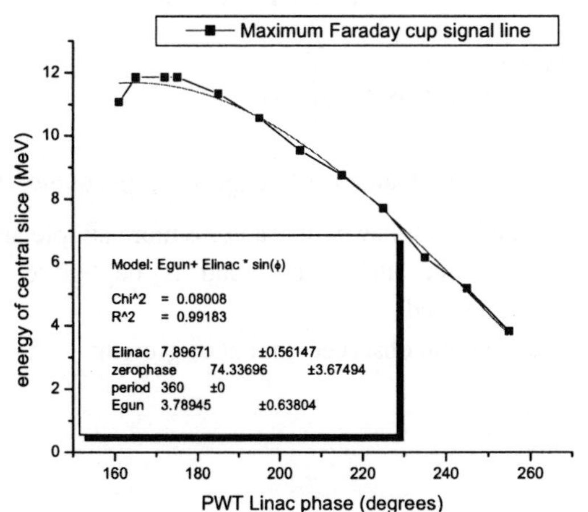

Figure. 4: Energy of central slice varying the phase of the PWT linac

4.2. *Quad scan measurement*

On the central reference beam slice we can perform the quad scan. Since the beam varies in size significantly going through the quad, an analysis that takes into account the thickness of the lens is needed.

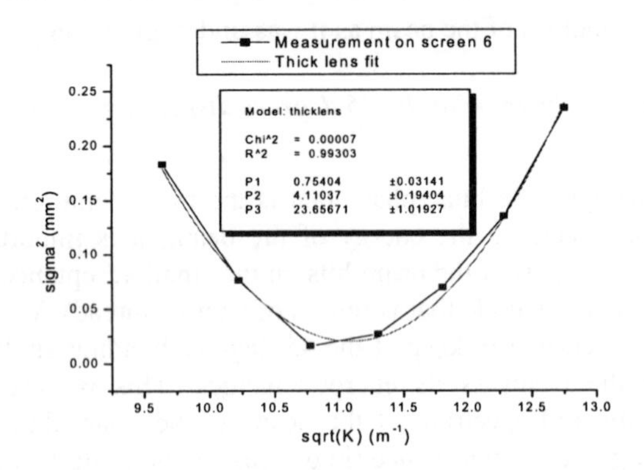

Figure 5: Thick lens treatment for the quad scan

The parameterization of the square of the measured beam size with respect to the quad strength ($K=qB'/p_0$) is:

$$\sigma^2(K) = \left[\cos\left(\sqrt{K}l_q\right) - \sqrt{K}l_d \sin\left(\sqrt{K}l_q\right)\right]^2 \sigma_{11} + \frac{2}{\sqrt{K}}\left[\sin\left(\sqrt{K}l_q\right) + l_d \cos\left(\sqrt{K}l_q\right)\right]\left[\cos\left(\sqrt{K}l_q\right) - \sqrt{K}l_d \sin\left(\sqrt{K}l_q\right)\right]\sigma_{12} + \left[\frac{\sin\left(\sqrt{K}l_q\right)}{\sqrt{K}} + l_d \cos\left(\sqrt{K}l_q\right)\right]^2 \sigma_{22} \quad (2)$$

where $l_{q,d}$ are the quad and drift lengths, respectively. Here we take into account the full thick lens matrix instead of a more simple analysis in which the betatron phase advance inside the quad is negligible and the thin lens approximation can be used.

In fig.6 is shown the observed emittance growth:

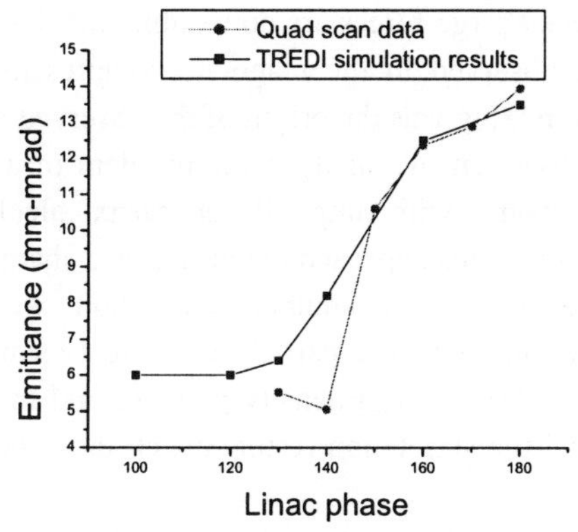

Figure 6: Emittance growth during compression, experimental results and simulations

5. Compression of a beam with chirped energy spectrum in a bending magnet

We expected to observe emittance growth for PWT phases for which the longitudinal waist of the beam occurs before the measurement screen due to the deleterious effect of a longitudinal cross-over in the transverse phase space. On the other hand, the data show an increase in emittance even for phases for which the beam can not be fully compressed by going through the small drift between the PWT Linac and the dispersing dipole. To explain this, it requires a deeper look into the dynamics of a chirped beam going through the bending magnet that we used as a slice selector in the measurement. For the vertical phase space the bending magnet is in fact just a drift, but looking at physical beam volume in x-s configuration space, we observe a strong compression if the beam is chirped in energy with particles with higher energy being in the tail of the beam. Note that a bending magnet has a negative R_{56} so that the compression is anomalous. The size of the projections of the beam density onto the curvilinear longitudinal axis s or onto the transverse dimension x and y do not get smaller. On the other hand, if the beam is chirped with higher energy particles in the tail and the transverse beam size is big in a way that we will quantify in the next section, the three-dimensional physical volume of the beam gets smaller (undercompression), goes through a waist, and then gets bigger again (overcompression). It is a three-dimensional effect that is ultimately due to the correlations in the configuration space introduced by the bending magnet.

The effect of this electronic density increase at the cross-over point is dramatic for the electron beam quality especially at moderately relativistic energies where space charge forces are dominating the dynamics of the beam. A similar effect was observed in the magnetic compression experiment, where configuration space mixing was the origin of the observed emittance growth [5].

In the next subsection, we analyze the problem of a chirped beam going through a bending magnet with simple linear matrix calculation to illustrate the compression dynamics. This approach neglects space charge forces and is only a first order approximation, in another subsection we show fully three-dimensional simulations with the particle tracking Lienard-Wiechert potential based code TREDI. The electron density increase and the configuration space mixing are clear and the simulation results match the experimental data for the emittance growth.

5.1. *Linear analysis*

A simple way to understand the physics of this process is the linear matrix-based approach to the beam dynamics. Because the vertical phase space is uncoupled in a bending magnet transport line, we study only the 4-dimension phase space (x,x',s,δ).

Knowing the initial beam matrix Σ

$$\mathbf{\Sigma} = \begin{pmatrix} \sigma_{xx} & \sigma_{xx'} & \sigma_{xs} & \sigma_{x\delta} \\ \sigma_{x'x} & \sigma_{x'x'} & \sigma_{x's} & \sigma_{x'\delta} \\ \sigma_{sx} & \sigma_{sx'} & \sigma_{ss} & \sigma_{s\delta} \\ \sigma_{\delta x} & \sigma_{\delta x'} & \sigma_{\delta s} & \sigma_{\delta\delta} \end{pmatrix} \tag{3}$$

where the σs are the second moment of the beam phase space distributions (for example $\sigma_{xx} = \langle x^2 \rangle$, $\sigma_{xx'} = \langle xx' \rangle$ and so on) and the transport matrix R for a bending magnet.

$$\mathbf{R}(s) = \begin{pmatrix} 1 & s & 0 & -R\cdot\left(1-\cos\left(\frac{s}{R}\right)\right) \\ 0 & 1 & 0 & -\sin\left(\frac{s}{R}\right) \\ \frac{s}{R} & \frac{s^2}{2\cdot R} & 1 & \frac{s}{\gamma^2} - s + R\cdot\sin\left(\frac{s}{R}\right) \\ 0 & 0 & 0 & 1 \end{pmatrix} \tag{4}$$

where s is the longitudinal coordinate along the beam path, R is the bending radius and γ is the design energy, it is easy to find out the final beam matrix , just by matrix multiplication:

$$\mathbf{\Sigma}_f = \mathbf{R} \cdot \mathbf{\Sigma}_i \cdot \mathbf{R}^{\mathrm{T}} \tag{5}$$

We can then calculate the configuration space volume taking into account, as we do when we calculate the rms volume of a phase space, possible correlations between x and the longitudinal coordinate s.

$$V = \sqrt{\sigma_{xx} \cdot \sigma_{ss} - \sigma_{xs}^{2}} \tag{6}$$

A good approximation is to neglect the transverse emittance contribution ($\sigma_{xx'}$, $\sigma_{x'x'} = 0$) because the RF non linearities make the longitudinal trace space contribution the most important one to the final volume. With this assumption we can write for the configuration space volume

$$V = \sqrt{\sigma_{xx}\left(\sigma_{ss} + 2\sigma_{s\delta} R^{*}{}_{56} + \sigma_{\delta\delta} \cdot R^{*}{}_{56}{}^{2}\right) + R_{16}{}^{2}\left(\sigma_{ss}\sigma_{\delta\delta} - \sigma_{s\delta}{}^{2}\right)} \tag{7}$$

where $R^{*}{}_{56}$ can be written as:

$$R^{*}{}_{56} = \frac{s}{\gamma^{2}} + R \cdot \sin\left(\frac{s}{R}\right) - s \cdot \cos\left(\frac{s}{R}\right) \tag{8}$$

a positive quantity and quickly increasing with distance (in the ultrarelativistic case, ~ s^3), and R_{16} is the regular dispersion function, also increasing with distance. The initial volume is $(\sigma_{xx} \cdot \sigma_{ss})^{1/2}$. If the beam initially has a negative energy-position correlation so that particles in the tail are more energetic than particles in the head, this correlation can be removed by the positive $R^{*}{}_{56}$ at the anomalous compression point. This is the minimum volume point. (fig.7)

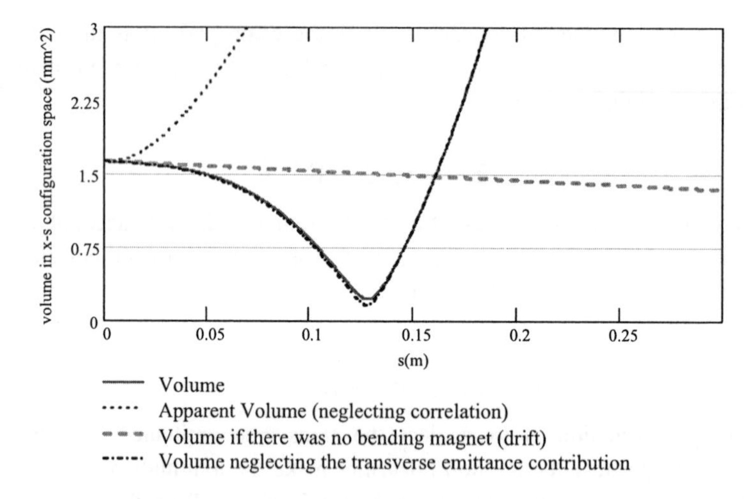

Figure 7: Anomalous compression for a beam chirped by going 70 degrees off crest in the PWT Linac. The initial beam sizes at the entrance of the dipole magnetic field are σ_x =2.8 mm, σ_s =0.6 mm. The anomalous compression point is 12 cm inside the magnet and the compression factor is 4.7.

Geometrically this minimum corresponds to a point along the beam line in which the more energetic particles in the tail of the beam overtake the less energetic ones. (fig.8)

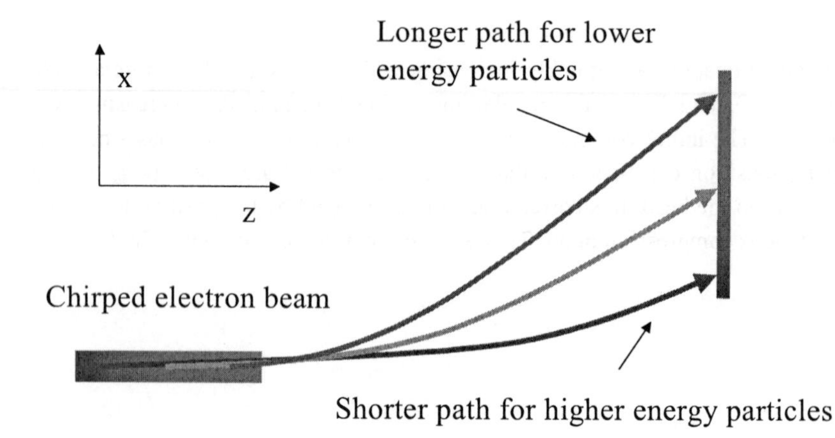

Figure 8: Geometrical description of anomalous compression.

The compression factor depends strongly on the initial transverse beam size and the longitudinal trace space emittance and is given by:

$$B = \frac{V_i}{V_f} = \sqrt{\frac{\sigma_{xx} \cdot \sigma_{ss}}{\varepsilon_s \cdot \left(\frac{\sigma_{xx}}{\sigma_{ss}} \left(R^*_{56}\right)^2 + \left(R_{16}\right)^2 \right)}} \tag{9}$$

where ε_s is the longitudinal trace space emittance, and R_{16} and R^*_{56} are evaluated at the anomalous compression point.

5.2. *Simulations*

The linear analysis is useful to understand at first order what happen to the beam going through the dipole. In order to fully understand our measurement we have performed fully 3dimensional simulations with the particle tracking Lienard-Wiechert potential based code TREDI [12].

The mixing between different sections of the beam that occurs at the anomalous compression point is ultimately the origin of the emittance growth observed in the measurement. The laminar flow for the space charge dominated beam that is a key to prevent emittance growth is broken by the configuration space crossover and the transverse phase space quality is degraded.

Looking in the configuration space from the particle tracking simulations it results that if the chirp on the beam is not too big (<10% when the beam is accelerated 30 degrees off crest), the crossover doesn't happen because the dispersion in the horizontal plane dilutes the beam before the anomalous compression point and the beam just bend inside the dipole without severe degradation of phase space quality. On the other side, for a beam with big energy spread that has been accelerated by the Linac far off crest the configuration space cross over takes place after less than 10 cm from the entrance of the dipole (fig. 9). The more energetic particles that are in the tail reach the less energetic ones that follow a trajectory with a bigger bending angle. This is the reason that we observe significant emittance growth at phases for which the compression should not be so severe in the rectilinear drift.

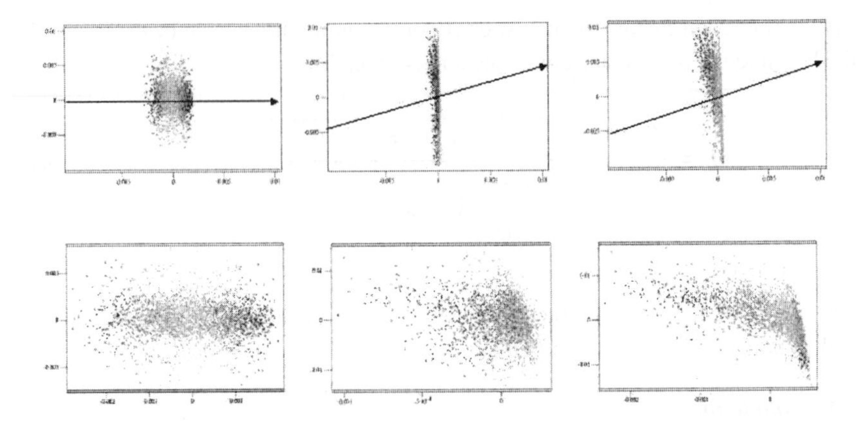

Figure 9. Cartesian configuration space for a beam accelerated 70 degrees off crest in the Linac at three different locations along the beam line, before the dipole (first column), at the anomalous compression point 12 cm inside the dipole (second column) and after the crossover (third column). The top line has equal scale on x and z axis and an arrow shows the beam propagation direction, the bottom line is a zoom-in of the first line. Particles with higher energy (blue) that were in the tail of the beam had overcome the lower energy particles (red) in the front.

6. ORION proposal

The Neptune configuration is a very particular one. It is not optimized for ballistic bunching compression experiment. The goal of our experiment was to explore as much as possible of the new scheme and compare with theory and simulations. The idea behind this approach is that once we understand what is going on in the Neptune experiment we would be ready to design a system in which velocity bunching actually increases the brightness of the electron beam.

For example in the proposed Orion injector, the injector is an S-band 1-6 cell gun and the booster accelerating cavities are two x-band structures. To limit the energy spread, the beam has to be compressed before being injected into the shorter wavelength cavities. One competitive proposal is to use the ballistic bunching. A short high gradient standing wave cavity will chirp the beam and the compression will happen in the following drift before the X-band cavity (fig.10).

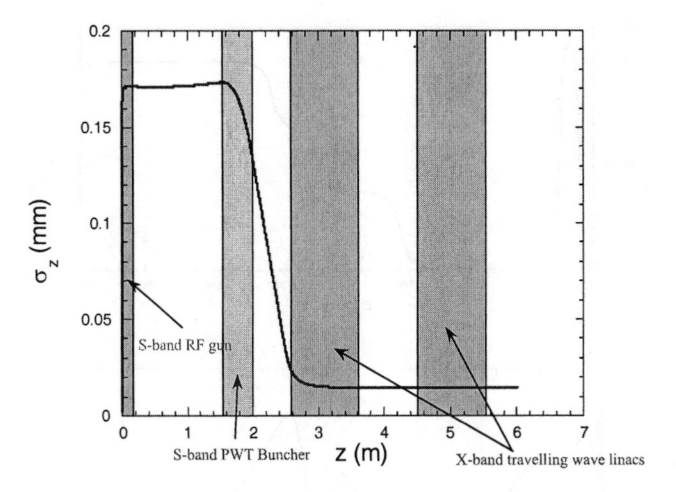

Figure 10. Orion Ballistic bunching. Longitudinal beam size

In order to maintain the transverse beam quality the beam is under-compressed in order to stay away from the deleterious effect of the longitudinal crossovers.

Another important element in a velocity bunching injector design is the solenoidal magnetic field to keep the beam under control while it is compressing. Because the beam is compressing and getting denser the plasma frequency of the transverse oscillation is increasing. Solenoid magnetic field keep the beam focused to control the beam size and the emittance oscillation. The increasing magnetic field to match the beam plasma frequency is given in the Orion case by properly tailoring the solenoids wrapped around the X-band RF structures (fig.11).

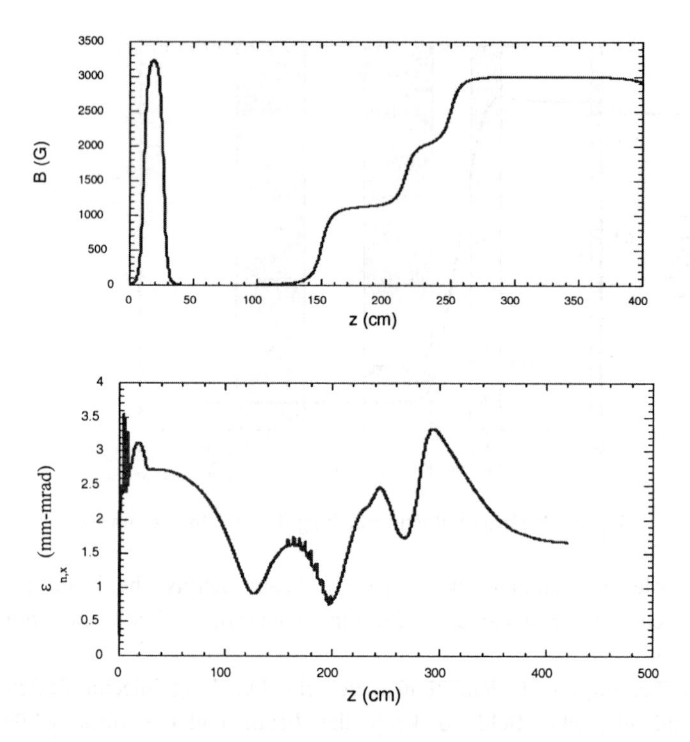

Figure 11. Magnetic field and emittance compensation in Orion case

7. Conclusion

The Neptune ballistic bunching experiment demonstrated the efficiency of the rectilinear RF compression. A compression ratio in excess of 10 was achieved due to the strong suppression of the effect of the RF-non linearities. Experimental investigation on the transverse phase space quality showed emittance degradation. A deeper look into the beam dynamics shows that the technique used to select a small energy spread slice affects the beam quality in a serious way. A first order linear analysis is performed to quantify the anomalous beam compression of a chirped beam going through bending magnet. Three-dimensional simulations are in reasonable agreement with the experimental data.

The Neptune experiment points out the deleterious effect of crossovers in a space charge dominated beam dynamics, but because of the lack of post-acceleration we can not make conclusive statements on the quality of a beam compressed with a velocity bunching scheme. Future experiments are needed to investigate the full potential of the velocity bunching method for increasing the brightness of photoinjector beams, and the use of the magnetic solenoids to keep the beam phase spaces under control. One important point to be addressed is to investigate the difference between the thin lens version 'ballistic' bunching and the long version of the rectilinear compressor. UCLA will be involved also in experiments on this last configuration both at the Pleiades Thomson source and at the INFN SPARC injector.

References

1. Linac Coherent Light Source (LCLS) Design Report, *Tech. Pub., SLAC-R-521* (1998).
2. C.E. Clayton, L. Serafini , *IEEE Trans. On Plasma Science*, **24**, 400 (1996)
3. W. J. Brown, et al., "Electron Beam Production and Characterization for the PLEIADES Thomson X-Ray Source", *Proc. Of Advanced Accelerator Concepts* 2002, Mandalay Beach, California
4. J.B. Rosenzweig, N. Barov, E. Colby, *IEEE Trans. On Plasma Science*, **24**, 409 (1996)
5. S. Anderson et al. "Commisioning of the Neptune Photo-injector" in *Proceedings of the 2001 Particle Accelerator Conference 2001, Chicago*, p. 89 (2001)
6. W. S. Graves et al. "Ultrashort electron bunch length measurement at DUVFEL" in *Proc. Of Particle Accelerator Conference 2001, Chicago*, p. 2224 (2001)
7. L. Serafini, M. Ferrario, "Velocity Bunching in Photo-Injectors" in *Physics of, and science with, the X-Ray Free-Electron Laser*, edited by S.Chattopadhyay et al., AIP Conference Proceedings 581, 19th Advanced ICFA Beam Dynamics Workshop, Arcidosso, pp. 87 (2001)
8. X.J. Wang, et al. "Experimental Observation of high brightness microbunching in RF gun" , *Phys. Rev. E*, **54**, 3121 (1996)
9. "Orion Research Facility: Technical Design Study" SLAC Tech. Rep. (2002)
10. R.J. England et al., "Negative R56 compressor", *Proc. Of Advanced Accelerator Concepts 2002*, Mandalay Beach, California
11. A. Murokh , J.B. Rosenzweig, A. Tremaine, "Coherent Transition Radiation based diagnosis of electron beam pulse shape" in *Proc. Of Advanced Accelerator Concepts 1998, Baltimore*, AIP Conference Proceedings **472**, 38 (1999)
12. L. Giannessi et al., *Nucl. Instr. Meth.*, **393**, 434 (1997)

COMMENTS ON OPTICAL STOCHASTIC COOLING

K.Y. NG
Fermi National Accelerator Laboratory, P.O. Box 500, Batavia, IL 60510, USA
E-mail: ng@fnal.gov

S.Y. LEE AND Y.K. ZHANG
Department of Physics, Indiana University, Bloomington, IN 47405, USA
E-mail: shylee@indiana.edu, yzhang2@indiana.edu

An important necessary condition for transverse phase space damping in the optical stochastic cooling with transit-time method is derived. The longitudinal and transverse damping dynamics for the optical stochastic cooling is studied. We also obtain an optimal laser focusing condition for laser-beam interaction in the correction undulator. The amplification factor and the output peak power of the laser amplifier are found to differ substantially from earlier publications. The required power is large for hadron colliders at very high energy.

1 Introduction

Transit-time optical stochastic cooling (OSC) was first introduced by Zolotorev and Zholents[1], where the optical frequency of $\sim 3\times 10^5$ GHz ($\lambda \sim 1$ μm) is used. This provides a bandwidth more than ten thousand times larger than microwave stochastic cooling[2]. The OSC can be used in low energy electron rings to provide high brightness beams. It can also be used in proton collider rings to increase luminosity by counteracting intra-beam scattering.

In Ref. 1, the horizontal cooling decrement is presented as

$$\alpha_x = \frac{1}{2}\left(\frac{\overline{\Delta(x^2)}}{\overline{x^2}} + \frac{\overline{\Delta(x'^2)}}{\overline{x'^2}}\right) = \frac{1}{2}\left[4GD_0\eta_0' k \exp\left\{-\frac{\overline{\Delta\phi_i}}{2}\right\} - \frac{G^2N_s}{2}\left(\eta_0'^2 + \frac{D_0^2}{\beta^2}\right)\frac{\beta}{\epsilon_x}\right], \tag{1}$$

where, in their notations, D_0 and $-\eta_0'$ represent the dispersion and dispersion gradient at the second undulator. In above, the first term on the right represents cooling while the second term represents heating. Thus, there is no horizontal cooling if $\eta_0' = 0$, which can hardly be correct. In order to understand OSC, we rederived all the equations of the dynamic again. The detail is given in Ref 3.

2 Transit-time OSC

In the transit-time OSC, the beam particles pass through a first undulator where photons are emitted. After amplification by a laser amplifier, these photons meet the beam particles again inside the second undulator and interact with them by correcting their momentum offset. In order to receive the right corrections, the phases of the beam particles are properly adjusted by allowing the beam particles traveling through a bypass as shown in Fig. 1.

Consider the i-th beam particle with a momentum deviation $\delta_i = \Delta P_i/P$, and the betatron phase space coordinates $(x_{i1}, x_{i1}', z_{i1}, z_{i1}')$ at the first undulator. In the

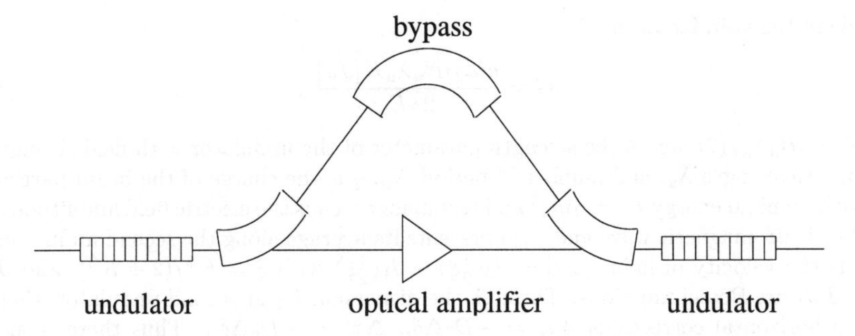

Figure 1. The module for transit time OSC, which consists of two undulators, a laser amplification system, and a by-pass for the beam particles.

Frenet-Serret coordinate system, the path length of the test particle in the bypass section is[4]

$$\ell_i = \int_{s_1}^{s_2} \sqrt{\tilde{x}_i'^2 + \tilde{z}_i'^2 + \left(1 + \frac{\tilde{x}_i}{\rho}\right)^2}\, ds \approx \int_{s_1}^{s_2} \left(1 + \frac{\tilde{x}_i}{\rho}\right) ds\,, \tag{2}$$

where $\hat{x}, \hat{s}$ and $\hat{z}$ form a curvilinear coordinate system with a horizontal bending radius ρ, the coordinates $\tilde{x}_i$ and $\tilde{z}_i$ are the deviations from a reference orbit, and s is the longitudinal coordinate along the reference orbit. We have also assumed $\tilde{x}_i' \ll 1$ and $\tilde{z}_i' \ll 1$ to obtain the last approximate equality. For a bypass with planar geometry, the transverse displacement of an orbiting particle is given by $\tilde{x}_i(s) = x_{\rm co}(s) + M_{11}(s,s_1)x_{i1} + M_{12}(s,s_1)x_{i1}' + D(s)\delta_i$, where $M_{11}(s,s_1)$ and $M_{12}(s,s_1)$ are transport matrix elements of the Hill's equation from the first undulator at s_1 to the second undulator at s_2 via the beam bypass, $x_{\rm co}(s)$ is the closed orbit around the reference orbit, and $D(s)$ is the dispersion function. The path length for the i-th particle in the bypass region becomes

$$\ell_i = \ell_0 + x_{i1}I_1 + x_{i1}'I_2 + \delta_i I_D, \tag{3}$$

where x_{i1}, x_{i1}' are the conjugate phase space coordinates for the i-th particle at the location s_1, and the integrals I_1, I_2, and I_D are

$$I_1 = \int_{s_1}^{s_2} \frac{M_{11}(s,s_1)\,ds}{\rho(s)}\,, \quad I_2 = \int_{s_1}^{s_2} \frac{M_{12}(s,s_1)\,ds}{\rho(s)}\,, \quad I_D = \int_{s_1}^{s_2} \frac{D(s)\,ds}{\rho(s)}\,. \tag{4}$$

The on-momentum particle arrives at the second undulator and sees exactly the zero crossing of the electric field of the photons. The i-th particle, however, arrives with a time delay and therefore at a phase

$$\Delta\phi_i = k(\ell_i - \ell_0) = k(x_{i1}I_1 + x_{i1}'I_2 + \delta_i I_D) \tag{5}$$

from the zero-crossing of the electric field, where k is the wavenumber of the photons. The correction through the electric field is[5]

$$\Delta\delta_i = -\,\mathrm{sgn}(I_D)G\sin(\Delta\phi_i)\,, \tag{6}$$

where the gain factor is

$$G = \frac{q\langle\mathcal{E}_2\rangle N_u \lambda_u K[JJ]}{2\gamma E_b} , \tag{7}$$

$K = qB_u\lambda_u/(2\pi mc)$ is the strength parameter of the undulator with field strength B_u, wavelength λ_u, and number of period N_u, q is the charge of the beam particle with nominal energy $E_b = \gamma mc^2$ and rest mass m, $\mathcal{E}_2$ is the electric field amplitude of the electromagnetic wave and $\langle\ \rangle$ represents its average along the second undulator, c is the velocity of light, $[JJ] = J_0(\frac{1}{2}\xi) - J_1(\frac{1}{2}\xi)$ with $\xi = K^2/(2+K^2)$, and J_0 and J_1 are Bessel functions. Through the dispersion D_2 at second undulator, there are horizontal corrections $\Delta x_i = -D_2\Delta\delta_i$, $\Delta x'_i = -D'_2\Delta\delta_i$. Thus there is also horizontal cooling.

3 Damping Decrements

The correction in Eq. (6) is the result of the interaction of the photons emitted by a beam particle on the same beam particle. This interaction is coherent and produces a damping. However, this particle also interacts with the photons emitted by other beam particles. This interaction is incoherent and produces a growth instead. Adding up cooling and anti-cooling, the resulting longitudinal damping decrement is

$$\alpha_\delta \equiv -\frac{\langle\delta_{ic}^2 - \delta_i^2\rangle}{\sigma_\delta^2} = 2GkI_D e^{-u} - \frac{G^2 N_s}{2\sigma_\delta^2} , \tag{8}$$

where

$$u = \frac{1}{2}k^2[(\beta_1 I_1^2 - 2\alpha_1 I_1 I_2 + \gamma_1 I_2^2)\epsilon_x + I_D^2\sigma_\delta^2] \tag{9}$$

is the total *thermal energy* of the system, with β_1, α_1, and γ_1 the corresponding Twiss parameters at the first undulator. Here, a Gaussian momentum distribution has been assumed for the beam particles at the first undulator with σ_δ being the rms momentum spread.

For the horizontal, we introduce the normalized coordinates $(x, P_x = \beta x' + \alpha x)$. The horizontal damping decrement is just

$$\alpha_x \equiv -\frac{\langle P_{x2c}^2 + x_{2c}^2 - (P_{x2}^2 + x_2^2)\rangle}{\sigma_{x2}^2} = 2GkI_\perp e^{-u} - \frac{G^2 N_s \mathcal{H}_2}{2\epsilon_x} , \tag{10}$$

where

$$\mathcal{H}_2 = \frac{D_2^2 + P_{D2}^2}{\beta_2} \tag{11}$$

is the $\mathcal{H}$-function at 2nd undulator, and

$$I_\perp = -\frac{\beta_1}{\beta_2}\Bigg\{P_{D2}\Bigg[\bigg((\beta_2 M_{21}+\alpha_2 M_{11}) - \frac{\alpha_1}{\beta_1}(\beta_2 M_{22}+\alpha_2 M_{12})\bigg)\bigg(I_1 - \frac{\alpha_1}{\beta_1}I_2\bigg) + \frac{\beta_2 M_{22}+\alpha_2 M_{12}}{\beta_1^2}I_2\Bigg] + D_2\Bigg[\bigg(M_{11} - \frac{\alpha_1}{\beta_1}M_{12}\bigg)\bigg(I_1 - \frac{\alpha_1}{\beta_1}I_2\bigg) + \frac{M_{12}}{\beta_1^2}I_2\Bigg]\Bigg\} . \tag{12}$$

We have introduced for the i-th particle at the second undulator the normalized betatron coordinate $P_{xi2} = \beta_2 x'_{i2} + \alpha_2 x_{i2}$ and the normalized dispersion phase space coordinate $P_{D2} = \beta_2 D'_2 + \alpha_2 D_2$. Thus, we arrive at the necessary condition of horizontal damping: $I_\perp > 0$. Notice that transverse cooling can still be possible even when $D_2 = 0$.

4 Cooling Dynamic

From the damping decrements, it is easy to compute the rates at which the transverse emittance and the momentum are damped:

$$\frac{d\epsilon_x}{dt} = -\frac{2GkI_\perp\epsilon_x}{T_0}e^{-u} + \frac{G^2 N_s \mathcal{H}_2}{2T_0} , \tag{13}$$

$$\frac{d\sigma_\delta^2}{dt} = -\frac{2GkI_D\sigma_\delta^2}{T_0}e^{-u} + \frac{G^2 N_s}{2T_0} . \tag{14}$$

In the special case of $I_D = I_\perp$, there is equal gain in both the transverse and longitudinal directions. We can then combine Eqs. (13) and (14) to obtain

$$\frac{du}{dt} = -\frac{2G_0kI_D}{T_0}ue^{-u} + \frac{G_0^2 N_s v}{2T_0} , \tag{15}$$

where

$$v = \frac{1}{2}k^2[(\beta_1 I_1^2 - 2\alpha_1 I_1 I_2 + \gamma_1 I_2^2)\mathcal{H}_2 + I_D^2] . \tag{16}$$

The optimum gain is

$$\frac{du}{dt} = -\frac{2k^2 I_D^2}{vN_sT_0}u^2 e^{-2u} , \tag{17}$$

which occurs when

$$G_{\text{opt}} = \frac{2kI_D}{vN_s}ue^{-u} . \tag{18}$$

The cooling of the thermal energy u at optimized gain is shown in Fig. 2 starting from the initial value of $u_0 = 3$. Because of the u^2e^{-2u} factor, the cooling slows down as time progresses. When $u \le 1$, the cooling process behaves like $u = \frac{1}{t}$ and becomes very inefficient. However, the OSC takes place through Eq. (6), which is proportional to $\sin(\Delta\phi_i)$, and the correction will be in the *wrong direction* if the phase shift $|\Delta\phi_i| > \pi/2$. Thus, for a large thermal energy, like $u_0 = 3$, only the part of the beam sufficiently close to the on-momentum particle will be cooled while the rest will be heated instead. To ensure OSC, we must make sure that all the particles in the beam (usually 95% is assumed) be within the $\pi/2$ phase shift. Since a bypass can be designed with very small I_1 and I_2, this phase shift requirement translates into

$$u = u_0 \approx \frac{1}{2}(kI_D\sigma_\delta)^2 \le \frac{\pi^2}{48}. \tag{19}$$

As a result, OSC at optimum gain factor is rather inefficient because the emittance of a cold beam decreases inversely with the cooling time. As will be seen below,

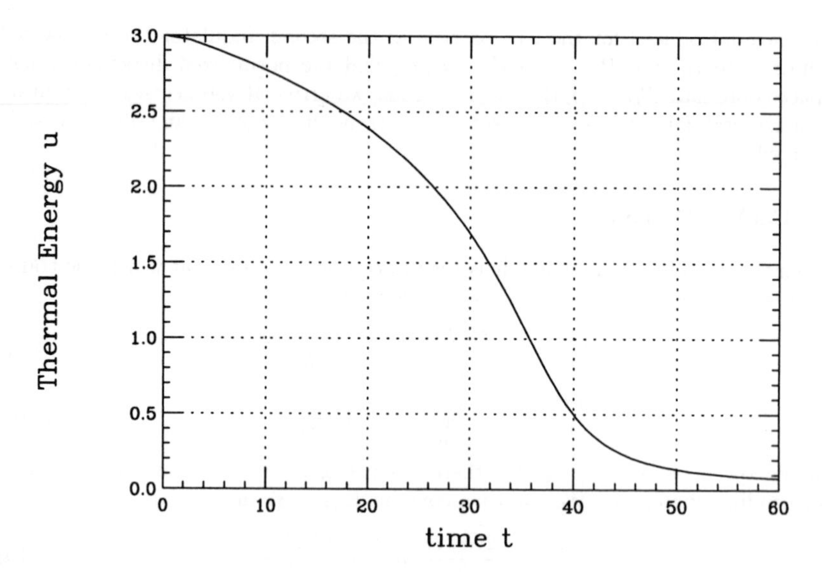

Figure 2. The cooling of the thermal energy u from the initial value $u_0 = 3$, obtained by solving the Eq. (17) is shown as a function of time t with optimal gain factor.

OSC at small gain turns out to be more efficient. Although the cooling represented by Eq. (17) is not exponential, an initial cooling time can nevertheless be defined by

$$\tau_{\text{cool}} = - \left. \frac{u}{du/dt} \right|_{u=u_0} \approx \frac{N_s T_0}{4} \frac{e^{2u_0}}{u_0} \tag{20}$$

for an optimum gain factor.

5 Peak Power

After being amplified by the laser amplifier and focused to the middle of the second undulator, the electromagnetic pulse of the emitted photons at the waist has a time duration of $\Delta t_R = N_u \lambda / c$, an electric field amplitude $\mathcal{E}_2$ and an area of cross section A_2. The total energy of the electromagnetic pulse is

$$W_2 = \frac{1}{2} \epsilon_0 \mathcal{E}_2^2 A_2 c \Delta t_R \ , \tag{21}$$

where ϵ_0 is the electric permittivity of free space. The output peak power of the laser amplifier is therefore

$$\hat{P}_2 = \frac{W_2}{\Delta t_R} N_s = \frac{1}{2} \epsilon_0 \mathcal{E}_2^2 A_2 c N_s \ , \tag{22}$$

where

$$N_s = N_B \frac{N_u \lambda}{2\sqrt{6} c \sigma_\tau} \tag{23}$$

is the number of particles in a sample within a bandwidth of $\Delta\omega|_{\rm FWHM} = \omega/N_u$, and σ_τ is the rms length of a bunch containing N_B beam particles to be cooled. Here, we have assumed 100% photon transmission in the optical amplifier, and assume that the bandwidth of the laser amplifier is larger than that of the undulator radiation.

To relate the peak output power to the gain factor G defined in Eq. (7), we need to compute the electric field amplitude averaged along the second undulator. Since the energy of the electromagnetic pulse must be the same as it travels down the second undulator, we must have

$$\mathcal{E}_2(s) A_2(s) = \mathcal{E}_2 A_2 \ . \tag{24}$$

On the other hand, diffraction requires the cross-sectional area

$$A_2(s) = A_2 \left(1 + \frac{s^2}{Z_R^2}\right) , \tag{25}$$

where $Z_R = \sigma_r/\sigma_{r'}$, the ratio of the transverse size of the electromagnetic wave to its divergence, is called the Rayleigh length, which just plays the role of β^*, the minimum betatron function in the situation of a particle beam. We therefore have

$$\langle \mathcal{E}_2 \rangle = \frac{2\mathcal{E}_2}{N_u \lambda_u} \int_0^{N_u \lambda_u /2} \frac{ds}{\sqrt{1 + (s/Z_R)^2}} , \tag{26}$$

where $N_u \lambda_u$ is the length of the second undulator. Combining Eqs. (7), (22), and (26), we arrive at the peak power for a given gain factor,

$$\hat{P}_2 = G^2 \frac{N_s (E_b/q)^2}{Z_0 \xi N_u [JJ]^2} \mathcal{F}_2 , \tag{27}$$

where Z_0 is the impedance of free space,

$$\mathcal{F}_2 = \frac{A_0/A_2}{8[\ln(A_0/A_2 + \sqrt{1 + (A_0/A_2)^2})]^2} , \tag{28}$$

$A_2 = 2\pi\sigma_r^2$ is the photon beam area at the waist of the second undulator[6] as defined in Eq. (21), and $A_0 = N_u \lambda_u \lambda/4$. Minimum laser amplifier power occurs[7] when $A_2 = 0.3012 A_0$, where $\mathcal{F}_2 = 0.1132$.

The average laser power is equal to the peak power multiplied by the duty factor, i.e.,

$$\langle P \rangle_2 = \hat{P}_2 \frac{n_b 2\sqrt{6} \sigma_\tau}{T_0} = G^2 \frac{(E_b/q)^2}{Z_0 \xi [JJ]^2} \frac{N_B n_b \lambda}{C} \mathcal{F}_2 , \tag{29}$$

where n_b is the number of bunches, σ_τ is the rms bunch length in time, and C is the circumference of the storage ring. Note that the average power is proportional to the total number of particles in the storage ring.

5.1 Peak Laser power for optimal gain

Substituting the optimal gain of Eq. (18) into Eq. (28), we obtain the output peak power of the laser amplifier given by

$$\hat{P}_2 = \frac{N_s (E_b/q)^2}{Z_0 \xi [JJ]^2 N_u} \left(\frac{2kI_D}{v\, N_s} u e^{-u} \right)^2 \mathcal{F}_2 \,, \tag{30}$$

Since the cooling rate is inversely proportional to N_s, the peak power for an optimized cooling of N_s particle is also inversely proportional to N_s. Because of the stability condition of $u \leq \pi^2/48$ in Eq. (19), the peak power is highly reduced.

Application to Hadron Machines

Figure 3 shows the peak power versus γ (beam energy) for proton storage rings at optimal gain. The laser wavelength is taken to be $\lambda = 1$ μm and each undulator has

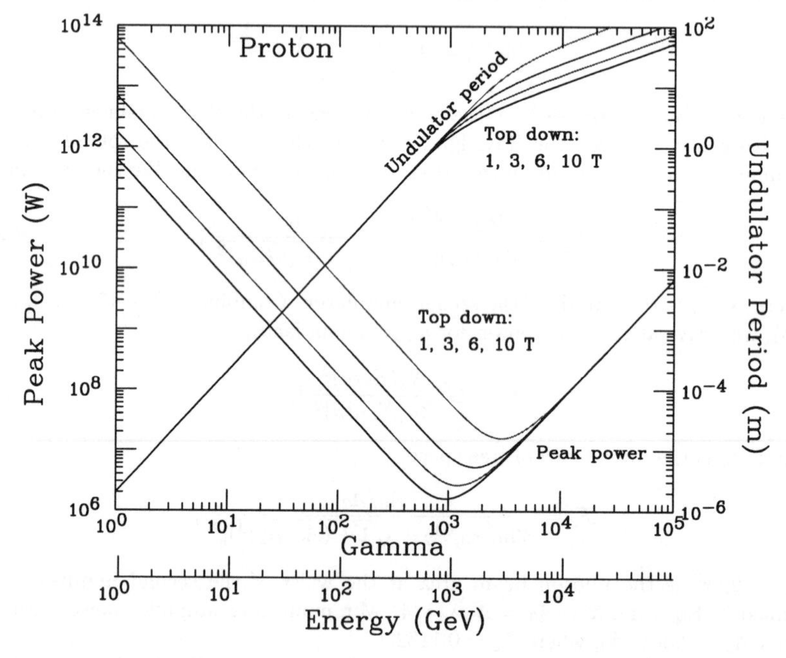

Figure 3. The peak laser amplifier power vs γ for an optimal gain in the optical stochastic cooling for a proton storage ring (TEVATRON). The parameters for the TEVATRON are $\sigma_\ell = 0.37$ m, $\sigma_\delta = 1.3 \times 10^{-4}$, $n_b = 36$ bunches, each containing $N_B = 2.7 \times 10^{11}$ particles, $E_b = 1$ TeV, the mean radius of the TEVATRON of 1000 m, and $B_u = 10$ T. The initial cooling time is given by Eq (20) with $u_0 = \pi^2/48$ or $\tau_{\rm cool} \approx 57$ s.

$N_u = 10$ periods. Most parameters correspond to the TEVATRON at Fermilab: $N_B = 2.7 \times 10^{11}$ particles, rms bunch length $\sigma_\ell = 0.37$ m, and $\sigma_\delta = 1.3 \times 10^{-4}$.

With the TEVATRON revolution period of $T_0 = 20.96\ \mu$s, the initial cooling time is 57 s given by Eq. (20). The magnetic field of the undulator varies from 1 to 10 T.

For a fixed laser wavelength and the undulator magnetic field, the undulator parameter is obtained by solving the cubic equation:

$$\lambda = \frac{\pi mc}{2qB_u\gamma^2} K(2+K^2)\ , \tag{31}$$

from which the undulator period λ_u can be solved and plotted in Fig. 3. The self-consistent solution gives $K \sim \gamma^2$ at low energies and $\hat{P}_2 \sim \frac{(E_b/q)^2}{\xi} \sim \frac{(E_b/q)^2}{K^2} \sim \frac{1}{\gamma^2}$; i.e., it requires a large laser power to compensate the small coherent radiation flux for hadron beams at low energies. At high energies, the particle beam is stiff and the number of photons emitted in the solid angle $\lambda/(N_u\lambda_u)$ becomes saturated ($\xi \to 1$). The output power increases as γ^2 instead. The position of the minimum laser power can be easily calculated to be

$$\gamma_{\min} = \sqrt{\frac{4\sqrt{2}\pi}{3\sqrt{3}}}\sqrt{\frac{mc}{qB_u\lambda}}\ . \tag{32}$$

The TEVATRON at 1 TeV happens to be near the minimum of the power-vs-gamma curve and is therefore favored[8] by OSC. The undulator period of $\lambda_u = 1.93$ m ($B_u = 6$ T) is long enough for superconducting undulators. The Relativistic Hadron Collider (RHIC) lies on the left side of the minimum and has its output amplifier power scale as $\gamma^{-2}(m/q)^4$. The Very Large Hadron Collider (VLHC) lies on the right side of the minimum and has its output power scale as $(m\gamma/q)^2$.

Application to Electron Machines

Figure 4 is a similar plot for electron rings. Because of the small electron mass, there is no need to consider high magnetic field undulators and we set the magnetic field at $B_u = 1$ T. The bunch parameters are $N_B = 1.0 \times 10^{11}$, $\sigma_\ell = 1$ cm, and $\sigma_\delta = 1.3 \times 10^{-4}$. Besides laser wavelength $\lambda = 1\ \mu$m, we also include $\lambda = 5$, 20, and 100 μm, where the corresponding numbers of sampling particle are $N_s = 2.0 \times 10^7$, 1.0×10^8, 4.1×10^8, and 2.0×10^9 respectively. The initial cooling time for the optimal gain given by Eq. (20) is $\tau_{\text{cool}} = 1.8 N_s T_0$, which depends on the revolution period T_0.

When $\lambda = 1\ \mu$m, the minimum peak power occurs at $\gamma_{\min} = 76.3$ or $E_b =$ 39.0 MeV; i.e., nearly all electron storage rings lie on the right side of the minimum. However, because of the $(m/q)^2$ factor, the output power of the amplifier is very much reduced. That does not imply that OSC favors electron rings of high energies because the radiation damping rate increases rapidly with energy. To be effective, the OSC cooling rate, discussed in the last paragraph, has to be faster than the radiation damping rate of the electron ring.

Now, we consider a possible example of converting the Cooler Ring at the Indiana University Cyclotron Facility (IUCF) to an electron ring and OSC is applied at the Ti-Saphire laser wavelength $\lambda = 0.78\ \mu$m with $N_u = 10$ and $\lambda_u = 5$ cm. Setting an initial cooling time of 0.10 s, we find $N_s = 1.92 \times 10^5$. Since the rms bunch length is 3.6 cm with the rf system, we find the number of particles in a bunch

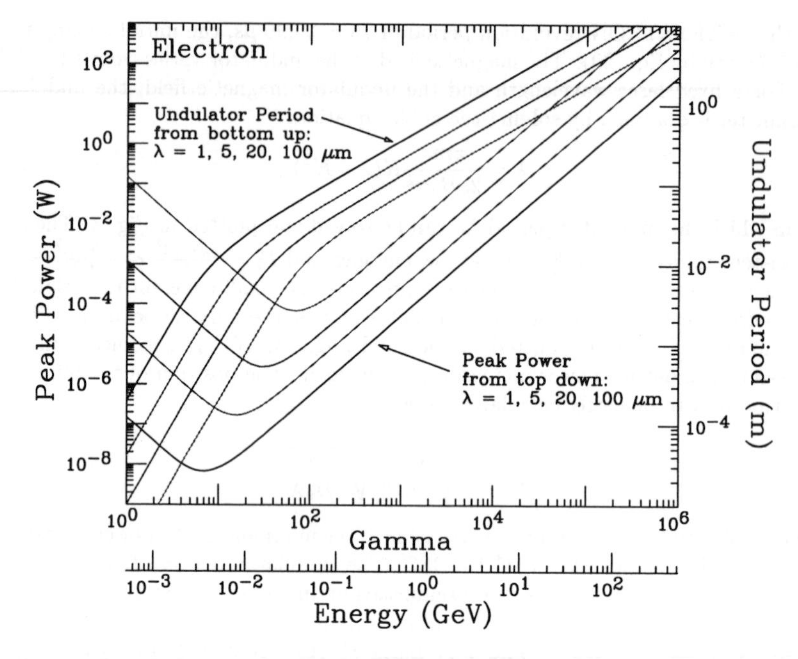

Figure 4. The peak laser amplifier power vs γ for optimal gain in the optical stochastic cooling for electron storage rings. The parameters for the electron storage ring are $\sigma_\ell = 1$ cm, $\sigma_\delta = 1.3\times 10^{-4}$, $N_B = 1.0 \times 10^{11}$, and $B_u = 1.0$ T.

$N_B = 4.36 \times 10^9$. At $E_b = 500$ MeV, the required laser peak power is $\hat{P} = 39$ W. The peak power is much larger than that of Fig. 4 because the number of the sampling particle is much smaller in this example. The natural horizontal emittance and the OSC-equilibrium emittance are plotted in Fig. 5 as functions of beam energy. Other parameters used in the plots are ring circumference $C = 85.03$ m, bending radius $\rho = 2.44$ m, momentum compaction $\alpha_c = 0.04938$, rf harmonic $h = 15$, and a bucket-to-bunch-height ratio of 40. We note that the emittance is OSC-damped by almost an order of magnitude or more when the electron energy is below 500 MeV. However, at higher energies, OSC damping is completely inefficient because of the rapidly increasing radiation damping rates. As a whole, applications of OSC to low energy electron storage rings can be useful for attaining high brightness electron beams.

6 Laser Power for Low-Gain Regime

At an optimal gain, the laser power requirement is usually high (see Fig. 3), and the damping dynamics is not necessarily the most favorable for beam cooling. It would be useful to consider the OSC in the low-gain regime. As an example, if the second undulator location is designed to be non-dispersive, i.e. $D_2 = P_{D2} = 0$,

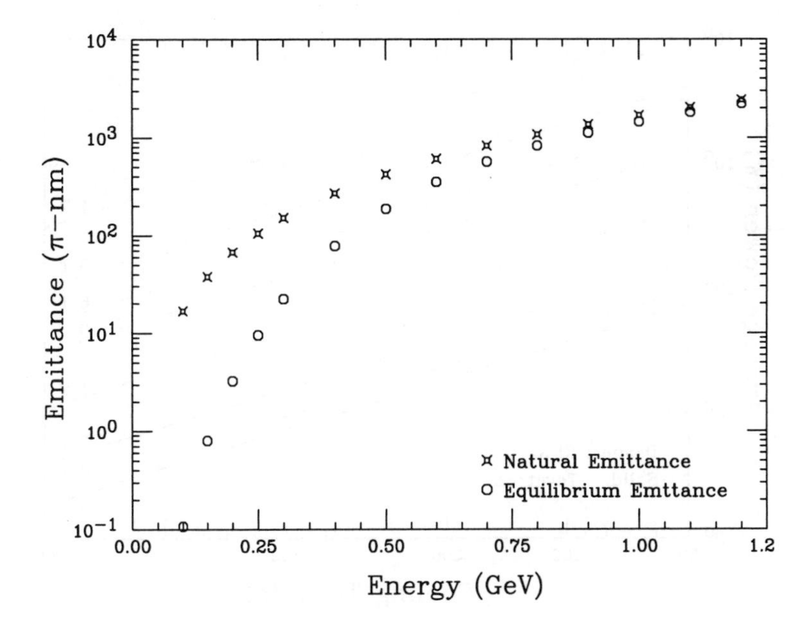

Figure 5. The equilibrium electron emittance for a cooling time of 0.1 s is shown as a function of the electron beam energy.

the betatron cooling and heating vanish. The OSC becomes a one-dimensional momentum cooling device and the cooling bypass design is simplified. Let $u_x = \frac{1}{2}k^2(\beta_1 I_1^2 - 2\alpha_1 I_1 I_2 + \gamma_1 I_2^2)\epsilon_x$, and $u_\delta = \frac{1}{2}k^2 I_D^2 \sigma_\delta^2$. In the low-gain regime, the incoherent heating term is now small and can be neglected. The damping equation becomes

$$\frac{du_\delta}{dt} = -\frac{2GkI_D}{T_0} e^{-u_x} u_\delta e^{-u_\delta} \, . \tag{33}$$

Since $u_\delta \leq \pi^2/48$ is small, the damping is almost exponential and becomes more so as the cooling proceeds and will continue until the cooling force is balanced by the heating forces coming from rf noise, intra-beam scattering, etc. This is highly in contrast with the cooling at optimum gain-factor discussed in Sec. III.C.2, where the cooling process becomes more and more inefficient as the beam is cooled. With $u_x = 0$, the cooling time is

$$\tau_{\rm cool} \approx \frac{e_\delta^u}{2GkI_D} T_0 \, . \tag{34}$$

The resulting peak power is

$$\hat{P}_2 = \left(\frac{T_0}{\tau_{\rm cool}}\right)^2 \frac{N_s \left(E_b/q\right)^2 e^{2u_\delta}}{Z_0 N_u \xi [\mathrm{JJ}]^2 (2kI_D)^2} \mathcal{F}_2 \, , \tag{35}$$

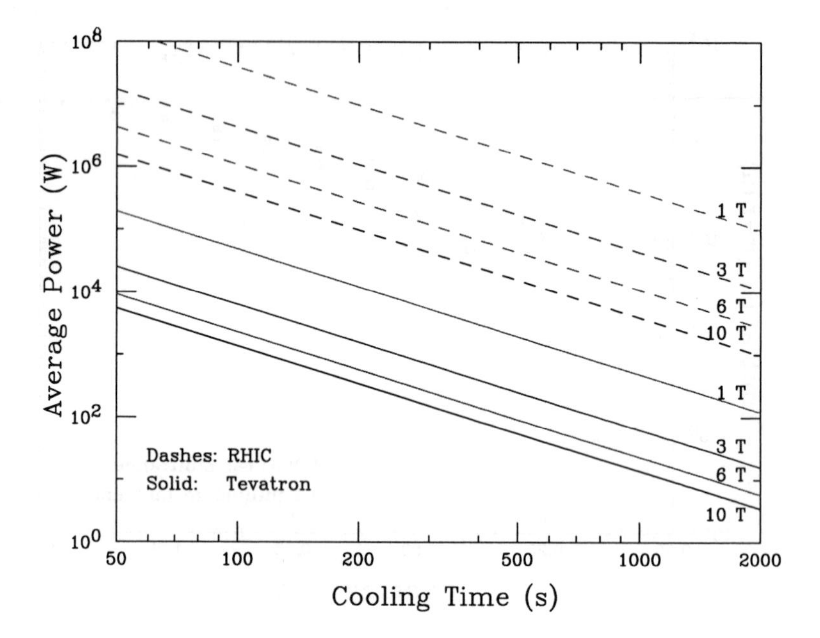

Figure 6. The laser amplifier power in the low gain regime for the TEVATRON at 1 TeV and the RHIC at 100 GeV/amu. The laser wavelength is $\lambda = 1\mu$, and the undulator parameters are $N_u = 10$ with the magnetic field strength B_u listed in the graph. The corresponding beam parameters are $\sigma_\ell = 0.37$ m, $\sigma_\delta = 1.3 \times 10^{-4}$, and $n_b = 36$ bunches, each containing $N_B = 2.7 \times 10^{11}$ particles, at $E_b = 1$ TeV for the TEVATRON; and $\sigma_\tau = 2$ ns, $\sigma_\delta = 1.0 \times 10^{-3}$, $n_b = 60$ bunches, each containing $N_B = 1.0 \times 10^9$ particles, $E_b = 100$ GeV/nucleon for gold ion, and the circumference of 3833.85 m for the RHIC.

The average power of the laser amplifier is

$$\langle P \rangle_2 = \left(\frac{T_0}{\tau_{\text{cool}}} \right)^2 \left(\frac{n_b N_B \lambda}{C} \right) \frac{(E_b/q)^2 \, e^{2u_\delta}}{Z_0 \xi [\text{JJ}]^2 (2kI_D)^2} \mathcal{F}_2 \,. \tag{36}$$

Note that the average power depends on the total number of particles $n_b N_B$ in the ring and the square of the energy over charge $(E_b/q)^2$.

Figure 6 shows the average power requirement versus cooling time in the low gain regime, where the undulator parameters are $\lambda = 1.0$ μm, $N_u = 10$, and the undulator magnetic field varying from 1 T to 10 T. The corresponding beam parameters are $\sigma_\ell = 0.37$ m, $\sigma_\delta = 1.3 \times 10^{-4}$, $n_b = 36$ bunches each containing $N_B = 2.7 \times 10^{11}$ protons at $E_b = 1$ TeV for the TEVATRON whose mean radius is 1 km, while $\sigma_\tau = 2.0$ ns, $\sigma_\delta = 1.0 \times 10^{-3}$, $n_b = 60$ bunches each containing $N_B = 1.0 \times 10^9$ gold ions ($A = 197$ and $Z = 79$) at $E_b = 100$ GeV/nucleon for the RHIC whose circumference is 3833.85 m. We see that for a cooling time of 1200 s which is fast enough to counteract intra beam scattering, the average output power for the TEVATRON is only 16 W when superconducting undulators at $B_u = 6$ T are used. On the other hand, the average output power for the RHIC is more than

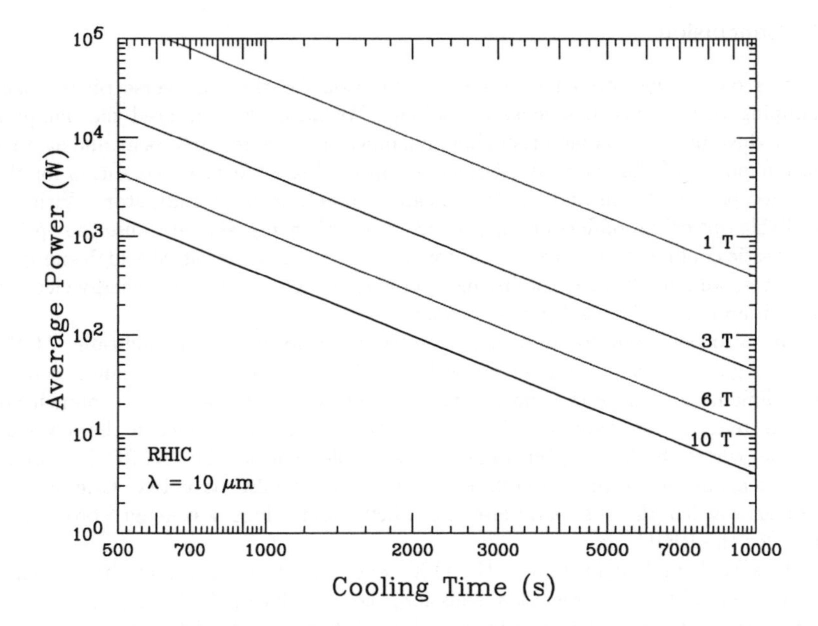

Figure 7. The laser amplifier power in the low gain regime for the RHIC at 100 GeV/amu with the laser wavelength of $\lambda = 10\mu$ and the undulator parameters are $N_u = 10$. The magnetic field strength B_u is listed in the graph. The corresponding beam parameters are $\sigma_\tau = 2$ ns, $\sigma_\delta = 1.0 \times 10^{-3}$, $n_b = 60$ bunches, each containing $N_B = 1.0 \times 10^9$ particles, $E_b = 100$ GeV/nucleon for gold ion, and the circumference of 3833.85 m.

1000 times larger. Because γ is one order of magnitude smaller than that of the TEVATRON, the undulator period becomes $\lambda_u = 2.3$ cm, two orders of magnitude smaller. This implies that superconducting undulators may not be used and only 1 T undulators are possible. The output power for the RHIC application is therefore increased at least one more order of magnitude.

Note that when the laser wavelength is chosen to be $\lambda = 1$ μm for the RHIC, the undulator period is $\lambda_u = 2.3$ cm, which may be difficult to attain a high field undulator magnet. The wiggler number becomes very small, and the required laser amplification power becomes very large (see Fig. 6). If there is a longer wavelength high bandwidth laser, e.g. $\lambda = 10\mu$m, the undulator period becomes 23 cm, and the required laser amplification power will be greatly reduced as shown in Fig. 7. Although it may still require 80 W of laser amplification power to attain a 1 hr cooling time (for $B_u = 6$ T), this is dramatically improved in comparison with the 1000 W requirement shown in Fig. 6.

7 Conclusion

In this paper, we derived a necessary condition for the transverse phase space damping in the optical stochastic cooling. We have also explored the damping rates, the amplification factor, cooling dynamics, and the required peak and average output power of the laser. We derived an optimal laser focusing condition for the charged particle beam and the laser beam interaction in an undulator. With the available optical amplifiers at the present, it is rather impractical to use the optical stochastic cooling method to cool proton and heavy ion beams at *very* high energies. However, we find that the cooling method may be beneficial to low energy electron beams, and around 1 TeV proton beams.

We also point out the difficulties of OSC with optimal gain condition. At the optimal gain, the required laser power is usually very large. As the beam is cooled, it is difficult to change the charged particle optics for a larger kI_D to compensate the decrease in emittances. The best solution is to cool beams in the low gain regime, where the heating term may be negligible. For the TEVATRON, it seems to be feasible to use the Ti-Saphire $\lambda = 0.78$ μm for OSC at 1 TeV. One needs a shorter wavelength broadband laser for VLHC, and a long wavelength broadband laser for the RHIC.

In actual implementation of the OSC, one should also consider the efficiency of laser pumping and optical transmission, the linearity of the laser amplification, noise, etc. These problems can be considered if there is a realistic project to carry out experimental tests.

References

1. M. Zolotorev and A. Zholents, Phys. Rev. E, **50**, p.3087 (1994).
2. S. van der Meer, *Stochastic damping of betatron oscillations*, CERN/ISR PO/72–31 (unpublished), (1972); D. Möhl, CERN Accelerator School Report, CERN 87-03, p.453 (1987).
3. S.Y. Lee, Y.K. Zhang, and K.Y. Ng, submitted to Phys. Rev. E for publication.
4. see for example, S.Y. Lee, *Accelerator Physics*, (World Scientific Inc., Singapore, 1999).
5. C. Pellegrini and I.E. Campisi, AIP conference Proceedings, **105**, p. 1062 (1983); E.D. Courant, C. Pellegrini, and W. Zakowicz, Phys. Rev. A **32**, 2813 (1985).
6. If we assume that the photon beam be distributed as *bi-Gaussian* radially but uniformly along the longitudinal s-direction, the total energy of the photons can be written as

$$W_0 = \int \frac{W_0/\Delta s}{2\pi\sigma_x\sigma_z} \exp\left(-\frac{x^2}{2\sigma_x^2} - \frac{z^2}{2\sigma_z^2}\right) dx dz \Delta s \ ,$$

where the energy density is $\frac{1}{2}\epsilon_0\mathcal{E}_2^2 = (W_0/\Delta s)/(2\pi\sigma_x\sigma_z)$ and σ_x and σ_z are the spreads in the transverse directions. Here $\mathcal{E}_2$ is the peak field at $x = 0$ and $z = 0$. Now, we can write $W_0 = \frac{1}{2}\epsilon_0\mathcal{E}_2^2 A_2 \Delta s$, i.e. the effective photon beam area is $A_2 = 2\pi\sigma_x\sigma_z$. For a photon beam with cylindrical symmetry, we find $A_2 = 2\pi\sigma_r^2$.

7. The emittance of the photon beam, $\lambda/(4\pi)$, may substantially differ from the emittance of the charged particle beams, e.g. 3.3 nm for the TEVATRON at 1 TeV and 16 nm for the RHIC beam at 100 GeV/amu. The efficiency of the cooling may be reduced by the overlap area between the charged particle and the photon beams. The optimal energy gain at the second undulator for the charged particle beams is equivalent to the minimum in the laser power.
8. At high energy, the self-consistent solution of $\lambda = \lambda_u(2+K^2)/(4\gamma^2)$ and $K = qB_u\lambda_u/(2\pi mc)$ leads to a conclusion that the peak power of the laser amplifier is proportional to γ^2 of Eq. (30). The minimum power requirement occurs at $\gamma_{\min} \sim 1.14\sqrt{\pi mc/(qB_u\lambda)}$.

SUB-PICOSECOND COMPRESSION BY VELOCITY BUNCHING IN A PHOTO-INJECTOR

PH. PIOT,
Deutsches Elektronen-Synchrotron (DESY), D-22607 Hamburg, Germany

L. CARR, W.S. GRAVES, H. LOOS
Brookhaven National Laboratory, Upton, NY 11973, USA

We present experimental results of a bunch compression scheme that uses a traveling wave accelerating structure as a bunch compressor. The bunch length issued from a laser-driven radio-frequency electron source was compressed by a factor >3 using an S-band traveling wave structure located immediately downstream of the electron source. Experimental data are found to be in good agreement with particle tracking simulations.

1 Introduction

In the recent years there has been an increasing demand on ultrashort electron bunches to drive short-wavelength free-electron lasers and novel accelerating techniques such as plasma-based accelerators [1,2]. Short bunches are commonly obtained using magnetic compression. In such a scheme, the compression is performed using a series of dipoles arranged in a chicane configuration to introduce an energy-dependent path-length. Thus an electron bunch can be shortened by introducing the proper time-energy correlation along the bunch prior to the chicane. However, problems inherent to magnetic compression such as transverse emittance dilution due to the bunch self-interaction via coherent synchrotron [3] has brought back the idea of bunching with radio-frequency (rf) structures [4]. It was recently proposed to incorporate the latter method (henceforth named velocity bunching) into the next photo-injectors designs [5]. The velocity bunching relies on the phase slippage between the electrons and the rf-wave that occurs during the acceleration of non ultra-relativistic electrons. In this paper after presenting a brief analysis of the velocity scheme, we report on its exploration at the deep ultraviolet free-electron laser (DUV-FEL) facility. The measurements are compared with tracking simulations performed with the computer program ASTRA [7].

2 Analysis of the velocity bunching technique

In this section we elaborate a simple model that describes how the velocity bunching works. A detailed discussion is provided in Reference [5].
An electron in a rf traveling wave accelerating structure experiences the longitudinal electric field:

$$E_z(z,t) = E_o \sin(\omega t - kz + \psi_o), \tag{1}$$

where E_o is the peak field, k the rf wavenumber and ψ_o the injection phase of the electron with respect to the rf wave. Let $\psi(z,t) = \omega t - kz + \psi_o$ be the relative phase of the electron w.r.t the wave. The evolution of $\psi(t,z)$ can be expressed as a function of z solely:

$$\frac{d\psi}{dz} = \omega\frac{dt}{dz} - k = \frac{\omega}{\beta c} - k = k\left(\frac{\gamma}{\sqrt{\gamma^2-1}} - 1\right). \tag{2}$$

Introducing the parameter $\alpha \doteq \frac{eE_o}{kmc^2}$, we write for the energy gradient [6]:

$$\frac{d\gamma}{dz} = \alpha k \sin(\psi). \tag{3}$$

The system of coupled differential equations (2) and (3) with the initial conditions $\gamma(z=0) = \gamma_o$ and $\psi(z=0) = \psi_o$ describes the longitudinal motion of an electron in the rf structure. Such a system is solved using the variable separation technique to yield:

$$\alpha\cos\psi + \gamma - \sqrt{\gamma^2-1} = \mathcal{C}. \tag{4}$$

Or, expliciting ψ as a function of γ:

$$\psi(\gamma) = \arccos\left(\frac{\mathcal{C} - \gamma + \sqrt{\gamma^2-1}}{\alpha}\right). \tag{5}$$

Here the constant of integration is set by the initial conditions of the problem: $\mathcal{C} = \alpha\cos\psi_o + \gamma_o - \sqrt{\gamma_o^2-1}$. The latter equation gives insights on the underlying mechanism that provides compression. In order to get a simpler model, we consider the limit: $\psi_\infty \doteq \lim_{\gamma\to\infty}\psi(\gamma) = \arccos\left(\cos(\psi_o) + \frac{1}{2\alpha\gamma_o}\right)$; we have assumed γ_o is larger than unity and did the approximation $\gamma_o - \sqrt{\gamma_o^2-1} \simeq 1/(2\gamma_o)$. After differentiation of Eq. (5), given an initial phase $d\psi_o$ and energy $d\gamma_o$ extent we have for the final phase extent $d\psi_\infty$:

$$d\psi_\infty = \frac{\sin(\psi_o)}{\sin(\psi_\infty)}d\psi_o + \frac{1}{2\alpha\gamma_o^2\sin(\psi_\infty)}d\gamma_o. \tag{6}$$

Hence depending upon the incoming energy and phase extents, the phase of injection in the rf structure ψ_o can be tuned to minimize the phase extent after extraction i.e. to ideally make $d\psi_\infty \to 0$. We note that there are two contributions to $d\psi_\infty$ in Eq. (6): the first term $\partial\psi_\infty/\partial\psi_o$ comes from the phase slippage (the injection and extraction phases are generally different). The second term $\partial\psi_\infty/\partial\gamma_o$ is the usual "ballistic" bunching. To illustrate the compression mechanism we consider a two macro-particles model. In Figure 1 we present results obtained by directly integrating the equation of motion for two non-interacting particles that are injected into a 3 m long ideal traveling wave structure. Given the incoming phase $\Delta\psi_o$ and energy spread $\Delta\gamma_o$ between the two particles and the accelerating gradient of the structure (taken to be 20 MV/m), we choose the injection phase to minimize the bunch length at the structure exit.

3 Experimental results

The measurement was carried out at the DUV-FEL facility of Brookhaven national laboratory [11]. A block diagram of the linear accelerator is given in Fig. 2. The

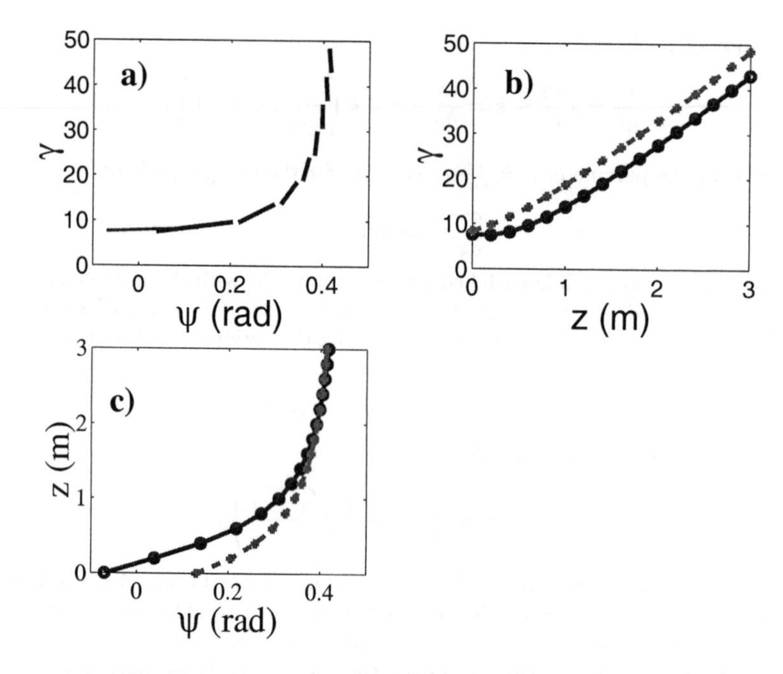

Figure 1. Simulation, using a two macro-particles model, of the velocity compression in a 3 m long traveling wave structure. The initial conditions are (ψ_o, γ_o)=(0.03,8) and the macro-particle spacing is $(\Delta\psi_o, \Delta\gamma_o)$=(0.1, 0.4). Plot **a)** shows snapshots at different z of the longitudinal phase space each segment extremities is determined by the two macro-particles positions. Plots **b)** and **c)** present the energy gain and phase evolution of the two macro-particles versus z. In these two latter plots, solid lines represent the leading particle and dashed lines the trailing one.

electron bunches of ~4 MeV, generated by a laser-driven rf electron source, are accelerated by a series of four linac sections. The linac sections consist of 2.856 GHz traveling wave structures operating on the $2\pi/3$ accelerating mode. The structures are approximately 3 m and can operate with an average accelerating voltage up to 20 MV/m. Nominally the bunch is shortened using a magnetic bunch compressor chicane located between the second and third linac sections. In this latter case, the linac sections L1, L3, L4 are ran on-crest while the linac L2 is operated to impart the proper time-energy correlation along the bunch to enable compression as the beam pass through the magnetic chicane.

To investigate the velocity bunching scheme, the linac section L1 was used as a buncher: its phase was varied and, for each phase setting, the section L2 was properly phased to maximize the beam energy with sections L3 and L4 turned off. The magnetic bunch compressor was turned off during the measurement. The nominal settings for the different rf and photo-cathode drive-laser parameters are gathered in Table 1.

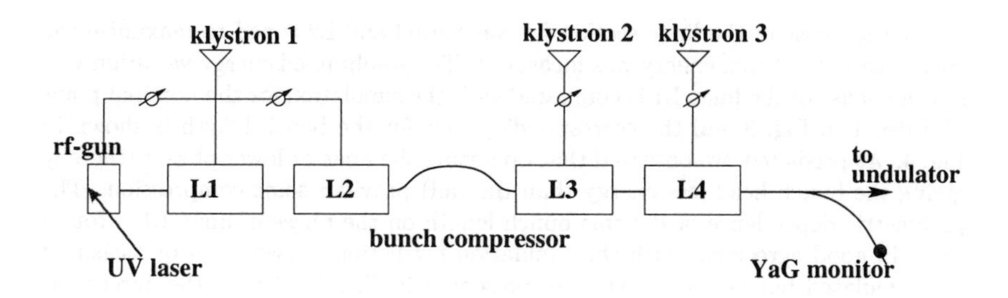

Figure 2. Overview of the Deep ultra-violet free-electron laser (DUV-FEL) accelerator. L1, L2, L3, and L4 are the four linac sections.

parameter	value	units
laser injection phase	40 ± 5	rf-deg
laser radius on cathode	0.75 ± 0.1	mm
laser rms length	1.15 ± 0.1	ps
E-peak on cathode	83 ± 1	MV/m
L1 average accelerating field	10.5 ± 0.1	MV/m
L2 average accelerating field	13.2 ± 0.1	MV/m

Table 1. *Rf and photo-cathode drive-laser setting. The values have been directly measured or inferred from beam properties.*

The measurements of bunch length that follow are compared with simulations using the tracking code ASTRA [7], a macro-particle code based on a rotational symmetric space charge algorithm. ASTRA incorporates a detailed model for the traveling wave accelerating structure [8,9]. For the simulation we used the settings of Table 1 for all the parameters. The laser transverse distribution was modeled by a uniform radial distribution with 0.75 mm radius, and the measured time profile, using a cross-correlation technique, was directly loaded into the simulation.

Both time- and frequency-domain techniques were used to characterize the bunching process as the phase of the linac L1 was varied.

The time-domain charge density was directly measured using the so-called zero-phasing method [12,13]. In the present case, we use the linac section L3 to cancel the incoming time-energy correlation, and operate the linac L4 at zero-crossing to introduce a controlled linear time-dependent energy chirp along the bunch (we investigate both zero-crossing points). The bunch is then directed to a beam viewer (YaG monitor in Fig. 2) downstream of a 72° angle spectrometer. The viewer, located at a dispersion of $\eta = 907$ mm, allows the measurement of the bunch energy distribution.

As the phase of the linac section L1 was varied and L2 tuned to maximize the energy gain, the beam energy was measured. The so-obtained energy variation versus the phase of the linac L1 is compared with the simulation for the nominal point of Table 1 in Fig. 3 and the corresponding plot for the bunch length is shown in Fig. 4. As predicted, we observed that operating the linac at lower phases (thereby giving the bunch head less energy than the tail) provides some compression. The parametric dependence of the rms bunch length on the phase of linac L1 is found to be in good agreement with the simulation predictions. Two cases of measured and simulated bunch time-profile are presented in Fig. 5. Again, the agreement between simulation and experiment is fairly good taking into account the uncertainties associated to the zero-phasing method. Noteworthy is the achieved peak current of 150 A.

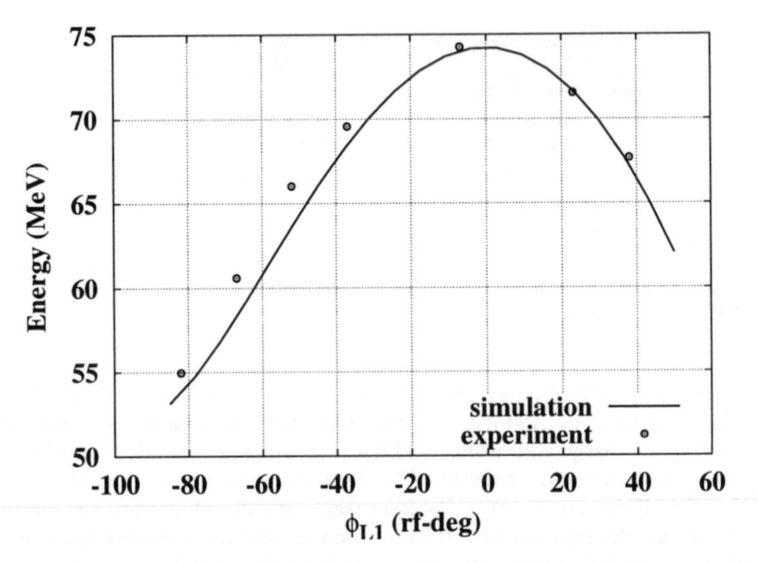

Figure 3. Total energy versus phase of linac section L1. The points are measurements obtained parasitically to the bunch length measurement. The solid line is a simulation result.

The frequency-domain technique is based on the measurement of the coherent radiation emitted by the electron bunch via some electromagnetic process. In the coherent regime (i.e. for frequencies $\omega \sim 2\pi/\sigma_t$ where σ_t is the rms bunch length) the radiated power scales with the squared charge and depends on the bunch form factor. Thus it provides indirect informations on the bunch time-profile. In DUV-FEL, we detect the far-field radiation associated to the geometric wake field caused by aperture variation along the accelerator (e.g. the irises of the rf-structure). The radiation shining out of a single-crystal quartz vacuum window, located prior to the linac section L3, was detected with a He-cooled bolometer. The detection system, composed of the bolometer and the vacuum extraction port, has a good frequency

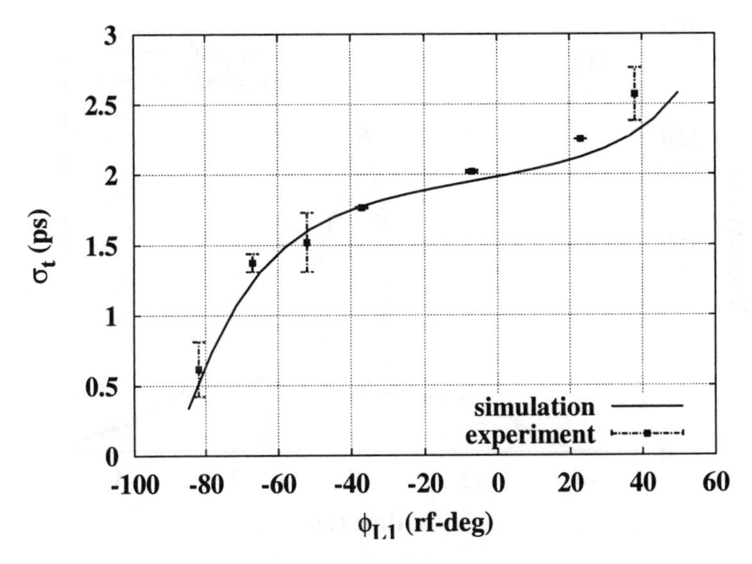

Figure 4. Rms bunch length versus phase of the linac section L1.

response within the range $[\omega_l, \omega_u] \simeq [1.2, 50]$ GHz. The lower and upper frequency limits being respectively due to diffraction effects related to the finite size of the detector and transmission function of the vacuum extraction port. Given the bunch charge Q and the Fourier transform of the bunch time-profile $\tilde{S}(\omega)$, the power is expected to scale as $P \sim Q^2 \int_{\omega_l}^{\omega_u} d\omega|\tilde{S}(\omega)|^2 \propto Q^2/\sigma_t$. The typical signal observed as the charge is varied is presented in Fig. 6: the nonlinear behavior observed confirms the emitted radiation is not incoherent. From simulation we expect the power to scale as $P \propto Q^{1.37}$ a number close to the one resulting from the fit of the data $P \propto Q^{1.57}$

In Figure 7, the measured bolometer output signal versus the phase of L1 is compared with the expectation (1) calculated from the simulated phase space density and (2) computed from the measured bunch time profile obtained by zero-phasing. As expected the increase of the coherent signal is an unambiguous signature of the bunch being compressed (the charge was monitored during the measurement and remained constant to 200±20 pC).

The data points computed from the measured time profile f_{meas} were obtained by numerically computing the Fourier transform of the bunch time-profile (using a FFT algorithm) and by performing the integration:

$$f_{meas} = \int_{\omega_l}^{\omega_u} d\omega|\tilde{S}(\omega) \times R(\omega)|^2. \tag{7}$$

where $R(\omega)$ stands for the frequency response of the detection system.

To generate the data points from the simulated phase space distributions

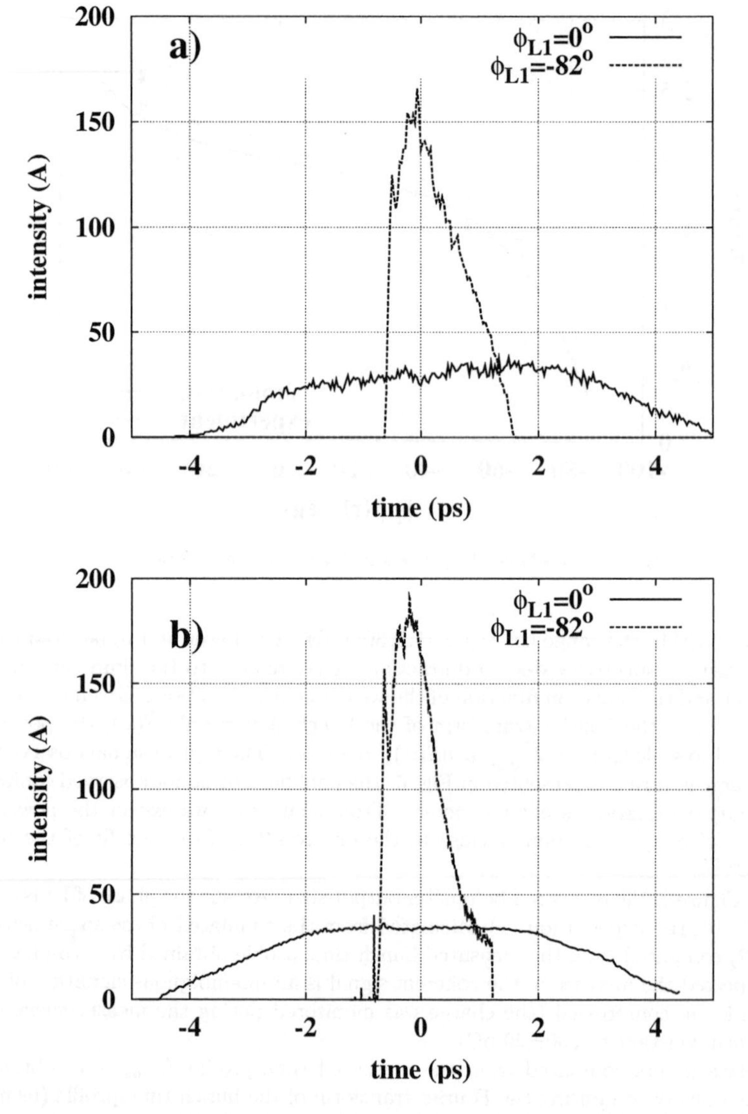

Figure 5. Comparison of the bunch time-profile for L1 on crest ($\phi_{L1} = 0°$), and $-82°$ off-crest. Plot **a)** was generated by tracking simulation; plot **b)** is a direct measurement using the zero-phasing method. The time > 0 corresponds to the bunch tail

f_{simu} we write the time-profile, $S(t)$ as a Klimontovitch distribution:

$$S(t) = \frac{1}{N} \sum_{i=1}^{N} \delta(t_i - t), \tag{8}$$

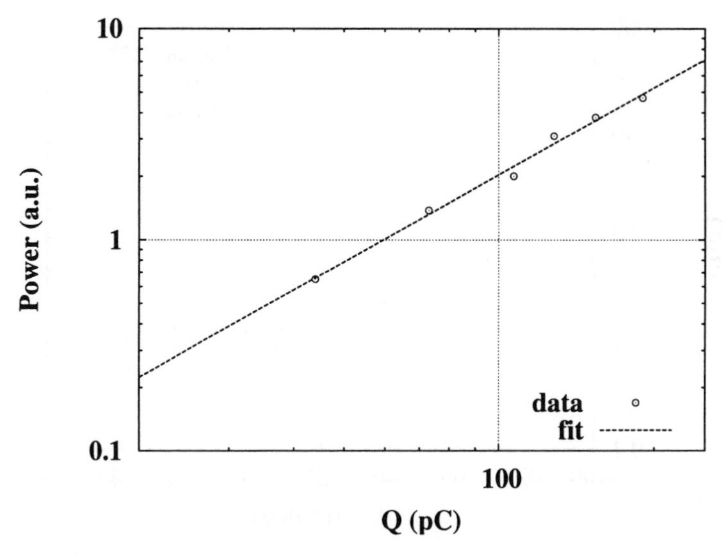

Figure 6. Dependence of bolometer signal versus bunch charge. The circles are measurement, the line is a fit of the measurement using a $\alpha \times Q^{\beta}$ law, the result gives $\beta = 1.37 \pm 0.06$.

N being the number of macro-particle used (50000 in the simulations presented hereafter) and t_i the time of arrival of the i-th macro-particle. Eq. (8) allows to write the integrated power as:

$$f_{simu} = \frac{1}{N^2} \int_{\omega_l}^{\omega_u} d\omega |R(\omega)|^2 \left(\left[\sum_{i=1}^{N} \cos(\omega t_i) \right]^2 + \left[\sum_{i=1}^{N} \sin(\omega t_i) \right]^2 \right). \quad (9)$$

Though Figure 7 shows the signal increase as the bunch is compressed, there are discrepancies between the measurement and the two calculation for the short bunch case, we believe this is due to the lack of precise knowledge of the transmission line.

4 Conclusion

We have measured the bunch length dependence on the phase of an accelerating traveling wave structure located just downstream of an rf electron source. We could compress the bunch by a factor ~3 down to ~0.5 ps for a bunch charge of approximately 200 pC. In our experimental setup, a stronger compression is currently difficult to achieve because of the interplay between longitudinal and transverse phase space. The linac section used for the compression also plays a crucial role in achieving low emittance since it quickly accelerates the beam at energies of approximately 60 MeV thereby freezing the transverse phase space. Hence operating the first linac far off-crest reduces the final energy and impacts

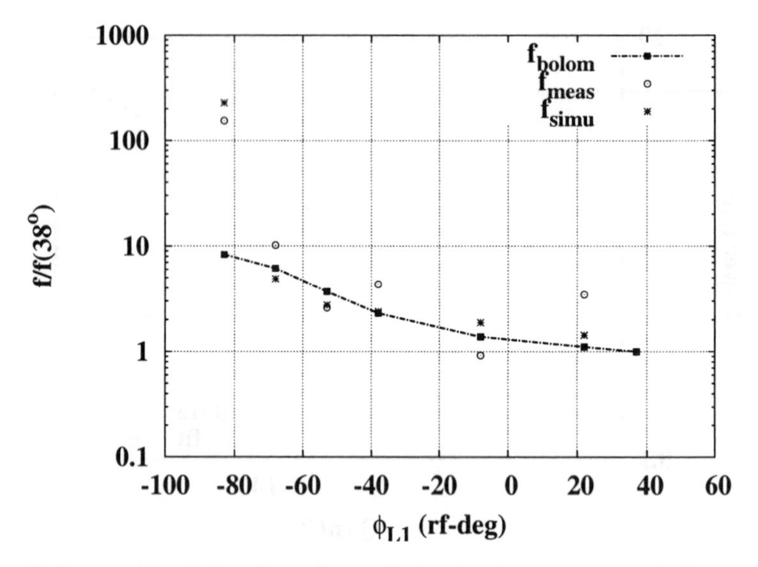

Figure 7. Integrated bunch form factor f normalized to its value at $\phi_{L1} = +38°$. f_{bolom}, f_{meas}, and f_{simu} correspond respectively to measurement with the bolometer, computation from the measured time-profiles and computation from the simulation-generated time profiles.

the emittance (since transverse space charge forces scale as $1/\gamma^2$). An improvement of our experiment could be to surround the accelerating structure used to compress with a solenoidal lens. Such a setup would enable a better control of the transverse beam envelope and emittance [14].

References

1. Ayvazian V., et al., *Eur. Phys. Jour.* **D20**:149-155 (2002)
2. Barov N., et al, *Phys. Rev. ST Accel. Beams* **3**:011301 (2000)
3. Derbenev., et al, report TESLA-FEL-95-05, DESY Hamburg (1995)
4. Haimson H., *Nucl. Instr. Meth.* **39**:13-34 (1966)
5. Serafini L., Ferrario M., "Velocity bunching in photo-injectors", in *Physics of, and science with, the X-ray free-electron laser* edited by S. Chattopadyay et al. AIP conference proceedings **581**:87-106 (2001)
6. Kim, K.-J., *Nucl. Instr. Meth.* **A275**:201-218 (1989)
7. Flöttmann K. *Astra user manual* DESY (2000)
8. Loew G.A., Miller R.H., Early R.A, Bane K.L. "Computer calculation of traveling-wave periodic structure properties", SLAC-PUB-2296 Stanford (1979)
9. Ferrario M., Clendemin J.E., Palmer D.T., Rosenzweig J.B., Serafini L., "HOMDYN study for the LCLS rf photo-injector", SLAC-PUB-9400 Stanford (2000) and report LNF-00/004 INFN-Frascati (2000)

10. Dowell D., Joly S., Loulergue A., proc. PAC 1997 Vancouver, 2684-2686 (1998)
11. Yu, L.-H. et al, proc. PAC 2001 Chicago, 2830-2832 (2002)
12. Wang D.X., Krafft G.A., and Sinclair C.K. *Phys. Rev.* **E57**(2):2283-2286 (1998)
13. Graves, W. et al, proc. PAC 2001 Chicago, 2224-2226 (2002)
14. Serafini L., Bacci A., and Ferrario M., proc. PAC 2001 Chicago, 2242-2244 (2002)

COMPARISON OF THE COHERENT RADIATION-INDUCED MICROBUNCHING INSTABILITY IN AN FEL AND A MAGNETIC CHICANE

S. REICHE

J.B. ROSENZWEIG

Department of Physics & Astronomy, University of California, Los Angeles

A self-amplified spontaneous emission free-electron laser (SASE FEL) is a device which is based on the creation of a very intense, relativistic electron beam which has very little temperature in all three phase planes. The beam in this system is described to as having "high brightness", and when it is bent repetitively in a magnetic undulator, undergoes a radiation-mediated microbunching instability. This instability can amplify the original radiation amplitude at a particular, resonant wavelength by many orders of magnitude. In order to create high brightness beams, it is necessary to compress them to create higher currents than available from the electron source. Compression is accomplished by use of magnetic chicanes, which are quite similar to, if much longer than, a single period of the undulator. It should not be surprising that such chicanes also support a radiation-mediated microbunching interaction, which has recently been investigated, and has been termed *coherent synchrotron radiation (CSR) instability*. The purpose of this paper is to compare and contrast the characteristics of the closely related FEL and CSR microbunching instabilities. We show that a high gain regime of the CSR instability exists which is formally similar to the FEL instability.

1. Introduction

A linear accelerator consists mainly of nominally straight accelerating structures and focusing elements, but also often relies on sections in the beamline where the electron beam is transversely deflected. In addition to bending magnets used to connect different beamlines that have a relative offset, there are dispersive sections, such as magnetic chicanes, in which the beam is finally placed along its original axis. One of the main purposes of such a bending configuration is to manipulate the bunch length. This manipulation requires a time-energy chirp to be applied to the electron bunch by the linear accelerator, with the correlation partially or fully removed by the beam dynamics in the chicane. Such a scheme is referred to as a bunch compressor. In the process of bending the beam in the chicane, broad-band synchrotron radiation is emitted. The amplitude of this radiation at wavelengths similar to, or longer than, the bunch dimension, can be greatly enhanced by constructive interference. This phenomenon, which is known as coherent synchrotron radiation (CSR), can lead to significant, undesirable distortions of the beam's longitudinal and transverse phase space distributions, as well as introducing correlations between phase planes.

Similar in the layout to the chicane, but much smaller in size, is a single period of an undulator or wiggler magnet. The periodic deflection of the electron beam also causes the spontaneous emission of synchrotron radiation. The advantages of undulators or wigglers, which consist of a large number of such periodic bending arrays, as radiation sources are two-fold: (1) the radiation power is accumulated over the entire undulator length in a compact angular region, and (2) the periodic nature of the radiative process narrows the bandwidth in the angular spectrum around resonant frequencies through repetitive interference effects. In addition, these resonant frequencies are tunable by simply changing the driving beam energy.

The spontaneous synchrotron radiation emitted in an undulator or wiggler can act back on the electron beam. Again, because of the repetitive nature of the interaction, the emitted radiation can modulate the beam energy on the scale of the resonant wavelength [1], by forcing the electrons to exchange energy with the field through motion transverse to the field propagation vector. The emitted radiation modulates the beam particles forward of the emitting electron, because the light travels essentially along the nominal beamline path (s) at speed c, which is faster than the electron velocity. Because the undulator is, like the magnetic chicane, a dispersive device, the energy modulation is eventually transferred into a temporal (current) modulation termed *microbunching.* Because the coherence of the radiation at the resonant wavelength is enhanced by this microbunching, more radiation power is emitted, and this feedback mechanism can drive a collective instability [2,3], where an initial bunching and the corresponding emitted radiation is amplified up to a point where maximum bunching (saturation) is achieved. This microbunching instability typically occurs on a scale typically much shorter than the bunch length, and is the operating principle of the free-electron laser [4] (FEL), where the instability is intended. To enhance the performance of an FEL, bunch compressors are often incorporated in the beam line to shorten the bunch length, increasing the driving beam current and beam brightness, and thus the FEL gain.

Although it is explicitly sought in an undulator or wiggler, the interaction of the electron bunch with the coherent, spontaneous radiation within a magnetic chicane can degrade the electron beam quality [5], reducing the performance of a succeeding FEL employing on the compressed beam. On their curved trajectory in bend magnets, the electrons emit radiation, which again moves forward in the beam, because the radiation propagates faster, and in a straight line. The radiation field and electron cross at an angle (typically much larger than in an FEL undulator) yielding a longitudinal electric field component, which changes the electron energy [6]. The difference between this energy exchange mechanism and that in an undulator is purely semantic. In the undulator, we define longitudinal as along z, the undulator axis, and thus energy exchange is possible because the electron travels transverse to the nominal radiation direction, along the electric field polarization direction. In the chicane magnet, we define longitudinal to be always locally along the direction of the electron s, so "transverse" motion of the beam's central trajectory is forbidden by definition. Thus the field propagation vector is now defined to have a transverse component, and the electric field to have a longitudinal component, allowing energy exchange. In analogy to an undulator the dispersion causes a growth in the beam current modulation and thus in the coherence level of the synchrotron radiations [7]. We shall see that the way in which microbunching due to energy modulation asserts itself in the chicane is qualitatively different than in the undulator case.

In the following sections we undertake a comparison of the collective beam dynamics induced by the beam-radiation interaction in an FEL and a magnetic chicane. For sake of simplicity we restrict ourselves to 1D models of this interaction. Because the FEL theory [8] is well understood, and because we will frequently refer to aspects of the FEL model, we begin our discussion with a brief summary of its basic results. In addition, there has recently been a tremendous improvement in understanding the model of the microbunching instability (CSR instability) in a bunch compressor [9,10]. Because this theory is rather complex, we present here a simplified, low gain model of the instability, based on a few, but well justified, assumptions. We identify first many points of comparison and contrast between the FEL and CSR instabilities. We then extend the model of CSR instability to the high gain limit, noting similarities in the theoretical analysis and underlying physics of both CSR and FEL cases.

2. Microbunch Instability in a FEL

Free-electron lasers employ the collective instability of an electron bunch in an undulator to increase the electromagnetic radiation output power level orders of magnitude beyond the level of spontaneous radiation. In addition, this radiation is at least partly coherent, and so the system has characteristics of more conventional lasers. We now review these characteristics through a discussion of the basic theory of the FEL.

The analytical framework for FEL theory is well understood and we only present here the results relevant to our discussion. Three dimensional effects [11] such as diffraction are ignored. A 1D treatment implies among other characteristics that the radiation does not, as in the case of the chicane system, leave the region of the beam transversely, either by diffraction, or cross-propagation. As we consider a long beam limit only, the radiation also does not leave the beam region longitudinally.

The FEL process is initiated by a pre-existing modulation in the electron bunch distribution at the undulator fundamental resonance wavelength (self-amplified spontaneous emission free-electron laser, or SASE FEL), which may exist initially only at the random noise level. For a planar undulator the velocity of the electron in an alternating B-field $B_y(s) = B_0 \sin(k_U s)$ of period $\lambda_U = 2\pi/k_U$ and longitudinal position s is in the xz-plane with

$$\begin{aligned}
\beta_x(s) &= \frac{K}{\gamma}\sin(k_U s) \\
\beta_y(s) &= 0 \\
\beta_z(s) &= 1 - \frac{1+K^2/2}{2\gamma^2} + \frac{K^2}{4\gamma^2}\cos(2k_U s) \quad .
\end{aligned}$$

The undulator parameter K is defined as $K = eB_0/mck_U$. Because the electrons have a transverse velocity component, they couple with a superimposed radiation field, propagating along with the beam.

The change in the energy depends on the phase between the motion and the electromagnetic field $E_x(s,t) = E_0 \cos(k(s-ct)+\phi)$. The FEL model assumes that at any phase the interaction over a single period is too weak to exhibit significant effects. As the energy transfer has to be accumulated over many periods, it requires that the phase remains almost constant over may undulator periods. One frequency fulfills this requirement, yielding the FEL resonance condition

$$< \beta_z >= \frac{k}{k+k_U} \quad . \tag{1}$$

This condition physically implies that the radiation slips one wavelength past the beam electrons per undulator period, as it must to allow optimal constructive interference in the emitted radiation. The restriction of examining only a small bandwidth [12] around the resonant frequency which obeys Eq. 1 is referred to as the resonance approximation.

It significantly simplifies the analytical framework of FEL theory if the equations of motion are averaged over the undulator period. The longitudinal variables of each electron – the ponderomotive phase $\theta_j = (k + k_U)s_j - ckt$ and the energy deviation from the mean energy $\delta_j = (\gamma_j - <\gamma>)/\rho <\gamma>$ – are slow-varying quantities in the co-moving frame of the ponderomotive wave. The scaling parameter[2] ρ is introduced to simplify the equation of motions. It is defined as

$$\rho = \left[\frac{K f_c <\gamma>^2 \omega_p}{4c\gamma_R^2 k_U} \right]^{\frac{2}{3}} \quad . \tag{2}$$

where $\omega_p = \sqrt{\mu_0 n_e e^2 c^2 / m\gamma_0^3}$ is the plasma frequency and γ_R the resonant energy, which fulfills the requirement of Eq. 1 for a given radiation and undulator wavenumber. In a planar undulator the electrons also performs a longitudinal oscillation (when the transverse velocity is large, the longitudinal velocity is diminished) which effectively smears out the particle position. This reduces the coupling and is expressed by the coupling factor $f_c = J_0(\eta) - J_1(\eta)$ with $\eta = K^2/(4+2K^2)$. With the normalized radiation field amplitude $A = -i(Kf_c/(2+K^2)\rho^2)(eE_0/mc^2k)e^{i\phi}$ the dynamics of an FEL is described by the FEL equations

$$\frac{d\theta_j}{d\hat{s}} = \Delta + \delta_j \quad , \tag{3}$$

$$\frac{d\delta_j}{d\hat{s}} = - \left[\left(A + i\sigma \left\langle e^{-i\theta j} \right\rangle \right) e^{i\theta_j} + c.c \right] \quad , \tag{4}$$

$$\frac{dA}{d\hat{s}} = \left\langle e^{-i\theta_j} \right\rangle \quad . \tag{5}$$

The detuning parameter $\Delta = (\gamma_0^2 - \gamma_R^2)/2\rho\gamma_R^2$ is the collective deviation of the beam energy from the resonant energy. Note that the change in longitudinal position θ is linear with the total momentum offset $\Delta + \delta$. Thus the longitudinal velocity responds without delay to changes in momentum. Note that this is true only because the equation of motions are averaged over a period - examination of delays in response shorter than one period are outside of the model's applicability.

We complete our description by noting that the space charge parameter $\sigma = (\omega_p^2 \gamma_0^2)/2c^2 k k_U \gamma_R^2$ represents the longitudinal repulsive forces which counter-act the induced bunching by the radiation field. The normalized position $\hat{s} = 2ck_u\rho t$ is the position in the undulator, measured in units of the characteristic length $1/2k_u\rho$.

Eqs. 3 – 5 are coupled and allow collective instability to occur. Assuming that the modulation in energy and position as well as the radiation field amplitude evolve exponentially as $\exp[i\Lambda\hat{s}]$ the FEL equations are reduced to the dispersion relation [13]

$$1 + \left(\frac{1}{\Lambda} - \sigma\right) \int \frac{\partial f_0}{\partial \delta} \frac{1}{\Lambda + \Delta + \delta} d\delta = 0 \tag{6}$$

where f_0 is the initial distribution in energy. In the case of no detuning, space charge and energy spread the dispersion relation reduces to the cubic equation $\Lambda^3 = -1$. One of the roots has an imaginary part, of amplitude $-\sqrt{3}/2$, which corresponds in an exponential growth of the field amplitude and the beam modulation. The characteristic scale of the field growth is the gain length $L_g = \lambda_U/\sqrt{12}\pi\rho$.

A limit is implicitly imposed on the FEL equations by the resonance approximation. It can be expressed using the FEL parameter ρ as

$$\rho \ll 1 \tag{7}$$

For all existing and planned FELs [14,15,16,17,18,19] the condition is sufficiently fulfilled. However, the conceptual consequences of violating the condition in Eq. 7 will be explored in the context of our comparison of the FEL and CSR instabilities.

In summary, the FEL instability has several unique features, which we point out here to aid in our comparison with the CSR microbunching instability in a magnetic chicane:

- Frequency components are located within a narrow bandwidth around a central, resonant wavelength Eq. 1. The relative bandwidth is of the order of ρ. If ρ were to attain values near unity, the resonance approximation would clearly be violated, and the physical picture we have deduced from the analysis above would qualitatively change.
- The amplitude of the transverse oscillation is typically small compared to the beam size. The same radiation field interacts with the electron bunch over the entire undulator length. This requires to include Maxwell equations for a self-consistent description of the instability process. Assuming validity of the resonance approximation and the resulting predominant direction of propagation of

the radiation field at a well-defined wavelength implies the paraxial approximation of the Maxwell equations, where the second order derivatives of the radiation amplitude and phase with respect to longitudinal position and time are ignored.

- The electron dynamics occuring during an undulator period is replaced by its averaged behavior. As with the paraxial approximation of the radiation field, this is valid as long as the resonance approximation holds. Again, if ρ is near unity, the paraxial approximation (and the related slowly-varying field amplitude approximation) would be violated, and the standard description of the FEL we have given is not valid.

3. Description of Particle Dynamics in a Magnetic Chicane

A magnetic chicane consists typically of bending magnets, drifts, and, optionally, focusing quadrupoles. For a simpler comparison to the FEL we exclude the latter two components from the discussion. The bend angle in a chicane dipole depends on the beam energy, and the orbit within the chicane is dispersive. If the electron energy is constant the path length difference in the bunch compressor can be expressed by the matrix element $R_{56} = \frac{\partial \zeta(s)}{\partial \delta(s_0)}$, with

$$\zeta(s) = \zeta(s_0) + R_{56}(s, s_0)\delta. \tag{8}$$

Here s and s_0 denote the final and initial position on the design orbit, respectively, $\zeta = ct$ the position in the frame of the moving bunch and $\delta = (\gamma - <\gamma>)/<\gamma>$ is the normalized deviation of the electron energy from the mean energy, and we are using the ultra-relativistic approximation $\beta = 1$. The matrix element R_{56} depends on the dispersion function η and the bend radius R and is defined as

$$R_{56}(s, s_0) = \int_{s_0}^{s} \frac{\eta(s')}{R(s')} ds'. \tag{9}$$

For the case where the electron beam interacts with the spontaneous radiation, the dispersion functions change with any change in the electron energy. We thus generalize the definition of R_{56} to

$$\zeta(s) = \zeta(s_0) + \int_{s_0}^{s} \delta(s')\tilde{R}_{56}(s, s')ds'. \tag{10}$$

In the case that δ is constant Eq. 10 reduces to Eq. 8 using the relation $R_{56}(s, s_0) = \int_{s_0}^{s} \tilde{R}_{56}(s, s')ds'$. The interpretation of Eq. 10 is that the change in ζ is given by the summation over all dispersion contributions between two arbitrary points s and s_0, wherever the beam energy changes. The longitudinal 'velocity' is easily obtained by taking the derivative of Eq. 10, yielding

$$\frac{d\zeta}{ds} = \int_{s_0}^{s} \delta(s') \frac{\partial \tilde{R}_{56}(s,s')}{\partial s} ds' \tag{11}$$

Note that the derivative with respect to the upper boundary of the integral of Eq. 10 is zero. The function $\tilde{R}_{56}$ implicitly depends on the bend field and radiated energy at all preceding positions in the magnetic chicane.

To describe a microbunched distribution of electrons, we begin by assuming a coasting beam with a small modulation

$$I(\zeta, s) = I_0 \left[1 + |b(s)| \cos(k\zeta + \phi(s))\right] \quad . \tag{12}$$

Because the magnitude and the relative position of the modulation can change within the bunch compressor, $|b|$ and ϕ depend on s. The emission of synchrotron radiation interacts with electrons in the forward direction. The resulting potential [6] seen by the electrons is

$$W(\zeta) = -\frac{2}{(3R^2)^{\frac{1}{3}}} \frac{1}{\zeta^{\frac{1}{3}}} \frac{\partial}{\partial \zeta} \tag{13}$$

for $\zeta < 0$ and zero otherwise. Note that in this model we exclude any transient effects, and assume that the interaction is instantaneous. The energy change of any given electron is

$$\begin{aligned} \frac{d\delta}{ds} &= -\frac{I_0}{I_A \gamma_0} \int_{-\infty}^{\infty} W(\zeta - \zeta') I(\zeta') d\zeta' \\ &= -\frac{I_0}{I_A \gamma_0} \frac{2k|b(s)|}{(3R^2)^{\frac{1}{3}}} \int_{\zeta}^{\infty} \frac{\sin(k\zeta' + \phi(s))}{(\zeta' - \zeta)^{\frac{1}{3}}} d\zeta' \\ &= -\frac{I_0}{I_A \gamma_0} \frac{2\Gamma(\frac{2}{3}) k^{\frac{1}{3}}}{(2R^2)^{\frac{1}{3}}} |b(s)| \sin\left(k\zeta + \phi(s) + \frac{\pi}{3}\right) \end{aligned} \tag{14}$$

The growth of the energy modulation scales linearly with the modulation of the current. Note also that the potential has a phase offset of $-\pi/6$ with respect to the modulation itself.

4. The Chicane Model

To perform a first analysis of the CSR instability, we consider a idealized chicane, which consists of three bending magnets with no drift space separating them. The outer magnets have a length of L while the inner one, which bends in the opposing direction of the outer two, is twice as long. This model, which describes many existing chicanes well, may not apply to certain proposed chicanes, which have long inter-magnet drifts.

To further simplify our calculations, the bend radius R is assumed to be much larger than L and the dispersion function is calculated based on the small deflection angle approximation $\theta_b \approx L/R << 1$. It is convenient to normalized the dispersion function η between any two arbitrarily chosen positions s and s' to the bend radius R. With the initial conditions $\eta(s') = 0$ and $\eta'(s') = 0$ the dispersion function is

$$\frac{\eta}{R} = \begin{cases} \frac{1}{2}\left(\frac{s-s'}{R}\right)^2 & \text{(I,I), (II,II),} \\ & \text{(III,III)} \\ \left(\frac{s-L}{R}\right)^2 - \frac{1}{2}\left(\frac{s-s'}{R}\right)^2 & \text{(II,I)} \\ \left(\frac{s-3L}{R}\right)^2 - \frac{1}{2}\left(\frac{s-s'}{R}\right)^2 & \text{(III,II)} \\ \frac{1}{2}\left(\frac{s-3L}{R}\right)^2 + \frac{1}{2}\left(\frac{3L-s'}{R}\right)^2 - \left(\frac{2L}{R}\right)^2 & \\ \quad - \left(\frac{s-3L}{R}\right)\left(\frac{L-s'}{R}\right) & \text{(III,I)} \end{cases} \tag{15}$$

where the pair of Roman numbers indicates in which dipole the ending and starting points, respectively, are located. The dispersion function is related to the modified matrix element $\tilde{R}_{56}$ as

$$\frac{\partial}{\partial s}\tilde{R}_{56} = -\frac{\partial}{\partial s'}\frac{\eta}{R} = \frac{s-s'}{R(s)R(s')} \tag{16}$$

We incorporated the bend direction of the dipoles into the sign of the bend radius as

$$R(s) = \begin{cases} R & s \text{ in 1st or 3rd dipole} \\ -R & s \text{ in 2nd dipole.} \end{cases} \tag{17}$$

With this convention, it can be easily seen that the sign of the differential matrix element changes when the electron enters the adjacent dipole. Physically this means that a particle that gains energy, becoming more rigid, has the tendency to fall behind the design trajectory in the magnet where the energy gain occurs. At a magnet boundary, however, the change in the dipole polarity reverses this effect, and the higher energy particle moves forward with respect to the design trajectory. In a chicane where the energy is held constant, the beam actually decompresses in the first and last dipoles, and compresses in the middle. Net compression is achieved in this case because the dispersion is much larger in the middle dipole.

In the present model, the differential equation for the longitudinal motion becomes

$$\frac{d\zeta}{ds} = \int_0^s \delta(s') \frac{s - s'}{R(s)R(s')} ds'. \tag{18}$$

5. The Low Gain Model

As indicated in Eq. 18 a change in the particle energy has a delayed effect on the particle's longitudinal position, which grows with the third power in s. In addition the change in the longitudinal position is also inhibited by the change in the polarity of the bending magnets. Particles with higher energy fall behind due to the larger bend radius, but catch up due to the shorter path length after a polarity change. Similar arguments are valid for lower energy particles. A third condition must be present before longitudinal motion and therefore instability can grow – the energy modulation must first grow significantly. Thus for short time scales, and low gain, $\delta(s')$ can be expected to change linearly in s'.

We recall for comparison that the resonance condition of the FEL theory averages over one period, and as a result the change in the longitudinal position due to an energy modulation is linear in s. Therefore the differential equations for phase and energy must be solved simultaneously. For the chicane, assuming low gain, the situation is qualitatively different. We can expect that the energy modulation is accumulated first before it affects the longitudinal position, which changes primarily in the last part of the chicane. The mechanism is very similar to that of a klystron [20].

This assumption of klystron-like behavior is invalid for high currents. In this case the delayed change in the longitudinal position is noticeable even in the first bending magnet. This instability is similar to the bunching instability in a storage ring [21], which can be regarded as a single long bending magnet. An initial bunching will be amplified exponentially till all particles are bunched and the beam breaks up. We discuss this limit further in the following section.

For many of the presently operating or proposed chicane compressors, the low gain limit is valid. We model the initial current by a equidistant distribution plus an added sinusoidal modulation in the positions. The amplitude and wavenumber of this modulation are $\Delta\zeta$ and k. The complex-value bunching factor is determined by

$$b = \frac{1}{N} \sum_{j=1}^{N} \exp[-ik\zeta_j + ik\Delta\zeta \sin(k\zeta_j + \phi)]$$

$$\frac{1}{N}\sum_{j=1}^{N}\exp[-ik\zeta_j]\left[\sum_{n=-\infty}^{\infty} J_n(k\Delta\zeta)\exp[in(k\zeta_j+\phi)]\right]$$
$$= J_1(k\Delta\zeta)e^{i\phi} \quad . \tag{19}$$

Because the effective radiation potential is harmonic in ζ the resulting modulation in the longitudinal position is harmonic as well with $\delta\zeta = Z(s)\sin(k\zeta_0+\psi)$, where ζ_0 is the initial position of a particle and $Z(s)$ vanishes at the entrance of the chicane. The resulting bunching factor becomes

$$b(s) = \sum_{n+m=1} J_n(k\Delta\zeta)J_m(kZ(s))e^{in\phi+im\psi}. \tag{20}$$

In our low gain model the initial offset $\Delta\zeta$ and the modulation amplitude $Z(s)$ are much smaller than the modulation wavelength. This implies that in the Taylor series expansion of the Bessel-functions in Eq. 20 the lowest order term is linear either in $\Delta\zeta$ or $Z(s)$. Further the initial amplitude of the longitudinal modulation is small compared to the modulation wavelength. As long as $Z(s)$ is comparable to the initial modulation amplitude $\Delta\zeta$, the bunching factor can be taken as constant and Eq. 14 can easily be integrated, giving a linear dependence on s. Inserting $\delta(s)$ into Eq. 18, the longitudinal position evolves in the chicane as

$$\zeta_j(s) = \zeta_j(0) - \frac{I_0}{I_A\gamma_0}\frac{2\Gamma\left(\frac{2}{3}\right)k^{\frac{1}{3}}}{(3R^2)^{\frac{1}{3}}}|b(0)|$$
$$\cdot\sin\left(k\zeta_j(0)+\phi+\frac{\pi}{3}\right)\Phi(s) \quad , \tag{21}$$

with

$$\Phi(s) = \int_0^s\left[\int_0^{s''}\frac{s'(s''-s')}{R(s'')R(s')}ds'\right]ds''$$
$$= \begin{cases} \frac{1}{24}\frac{s^4}{R^2} & \text{(I)} \\ \frac{1}{24}\frac{s^4}{R^2} - \frac{1}{2}\frac{L^2}{R^2}s^2 + \frac{2}{3}\frac{L^3}{R^2}s - \frac{1}{6}\frac{L^4}{R^2} & \text{(II)} \\ \frac{1}{24}\frac{s^4}{R^2} - 4\frac{L^2}{R^2}s^2 + \frac{52}{3}\frac{L^3}{R^2}s - \frac{56}{3}\frac{L^4}{R^2} & \text{(III).} \end{cases} \tag{22}$$

The Roman numerals indicate the dipole in which the position s lies. Fig. 1 shows the universal function $\Phi(s)$. The function can be characterized differently in four sections. In section 1 (first dipole, I) an electron, which gains energy, is slowed down because the bend radius and thus the path length gets larger. Section 2 is the first half of the second dipole (II), where the longitudinal motion is inverted as the polarity of the magnetic field

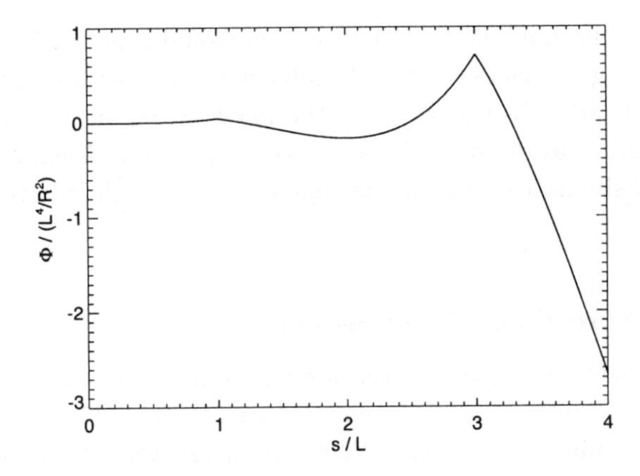

Figure 1. The function Φ (Eq. 22).

changes. In the second half of the second dipole the 'slow-down' effect of the increasing bend radius takes over. Finally in the third and last magnet (section 4, dipole III) the path length changes originating in the first and second dipoles add up with the same sign. Here the strongest change in the longitudinal position occurs. Somewhat surprisingly, the final offset in ζ is identical to that calculated if the accumulated energy change is applied directly at the entrance of the chicane and the electron is tracked by using the standard expression for the matrix element R_{56}!

The value of Φ at the end of the chicane is $-(8/3)(L^4/R^2)$, which also defines the maximum offset in longitudinal direction a particle will see. The final, normalized amplitude is

$$\xi = \frac{I_0 \Gamma\left(\frac{2}{3}\right)}{2 I_A \gamma_0} \left(\frac{8L^3 k}{3R^2}\right)^{\frac{4}{3}} . \tag{23}$$

With $kZ(4L) = \xi|b(0)| \ll 1$ the gain, defined as the ratio between final and initial amplitude of modulation, becomes

$$G = \frac{|b(4L)|}{|b(0)|} = \sqrt{1 + \xi + \xi^2}. \tag{24}$$

Note that the gain is broadband, not resonant. This stands in contrast to the FEL instability, where resonant behavior is derived from the periodic nature of the beam-undulator-radiation interaction.

As an example of a generic magnetic chicane, modeling the first LCLS bunch compressor (γ = 500, I_0 = 100 A, R = 12 m and L = 1.5 m), an initial modulation with a period of 5μm would grow by a factor of 25. Note that the gain scales roughly linearly with the beam current. This is identical with an FEL operating in the low-gain regime [22]. In the FEL high-gain regime, as discussed in Sec. 2, the dependence is exponential. We will also find similar exponential dependence in the high-gain regime of the CSR instability.

6. Energy Spread and Compression

The gain growth, discussed above, has a singularity at a zero modulation period length. This artifact is removed if energy spread is included in the model. With finite uncorrelated (thermal) energy spread, any modulation in the beam current is spread out and the emission level of the synchrotron radiation is reduced. This effect is particularly strong at short modulation wavelengths. Thus one expects that there is an optimum wavelength for gain of the CSR-induced modulation, where the natural tendency of the system to amplify shorter wavelengths is not yet damped by thermal effects. This type of instability suppression is often designated as Landau damping.

We now provide an analysis of energy spread effects. Including energy spread, the current profiles evolves as

$$I = I_0 \left[1 + \int f(\delta)|b(0)| \cos(k(\zeta + R_{56}\delta) + \phi) d\delta\right] \tag{25}$$

where $f(\delta)$ the normalized energy distribution. We assume no correlation between energy and longitudinal position on the scale of the modulation. For a Gaussian energy distribution with rms spread σ_δ the integration of this equation yields

$$I = I_0 \left[1 + e^{-\frac{1}{2}(\sigma_\delta R_{56} k)^2} |b(0)| \cos(k\zeta + \phi)\right] \tag{26}$$

The modulation decays if the spread in the longitudinal position $\sigma_\delta R_{56}$ is comparable with the modulation period length. The initial modulation is sheared mainly in the second dipole, where the value of matrix element R_{56} changes significantly. Because the seed for the microbunch instability – the accumulated change in the electron energy – occurs before that, the change in the longitudinal position (Eq. 22) is hardly effected. We can just apply the damping factor due to the energy spread to the previous results of Eq. 24. With $R_{56}(4L, 0) = -(4/3)L^3/R^2$ the final gain is

$$G = e^{-\alpha|\xi|^{\frac{3}{2}}} \sqrt{1 + \xi + \xi^2}, \tag{27}$$

defining the normalized energy spread as

$$\alpha = \sqrt{\left(\frac{I_A \gamma_0}{2 I_0 \Gamma\left(\frac{2}{3}\right)}\right)^3 \sigma_\delta^2}. \tag{28}$$

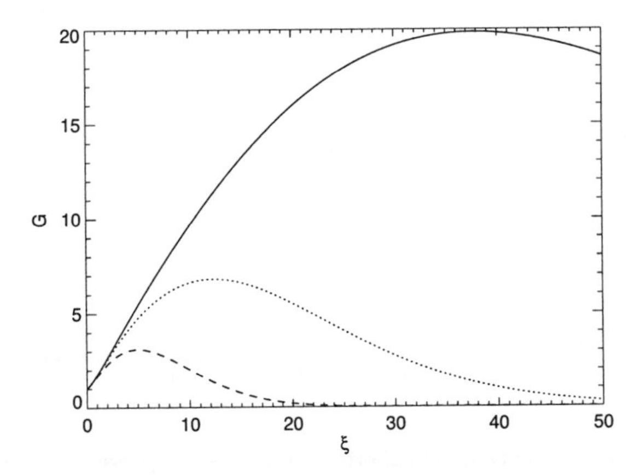

Figure 2. Gain curve for energy spreads of $\alpha = 0.003$, 0.015 and 0.05 (solid, dotted and dashed line, respectively).

For the LCLS case with an energy spread of 0.01% ($\alpha = 0.05$), the gain at 5 μm would be reduced by 7 order of magnitudes. Fig. 2 shows the achievable gains for different energy spreads.

In the case that a linear chirp is imposed on the beam to compress it, the path length differences in the chicane change the period length of the modulation. With the compression factor

$$C(s) = \frac{1}{1 + hR_{56}(s,0)} \tag{29}$$

and the chirp gradient $h = d\delta/cdt$, the wavenumber k scales along the chicane as $k(s) = C(s)k(0)$. This results in a larger wakefield amplitude although the dependence is rather weak due to the cubic root of the wavenumber in Eq. 14. Further, this wakefield enhancement is mainly visible in the last dipole. As in our discussion of the energy spread, the induced energy modulation in the first half of the chicane drives the longitudinal motion. Fig. 3 shows the change in the function Φ for a strong compression by a factor of 10, solving Eq. 14 and 18 self-consistently. The maximum change is less than 10%. The remaining effect of compression is that the electrons move further with respect to the reduced modulation wavelength. The normalized amplitude and energy spread scales as

$$\xi \to C\xi \qquad \text{and} \qquad \alpha \to \sqrt{C}\alpha \quad .$$

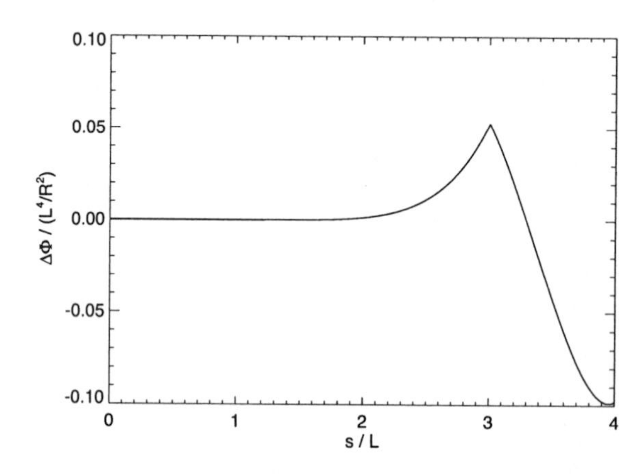

Figure 3. Change in the longitudinal orbit Φ due to a compression by a factor of 10.

7. The High Gain Model

In our low gain model we assumed that the CSR microbunching instability does not drive any bunching within the first dipole. To qualitatively place a limit on this assumed scenario, we develop here a high-gain, exponential growth model as well. There may be situations where the exponential gain does not assert itself in the first dipole, but may, by compression and thus higher current, become notable in the last dipole. We shall see that the exponentially growing regime of the CSR instability has many similarities with the high-gain FEL.

To begin our analysis, we define the collective variables

$$B = -ik < e^{-i\Psi}\zeta > \tag{30}$$

$$\Delta = < e^{-i\Psi}\delta > \quad , \tag{31}$$

where $\Psi_j = k\zeta_{0,j}$ is the initial phase of the jth electron of a uniform distribution. Note that B is the Taylor series expansion of the bunching factor b (Eq. 20). The equations of motion for a cold beam become

$$\frac{d\Delta}{ds} = -\frac{\rho_{CSR}^4}{kR^2} e^{i\frac{\pi}{3}} B \tag{32}$$

$$\frac{dB}{ds} = -i\frac{k}{R^2}\int_0^s \Delta(s-s')ds', \tag{33}$$

with the definition (in direct analogy with the FEL case) of the dimensionless ρ_{CSR}-parameter

$$\rho_{CSR} = \left[\frac{I_0}{I_A\gamma_0}4^{\frac{1}{3}}\Gamma\left(\frac{2}{3}\right)\right]^{\frac{1}{4}}(kR)^{\frac{1}{3}}, \tag{34}$$

The equations can be combined into a forth order differential equation. Using the ansatz $B \propto \exp[i\Lambda s]$ we obtain the dispersion relation

$$\Lambda^4 = \left(\frac{\rho_{CSR}}{R}\right)^4 e^{i\frac{5\pi}{6}}. \tag{35}$$

Four solutions exist, where two of the roots have a negative imaginary part, corresponding to an exponentially growing instability. The growth rates are $(\rho_{CSR}/R)\sin(7\pi/24)$ and $(\rho_{CSR}/R)\sin(5\pi/24)$, respectively. Just as the gain length of the high-gain FEL scales as λ_U/ρ, the gain length of the high-gain CSR instability is roughly R/ρ_{CSR}. Because the calculations are based on a relatively small deflection angle ($L \ll R$) exponential gain within a single dipole becomes significant only for

$$\rho_{CSR} > \frac{R}{L} \gg 1 \quad . \tag{36}$$

In comparison to the FEL the dispersion relation for CSR instability is of 4th order, because the change in the bunching factor is of third order in energy, while the energy modulation is linear in the bunching. For the FEL model, all equations (energy, phase and radiation field) are linear, resulting in a third order expression.

The start-up regime of the high gain CSR instability is of particular interest because it determines after how many gain lengths the exponential growth becomes dominant. In the case of an FEL this is typically achieved after the first or second gain length. Here the situation is different because the number of modes is higher. In addition the two growing modes have similar growth rate but different phase slippages (real part of Λ), so that the interference between these two is still noticeable after several gain length. Fig. 4 shows the growth rate normalized to $|\Im m(\Lambda_1)| = (\rho_{CSR}/R)\sin(7\pi/24)$. It takes at least 5 gain lengths before the bunching factor has grown by one order of magnitude, compensating an initial amplitude drop. The interference with the other growing mode is still notable after 20 gain lengths.

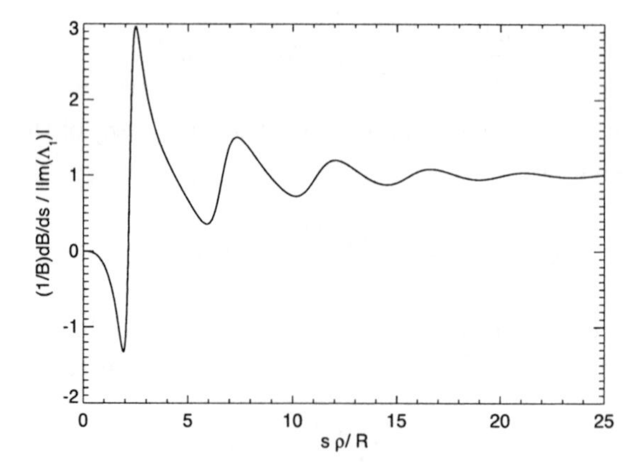

Figure 4. Growth rate of the bunching factor, normalized to the growth rate of the dominant growing mode.

The integral in Eq. 33 prevents an easy solution of the Vaslov equation to include energy spread effects. We conclude this section with a qualitative analysis. For a characteristic spread σ_δ in energy the momentum dispersion couples it to a phase spread of approximately $(k/6R^2)\sigma_\delta s^3$. In units of the gain length ($\hat{s} = s\rho/R$) the normalized phase spread $\hat{\sigma}_\delta = (kR/6\rho^3)\sigma_\delta$ is independent on the bend radius and wavelength. An estimate on the mitigation of the instability by the energy spread can be obtained by the condition that for a Gaussian-spectrum beam the growth over 5 gain lengths is canceled by the phase spread. The threshold for the normalized energy spread is $\hat{\sigma}_\delta < 0.02$ by this criterion – any value larger than 0.02 would completely suppress the exponential growth of this instability. Note that if even the constraint is fulfilled the energy spread may affect the instability at a certain point due to its strong dependence on s. For a Gaussian spread this occurs after $\hat{\sigma}_\delta^{-\frac{1}{3}}$ gain lengths. Because the gain length depends on k, longer wavelengths are less effected by the energy spread, as we have seen in the low gain analysis already.

8. Conclusions

Because of the mechanism of coherent synchrotron radiation-induced energy feedback, any transverse deflecting beam system may support a microbunching instability. The underlying dynamics of these instabilities may be quite different, however. In the cases we have considered, the undulator and the magnetic chicane, these differences are particularly strong. In

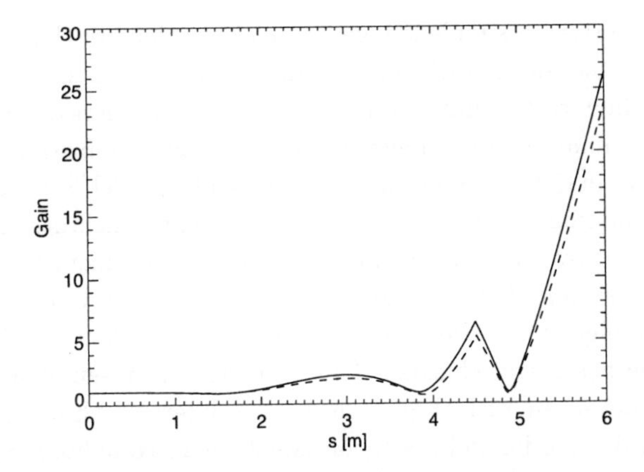

Figure 5. Evolution of the gain along the chicane, using the low gain model (Eq. 24) and a self-consistent numerical simulation (solid and dashed line, respectively). The input parameters are $I_0 = 100$ A, $\gamma_0 = 500$, $R = 12$ m and $L = 1.5$ m.

an FEL the local interaction is rather weak over one period, and must be accumulated over many bending oscillations to yield an instability. It is quite important to the FEL instability that the emitted radiation stay in the electron beam region, therefore, to allow a significant build-up of the beam-radiation interaction. The relative weakness of this interaction is expressed by the value of the ρ-parameter, which is much smaller than unity, indicating a slowly varying system.

Theoretically, an important measure of the transition between the FEL instability and the microbunching instability in a chicane would be the extrapolation of ρ to unity and beyond. It should emphasized that the use of the standard analytical model in this case would not yield reasonable results, as the assumptions (slowly varying system, paraxial radiation) would be violated. A valid theory in this limit must include the entire radiation field in all directions, as well as the explicit longitudinal motion variations of the electrons. Some of these characteristics are present in our chicane-based analysis, which is not fully general. In a chicane, the orbit offset is large compared to the beam size so that an emitted radiation field interacts only briefly with the electrons before escaping the beam. Under this assumption the radiation field dynamics have been simplified to a wake potential which acts instantaneously on the beam. A general treatment, covering both FEL-like and chicane-like cases, would demand a much more sophisticated model of the radiation emission.

We have derived a simple low-gain model to calculate the growth of an initial current modulation within a magnetic chicane. As in a klystron the physics is split into the initial modulation in beam energy, and a change in the longitudinal position (followed by an enhanced emission level of radiation). The major assumption in our low-gain model is that the beam current modulation, which generates the coherent synchrotron radiation, is held constant over the entire chicane. The klystron-like assumption implies that the system does not have an exponential gain, but rather acts as a linear amplifier where the gain coefficient is dependent on the physical details of the beam and chicane. This is in striking resemblance with the low-gain model of an FEL. Fig. 5 shows the difference, however, between our klystron-like model and a self-consistent, more complex model [9].

In order to extend the low-gain model to account for self-consistent behavior, we have developed a high gain model that has notable similarities with high-gain FEL theory. In particular, we obtained a quartic dispersion relation which allows us to scale the gain through a ρ-parameter (ρ_{CSR}), with a gain-length proportional to R/ρ_{CSR}. In contrast to the FEL case, with the CSR instability, the parameter ρ_{CSR} must have a value in excess of unity, so that exponential gain occurs within a single dipole.

In conclusion, a microbunching instability, displaying familiar behaviors such as klystron-like or exponentially growing, occurs in transverse deflecting magnetic devices. Different assumptions and approximations allows this instability to be investigated analytically, within the confines of physically relevant models, such as the FEL and the compressor chicane. The FEL analysis relies on simplified, period-averaged view of the electron motion, allowing an exponentially growing instability. In the case of the magnetic chicane, on the other hand, electron motion must be treated in a more intricate way, but the radiation field can be simply expressed by a continuous force acting to change the beam energy. The transition between FEL and the CSR instability is controlled by strength of the ρ-parameter; when it nears unity, a CSR-like case would occur. As the ρ-parameter is a valid number only when it is small, it is really only of value for FEL-like (resonance approximation, paraxial radiation) case. For the CSR instability, it is more valuable to define the ρ_{CSR} parameter in the context of exponential growth in a single bend magnet. With these issues in mind, we can classify bending systems in a hierarchy of gain strength, from lowest to highest: (1) standard FEL instability, where $\rho \ll 1$; (2) ultra-high gain FEL $\rho \simeq 1$, where the standard theory is no longer valid; (3) CSR instability in the low-gain limit, where ρ_{CSR} is not greater than 1, and the system is klystron-like, requiring the entire chicane to develop; (4) high-gain CSR instability, where $\rho_{CSR} > 1$ and the beam microbunches in one chicane magnet. At present, the only systems which are commonly encountered are (1) and (3). The relevance of analysis of systems (2) and (4) awaits further progress in high brightness electron beam production.

References

1. J.M.J. Madey. *J. App. Phys.*, 42:1906, 1971.
2. R. Bonifacio, C. Pellegrini, and L.M. Narducci. *Opt. Comm.*, 50:373, 1984.
3. A.M. Kondratenko and E.L. Saldin. *Par. Acc.*, 10:1980, 1980.
4. W.B. Colson. *Phys. Rev. Lett.*, 36:717, 1976.
5. L.V. Iogansen and M.S. Rabinovich. *Sov. Phys. JETP*, 37(10):83, 1960.
6. Y.S. Derbenev, J. Rossbach, E.L. Saldin, and V.D. Shiltsev. *TESLA-FEL 95-05*, 1995. Deutsches Elektronen-Synchrotron, Hamburg, Germany.
7. H. Braun. *Phys. Rev. Lett.*, 84:658, 2000.
8. E.L. Saldin, E.A. Schneidmiller, and M.V. Yurkov. *The Physics of Free Electron Lasers.* Springer-Verlag, Berlin, 2000.
9. S. Heifets, S. Krinsky, and G. Stupakov. *SLAC-PUB-9165*, 2002. Stanford Linear Accelerator Center, Stanford, USA.
10. Z. Huang and K.-J. Kim. Formulae for csr microbunching in a bunch compressor chicane. In *24th Advanced ICFA beam Dynamics Workshop on Future Light Sources*, Japan, 2002. SPRING-8.
11. G.T. Moore. *Opt. Comm.*, 52:46, 1984.
12. J.M. Wang and L.-H. Yu. *Nucl. Instr. & Meth.*, A250:484, 1986.
13. J.B. Murphy and C. Pellegrini. Introduction to the physics of the fel. In *Proc.of the South Padre Island Conference.* Springer, 1986.
14. M. Hogan et al. *Phys. Rev. Lett.*, 81:4867, 1998.
15. A. Tremaine et al. *Nucl. Instr. & Meth.*, A483:24, 2002.
16. S.M. Milton et al. *Science*, 292:2037, 2001.
17. V. Ayvazyan et al. *Phys. Rev. Lett.*, 88:104802, 2002.
18. Linac Coherent Light Source (LCLS) Design Study Report. *SLAC-R-521*, UC-414, 1998.
19. TESLA - Technical Design Report. *TESLA-FEL 2001-05*, 2001. Deutsches Elektronen-Synchrotron, Hamburg, Germany.
20. E.L. Saldin, E.A. Schneidmiller, and M.V. Yurkov. *TESLA-FEL 2002-02*, 2002. Deutsches Elektronen-Synchrotron, Hamburg, Germany.
21. S. Heifets and G. Stupakov. *SLAC-PUB-8761*, 2001. Stanford Linear Accelerator Center, Stanford, USA.
22. W.B. Colson. *Phys. Quantum Electron.*, 5:157, 1978.

COHERENT SYNCHROTRON RADIATION IN STORAGE RINGS

M. VENTURINI, R. WARNOCK, AND R.RUTH

Stanford Linear Accelerator Center
Stanford University,
Stanford, CA 94309, USA
E-mail: venturin@slac.stanford.edu

We take a detour from the main theme of this volume and present a discussion of coherent synchrotron radiation (CSR) in the context of storage rings rather than single-pass systems. Interest in this topic has been revived by a series of measurements carried out at several light source facilities. There is strong evidence that the observed coherent signal is accompanied by a beam instability, possibly driven by CSR itself. In this paper we review a "self-consistent" model of longitudinal beam dynamics in which CSR is the only agent of collective forces. The model yields numerical solutions that appear to reproduce the main features of the observations.

1. Introduction

The realization of a possible role for coherent synchrotron radiation (CSR) in cyclic electron machines is old, preceding the construction of the first electron synchrotrons, and dates back at least to an unpublished paper by Schwinger[1] in 1945. Schwinger's work was motivated by concerns about energy efficiency. When passing through bending magnets charged particles radiate incoherently, with power proportional to the number of particles per bunch N, and also coherently at longer wavelengths, with power proportional to N^2. However, Schwinger[1] and others[2,3] pointed out that unless the bunch length were exceedingly small, coherent radiation would be effectively suppressed by shielding from the metallic vacuum chamber. In spite of its more unfavorable scaling with N the coherent part of the radiation is typically a very small fraction of the overall dissipated power, and has little consequence in machine operation. As a result, interest in CSR faded somewhat after those early papers, but the subject was kept alive in several theoretical studies through the 1960's and 1970's (see the bibliography in the first paper of Ref.[4]). There was interest at Berkeley [5], Dubna [6], and elsewhere in connection with the "smoke ring" acceleration concept. It was not until the mid 1980's that the first experimental indications[7,8] of CSR were reported, and 1989 that the first clear observation was made, by Nakazato *et al.* [9]. The latter involved a linac and a bending magnet, rather than a circular machine. Only in recent years detection of CSR from storage rings has been more conclusively established through a series of measurements carried out at NIST[10] (Maryland), NSLS[11,12,13] (Brookhaven), MAX Laboratory[14] (Lund, Sweden), BESSY[15] (Berlin), and ALS[16] (Berkeley). The renewed attention to CSR stems in part from the prospect of exploiting the process to create a new class of high-power light sources in the infrared region. A first suggestion of the practical implications of CSR was contained in a paper by Michel[18] in 1982. Later, a detailed proposal was made by Murphy and Krinsky[19], and the design of a dedicated CSR source is currently being explored at LBNL[17] (Berkeley).

As this volume well illustrates, CSR has been gaining increasing attention because of its potential role as a mechanism for driving collective instabilities. While at present not a factor limiting beam quality in storage rings (a marked difference from the case of bunch compressors in FEL applications), CSR instabilities are nonetheless important as they appear to be connected with most of the above observations[10–16]. Two features in these measurements indicate a connection: the existence of a current threshold for detecting a coherent signal and a radiation wavelength considerably shorter than the nominal bunch length. In particular, the latter implies that the bunches carry a modulation in the longitudinal distribution. This is because, in general, coherent radiation at a wavelength λ can be emitted only if the Fourier spectrum of the longitudinal charge density in the bunch is significant at that λ (if λ is sufficiently small the shielding by the vacuum chamber becomes ineffective and a coherent signal can be detected). The required modulation could naturally be provided by a collective instability. There is an on-going debate about whether the cause of such an instability might be the machine geometric impedance or, as recently suggested by Heifets and Stupakov[20], the CSR process itself. Perhaps the most convincing argument in favor of the latter hypothesis is the apparent generality of coherent emissions, which have been detected over a number of very different machines. In addition, there appears to be a substantial agreement so far between predictions by the standard linear theory of collective instabilities applied to CSR and observations.

In this paper we review some recent work[21,22] that we have undertaken to study the interplay between the coherent radiation process, the longitudinal beam dynamics, and related instabilities. The goal is to go beyond linear theory, hoping to explain aspects of the observations that cannot be captured otherwise. For example, in most of the current observations of CSR in storage rings, a typical signal presents a bursting time structure. Radiation appears in recurrent short peaks separated by relatively long intervals, some fraction of the damping time. Details like the separation between the peaks and the amount of radiation released at each burst depend on the beam current. Smaller currents are typically associated with more regular patterns of emission. A linear theory can explain the onset of the instability generating the emission but not the time structure of the ensuing signal.

We proceed to a numerical solution of the fully nonlinear 1-D Vlasov-Fokker-Planck (VFP) equation for the bunch distribution in phase space. The model is "self-consistent", as it includes the distribution-dependent collective force associated with CSR. With the representation of the CSR-induced collective force that will be discussed in the next Section, the resulting solutions appear to reproduce all the qualitative features of the coherent signal found experimentally. Since the only collective force in the model is from CSR, the result is an argument in favor of the idea that the machine geometric impedance plays at most a secondary role. Nevertheless, it will still be interesting in further work to assess the role of the geometric impedance, which may vary from one machine to another.

2. The model

Our model is based on the usual one-dimensional longitudinal motion under linear r.f. focusing, with radiation damping and quantum fluctuations from incoherent emission of photons [24]. To this we add a "self-consistent" account of the nonlinear interplay of CSR and particle dynamics, based on the Vlasov-Fokker-Planck (VFP) equation for the phase space distribution.

In the ultra-relativistic limit, which is of interest here, the Lorentz factor γ is much larger than unity and $\alpha \gg 1/\gamma^2$, so that the momentum compaction α is about the same as the slippage factor. It is convenient to work with the dimensionless phase space variables, $q = z/\sigma_z$ and $p = -\Delta E/\sigma_E$, where z is the distance from the test particle to the synchronous particle (positive when the test particle leads), and $\Delta E = E - E_0$ is the deviation of energy from the design energy. Normalization is by the low-current r.m.s. bunch length and energy spread, which are related by the equation $\omega_s \sigma_z / c = \alpha \sigma_E / E_0$, where ω_s is the angular synchrotron frequency. In these coordinates the unperturbed equations of motion are $dq/d\tau = p$, $dp/d\tau = -q$, with time variable $\tau = \omega_s t$.

The VFP equation for the phase-space distribution function $f(q, p, \tau)$ is

$$\frac{\partial f}{\partial \tau} + p\frac{\partial f}{\partial q} - [q + I_c F(q, f, \tau)]\frac{\partial f}{\partial p} = \frac{2}{\omega_s t_d}\frac{\partial}{\partial p}\left(pf + \frac{\partial f}{\partial p}\right), \tag{1}$$

where $-I_c F(q, f, \tau)$ is the collective force due to CSR, in principle the longitudinal electric field obtained from Maxwell's equations with charge/current densities derived from f itself. The nonlinear Vlasov operator on the left side accounts for the complicated short term dynamics, while the Fokker-Planck operator on the right side accounts for long-term effects of incoherent radiation: damping, and diffusion due to quantum fluctuations. The longitudinal damping time is t_d. We normalize F so that the current parameter is $I_c = e^2 N/(\omega_s T_0 \sigma_E)$, where N is the bunch population and T_0 is the revolution time.

Since it is difficult to solve the Maxwell equations with a realistic representation of particle orbits and vacuum chamber walls, we compute the collective force as though it came from a simple model which is meant to express the essential features. The vacuum chamber is represented by infinite parallel plates, perfectly conducting, with vertical separation h. The particles move on circular orbits of fixed radius R. In cylindrical coordinates (r, θ, y), with y-axis normal to the plates and origin in the midplane, the charge density has the form $\rho(r, \theta, y, t) = eN\lambda(\theta - \omega_0 t, t)H(y)\delta(r - R)/R$, where $\omega_0 = \beta_0 c/R$ is the revolution frequency of the circular model (not of the actual ring). The vertical density $H(y)$ is fixed; we choose H to be constant for $|y| < \delta h/2$, and 0 otherwise. The longitudinal density in the beam frame evolves by VFP dynamics through the relation $\lambda(\theta, t) = (R/\sigma_z)\int f(R\theta/\sigma_z,\ p,\ \omega_s t)dp$.

The radius R is identified as the radius of curvature in the bending magnets, not the average geometrical radius, of the actual ring. Thus, we effectively neglect transient effects as the particles enter and leave bends, hoping that at least the total work done by the CSR force over a turn will be approximated by the model. The plate separation is taken to be the average height of the actual vacuum chamber in the bends. The parameters entering the unperturbed equations of motion will be those of the actual ring. Only the CSR force is computed as though the trajectory were circular.

We define $E(\theta, t) = \int e_\theta(\theta, R, y, t)H(y)dy$ to be the longitudinal electric field averaged over the transverse distribution. The double Fourier transform (FT) of the field is $\hat{E}(n, \omega) = (2\pi)^{-2}\int d\theta \int dt \exp(-in\theta + i\omega t)E(\theta, t)$,

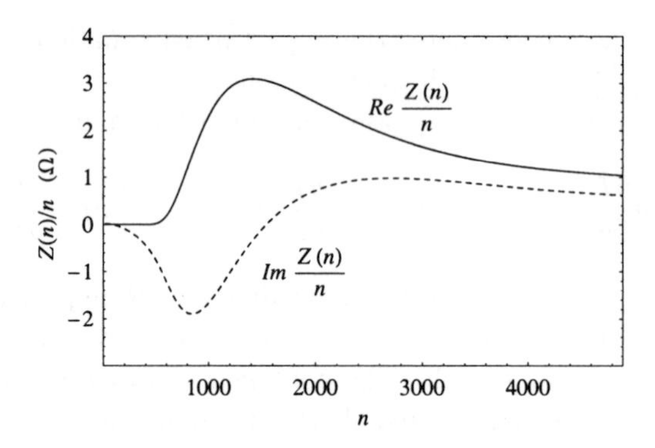

Figure 1. Real (solid line) and imaginary (dashed line) parts of $Z(n)/n$ in ohms. Parallel-plate model with $h = 4.2$ cm, $R = 1.9$ m, and $E_0 = 737$ MeV.

which is related to the corresponding FT of the current I through the impedance: $-2\pi R\hat{E}(n,\omega) = Z(n,\omega)\hat{I}(n,\omega)$. The impedance is given by [4]

$$\frac{Z(n,\omega)}{Z_0} = \frac{(\pi R)^2}{\beta_0 h} \sum_{p=1,3,\cdots} \Lambda_p \left[\frac{\omega\beta_0}{c} J_n{}' H_n^{(1)\prime} + \left(\frac{\alpha_p}{\gamma_p}\right)^2 \frac{n}{R} J_n H_n^{(1)} \right]. \qquad (2)$$

Here $H_n^{(1)} = J_n + iY_n$, where J_n and Y_n are Bessel functions of the first and second kinds, respectively, evaluated at $\gamma_p R$, with $\alpha_p = \pi p/h$, $\gamma_p^2 = (\omega/c)^2 - \alpha_p^2$, $\Lambda_p = 2(\sin x/x)^2$ and $x = \alpha_p \delta h/2$. In MKS units $Z_0 = 120\pi\ \Omega$. The sum over positive odd integers p arises from a Fourier expansion with respect to y.

We suppose that during the i-th time step $t_i \rightarrow t_i + \delta t$ in integration of (1) the bunch can be considered as rigid. Next, we assume that during that time step the CSR force can be computed as though the bunch had its present form for all time. In that case we get the field from the source $\hat{I}(n,\omega) = eN\omega_0\lambda_n(t_i)\delta(\omega - n\omega_0)$, where $\lambda_n(t_i) = (1/2\pi)\int d\theta \exp(-in\theta)\lambda(\theta, t_i)$. Then only the "diagonal" part of the impedance, $Z(n) = Z(n, n\omega_0)$, enters the picture. The inverse FT gives the collective force for (1) through $F(q, f(\tau_i)) = -\omega_0 \sum_n \exp(inq\sigma_z/R)Z(n)\lambda_n(t_i)$. The real part of $Z(n)/n$ has a peak value of about $132h/R\ \Omega$ and is negligible for $n < n_0 = \pi(R/h)^{3/2}$. Fig. 1 shows the real and imaginary parts of $Z(n)/n$ for a choice of parameters meant to model the NSLS VUV Storage Ring.

A more exact treatment of bunch deformation in the impedance picture, accounting strictly for causality and retardation, involves off-diagonal contributions of $Z(n,\omega)$. This matter will be discussed elsewhere [22], as will our procedure for fast evaluation of the FT defining $\lambda_n(t_i)$.

3. A Case Study: the NSLS VUV Storage Ring

In the following we show examples of solutions to Eq. (1) that are meant to model the beam dynamics for some typical setting of the Brookhaven NSLS VUV Storage Ring. We chose to refer to this machine mostly because of the extensive measurements of CSR that have been carried out over the past few years [11,12,13]. The VUV Ring has a double-bend achromat lattice with a local radius of curvature $R = 1.9$ m and a vacuum chamber size $h = 4.2$ cm. The list of other relevant parameters includes a synchrotron frequency $\omega_s/2\pi = 12$ kHz; revolution frequency $1/T_0 = 5.9$ MHz; damping time $t_d = 10$ ms; energy $E_0 = 737$ MeV. The rms bunch length corresponding to the natural energy spread of $\sigma_E/E_0 = 5 \times 10^{-4}$ is $\sigma_z = 5$ cm.[a] During CSR measurements the VUV ring was operated in a single-bunch mode. In this regime the ring supports a current up to 400 mA, corresponding to $N = 4 \times 10^{11}$ particles. The experimental current threshold for detection of coherent signal for the particular setting under consideration here is 100 mA. This can be changed considerably through variations of the machine momentum compaction[11].

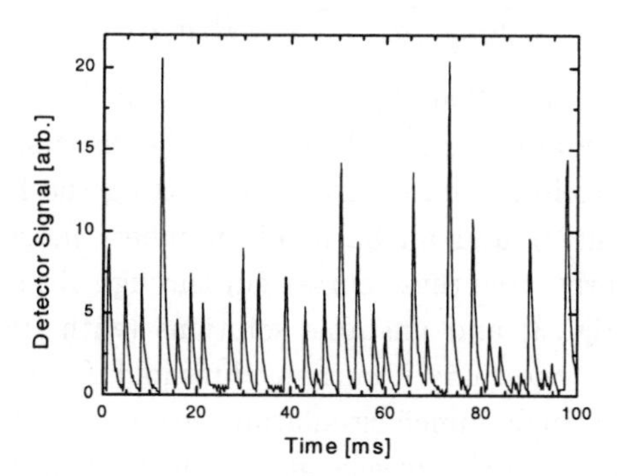

Figure 2. Far infrared detector output with emission bursts at the NLSL VUV Ring. (Courtesy of G. Carr.)

Figure 2 shows an example of a coherent signal in the far infrared with a

[a] For beam height δh, which is not a critical parameter, we take 0.1 mm.

characteristic bursting structure (from G. Carr *et al.*[11]). In this particular instance the peaks are separated by a few msec and appear to have a fairly regular spacing. The duration of the peaks is dominated by the thermal time constant of the detector (about $200\mu s$[11]).

4. Linear Theory

Linearization of Eq. (1) can be used to obtain useful information regarding the conditions for the onset of the instability. As we neglect the effects of the geometric machine impedance, a Gauss distribution $f_0 = e^{-(p^2+q^2)/2}/(2\pi)$ in energy spread and spatial variable is a very good approximation to an equilibrium. For bunched beams the linear equation obeyed by f_1, a small deviation about equilibrium, is still too complicated to have a solution in closed analytical form. However, under the conditions that the instability be fast and the wavelength of the unstable mode small compared to bunch size one can use the linearized Vlasov equation for a coasting beam (*i.e.* neglect the rf focusing term proportional to q) . The current carried by the coasting beam should be the same as the peak current for the bunched beam (Boussard criterion). The modified linear equation admits wave solutions with space-time dependence $\exp[i(nq\sigma_z/R - \nu\tau)]$ yielding the dispersion relation

$$\frac{I_c \omega_0 R^2}{\sqrt{2\pi}\sigma_z^2} \frac{Z(n)}{n} = \frac{i}{D(\nu R/\sigma_z n)}, \tag{3}$$

where $D(z) = 1 + iz\sqrt{\pi/2}w(z/\sqrt{2})$ and $w(z) \equiv e^{-z^2}\mathrm{erfc}(-iz)$ is the error function of complex argument[24]. Analysis of the dispersion relation is best represented on a Keil-Schnell diagram by plotting the LHS part of Eq. 3 for a given current I_c and all harmonic numbers n; see Fig. 3. If the current is sufficiently large these curves cut through the stability boundary (dashed line in Fig. 3) and unstable solutions (with Im $\nu > 0$) emerge. From this analysis one finds a current threshold $I_c > I_c^{th} = 6.2$ pC/V, corresponding to a single-bunch circulating current of 168 mA or $N = 1.8 \times 10^{11}$. Close to threshold the wavelength of the most unstable mode is $\lambda = 2\pi R/n = 6.8$ mm with $n = 1764$. These values are reasonably close to the observed wavelength $\lambda = 7$ mm and critical current 100 mA for detection of a coherent signal [11]. The linear theory also indicates that the instability is very fast: the exponential growth-time of the most unstable mode is as low as one tenth of synchrotron period even for a current only 5% above threshold – validating the use of the Boussard criterion. This linear analysis is essentially the same as the one worked out by Heifets and Stupakov[20].

The only difference is that there the radiation impedance is relative to free space – implying that the result is meaningful only when the calculated unstable wavelength is below the shielding cutoff. A linear analysis for a bunched beam [23], with the radiation impedance for a resitive toroidal vacuum chamber, indicated some time ago that CSR alone could cause an instability at plausible current, at least for parameters of a compact storage ring [19].

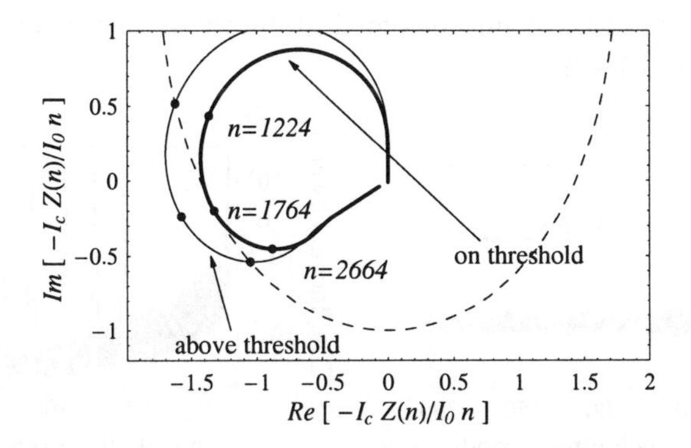

Figure 3. Keil-Schnell diagram for the linear motion of the NSLS VUV Ring. The dashed line is the familiar onion-shaped stability boundary for Gaussian equilibria. The thick line corresponds to the current (parameter) threshold $I_c = 6.2$ pC/V; $I_0 = \sqrt{2\pi}\sigma_z^2/\omega_0 R^2$.

5. Numerical Results

Above current threshold for instability, the numerical solutions of Eq. (1) present a characteristic sawtooth behavior that sets in after a transient depending on the initial condition. The pattern is evident from the plot of the rms bunch length versus time shown in Fig. 4 (picture on the left). The corresponding solution was obtained for a value $I_c = 10.5$ pC/V of the current parameter (equivalent to 3×10^{11} particles/bunch) starting from a slightly perturbed gaussian distribution. The thickness of the curve is due to the fast oscillation of the quadrupole mode.

The cycle of the sawtooth pattern follows a sequence: instability → saturation → damping. Close inspection of the solutions shows that where the envelope of the bunch length oscillation is minimum, a ripple (microbunching) appears on top of the charge distribution, see Fig. 5, right picture. This is the point where the density of the distribution in phase space is the largest and one may expect that the conditions for the collective instability are met. As predicted by the linear theory, the amplitude of this modulation grows rapidly. In the process the distribution function experiences a sudden enlargement as reflected by jumps in the evolution of either the rms bunch length (see picture) or energy spread. Next, as the density of the distribution in phase space decreases and reaches some critical value, saturation of the instability follows. This is where the effect of radiation damping starts to become apparent. It causes the bunch distribution to slowly shrink until the conditions for instability are met again so that the cycle can repeat itself.

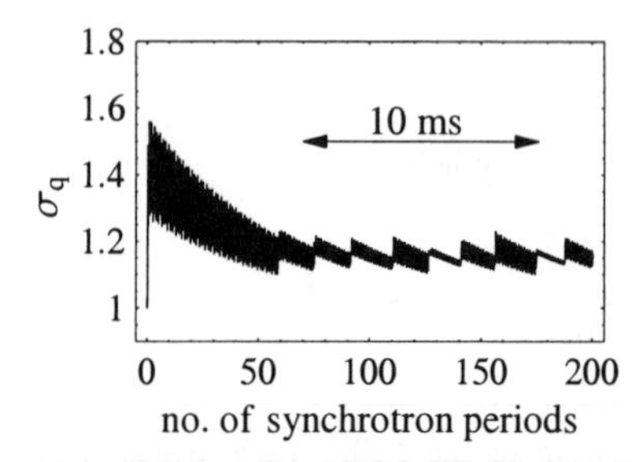

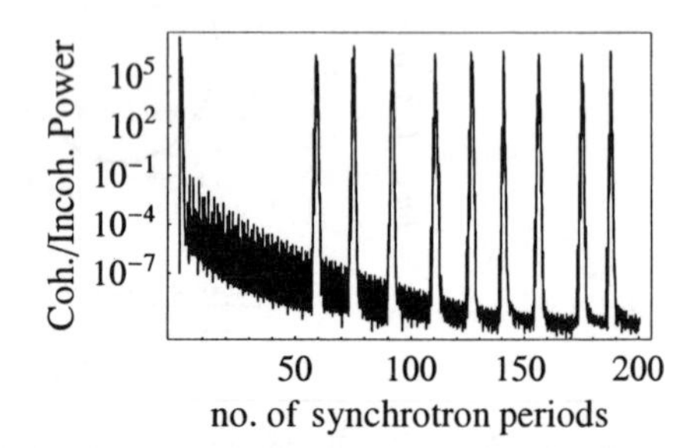

Figure 4. On the left: rms bunch length (normalized coordinates) vs. number of synchrotron periods. On the right: ratio of coherent radiation power to incoherent power vs. number of synchrotron periods for a narrow band of modes about a wavelength $\lambda = 7$ mm. In both cases $I_c = 10.5$.

Emission of coherent radiation takes place in correspondence to the appearance of microbunches, at the notches of the bunch length envelope. The time structure of the signal follows naturally that of the sawtooth pattern. The expected coherent radiation emitted at the n–th harmonic is given by

$$P_n^{coh}(t) = 2(eN\omega_0)^2 \mathrm{Re}Z(n)|\lambda_n(t)|^2 \tag{4}$$

where λ_n is the Fourier component of the (normalized) charge density (see Section 2). By comparison the incoherent part of radiation is expressed by $P_n^{incoh} = 2N(e\omega_0)^2\mathrm{Re}Z(n)/(2\pi)^2$. Plot of the coherent-to-incoherent-power ratio for a narrow band of modes around the wavelength $\lambda = 7$ mm is reported in Fig. 4 (picture to the right).

In qualitative agreement with the experimental observations the pattern of bursts is fairly regular for moderate currents while for larger current more stochastic features start to appear – both in the occurrence of the bursts as well in their peak values. Moreover, we found a dependence of the average separation between bursts on the current, again in qualitative agreement with observations[25]. A plot of the average separation between bursts vs the current parameter I_c is reported in Fig. 5 (picture to the right).

In solving Eq. (1) as a time-domain initial-value problem we used a variant of the Perron-Frobenius method presented by R. Warnock and J. Ellison[26]. This involves a representation of the distribution function f on a cartesian grid and propagation of the derivatives $\partial_q f$ and $\partial_p f$ along with f. A more detailed description will be found a forthcoming publication[22].

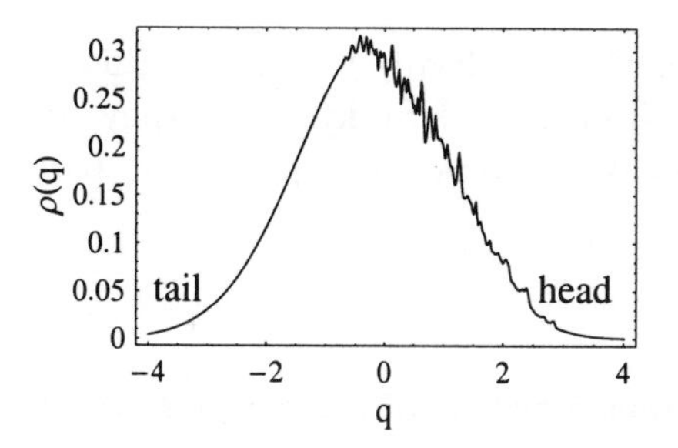

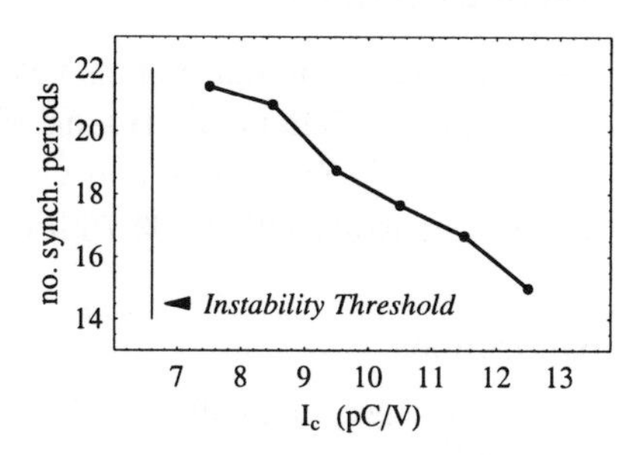

Figure 5. On the left: snapshot of the charge distribution with microbunching taken during a radiation burst ($I_c = 12.5$ pC/V). On the right: plot of average burst separation vs. current parameter.

6. Conclusions

CSR has become a very active subject of research over the past few years. While most of the attention is presently devoted to CSR effects in single-pass systems, we hope to have shown their relevance in storage rings as well. The basic physics of the instability caused by CSR is identical in both cases and its basic signature, *i.e.* the existence of microbunching, is also the same. However, while CSR-driven microbunching has been observed in bunch compressors both in simulations and in direct measurements, at this time there is no direct experimental evidence of it in storage rings (*e.g.* from streak camera measurements). The evidence is indirect through the detection of a coherent signal.

Numerical integration of our dynamical model supports the notion that the collective instability caused by CSR alone is sufficient to account for microbunching. In addition, the model appears to reproduce at least qualitatively the main features of the observations: the existence of a current threshold for detection of CSR, the wavelength at the peak of the coherent radiation spectrum, and the time structure of the signal characterized by short recurrent bursts separated by a substantial fraction of the damping time.

It is in our plans for the future to explore extension of the VFP solver to include the horizontal degree of freedom. If successful, such an extension could provide an interesting alternative to the macroparticle methods currently used to study beam propagation through bunch compressors.

Acknowledgments

We benefitted from discussions with K. Bane, J. Byrd, A. Chao, J. Ellison, P. Emma, S. Heifets, Z. Huang, S. Kramer, S. Krinsky, J. Murphy, B. Podobedov, F. Sannibale, and G. Stupakov. Work was supported in part by DOE contract DE-AC03-76SF0051.

References

1. J. Schwinger, unpublished (1945) [transcribed in LBL report LBL-39088 (1996)].
2. L. Schiff, Rev. Sci. Instr., **17** no. 1, 6 (1945).
3. J. Nodvick and D. Saxon, Phys. Rev. **96**, 180 (1954).
4. R. Warnock and P. Morton, Part. Accel. **25**, 113 (1990); R. Warnock, SLAC reports SLAC-PUB-5375 (1990), -5417 (1991), -5523 (1991).
5. D. Keefe *et al.*, Phys. Rev. Lett. **552**, 558 (1969); A. Faltens and L. J. Laslett, Part. Accel. **4**, 151 (1973); V. Brady, A.Faltens, and L. J. Laslett, LBL report LBID-536, 1981.
6. A. G. Bonch-Osmolovskii and E. A. Perel'shtein, Izv. Rad. Phys. **8**, 1080 (1970); *ibid.* **8**, 1089 (1970); B. G. Shchinov *et al.*, Plasma Physics **15**, 211 (1973).
7. J. Yarwood *et al.*, *Nature* **312**, 742 (1984).
8. E. Schweizer, *et al.*, Phys. Res. Sect. A **239**, 630 (1985).
9. T. Nakazato *et al.*, Phys. Rev. Lett. **63**, 1245 (1989).
10. U. Arp *et al.* Phys. Rev. ST Accel. Beams **4**, 054401 (2001).
11. G. Carr, S. Kramer, J. Murphy, R. Lobo, and D. Tanner, Nucl. Instr. Meth. Phys. Res. Sect. A **463** 387-392 (2001).
12. B. Podobedov *et al.*, Proc. 2001 IEEE Part. Accel. Conf., p. 1921 (IEEE, Piscataway NJ, 2002).
13. S. Kramer *et al.*, Proc. 2002 Euro. Part. Accel. Conf., to be published; S. Kramer, Phys. Rev. ST Accel. Beams **5**, 112001 (2002).
14. Å. Anderson, M. Johnson, and B. Nelander, Opt. Eng. **39**, 3099 (2000).
15. M. Abo-Bakr, J. Feikes, K. Holldack, G. Wüstefeld, and H.-W. Hübers, Phys. Rev. Lett. **88**, 254801-1 (2002).
16. J. M. Byrd *et al.*, Proc. 2002 Euro. Part. Accel. Conf., to be published.

17. W. C. Barry *et al.*, Proc. 2002 Euro. Part. Accel. Conf., to be published.
18. F. Michel, Phys. Rev. Lett. **48** 580 (1982).
19. J. B. Murphy and S. Krinsky, Nucl. Instr. and Methods in Phys. Res. A **346** 571 (1994)
20. S. Heifets and G. Stupakov, Phys. Rev. ST Accel. Beams **5**, 054402 (2002).
21. M. Venturini and R. Warnock, Phys. Rev. Lett. **89**, 224802 (2002).
22. M. Venturini, R. Warnock, and R. Ruth, to be submitted to Phys. Rev. E.
23. R. Warnock and K. Bane, SLAC report SLAC-PUB-95-6837 (1995).
24. S. Y. Lee, *Accelerator Physics*, Chap. 3 (World Scientific, Singapore, 1999).
25. B. Podobedov, Private communication.
26. R. Warnock and J. Ellison, in *The Physics of High Brightness Beams*, (World Scientific, Singapore, 2000).

WORKING GROUP C

Application to FELs

WORKING GROUP C SUMMARY*

H.-D. NUHN
Stanford Linear Accelerator Center
2575 Sand Hill Rd,
Menlo Park, CA 94325, USA
E-mail: nuhn@slac.stanford.edu

Working group C, "Application to FELs," of the Joint ICFA Advanced Accelerator and Beam Dynamics Workshop on July 1-6, 2002 in Chia Laguna, Sardinia, Italy addressed a total of nine topics. This summary will discuss the topics that were addressed in the stand-alone sessions, including Start-To-End Simulations, SASE Experiment, PERSEO, "Optics Free" FEL Oscillators, and VISA II.

1. Introduction

This is the summary of the activities in working group C, "Application to FELs," which was based in the Bithia room at the Joint ICFA Advanced Accelerator and Beam Dynamics Workshop on July 1-6, 2002 in Chia Laguna, Sardinia, Italy. Working group C was small in relation to the other working groups at that workshop. Attendees include Enrica Chiadroni, University of Rome "La Sapienza", Luca Giannessi, ENEA, Steve Lidia, LBNL, Vladimir Litvinenko, Duke University, Patrick Muggli, UCLA, Alex Murokh, UCLA, Heinz-Dieter Nuhn, SLAC, Sven Reiche, UCLA, Jamie Rosenzweig, UCLA, Claudio Pellegrini, UCLA, Susan Smith, Daresbury Laboratory, Matthew Thompson, UCLA, Alexander Varfolomeev, Russian Research Center, plus a small number of occasional visitors.

The working group addressed a total of nine topics. Each topic was introduced by a presentation, which initiated a discussion of the topic during and after the presentation. The speaker of the introductory presentation facilitated the discussion.

There were six topics that were treated in stand-alone sessions of working group C. In addition, there were two joint sessions, one with working group B,

* This work is supported by U.S. Department of Energy, Office of Basic Energy Sciences, Division of Material Sciences, under Contract No. DE-AC03-76SF00515.

which included one topic, and one with working group C, which included two topics. The presentations that were given in the joint sessions are summarized in the working group summary reports for groups B and D, respectively. This summary will only discuss the topics that were addressed in the stand-alone sessions, including Start-To-End Simulations, SASE Experiment, PERSEO, "Optics Free" FEL Oscillators, and VISA II.

1.1. *Start-To-End Simulations*

The first presentation in working group C was by Sven Reiche about the status of start-to-end simulations. With the development of high-brightness electron beam sources and short wavelength SASE free-electron lasers, the need arises to parameterize the electron beam on a much more detailed level than projected beam quantities such as energy spread or emittance. This is done by start-to-end simulations, where an electron bunch is tracked from its generation in an RF photocathode gun to the exit of the undulator. The presentation summarized the methods that used for the recent start-end simulations and gave an outlook on the trends for the near future.

Sven Reiche pointed out that a SASE FEL system consists of quite a number of different subsystems in which different processes take place. These subsystems are: the drive laser, the RF gun, the linac sections, the bunch compressors, the FEL undulators, and the FEL beam optics. There is no single code that covers the entire system and it is felt that it is not necessary or desirable to create such a code. Instead, there are different codes for each subsystem that handle the different processes in that subsystem. The codes are modules in a modular system. All modules, except the first, work on the output of the preceding module, which is stored in data files. The results of the computations are in-turn stored in data files, again, to be picked up by the following module. So far, the start-to-end simulations have been done with modules for the gun, the linac sections, the bunch-compressors, and the FEL undulator. It is desirable to add modules for gun laser, multi-stage FELs and FEL optics components.

The actual computer codes used as modules are listed in the table:

Table 1 Start-to-end simulations codes

Drive Laser	None
Gun	Astra, Parmela
Linac	Elegant
Bunch Compressor	Traffic4, (Elegant)
FEL	Genesis 1.3, Ginger, Fast
FEL Optics	SRW, Phase, R.Bionta@LLNL, Shadow

No code is available yet that would allow simulating the drive laser. For the RF photocathode gun there are Astra by a group at DESY and Parmela by Lloyd Young at Los Alamos. For the linac, the code Elegant by Michael Borland from Argonne is used. For the bunch-compressor calculations, which are mostly concerned with the handling of coherent synchrotron radiation (CSR), there is Traffic[4] from DESY and also Elegant contains a 1-D treatment of CSR. The predominant time-dependent FEL codes are Genesis 1.3 by Sven Reiche from UCLA, Ginger by William M. Fawley from LBNL, and FAST by Evgeny Saldin, Evgeny Schneidmiller and Michael Yurkov from DESY.

Several codes exist or are being developed for the FEL optics, some at BESSY and the ESRF and some at LLNL by Richard Bionta and co-workers. Those codes are not yet directly coupled to the rest of the modules but there are plans to do this.

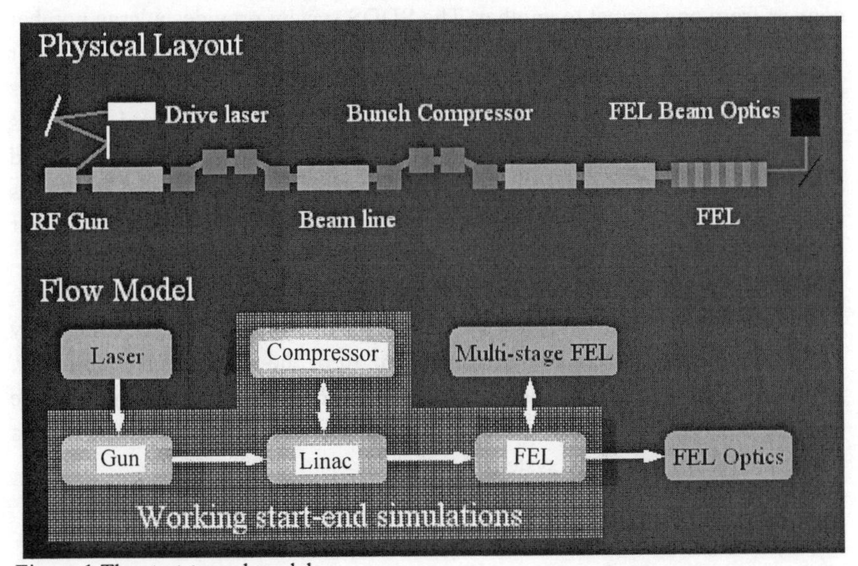

Figure 1 The start-to-end model.

Important is the file format used to exchange results between the various modules. In the past, applications have been written in native file formats. This means that the authors just invented a new set of input and output formats every time they wrote a new code. Of course, different programs used completely different formats. In order to be able to exchange data from one code to another one needed to write a translator code to convert between the formats. These special formats are difficult to maintain. Often, it is even difficult for the authors themselves to remember what the exact input format was. The format is normally fixed, meaning that, for instance, the numbers have to be put into a certain range of columns in the input line or either must or must not contain a decimal point. If errors are made the programs may read the numbers wrongly, which is often difficult to catch by the user. The input and output files are normally so specialized that preprocessor and postprocessor programs have to be distributed together with the module code itself in order to produce and/or read these files. No standard software is able to read these files.

This is different with the new approach of using Self Describing Data Formats. The start-to-end effort is using the SDDS (Self Describing Data Set) system developed by Michael Borland at Argonne. The files that are produced, maintained, and evaluated by this package do not only contain the data themselves but also a description of the data, i.e., the names, the physical units and even the format in which the data is coded. The programs are storing the data in ASCII as well as in binary format. There are functions available that can convert from one format to another. The SDDS system provides a large number of library functions to be used by code developers and there is a toolkit for processing and plotting the data. Using Self Describing Data Format tools greatly simplifies the flow in start-to-end simulations and Michael Borland's SDDS set is particularly good for homogeneous data sets. Sven Reiche pointed out that there are desirable aspects that are still missing in Michael Borland's SDDS system. In particular, SDDS does not validate files against a given format

Over the last several years, a more general Self Describing Data Format, which is called the Extensible Markup Language (XML), has been developed by international groups completely independent from the accelerator community. XML provides significantly better control over the data and makes it possible that the data validity can be checked by generic codes. Sven Reiche made a strong pledge for using this system.

Under XML, the files are well formatted, validated, and platform independent. Various software libraries and editors are available. The system is widely used, outside of the accelerator community. The data organization in the files is so that generic programs can handle the data. The system allows for flexible transformation between various user-defined formats. One of the drawbacks is that there is a large overhead for long, similar data sets, such as large lists of particle phase space information. When storing these kinds of data, the data files would be of an enormous size because the storage is in ASCII and there is a description for each number. XML seems not to support binary formats in a controlled way. The SDDS system does a better job for those large data sets.

Sven Reiche believes that the equivalent for XML for binary data sets would be the Hierarchical Data Format Version 5 (HDF5). The HDF5 software is developed and supported by NCSA and is freely available. It is used world-wide in many fields, including Environmental Science, Neutron Scattering, Non-Destructive Testing, and Aerospace, to name a few. Scientific projects that use HDF5 include NASA's HDF-EOS project, and the DOE's Advanced Simulation and Computing Program. Sven Reiche will examine this option.

The next step in the development of the start-to-end simulation should be the addition of the FEL optics module, the improvement of the support for a standardized format, extension of the common file format to all major data streams, automation and automatic checking of the input data, and support for scripted execution of the entire system. The latter is important because the separate modules have long execution times. Scripted execution allows for them to run in the background without need for intermittent attention from the user.

In addition, the development of user friendly GUIs, built on top of scripted shells, are desirable because they would make it easier to distribute the packages to other users.

1.2. *SASE Experiments*

The next topic was introduced by Alex Murokh, who summarized what has been learned from the SASE experiments.

Three major SASE experiments have been successfully carried out in the last few years: LEUTL at Argonne, the TTF-FEL at DESY and VISA at the ATF at Brookhaven. VISA was carried out by a collaboration of a number of laboratories including the LCLS.

Alex Murokh found that there was a lack of detailed measurements of particle distributions at the input to the undulator. The VISA experiment had done the best job in simulating the distribution. The other two experiments had large uncertainties for average parameters such as emittance.

During the evaluation of the experimental data it became clear that one could only fully analyze the FEL experiments when using simulation codes, because the particle phase space distributions, used in the experiments, are too complicated to be reasonably describable by analytic theory. It is still necessary that detailed measurements of particle distributions be made in order to check the simulations at a number of positions along the system.

All three experiments worked with rather complicated distributions, which, in the case of VISA, could be simulated, and the measured results agreed with the simulations to an amazing degree.

If the experiments could be made with simpler distributions it would be easier to measure the distributions experimentally. This would also make these results more available for the comparison with analytical theory.

Alex Murokh looked at 3-D scaling laws to test if parameters that were measured at VISA scaled as the theory predicts. He is planning to do this also for the other two experiments as soon as he gets sufficient data. In order to check the 3-D scaling laws, Alex Murokh estimated the effective FEL parameter, $\tilde{\rho}$, from the measured gain length According to

$$\tilde{L}_g \equiv \frac{\lambda_u}{4\pi\sqrt{3}\tilde{\rho}} . \tag{1}$$

Once $\tilde{\rho}$ is estimated from the measured gain length, a number of parameters can be calculated such as the saturation length, the spread of the saturation lengths over many pulses, the saturation power, the bandwidth, the corporation length, the gain at saturation, and the noise bandwidth. These parameters can also all be obtained independently from the experimental results. Alex Murokh compared the results of these two approaches and came to the conclusion that the agreement is rather good.

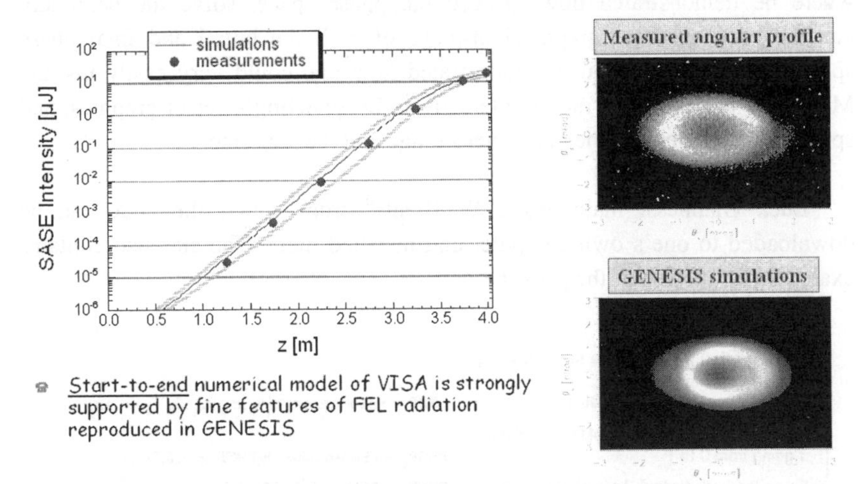

Figure 2 Results of start-to-end simulations for the VISA experiment.

As already mentioned, the main part of the VISA experimental analysis was based on intense start-to-end simulations. Results from these simulations include the prediction of the gain development along the undulator. The measurements agreed very well with the predicted gain curve.

One of the results, caused by the complicated phase space distribution in the VISA experiment, was a donut shaped distribution of the FEL radiation at the exit of the undulator. It is one of the satisfying successes of start-to-end simulations that even the radiation pattern of such a complicated phase space distribution can be predicted and explained by start-to-end simulations. This and a number of other points give very good confidence to the simulation codes.

1.3. *PERSEO*

The next contribution was by Luca Giannessi. He introduced his tool set, PERSEO, for FEL simulations using the commercial MathCAD package[c]. Luca

[c] http://www.mathcad.com/products/mathcad.asp

Giannessi wrote a suite of complex functions in the C language that can be called from MathCAD and allow the 1-dimensional simulation of SASE FEL configurations. He also wrote a number of MathCAD scripts that use these functions and provide more complex functionality. He showed an example where he demonstrated how to seed the phase space, solve the pendulum equation, and produce graphical displays of both the initial and final phase space distributions. He also demonstrated how one could very easily instruct MathCAD to generate a movie interactively by selecting a set of graphics, and specifying a parameter, the parameter's range and the step size.

Luca Giannessi maintains a WEB site[d] from which the code can be downloaded to one's own computer and installed there. The site also contains examples for the use of the package.

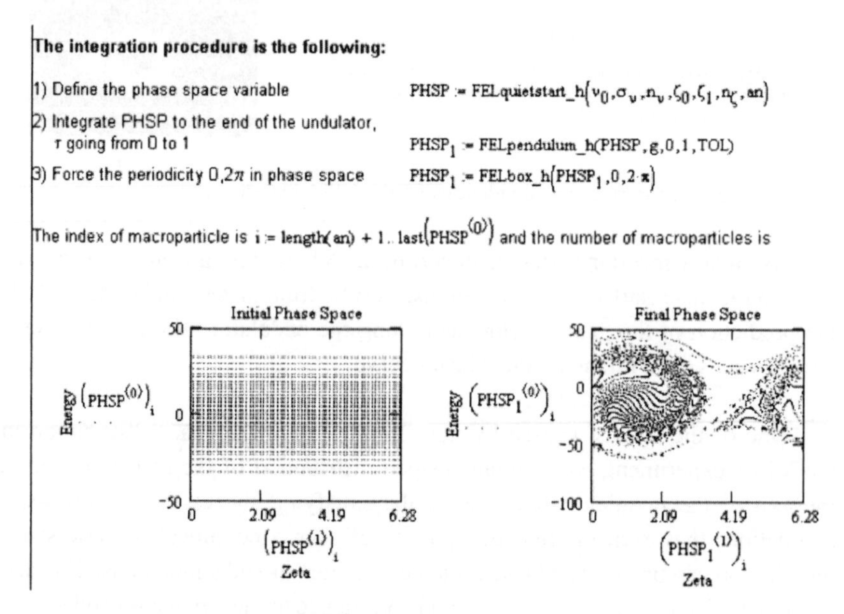

The integration procedure is the following:
1) Define the phase space variable
PHSP := FELquietstart_h(ν0,σν,nν,ζ0,ζ1,nζ,an)
2) Integrate PHSP to the end of the undulator, τ going from 0 to 1
PHSP1 := FELpendulum_h(PHSP,g,0,1,TOL)
3) Force the periodicity 0,2π in phase space
PHSP1 := FELbox_h(PHSP1,0,2·π)
The index of macroparticle is i := length(an) + 1 .. last(PHSP⟨0⟩) and the number of macroparticles is
Initial Phase Space
Final Phase Space
Energy
Zeta
50
0
-50
-100
2.09
4.19
6.28

Figure 3 Example for using the PERSEO system.

[d] http://www.afs.enea.it/gianness/perseo/

1.4. *"Optics Free" FEL Oscillators*

The next presentation was by Vladimir Litvinenko about "Optics Free" FEL Oscillators. Vladimir Litvinenko addressed the problem that the spectral distribution of radiation from SASE FELs is rather wide, i.e., on the order of 0.3-0.5 % while typical VUV FEL oscillators, such as the OK-4 at Duke University, have a much smaller bandwidth of 0.01-0.0003%. Of course, the absence of good optics in the VUV and X-Ray spectral ranges makes traditional oscillator schemes prohibitory complex or impossible. That is why Vladimir Litvinenko introduces the concept of optics-free FEL oscillators. The basic idea is to use a high gain FEL, which produces a radiation pulse that can be used for experiments and, at the same time, can also be used to add a momentum modulation to a low energy feedback electron beam that travels through a small modulator undulator together with the radiation pulse. The feedback beam is produced by an independent electron source and is longer than the radiation pulse. Only the part of the feedback beam that overlaps with the radiation pulse will be modulated. The modulated feedback beam is then transported through an isochronous beamline to the beginning of the high-gain FEL where it enters a small radiator undulator to produce seed light for seeding the FEL process for the next arriving electron bunch entering the high-gain FEL (See Figure 4).

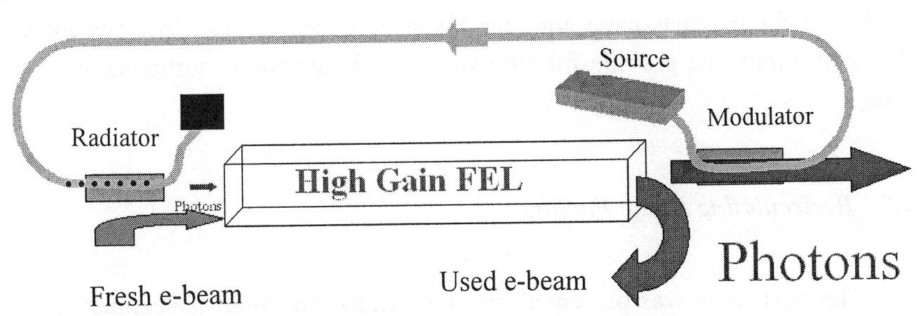

Figure 4 System diagram of the "Optics Free" FEL Oscillator.

Of course, the seed light amplitude needs to be larger than the noise power produced by the next bunch at the beginning of the high-gain undulator and there has to be a large temporal overlap between the seed light and the modulated part of the feedback beam. If the temporal jitter of the incoming electron bunch is larger than its width the scheme will not work. To overcome this problem one could work with a train of bunches that are temporally spaced by the same time that it takes for the electron bunch to travel through the feedback undulator and for the feedback beam to be transported back to the radiator undulator. These bunches would have very little time jitter with respect to each other. This might make such a scheme possible.

In order for the system to function, the momentum modulation in the feedback beam must be preserved during the transport from the modulator undulator to the radiator undulator. One condition is that the beamline needs to be isochronous, i.e., the momentum compaction factor needs to be fairly small up to higher order. Vladimir Litvinenko calculated limit values of 10^{-8} for the first order, 10^{-4} for the second order and 1 for the third order of the momentum compaction factor. Effects of emittance increase need to be compensated by sextupoles that are positioned inside the achromatic arcs.

The main limiting factor are quantum fluctuations due to synchrotron radiation in the transport bending magnets, setting an upper limit to the energy of the feedback beam. Vladimir Litvinenko believes that the scheme is still doable. He calculated that the dimensions of the radiator undulator could be 6 cm in length for 100 periods with a period length of 0.6 mm. With an energy of 500 MeV, an intensity of about 1 A and a normalized emittance of 0.1-0.3 mm mrad the oscillator gain could be as high as 10^3-10^4.

So far, effects such as CSR, the stability of power supplies, and the limited quality of the magnets have not been taken into account. The low current and the small emittance provide for "toy sizes" of the magnetic components in the arcs.

1.5. *Recirculating Linac Facility*

The next topic was presented by Steve Lidia, an initial conceptual design for a femto-scale pulse length synchrotron radiation facility ('femtosource'). The facility, planned at LBNL, is based upon a recirculating linac, which is being designed to generate flat electron beams of 2 ps duration in a series of conventional undulators and bend magnets. There is a strong scientific case for time resolved experiments at time scales of the order of atomic vibration levels of about 100 fs.

Starting from an injector system the beam makes several passes through a number of linac sections as shown in Figure 5. Between passes, the beam is transported by several beam transport lines with increasing radii. After the fourth pass through the linac the electron beam goes through a sequence of radiators, i.e., an alternate array of insertion devices and bending magnets that produce radiation that can be adjusted over a range between 1 and 10 keV. The x-ray pulses are expected to be shorter than 100 fs FWHM at 10 keV. The proposed repetition rate is quite high at 10 kHz. The goal for the photon flux is 10^7 photons/pulse/.1% BW at 10 keV. The proposed level of synchronization is 10 fs.

The system is to be incorporated in the LBNL laboratory close to the Advance Light Source.

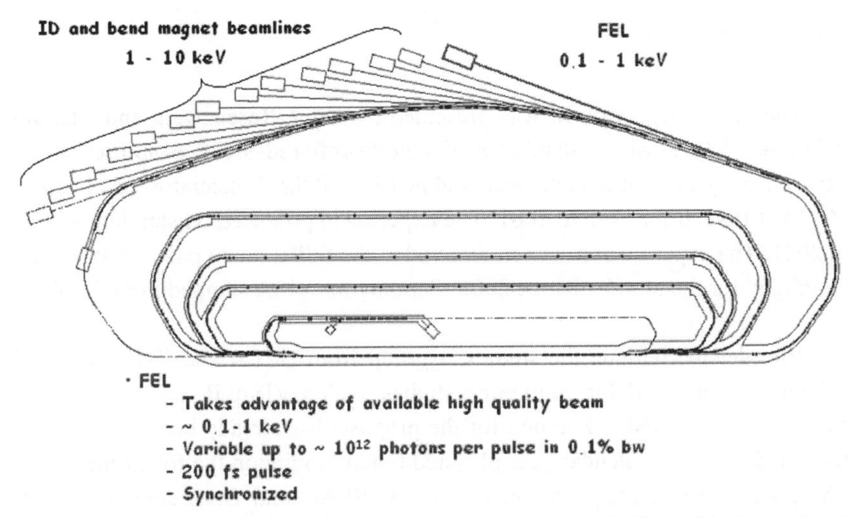

Figure 5 The components of the planned recirculating linac facility with ID and bend magnet beamlines as well as a soft x-ray FEL.

The presentation included a description of the general parameters of the facility, emphasizing the use of existing technologies along with those now in development. The generation of the flat electron beam from a novel RF photoinjector and a skew-quadrupole transport line were described, and recent experimental results, confirming the basic theory, were presented. Emittance preservation techniques for the linac and the recirculating arcs, by proper betatron tune phasing against the single bunch transverse wakefields, were mentioned. Facility infrastructure, cryogenic plant and RF power requirements were presented.

Of particular interest for the SASE FEL community is the development of the photocathode RF gun with pulse repetition rates of 10-100 kHz producing an 8-MeV electron beam. Through a workshop on New Opportunities in Ultrafast Science using x-rays, the group at LBNL established that there is strong scientific interest for the facility. Their outline of the machine feasibility was backed by a Machine Technical Advisory Committee. They plan to document the scientific case and the machine feasibility this year to put the Femtosource on the BESAC agenda. They need to develop mastery of technologies outside the present core competencies of the lab.

1.6. *VISA II*

The last topic, VISA II, was presented by Jamie Rosenzweig and Claudio Pellegrini. VISA, which stands for Visible-to-Infrared-SASE-Amplifier, is an experiment that has been developed and installed at the Accelerator Test Facility (ATF) at BNL from 1998 to 2001. The experiment produced outstanding results in 2001. The experiment has been featured at the SSRL home page[e] as Research Highlight[f] for the month of June 2002. Funding for VISA stopped in mid-2001.

In order to continue the studies, a group from UCLA and BNL joint and submitted a proposal for continuing studies, called VISA II, to the Office of Naval Research (ONL). Funding for the proposal has already been granted at a level of $300k this calendar year plus additional funding in the following years. The plan is that a collaboration of UCLA, SLAC and BNL continue VISA experiments.

VISA measured a large range of phenomena, including exponential gain at saturation, photon statistics, radiation spectra, and micro-bunch using CTR. The operating point was somewhat pathological, making use of strong bunch compression due to the non-linear properties of ATF beamline 3.

---

[e] http://www-ssrl.slac.stanford.edu/

[f] http://www-ssrl.slac.stanford.edu/research/highlights_archive/visa.html

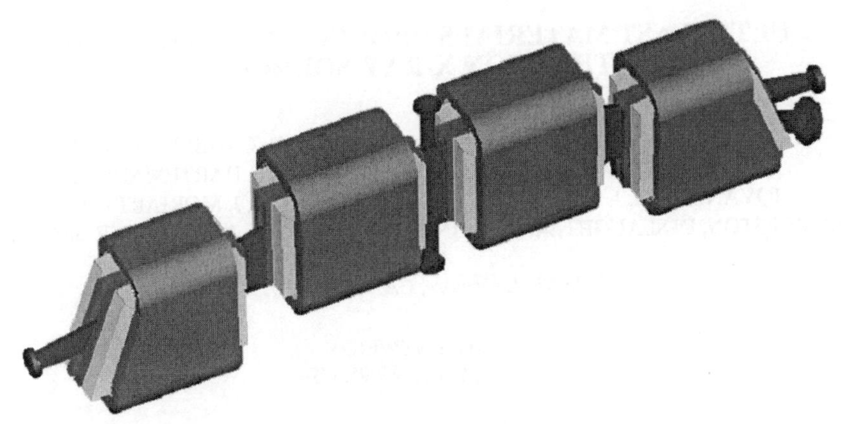

Figure 6 3-D Rendering of the new bunch compressor chicane recently installed in the ATF.

The system was extremely sensitive to RF phase jitter and the information of the initial longitudinal phase space distribution from the gun was completely lost in the non-linear beamline. There is interest of doing a more linear experiment using a desired longitudinal phase space distribution. This is where VISA II comes in.

Part of VISA II is the commissioning of the recently installed linear ATF bunch compressor (see Figure 6). The next step will then be to mitigate the severe nonlinearities in the existing beam by inserting sextupoles (2 might be enough, the study is on-going). And finally add more diagnostics to the beginning of the undulator to better characterize the beam.

Once the system is running, studies can be done with compressed and with uncompressed but chirped beams. Using compressed beams, the system can be used to diagnose the new bunch compressor. With uncompressed but chirped beams the impact of chirped beams on FEL gain can be studied. This will be in support of the LCLS, for which schemes have been proposed that use momentum chirping to produce short x-ray pulses. The chirped beam can also be used as direct diagnosis of FEL longitudinal pulse characteristics.

ULTRAFAST MATERIALS PROBING WITH THE LLNL THOMSON X-RAY SOURCE

G. LE SAGE, S. ANDERSON, W. BROWN, C. BARTY, R. CAUBLE, J CRANE, H. CYNN, C. EBBERS, D. FITTINGHOFF, D. GIBSON, F. HARTEMANN, I. JOVANOVICH, J. KUBA, A. MCMAHAN, R. MINICH, J. MORIARTY, B. REMINGTON, D. SLAUGHTER, P. SPRINGER, F. H. STEITZ, A. TREMAINE, and C-S. YOO

LLNL, Livermore, CA, 94550, USA

J. ROSENZWEIG

UCLA, LA., CA, 90095, USA

T. DITMIRE

University of Texas, Austin, TX, 85721, USA

The use of short laser pulses to generate high peak intensity, ultra-short x-ray pulses enables experimental capabilities that are otherwise unattainable. In principle, femtosecond-scale pump-probe experiments can be used to temporally resolve structural dynamics of materials on the time scale of atomic motion. Current research at Lawrence Livermore National Laboratory is leading toward such a novel x-ray source. The system is based on a low emittance photoinjector, a 100 MeV electron RF linac, and a 300 mJ, 35 fs solid-state laser system. The Thomson source will produce ultra-fast pulses with x-ray energies capable of probing into high-Z metals. A wide range of material and plasma physics studies with unprecedented time resolution will become possible.

1. Introduction

Initial fast dynamic studies of laser excited materials [1] have been pursued in recent years using time-resolved x-ray detectors at synchrotron radiation sources, and laser-plasma x-ray sources in pump-probe experiments [2,3]. Third-generation synchrotrons have been critical for x-ray studies in materials, but have a major limitation: although they probe structure on the atomic length scale, their time resolution (~ 100 ps) is not well-matched to the natural dynamics of elemental processes in solids, such as the time scale for atomic motion (~10-50 fs). Laser melting in semiconductors may involve excitation, electron-phonon coupling, melt front motion, and shock waves, all with relevant timescales less than 30 ps. X-rays produced from hot plasma sources can achieve very short pulse lengths, but the low peak brightness forces multiple shot measurements. In order to study irreversible changes in chemical, biological, or material systems (e.g., shock wave propagation), single-shot measurements are required. Additionally, strong x-ray and laser probe pulses may lead to target destruction in some systems. For single-shot pump-probe measurements, pulse length and peak brightness are the key parameters.

We have undertaken the development of an x-ray source based on laser Thomson scattering from a relativistic electron bunch. This approach offers the potential to produce x-rays in a unique regime. The LLNL source will exceed peak brilliance currently offered at third generation synchrotron light sources while delivering pulse lengths below one picosecond. In order to produce an ultra-short x-ray pulse, a temporally compressed laser pulse is focused onto a short, relativistic electron pulse. Figure 1 shows the schematic interaction of the laser pulse and electrons.

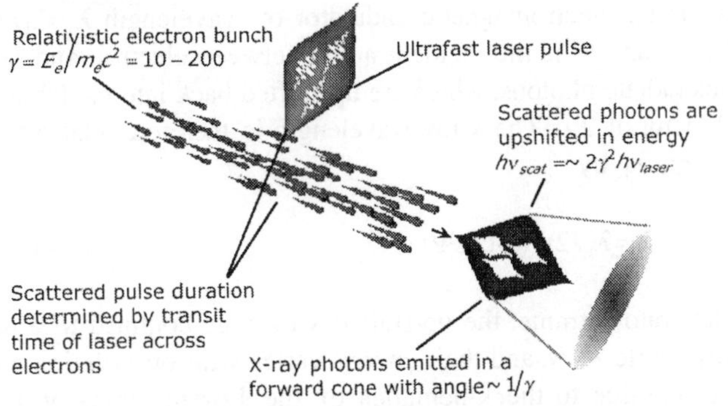

Figure 1: Thomson scattering geometry

The resulting hard x-rays can then be used for laser-x-ray pump-probe experiments, including extended x-ray absorption fine structure (EXAFS) spectroscopy [1]. As shown in Fig. 2, a variety of experiments will take advantage of laser and x-ray pulse lengths, as well as their relative synchronization in order to probe the evolution of sub-picosecond phenomena.

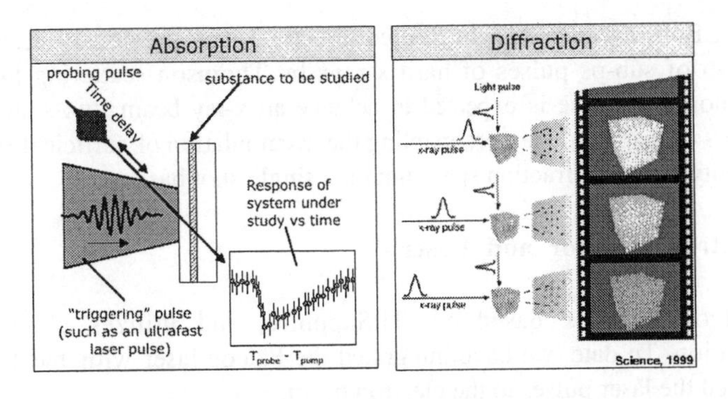

Figure 2: Laser - x-ray pump-probe experiments

2. Experimental design

The high photon energy of Thomson scattered x-rays is enabled by relativistic electrons produced by the LLNL 100 MeV electron Linac. The scattered laser photons are relativistically up-shifted in frequency into the hard x-ray range, and are emitted in a narrow cone about the electron beam direction.

As viewed in the frame of the moving electrons, the incident laser pulse train represents an electromagnetic undulator of wavelength $\lambda_u=\lambda_L/\gamma(1-\cos\Psi)$, where $\gamma=E/m_oc^2$ and Ψ is the incident angle between electron and laser beams. The electrons radiate photons, which are up shifted back into the laboratory frame by a second factor of 2γ. The x-ray wavelength is therefore related to the initial laser wavelength λ_L by

$$\lambda_x=\lambda_L/2\gamma^2(1\text{-}\cos\Psi) \qquad (1)$$

In the laboratory frame, the up-shifted x-rays are confined to a narrow cone with opening angle $\sim 1/\gamma$, and their energy varies with observation angle in the laboratory frame due to the kinematics of the Lorentz transformation. The laboratory-frame x-ray energy (for scattering of the fundamental) is given by

$$E_x=E_L\, 2\gamma^2(1\text{-}\cos\Psi)/(1 + \gamma^2\theta^2 + a_o^2) \qquad (2)$$

where E_L is the laser photon energy, θ is the observation angle and a_o is the usual normalized vector potential of the laser field, which is analogous to the K parameter of a static field undulator.

The LBNL Advanced Light Source injector Linac previously demonstrated generation of sub-ps pulses of hard x-rays by Thomson scattering [4,5]. The LLNL Thomson source is expected to achieve an x-ray beam flux some four to five orders of magnitude larger, enabling the accumulation of sufficient data for a high quality Bragg diffraction spectrum on a single shot basis.

3. Electron Injector and Laser

The Falcon laser is based on Ti:Sapphire, and utilizes chirped pulse amplification. To date, we have integrated the Falcon laser with the Linac and transported the laser pulses to the electron beam.

The LLNL Falcon laser [8,9] currently produces 300 mJ pulses at a 10 Hz repetition rate. Each pulse can have width as short as 35 fs, giving nearly three decades higher brightness than in previous scattering experiments. Anticipated laser upgrades to 4 J coupled with increases in electron bunch charge will extend this advantage to 4-5 decades.

An RF photocathode electron injector [10] produces the high brightness, low emittance electron beam necessary for interaction with focused laser pulses. This photoinjector, shown in Fig. 3, has been designed, constructed, and fully characterized.

Figure 3: PLEIADES Photoinjector

A pulse of S-band (2.8545 GHz) RF input with 7 MW peak power and 3 μs pulse length produces a peak standing wave electric field of 100 MV/m that accelerates electrons to 5 MeV in a distance less than 10 cm. A laser pulse, split from the Falcon laser oscillator is amplified and frequency tripled before striking a copper photocathode near the peak RF field. The electrons bunches produced have 1-10 nC of charge with pulse lengths of about 10ps. Focusing solenoids are employed to preserve the transverse emittance [11] of the electron beam, and to help match the electron beam into four SLAC-type linac sections, which increases the beam energy to 40-100 MeV.

Initial production of Thomson x-rays at LLNL utilized 20 mJ laser pulses and 5 MeV electron bunches produced directly by the photoinjector at a 10 Hz repetition rate. Successful overlap of the two beams demonstrated 0.6 keV x-ray production [12]. This result demonstrated the ability to synchronize and overlap both beams. For high-flux x-rays, the electron beam, with energy of 100 MeV, and falcon laser will both need to be focused to under 40 μm with synchronization better than 3ps.

At high energy (100 MeV) the minimum focal spot obtainable for the electron beam is emittance dominated for bunch charges in the nC range. Simulations of the electron beam emittance have been carried out using PARMELLA and results have been used to predict beam performance for the Thomson scattering experiments. The results predict a low emittance beam that can be focused to a 20 μm spot with a convergence angle of 5 mrad. To accomplish this, the required normalized transverse emittance must be less than 10 mm–mrad, requiring the use of a photoinjector.

A high-gradient quadrupole triplet will provide the final focusing of the electron beam before interaction with the Falcon laser. In a strong focusing field, low energy spread is also important to maintain a short longitudinal width at the focus, which the photoinjector can produce. A spectrometer magnet, shown in Fig. 4 is used both as a beam dump, and as an energy spectrometer. The interaction point of the FALCON beam and focused electrons is located between the final triplet and the spectrometer magnet. This spectrometer has been used to measure an energy spread on 0.2% at a beam energy of 60 MeV.

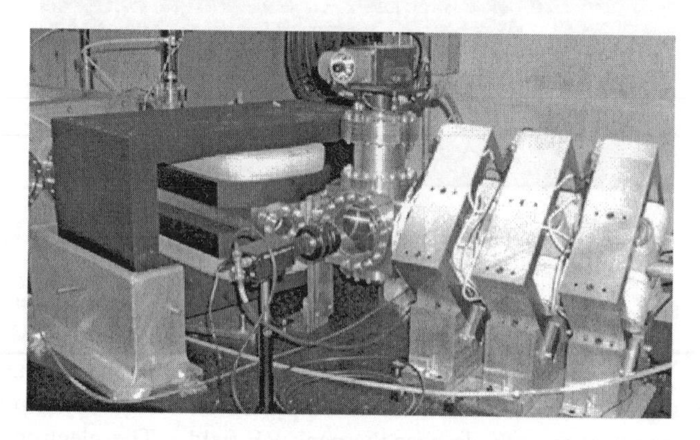

Figure 4: Final focus triplet, interaction, and spectrometer

The ultimate goal of the beam delivery system is to place as many monoenergetic, collimated electrons as possible into the FALCON laser focus, maximizing the production of up–shifted x-ray photons. Achieving a very high peak current is also critical, especially in the case of transverse interaction between the Falcon laser pulse and the electron beam. To attain the high current, the electron energy in the bunch will be chirped by a dephased Linac section, allowing for a ballistic compression of the electron beam from 10ps to 2ps [13]. A complementary dephased Linac section removes the energy chirp after compression for maximum beam peak current. Compression is optimized with respect to emittance and the energy spread, which increase with decreasing bunch length. A series of simulations were performed to optimize the bunch compression system. Simulation results are shown in Fig. 5. At the interaction point for this example, the 250 Amp electron bunch has been compressed to 2 ps, with a transverse width of 15 μm. In this case, high gradiant permanent magnet quadrupoles are used very near to the final interaction point. These new magnets are currently being developed.

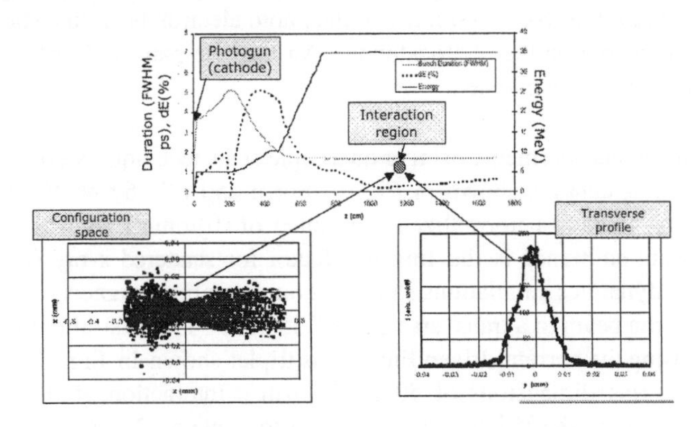

Figure 5: PARMELA simulation of compressed bunch

A three-dimensional simulation of the electron-photon interaction was performed using our laser and electron parameters to determine the x-ray photon flux. The result is shown in Fig. 6, including the spatial profile and spectrum produced. We expect to produce up to 10^8 x-ray photons in a ~100 fs pulse, and up to 10^{10} x-rays in a 1-10 ps pulse, by scattering the laser respectively either across the electron beam ($\Psi = 90^o$) which minimizes the temporal overlap, or in a head-on ($\Psi = 180^o$) geometry [6,7]. To achieve enhancements over other Thomson sources, we are making two straightforward but critical improvements: a low emittance electron beam derived from a photoinjector and a high power, 10 TW Ti:Sapphire laser with an future upgrade to 100 TW.

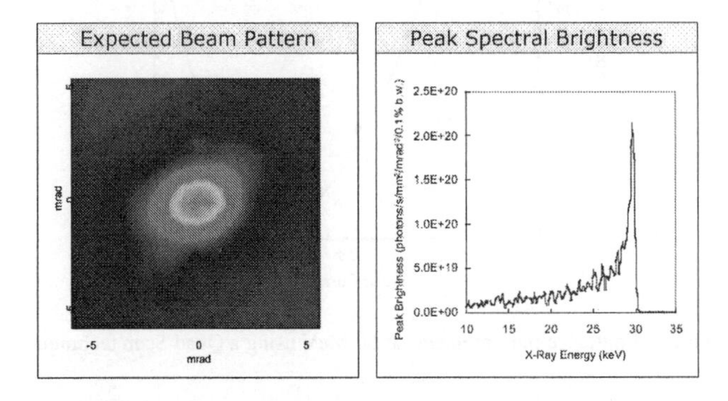

Figure 6: Three-dimensional Thomson x-ray simulation

4. Photo Electron Beam measurements

Initial Thomson scattering experiments were completed using the photoinjector alone. The photo-injector was then integrated with the Linac beamline, enabling 100 MeV acceleration. The photo-cathode UV drive laser transport was completed, and initial commissioning of the photo electron beamline started with a 110 μJ, 1mm spot on the photocathode. An upgrade path to 1 mJ of UV and uniform spots has been planned.

Emittance measurements using a quadruple scan technique were performed, and the resulting data are shown in Fig. 7. For a 250 pC, 60 MeV bunch with 0.2% energy spread, a normalized emittance of 10mm-mrad was measured. Electron beam emittance is the limiting factor for scattered x-ray flux in this case. Beam dynamics simulations show that our beamline magnets can focus the present electron beam to a transverse size of 35 μm. The first attempt to focus this beam using the current-driven final focus triplet shown in Fig. 4 resulted in a beam with rms dimensions of 50 by 70 μm. Interaction of this measured electron beam with the Falcon laser is expect to generate $2*10^6$ x-ray photons. Several straight-forward upgrades can be made, and the emittance should reduce to the desired range of less than 10 mm-mrad. The UV laser spot was not uniform during this measurement, and thus not optimized for minimum emittance. The measured emittance represents an upper bound for our system. Our initial measurements are encouraging, and we look forward to initial high-energy x-ray production.

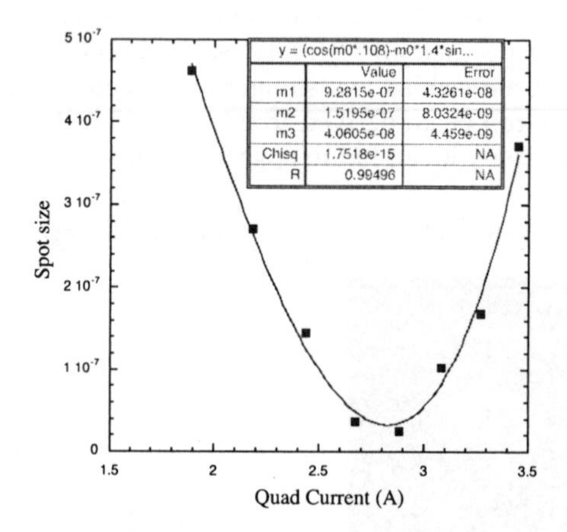

Figure 7: Emittance measurements at 60 MeV using a Quad Scan technique

5. Summary

A high power (10 TW) short pulse (35 fs) laser has been integrated with a 100 MeV electron Linac to provide a Thomson scattering source. The x-rays produced will be tunable in wavelength over a range of 10-200 keV. By driving the photoinjector and Linac with the same short pulse used for beam and target interaction, synchronization of 1-2 ps should be possible. Very low emittance electron bunches allow for tightly focused spots at the interaction point. The optimized system is predicted to provide an unprecedented high brightness (up to 10^{11} x–rays per pulse) in durations as short as 1-2 ps. The system should also be able to produce lower intensity pulses with duration as short as 100 fs. The 10% bandwidth, tunable wavelength range, and high per–pulse flux of the Thomson source will make it ideal for performing absorption edge spectroscopy of metals at a variety of wavelengths. Single-shot absorption and diffraction measurements at sub-picosecond timescales will allow unprecedented measurement capability.

Acknowledgments

This work was performed under the auspices of the U.S. Department of Energy by Lawrence Livermore National Laboratory under contract no. W-7405-Eng-48.

References

1. P. A. Lee, P. H. Citrin, P. Eisenberger, B. M. Kincaid, *Rev. Mod. Phys* **53**, 769-806 (1981).
2. C. Rose-Petruck, *et al.*, *Nature* **398**, 310-312 (1999).
3. C. W. Siders, *et al.*, *Science* **286**, 1340-1342 (1999).
4. R. W. Schoenlein, *et al.*, *Science* **274**, 236-238 (1996).
5. W. P. Leemans, *et al.*, *IEEE J. Quant. Elec.* **33**, 1925-1933 (1997).
6. E. Esarey, S. K. Ride, P. Sprangle, *Phys. Rev.* **E 48**, 3003-3021 (1993).
7. F. V. Hartemann, *et al.*, *Phys. Rev.* **E 64**, (2001).
8. V. P. Yanovsky, *et al.*, Conference on Lasers and Electron Optics, 410, (1999), Washington, DC, OSA Technical Digest.
9. T. Ditmire, M. D. Perry, Lawrence Livermore National Laboratory, **UCRL-ID-133293**, 1999.
10. G. Lesage, *et al.*, Proceedings of the 2000 Particle Accelerator Conference, Chicago, IL., 2000.
11. B.E Carlsten, *Nulc. Instrum Methods Phys. Res.* , **Sect. A**, 285, P. 313 (1989)
12. J. Crane, Private Communication.
13. L. Serafini, A. Bacci, and M. Ferrario, Proceedings of the 2001 Particle Accelerator Conference, p. 2242.

REVIEW OF SURFACE ROUGHNESS EFFECT ON BEAM QUALITY

A. MOSTACCI, L. PALUMBO
Dip. Energetica, Università degli Studi di Roma "La Sapienza", Roma, Italy

D. ALESINI
LNF-INFN, Frascati (Roma), Italy

In recent years a strong attention arose around the problem of the e.m. interaction of an ultra-relativistic beam with the residual roughness inside a beam tube, in particular in the framework of future 4th generation coherent light sources. The main concern was the effect of the wake-fields on the relative energy spread of the beam which has to be of the order of 10^{-4}, as for example in the LCLS and TESLA case. Although the real roughness has a stochastic feature, most studies dealt with periodic structure, or dielectric-equivalent layer which are considered to be conservative with respect the stochastic case. In this paper we will review the main theoretical models, and the most significant measurements trying to provide to the reader a complete picture of the present status of understanding.

1. Introduction

Looking at the literature on this subject, one finds that the problem of "beam tube roughness" is addressed for both "artificial" roughness, like for the LHC beam pipe,[1] and for "natural" roughness as in the case of ultra-short bunches machines. This problem was been first addressed in the framework of the Linac Coherent Light Source (LCLS[2]) design studies. An exhaustive phenomenological analysis of a rough surface features (in our accelerator contest) can be found in the paper by G. Stupakov (see Ref. 3 and related work). Very interesting pictures of scanned surfaces for different type of machining can be found in surface metrology books which show typical peak-to-valley height h of the roughness much smaller than the spacing between the crests g with the aspect ratio g/h that can easily exceed a hundred for smooth surfaces. Small corrugation in the beam pipe were already treated as perturbation of the wall profile in Ref. 4 and estimates were given for electromagnetic field and for the total energy loss up to the second order in the perturbation parameter; Ref. 5 reformulated such an expansion with special focus on the resonant frequencies and on a square-wave like wall profile. Recently, many authors focused their attention to the problem proposing different methods of solution. For each model presented here, we shall discuss only the longitudinal interaction with the structure (i.e. charge traveling on axis) being responsible of energy loss and energy spread in the beam. We will see that there can be two possible effects on the beam: a low frequency "inductive" field, mainly affecting the energy spread, and a high frequency synchronous field (slow wave) which causes both energy loss and energy spread. In Section 2, we give the basic definitions of the relevant quantities, impedance, wake and loss factor, expressed in such a way to make simpler the comparison among the various models. Section 3 is devoted to the low frequency interaction due to single and random bumps, causing an inductive impedance. In Section 4 we deal with the synchronous modes excited on a surface with periodic roughness. In Section 5 we analyze the experimental results obtained in two interesting measurements made with an electron beam while in the last Section we present our conclusions and the final remarks.

2. Meaningful quantities

To describe the beam-wall interaction one can follow a very standard approach which is discussed in details in Ref. 6 and its references. In this class of problems, one usually looks for the steady state solution which can be obtained from the field generated by a point charge traveling in the structure (a Green function approach). We limit our attention to the "longitudinal" problem , where all the charges involved are traveling on the longitudinal axis of a structure with rectangular or circular symmetry. Furthermore, since we will describe infinitely long structures, only quantities per unit length will be considered. For an ultra-relativistic particle, the longitudinal coupling impedance per unit length is given by

$$\frac{dZ(\omega)}{dz} = -\frac{1}{q} E_z \left(x=0, y=0, z, \omega\right) \exp\left(jz\omega/c\right), \tag{1}$$

where E_z is the z-component of the electric field in the frequency domain and q the particle charge. If we consider two particles traveling in a beam pipe, they can interact because of the presence of this field. The energy lost by the the trailing charge per unit of both charges (leading and trailing ones) is called longitudinal wake function and it is measured in Volt/Coulomb. The wake function per unit length is strictly related to the coupling impedance, i.e. (for an ultra-relativistic particle)

$$\frac{dw(\tau)}{dz} = \frac{H(\tau)}{\pi} \int_{-\infty}^{\infty} \frac{dZ_r(\omega)}{dz} e^{j\omega\tau} d\omega, \tag{2}$$

where τ is the time distance of the trailing charge from the leading one, $H(\tau)$ is the Heaviside function and Z_r is the real part of the coupling impedance. The energy lost by the leading particle (per unit charge and per unit length) is then computed from the beam loading theorem:

$$\frac{dk}{dz} = \frac{1}{2}\frac{d}{dz}\left[w\left(\tau \to 0^+\right)\right] = \frac{1}{\pi} \int_0^{\infty} \frac{dZ_r(\omega)}{dz} d\omega, \tag{3}$$

assuming an ultra-relativistic particle. k is usually referred to as the point charge loss factor, it is measured in Volt/Coulomb and it depends only on the real part of the coupling impedance (the imaginary part gives the odd part of the wake-function which vanishes in the origin). The bunch loss factor K can be easily computed from the coupling impedance. Being $I(\omega)$ the Fourier transform of the intensity of the bunch current

$$\frac{dK}{dz} = \frac{1}{Q^2\pi} \int_0^{\infty} \frac{dZ_r(\omega)}{dz} \left|I(\omega)\right|^2 d\omega, \tag{4}$$

where $Q = I(0)$ is the total bunch charge ($I(\omega)$ has the dimensions of Coulomb).

When the wake field is resonating only at one angular frequency ω_0, the above quantities have very simple expressions. Being w_0 the wake function amplitude (per unit length):

$$\frac{dw(\tau)}{dz} = w_0 H(\tau) \cos\left(\omega_0 \tau\right) \longleftrightarrow \frac{dZ_r(\omega)}{dz} = \frac{\pi}{2} w_0 \left[\delta\left(\omega - \omega_0\right) + \delta\left(\omega + \omega_0\right)\right] \tag{5}$$

and the bunch loss factor per unit length

$$\frac{dK}{dz} = \frac{w_0}{2} \frac{\left|I\left(\omega_0\right)\right|^2}{Q^2}. \tag{6}$$

An ultra-relativistic charge traveling on the axis of a smooth perfectly conducting pipe doesn't loose energy and can produce only fields vanishing as $1/\gamma^2$. Non zero wake fields are due to a "perturbation" of the beam pipe from the ideal lossless case (i.e. uniform cross section and perfectly conducting).

In the ultra-relativistic limit and cylindrical symmetry, the monopole longitudinal wake does not depend on the radial displacement of both leading and trailing charge. Numerical codes computing that wake in structures perform the integration along trajectories at the radius of the beam pipe since the longitudinal electric field component vanishes at the perfectly conducting walls. When the beam is interacting with one single mode (e.g. in cavities), the loss factor is simply the ratio between the peak voltage V and the energy stored U in that mode,[6,7] that is

$$k = \frac{|V|^2}{4U}. \tag{7}$$

For periodic structures, the loss factor per unit length is computed dividing the previous loss factors by the period L. In finite length traveling wave (multi-cells) structures, it was found that the point charge loss factor depends on the group velocity v_g as well and Eq. (7) has to be modified as[8,9]

$$k = \frac{|V|^2}{4U\,(1 - v_g/c)}. \tag{8}$$

This correction derives from the finite interaction length of the beam with the structure which broadens the spectrum of excited space harmonics. Therefore, not only the wave with phase velocity equal to c, but also other harmonics can exchange energy with the beam.

For an infinite structure, and steady state field, only a single space harmonic of the field is selected, and the interaction between the particle and a resonating mode depends only on the phase velocity.[10] A more recent paper[11] has discussed the effect of the group velocity in the case of final length structure, showing that the classical result of Eq. (7) holds for steady state cases while for transient cases one should use Eq. (8).

In the next section we discuss different theoretical models, which are divided in two main groups. First we review the inductive effects, which affect bunches of any length (Sec. 3); a second potential source of bunch losses is the interaction with synchronous modes, addressed in Sec. 4, relevant for short enough bunches.

3. Theoretical models in the long bunch limit: inductive effects

The longitudinal impedance at low frequencies (or equivalently for long bunches) of small perturbations on a vacuum chamber pipe has been studied for long time (see e.g. Refs. 12–14). In this regime, each perturbation (bump) acts independently from the others and total effect is given by the sum of all the bumps. The coupling impedance is purely inductive and therefore only the beam energy spread is affected. In particular, being Z_i the imaginary coupling impedance, the inductance $\mathcal{L}$ is such that

$$Z_i(\omega) = -j\omega\mathcal{L}$$

and, assuming a Gaussian bunch of r.m.s length σ_z, the relative energy spread $\delta E/E$ reads[15]

$$\frac{\delta E}{E} = \frac{1}{3^{3/4}\sqrt{2\pi}} \frac{Ne^2c^2\mathcal{L}}{E\sigma_z^2} L_{tot}, \tag{9}$$

if the bunch r.m.s length is much bigger than a typical size of the beam pipe (long bunch limit). E is the energy of the N particles of charge e traveling in a pipe of length L_{tot} at the speed of light c. The two main models are reviewed and compared in Ref. 15. For readers' convenience we will summarize briefly their assumptions and their main conclusions.

3.1. *The inductive impedance model*

This model[16] assumes that the a rough surface can be represented as a random distribution of small bumps and cavities of a certain size (the granularity size) on a smooth surface. Since the impedance of a small bump tends to be significantly larger than that of a cavity of similar size, the effect of cavity-like features was neglected and the rough surface was simply considered as a collection of small bumps of radius r on the surface of a tube of radius r_0. For many bumps, assuming that they are separated by at least their size, the total impedance is approximated by the sum of the impedances of individual bumps. At low frequencies ($\omega \ll r/r_0$), the total impedance per unit length becomes

$$\frac{Z_i(\omega)}{L_{tot}} = -j\alpha f \frac{Z_0\omega}{2\pi c} \frac{r}{r_0} \tag{10}$$

with α the packing factor equal to the relative area on the surface occupied by the bumps, f a form factor accounting for the shape of the bumps ($f = 1$ for hemispherical bumps) and Z_0 is the vacuum impedance. Numerical values for f are reported in literature.[15,16] In particular Ref. 15 estimates with Eq. (10) the effect of roughness wake in LCLS[2] and predicts a severe requirement on the smoothness of the beam tube surface to allow the designed energy spread.

3.2. *Small angle approximation*

A further important step in describing the surface roughness effect is done in Refs. 15, 17 and 18 by G. Stupakov who introduces the "small angle approximation". Interested reader will find in Ref. 17 all the details of the derivation and in Ref. 18 the report of some measurements of surface roughness of a tube similar to the one that will be used for the fabrication of the LCLS undulator. This model seems to be closer to realistic beam pipe materials and, moreover, its predictions are much less sever than those based on the previous model of Sec. 3.1.

The approach is based on the assumption that the angle between the normal to the rough surface and the radial direction is small compared to unity. Thus if the rough surface is given by the equation $y = h(x, z)$, where x, y and z are the Cartesian coordinates, and h is the local height of the surface, then the small angle approximation means that

$$|\nabla h| \ll 1. \tag{11}$$

The other assumptions are that the height and the characteristic width t of the bumps are small compared to the radius of the tube r_0 and the characteristic frequency of interest ($\omega \approx c/\sigma_z$) is small compared to c/t, that is

$$t, |h| \ll r_0 \quad \text{and} \quad \omega \ll c/t. \tag{12}$$

Therefore this model requires that the size of the bumps t is smaller than the bunch length σ_z. Typical surface measurements confirm the applicability of those assumptions.[18]

Detailed formulas of the inductance are given in Ref. 17 both for the case of a single bump and for rough surface with random profile; there, the validity limits of the small angle approximation are discussed in comparison to well-established results of Ref. 14.

In Ref. 18, measurements with atomic force microscope of surface profile are used to give realistic estimates of the inductance for different sizes of samples. It is shown that the additional energy spread that such a roughness would produce, is several times smaller than the design allowable value for the LCLS case. Although these positive results there are ongoing efforts to improve the surface roughness of commercially available high quality stainless steel tube. In fact the above results rely on the assumption of bunches much larger than the horizontal size of the roughness (marginally satisfied for future undulators) and they don't include the existence of a single synchronous mode (see Sec. 4).

4. Theoretical models: synchronous mode

While the inductive effects (imaginary impedances) seem to be well understood, it is not the same for the real part of the coupling impedance. Although there is a general agreement on the mechanism by which particles may loose energy, a series of concurrent models give different estimations. We will first discuss briefly the physics and then we will try to summarize the different models features. Unless explicitly declared, all the models assume a skin depth small with respect to the size of the perturbation, thus Ohmic losses are neglected for the moment.

The presence of a synchronous mode can be easily understood in a periodically perturbed surface. Although it is a very simplified model for surface roughness, this problem has often been addressed as a "worst case" for realistic surfaces and it can give some insights on the physical behavior. Moreover it is a good benchmark to compare various theoretical approaches and experimental observations; all the models that we will review, investigates also this case.

The main difficulties of a numerical approach is that the size of the bump is so small with respect to other dimensions that reliable simulation based studies are very difficult for any realistic rough surface model (the effects are very small). Also experimental observations are difficult because in existing machines and available beams, the rough surface effect is very small or negligible and the beam tube used are short. Analytical results are therefore necessary to guide the numerical simulations and propose meaningful experiments.

It is well known, that any regularly corrugated surface can guide a wave. Generally (rectangular and plane geometries) the field is exponentially confined to the rough surface which actually guides those "surface" waves. Such waves have a phase velocity which depends on the height of the corrugation. As an example, Fig. 1 shows the dispersion curve of a TM mode in a squared waveguide of side a for different value of height h of the corrugations in two opposite sides. The dash-dotted line represent the locus of points having the phase velocity equal to the speed of light: any point above (below) has a phase velocity $v = \xi_z/\omega$ smaller (greater) than c and it gives a slow (fast) wave. Normal modes in a smooth (perfectly conducting) guide are faster than light and, in fact, their dispersion curve (dashed line) lies all in the fast wave region crossing the $v = c$ line only at infinite frequencies. Increasing the corrugations height, the dispersion curve gradually bends entering the slow wave region of the dispersion diagram and there are frequencies at which the phase velocity is smaller than c. Particularly important are the frequencies where the phase velocity equals c (the dispersion curve crosses the $v = c$ line), since the slow wave will have the same velocity of an ultra-relativistic particle and thus will continuously exchange energy with it. We will refer to these frequencies as the "synchronous frequencies" f_{sc} ($\omega_{sc} = 2\pi f_{sc}$); already from the plot we notice that they must be inversely proportional to some power of h and most of the models that we will show later predict $f_{sc} \propto 1/\sqrt{h}$. Similar curves can be obtained for circular cross-section rough waveguides.

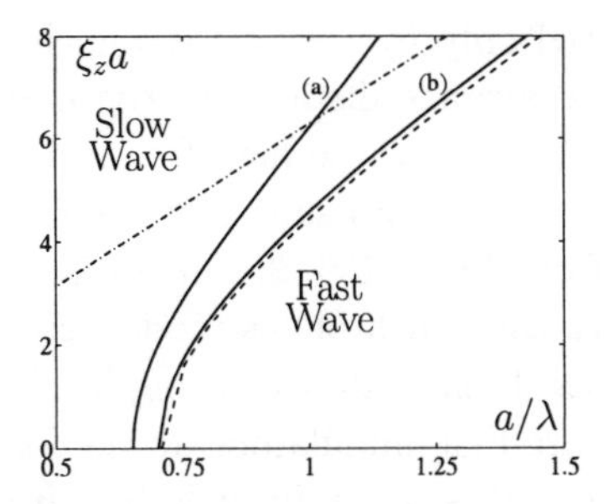

Figure 1. Dispersion diagram for a square waveguide of side a with corrugations on two opposite sides (ξ_z is the propagation constant). The solid lines are obtained from the dispersion equation: line (a) is for $h/a = 0.1$, while line (b) is for $h/a = 0.01$. The dashed line corresponds to the unperturbed TM_{11} mode of the smooth waveguide and the dot-dashed line represents an ultra-relativistic beam.

The possibility of transferring particle energy to a slow surface mode (and vice-versa) has interesting applications from old stochastic cooling pick-ups[19] to novel particle acceleration schemes for future linear colliders. These applications rely on the periodicity of the corrugated surfaces.

Figure 2 defines the symbols used through the paper concerning the relevant geometry. The problem has been studied since long time, both for circular and rectangular geometries. For a tube with shallow periodic oscillations, the monopole coupling impedance was found by a perturbation calculation[4] to be composed of many weak, closely spaced modes at frequencies such that $\omega L/c \gtrsim \pi$. Note, however, that it assumes both the perturbations and the slope of the perturbations to be everywhere small and this is not the case of Fig. 2 and of real roughness.

The method proposed in Ref. 4, has been somehow generalized by Ref. 3 for a sinusoidal corrugation, again in the framework of a small angle ap-

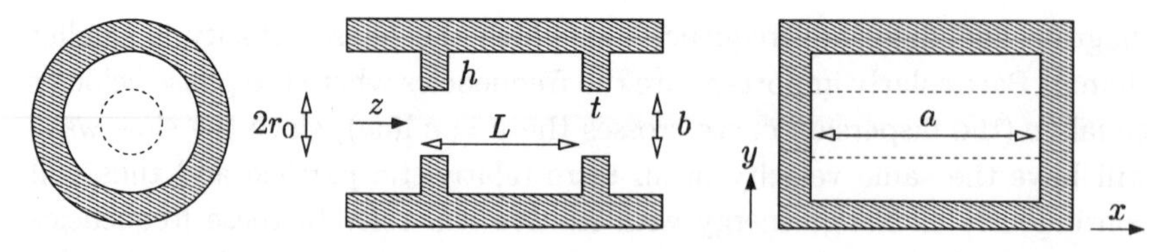

Figure 2. Relevant geometries concerning the corrugated waveguide model. Such model deals with circular or rectangular cross section waveguides; the longitudinal section is identical for the two models when changing b with $2r_0$.

proximation. Synchronous modes are found at very high frequency, so high that they can be excited only by very short bunches (with r.m.s. bunch length smaller than the average distance of the bumps). Using notation of Fig. 2, the wavelengths λ of those high frequency modes are such that $\lambda/(2\pi L) \leq 1$. We will not discuss such high frequency modes and interested reader can find formulas in Ref. 3. On the contrary, more dangerous for practical beams are the synchronous modes excited at lower frequencies, namely with $\lambda/L \gg 1$, since they can couple efficiently with the short bunches foreseen in future FEL undulators. In the following we will discuss the presently available models to study such "low frequency" synchronous modes.

4.1. *The "thin dielectric layer" model*

The oldest model investigating the possible effect of a synchronous mode in rough beam pipe, was proposed by A. Mosnier and A. Novohatski.[20] Sometimes it is referred to as the "resonator impedance model" (e.g. Ref. 21) since it was the first resonator impedance model in the contest of rough surface studies; for readers' convenience we will call it "thin dielectric layer" model.

As it is well summarized in Ref. 21, the authors of Ref. 20 began by representing a rough surface by a tube with periodic, shallow corrugations. Through time-domain simulation they find that the resulting wake-field is approximated by a single, loss-free resonator wake-field. The authors noted that the same kind of impedance is found for a beam tube with a thin dielectric layer.[22] One important result of Ref. 20 is that when the surface with the periodic perturbations is replaced by one with random perturbations, the character of the wake field experienced by the beam is left unchanged and the thin dielectric layer model is still valid.

According to Ref. 20, the wake-function amplitude per unit length (as defined in Sec. 2) for a slightly corrugated circular pipe (radius $r_1 = r_0 + h$) is

$$w_0 \approx \frac{cZ_0}{\pi r_1^2}, \tag{13}$$

which is the result given in Ref. 22 for a dielectric lined beam pipe. The synchronous frequency is

$$\omega_{sc} \approx c\sqrt{\frac{2}{r_1 h}\frac{\varepsilon}{\varepsilon - 1}}, \tag{14}$$

with ε being an "equivalent dielectric" constant which depends on the details of the corrugation.

Equation (13) states that the amplitude of the field generated by a point charge in a rough beam pipe does not depend on the roughness properties (and in particular it doesn't vanish for vanishing roughness).

On the contrary the synchronous frequency depends on the corrugation because the equivalent dielectric constant ε depends on it. The thin dielectric layer model has been partly updated from its first version and different values of ε have proposed on the basis of always refined numerical simulations. Those simulations were done for limited length structures, finite bunch sizes and choosing t, h, L of the same order. At the beginning[20] numerical simulations were suggesting dielectric constant in the range of 2; later developments can be found in Refs. 23–25. Anyway, periodic and random corrugations with bumps of the same sizes, exhibit the same value of ε.

The first attempt to have a formal derivation of the equivalent dielectric constant is in Ref. 26 for field wavelength $\lambda \gg L$. The corrugation slots are treated as parallel plate capacitors (i.e. assuming the field is constant in the radial direction) deriving an equivalent dielectric constant which depends only on t and L. The resulting synchronous mode angular frequency is

$$\omega_{sc} \approx c\sqrt{\frac{2}{r_1 h}\frac{L}{L - t}}. \tag{15}$$

This last equation is as well applied to random surfaces, using average values for t, L and h.

To the best of our knowledge, all the electromagnetic analysis is always limited to the computation of the propagation constant. The rest of the solution, such as field generated by the charge and then beam related quantities, are assumed as in Ref. 22.

Other results have been derived in the framework of this model and the most comprehensive reference is Ref. 27 and related publications. The synchronous mode contribution to energy spread is significant compared to the purely inductive contribution (i.e. the "non-interacting" bumps).[25] There is a kind of additivity of the effect of single walls on the longitudinal wake-field: the amplitude of the wake-field in the center of a 2-sided rough square guide is one half of the 4-side case.[25] The rough surface wake-field effect is dominant with respect to the finite resistivity effect provided that[28]

$$h > \frac{4}{3}\left(\frac{cr_1^2}{2\pi\sigma}\right)^{2/3} . \tag{16}$$

The effect of the group velocity, which can be significant for the electric coating is not included in the model.[22]

4.2. *The "corrugated waveguide" model*

This approach proposes an analogy between the surface roughness wake-field and the wake-field of ultra-relativistic charges in a corrugated waveguide; it has been applied to beam pipes with circular cross section (radius r_0) or a rectangular one (sides a, b) as well (Fig. 2). The model deals with periodic corrugations in the limit of $h \ll a, b, r_0$ and $t \ll L \ll \lambda$; the structure length is assumed infinite.

The derivation details and the results are reported in Ref. 29 for the rectangular cross section waveguides and in Ref. 30 for the circular cross section case.

Following Ref. 31, the field in the corrugation slots are assumed to be the TE modes which can propagate and reflect at the bottom of the slots. Higher-order evanescent modes, though present, are not used in the theory and, provided that the slot length L is a small fraction of the wavelength λ, the assumption is justified. In particular the only propagating mode in the slot is the fundamental TE mode which reflects at the wall producing a standing wave in the slot. This assumption is justified by experimental results in the field of corrugated antennas.

Exploiting the periodicity of the geometry (Floquet's theorem), the field generated by a point charge in the corrugated guide is analytically derived through a simplified field matching technique (matching only the fundamental mode[31]) and with the Lorenz reciprocity principle.[32] From the assumption of an infinitely long structure, it descends a spectrum of the radiated field described by Dirac delta functions $\delta(\omega \pm \omega_{sc})$, since only field synchronous to the driving particle can be generated. Once the scattered field is known, the beam related quantities are given by Eqs. (5–6) in Sec. 2.

The synchronous frequency and the loss factor amplitude depends on the geometry and it has being computed in Refs. 29–30 for rectangular and circular cross sections respectively. The synchronous radial frequencies are (in the limit of $h \ll a, b, r_0$ and $t \ll L \ll \lambda$)

$$\omega_{sc}^{rect} \approx c\sqrt{\frac{\pi}{ah}\coth\left(\frac{\pi}{2}\frac{b}{a}\right)} \quad \text{and} \quad \omega_{sc}^{circ} \approx c\sqrt{\frac{2}{r_0 h}} \tag{17}$$

where the superscript "rect" stands for "rectangular" and "circ" for "circular". The loss factor per unit length amplitude are then [a]

$$w_0^{rect} \approx \frac{Z_0\, c^3}{a^3 b\, (\omega_{sc}^{rect})^2}\, 8\pi^2 \left[\frac{\sinh(\pi b/a)}{\pi b/a} - 1\right]^{-1} \quad \text{and} \quad w_0^{circ} \approx \frac{Z_0\, c^3}{r_0^4\, (\omega_{sc}^{circ})^2}\, \frac{16}{\pi}. \tag{18}$$

It is worth noting that, according to this model, the point charge loss factor vanishes for vanishing corrugations. In fact its amplitude is, in both cases, proportional to h since $(1/\omega_{sc})^2 \propto h$. Only the lowest mode has been considered, being the only relevant one for the interaction with realistic finite length beams; obviously the point charge excites an infinite number of modes synchronous with the beam.

There is an interesting difference between the two geometries (circular and rectangular). In the rectangular case (where only two opposite sides are corrugated) the longitudinal electric field component (E_z) decreases exponentially going from the rough wall to the pipe center: it has a typical "surface wave" radial dependency.[29] On the contrary in the circular cross section pipe (where the corrugations are like irises), the beam generated E_z is constant with the radius (at least outside the corrugations) as shown in Eq. (A.5). Such a behavior is in agreement with numerical simulation results.[25]

To properly compare the results for the two geometries one should compare the circular case (radius r_0) to the square case (side a) keeping in mind that in the rectangular model the corrugation are assumed only in two opposite sides (and not on the four of them). Thus in the ratio between the wake amplitudes, the square one is multiplied by a factor two:

$$\frac{w_0^{circ}}{2w_0^{rect}} = \frac{a^4\, (\omega_{sc}^{rect})^2}{r_0^4\, (\omega_{sc}^{circ})^2}\, \frac{1}{\pi^3} \left[\frac{\sinh(\pi)}{\pi} - 1\right]. \tag{19}$$

[a] For readers' convenience, some steps in the derivation of the circular cross section case have been repeated in Appendix A which also solves some minor inconsistencies in the derivation of Ref. 30.

Being the structure infinite, the effect of the group velocity has not been included in the model.

4.3. *Shallow periodic corrugations: a field matching approach*

This model[21,33] studies again a cylindrical pipe with periodic corrugations and it was developed in the framework of the LCLS deign studies. Differently from Ref. 20 which uses a time-domain approach, this study proposes a frequency domain approach. The geometry is shown in Fig. 2 and "shallow corrugations" means $h/r_0 \ll 1$ and $L/h \lesssim 1$.

The approach is conceptually similar to the corrugated waveguide model since they both uses the field matching to find the homogeneous field solution. Using the numerical code TRANSVERSE,[7] the field matching matrix can be truncated at (virtually) any order and therefore the effect of many modes can be included in the simulations.

In the parameter regime of shallow corrugations (i.e $h/r_0 \ll 1$ and $L/h \lesssim 1$), the system matrix can be reduced to dimension 1 and, assuming kh small, one gets the synchronous frequency of the lowest mode[21]

$$\omega_{sc} \approx \sqrt{\frac{2}{r_0 h}\frac{L}{L-t}}. \tag{20}$$

Solving numerically the system matrix, the resonant frequencies can be determined (virtually) at any accuracy. It is shown that the first order result is accurate at least in the shallow corrugation assumption (and with L of the same order of t).

Once mode frequency is known, the loss factor can be computed according to the classical formula of Eq. (7),[7,21] with the voltage V and for the energy U such that

$$V = A_0 L \quad \text{and} \quad U = \frac{1}{2Z_0 c}\int_V \mathbf{E}\cdot\mathbf{E}^* dv = \frac{\pi A_0^2 r_0^2 L}{2Z_0 c} \tag{21}$$

where A_0 is the synchronous space harmonic and U is obtained with a volume integration. Accordingly the loss factor becomes

$$\frac{dk}{dz} = \frac{w_0}{2} \approx \frac{cZ_0}{2\pi r_0^2}, \tag{22}$$

which is again identical to the results of the thin dielectric model in Sec. 4.1. The last equation is valid as long as h/r_0 is small and L/h is not large.

In this approximation, one can show that the group velocity v_g is also close to the speed of light since

$$1 - \frac{v_g}{c} = \frac{4h}{r_0}\frac{L-t}{L}. \tag{23}$$

More recently[33] such expression of the group velocity was used in Eq. (8) to compute the loss factor. The energy stored in the mode is corrected as

$$U = \frac{1}{2Z_0 c}\int_V \boldsymbol{E}\cdot\boldsymbol{E}^* dv = \frac{\pi A_0^2 r_0^2 L}{2Z_0 c}\left(1 + \frac{\omega_{sc}^2 r_0^2}{8c^2}\right) \tag{24}$$

and the resulting loss factor is still as in Eq. (22). In conclusion, at least in the parameter range where h/r_0 is small and L/h is not large, this model gives a point charge loss factor which does not decrease if $h \to 0$, in agreement with the thin dielectric layer model.

In Ref. 21 it is pointed out that the previous formulas (based on the first order solution of the mode matching matrix) are not valid in the case when L/h is sufficiently large. Numerical studies[21] have been performed to explore such range of large L/h (but t of the same order of L) where the matrix can not be truncated at the first order. Interestingly enough, the loss factor starts decreasing while h decreases (and L/h is not small), i.e. the fundamental mode is suppressed. This agrees with predictions from the corrugated waveguide model.

4.4. *Sinusoidal corrugations: a small angle approach*

An interesting analytical model has been developed in the framework of a small angle approximation[3,33,34] for sinusoidal corrugations. The (circular) pipe surface is represented by

$$r = r_0 - h\sin\left(\frac{2\pi}{L}z\right),$$

and both the amplitude and the wavelength are small ($h, L \ll r_0$). This allows to neglect the curvature effects and to consider the surface locally as a plane one. The amplitude of the corrugations is also assumed to be smaller than the period, i.e. $h \ll L$.

The surface impedance, that is the ratio between the longitudinal electric field and the tangential magnetic field at the surface, can be found with a perturbation theory in the (small) parameter h/r_0[33,34]. Once the surface impedance is known, the general solution of the Maxwell equations is derived (see Ref. 34 for details).

It turns out that in the region of wavelength $\lambda > 2L$ there exists a synchronous solution. The longitudinal wake generated by this mode is given by

$$\frac{dw(\tau)}{dz} = \frac{2Z_0 c}{\pi r_0^2} U H(\tau) \cos(\omega_0 \tau), \tag{25}$$

where the dimensionless factor U and the frequency of the mode ω_0 depend on the parameter ℓ

$$\ell \equiv \frac{h}{r_0} \left(\frac{r_0}{L}\right)^{3/2}.$$

The limiting behavior of U and the frequency are:[3]

$$\omega_0 \to \pi \frac{c}{L}, \quad U \approx \frac{\pi^6}{8} \frac{h^4 r_0^2}{L^6} \quad \text{for} \quad h \to 0. \tag{26}$$

while

$$\omega_0 \approx \sqrt{\frac{2}{\pi} \frac{L}{r_0} \frac{c}{h}}, \quad U \to \frac{1}{2} \quad \text{for} \quad \frac{h}{r_0} \gg \left(\frac{L}{r_0}.\right)^{3/2}. \tag{27}$$

It is interesting to note that, in the limit of small corrugations, the amplitude of the wake function (i.e. also the loss factor) vanishes for vanishing corrugations (as for the corrugated waveguide model, but much faster). On the contrary in the limit of large ℓ the amplitude of the wake becomes equal to the prediction of the dielectric thin layer model in Eq. (13).

The effect of the group velocity has not been included in the model.

4.5. *Comparison and open questions*

The variety of models presently available may confuse the reader. For this reason we need to make a comparison among them to highlight the main similarities and the main differences. We will discuss only the Thin Dielectric Layer model (TDL, see Sec. 4.1), the Corrugated Waveguide model for circular geometries (CW, see Sec. 4.2) and the Field Matching model (FM, see Sec. 4.3) since they all deal with similar geometries (periodic corrugation in infinitely long pipes).

The first noticeable result is that all the models agree on the frequency of the synchronous mode with the general formula given by Eq. (20). In particular the synchronous frequency is proportional to $1/\sqrt{r_0 h}$.

Both the CW and FM models consider only the lowest synchronous mode excited in the structure. However they put different constraints that limit the theoretical results. The CW models requires $\lambda \gg L \gg t$ while the FM model seems to consider a "square wave"-like shape of the periodic roughness. Therefore FM model keeps in the theoretical study and in the simulation always the assumption $L \approx t \approx h$. This last assumption is considered also in all the numerical studies in the framework of the TDL model.

The important disagreements among the models is in the loss factor amplitude. The TDL and FM models predict a field and loss factor amplitude constant even in the case of $h \to 0$. On the contrary the CW model predicts a loss factor whose amplitude is proportional to h/r_0, that is it vanishes for vanishing corrugations. The intuitive behavior of a zero loss factor for a smooth perfectly conducting beam pipe is not fulfilled by TDL and FM models since the loss factor remains constant when the corrugation disappears.

Explicit derivation of the loss factor can be found only for the CW and the FM models. The relevant difference between those two is in the equations used to compute the loss factor. The former uses the classical approach where the interaction between the particle and the corrugated pipe mode depends only on the phase velocity. The latter states that the loss factor depends also on the group velocity (see Eq. (8)), also for infinitely long structures. Such a dependency is usually considered for traveling wave LINAC[9,8] and applies to finite length accelerating structures. In a recent paper,[11] A. Smirnov analyzes this case and distinguishes a transient regime which is characterized by a loss factor dependent on the group velocity as in Eq. (8) and a steady state regime where the classical formula of Eq. (7) should be used.

Also the TDL model predicts a loss factor not depending on h on the basis of the available results for a dielectric lined beam pipe.[22] It should be noticed that the author of Ref. 22 expects vanishing wake forces for vanishing height of dielectric layer. The derivation is based on the assumption $\gamma \gg \sqrt{2\varepsilon a/h(\varepsilon - 1)}$ which may not be fulfilled in the limit $h \to 0$. Therefore the author of Ref. 22 concludes that *"... (its) calculation cannot lead to the situation of $h = 0$, or the removal of the dielectric lining."*. Realistic pipe surfaces ($L \gg t$) may be characterized by an equivalent dielectric constant (computed as in Ref. 26) close to unity ($\varepsilon \approx 1$) limiting the applicability of results of Ref. 22. Furthermore this model, despite the agreement with Ref. 21, doesn't make use of the group velocity correction which can be not negligible in dielectric coated waveguide.

An interesting point of discussion comes from the observation that the synchronous modes excited in the corrugated waveguide correspond to modes supported by the perfectly conducting pipe deformed because of the corrugations. It is not surprising that, when $h \to 0$, the synchronous mode is excited at higher frequencies, where the "deformed" dispersion curve crosses the line of the speed of light (see Fig. 1) and that at this frequency the group velocity approaches the speed of light as well. However for $h \to 0$ the whole dispersion curve of the mode approaches that of a perfectly conducting smooth pipe which does not couple to the charge in term of energy exchange (radiation), as it is well known.

A vanishing loss factor with vanishing h is also found in the framework of the FM model when they investigate (numerically) the regime $L/h \gg 1$, as reported in Ref. 21.

Another interesting point arises when one considers that in the corrugated waveguide there can be an infinite number of synchronous waves excited at high frequencies whose frequency separation decreases for $h \to 0$. According to Ref. 21 the total loss factor, i.e. the sum of the loss factor for each mode, must equal $Z_0 c/(2\pi r_0^2)$ (see also Ref. 35). According to FM and TDL model, this amount of energy is condensed in a single mode, the fundamental one at lowest frequency. In particular what is the contribution of the other (infinite) modes and what is the overall loss for $h \to 0$?

In a recent paper,[36] results analogous to the TDL model ones have been obtained. According to Ref. 36, the difference between the results of CW model on one side and TDL and FM models on the other side can be explained with a different calculation method of the field amplitude and loss factor. Both CW model and Ref. 36 use the reciprocity principle for a point charge current distribution. However while in the CW model,[29] the integration is performed from $-\infty$ to $+\infty$ to obtain the field, in Ref. 36 a similar integration is performed in a semi-infinite region. The limited integration interval causes the dependency on the group velocity of the loss factor (as in Eq. (8)) which is not found in the CW model whose result depends only on the phase velocity.

In conclusion, there are two different theoretical approaches, relying basically on two different expression of the loss factor, i.e. Eq. (7) and Eq. (8), and giving significantly different scale for the amplitude of the wake function and loss factor.

Physically speaking, according to the first approach (CW model) the corrugation of the walls modifies the smooth waveguide modes, i.e. changing their phase velocity and field distribution. An infinite number of synchronous modes can be excited, the fundamental one being the mode with ω_{sc}^{circ}, w_0^{circ}. The amplitude of the modes depends on the height of the corrugation and the fundamental one scales as h/r_0 . This behavior is in agreement also with results from the code TRANSVERSE[21] which predicts a dependence on the parameter h for the loss factor of all the modes and it is not in contradiction with the sum rule reported in Refs. 21 and 35. When $h \to 0$, all the field amplitudes vanishes and the total loss factor goes to zero.

The second approach (TDL and FM models) predicts synchronous waves with same frequency as before, but with constant amplitude (depending only on the pipe radius). The fundamental mode brings an amount of energy equal to the total energy foreseen by Refs. 21 and 35. When the corrugation vanishes ($h \to 0$) the recovery of the classical case of loss-free beam tube is possible for finite bunch length (since the synchronous frequency goes to ∞) or assuming an exponential roll off of the relativistic point charge spectrum at angular frequencies above $\gamma c/r_0$.

4.6. *Synchronous modes and random roughness*

The applicability of the concept of beam coupling impedance to (statistically) rough beam pipes has been questioned by Ref. 37. Waves propagating in the system with random perturbations of the wall have continuous spectrum with respect to the wave number, because reflections from individual perturbations are not coherent, that is variations of particle energy are not regular in general cases. According to the authors of Ref. 37 one can calculate such energy variation using the straight way from classical electrodynamics without any artificial interpretation of the power of radiation losses as a real part of an impedance satisfying Kramers-Kronig relations. The total field generated by the ultra-relativistic bunch is a sum of the primary field (as in a smooth pipe) and the radiated one which forces variation of particle energy. Once the radiated fields are computed, the energy gain of a test charge at a distance ζ from the center of a Gaussian bunch (transversely uniform) can be calculated. The resulting standard deviation σ_E is

$$\sigma_E\left(\frac{\zeta}{r_0}\right) = \sigma_\delta\sqrt{l_c\int_0^L\left|G\left(\frac{\zeta}{r_0},\xi\right)\right|^2 d\xi}, \tag{28}$$

where σ_δ and l_c are respectively the standard deviation and correlation length of the statistical surface roughness. The function G depends in a complicated way on the beam pipe radius r_0 and on the beam energy. Applying their result to LCLS case, they conclude that the resulting energy spread is within the specification limits.

Another theoretical investigation[38] uses the Green function method to describe the effect of statistical roughness on the propagation of electromagnetic field in a (circular) guide. The surface is again described within a small angle approximation. The Helmholtz equation for the Green function is solved with the so called "effective boundary conditions" which are the first order term of the expansion of the general boundary condition considering small deviations of the pipe walls from an ideal surface. They estimate the shift of the propagation constant in the case of longitudinal (that is our case) and azimuthal roughness. In both cases, they see decreasing phase velocity (depending on the roughness properties) but the phase velocity shift is not enough to excite slow waves, at least at the first order. At the moment they have investigated only the propagation characteristics. Ref. 38 introduces a new theoretical method but contains only preliminary results which need further extensions.

These models don't predict the excitation of not negligible synchronous wave. The main reason being the assumption of very smooth approximation of the wall corrugations. In general one should expect the excitation of slow waves for real roughness provided that the slow wave wavelength is much larger than average roughness geometrical parameters, in particular the distance between peaks.

5. Experimental studies

It is clear that with such an uncertainty between the models, an experimental result is necessary to validate one or the other model. Unfortunately, as it will be shown in this section, there are few experiments done so far, from which is hard to derive definitive conclusions.

From a practical point of view, the most important question is whether a random bump pattern can support a synchronous wave. Two experiments been performed respectively in the Deutsches Elektronen-SYnchrotron (DESY) laboratory and in the Brookhaven National Laboratory (BNL). They are quite different, both in the properties of the corrugation under test and in the measured quantities. The results were analyzed in both cases with the thin dielectric layer model. We will first explain the experimental set-ups and then compare and discuss the results.

5.1. *The DESY-TTF experiment*

The DESY experiment was carried out at the linear electron accelerator of the TESLA Test Facility (TTF). The experimental set-up and the main result are reported in Ref. 39 while a more detailed description can be found in Ref. 40.

Using the 100nm FEL radiation production line, a properly shaped beam is delivered in a special ultrahigh vacuum chamber, containing six 800-mm-long test pipes of radii between 3 and 5mm and with different surface preparations, which is mounted behind the undulator. By remote control any of the pipes can be moved into the electron beam line and centered with an accuracy of 0.1mm. The pipes are composed of half cylinders machined into two flat aluminum plates. The surfaces were prepared by sandblasting or grooving and the resulting roughness was measured with a tracer-type measuring device featuring a resolution of 0.02μm. The results discussed concern three sandblasted pipes with different radius (3, 4, 5mm) and the same r.m.s. value of the corrugation height (8.3μm); a reference "smooth" pipe (r.m.s height 1.4μm) is used for comparisons. Resistive wall wake fields become important when the r.m.s. roughness is comparable with the skin depth of the metal (about 0.1μm in aluminum at 500GHz) and therefore they can be neglected in the rough test pipes.

The bunches were shaped by a bunch compressor such that they had a steep rising edge ($\approx$100fs) and a long tail ($\approx$10ps) and a further off-crest acceleration generates a correlated energy position distribution in the tail of the bunch. The synchronous mode wake fields, which are mainly produced by the sharp front peak of the bunch, can then be observed via the imposed energy modulation in the long tail. This set-up is very interesting because it allows direct measurements of the synchronous mode wake-field and other spurious effects (e.g. resistive wall, cross section change) have little influence.

The energy profile of the bunch tail, after the rough test pipes, showed a regular peak structure in the energy distribution which can be justified with a harmonic wake-field[39] whose frequency agrees within measurement error with the $1/\sqrt{r_0}$ behavior predicted by the thin dielectric layer model in Eq. (14); the frequencies measured are in the 500GHz range. The method adopted to measure the frequencies is independent from the initial energy distribution and from any specific wake-field model. Complementary experiments measuring the autocorrelation function of the radiation pulses produced by the bunches gave similar results.[41]

To estimate the effective dielectric constant ε_{eff} to be used in Eq. (14), they compare the measured frequencies to Eq. (14) itself. It result with a $\varepsilon_{eff} \approx 1.75$ while numerical calculations of Ref. 24 prefer $\varepsilon_{eff} \approx 2$, corresponding to $\approx$8% lower frequencies. Anyway such a deviation is still within the measurement uncertainty. The method proposed in Ref. 26 to determine *a priori* the effective dielectric constant, was not used in this case.

A numerical simulation of the whole experiment has been carried out imposing a wake-field in the rough test pipe according to dielectric layer model and using the measured frequency. The simulation parameter of each element (e.g. initial charge distribution or machine parameter of the sections) are allowed to vary in the experimental uncertainty interval. The predictions for an optimized parameter set is in a satisfactory agreement with the measured profile.[39]

5.2. *The BNL-ATF experiment*

A second interesting experiment was carried out at BNL with the high brightness electron beam of the Accelerator Test Facility (ATF).[42,43,44] They aimed to solve the controversy whether a synchronous mode exists in the real surface roughness. The 1ps electron beam pulse length (relatively long compared to the 100fs length planned for X-rays FELs) and the short beam tube length, obliged them to maximize the surface roughness wake-field in order to observe a measurable effect. Therefore they artificially created large bump beam tubes whose corrugations were distributed regularly as well as randomly. A smooth pipe was measured as reference too. The measurements were carried out at relatively low energy E=40MeV.

Using a high resolution electron beam spectrometer (0.05% resolutions), they measured the electron beam energy loss and energy spread as a function of electron beam pulse length for four different beam tubes (with corrugations of two different heights). The energy loss measurement is very useful to check theoretical model predictions because it depends only on the synchronous mode while the energy spread depends also on the inductive effects (non interacting bumps).

The corrugated pipes used in the BNL-ATF experiment have an aspect ratio bigger (i.e. closer to the real roughness case) than the sandblasted pipe of the DESY-TTF experiment. Moreover the average separation L between the bumps is comparable to the wavelength of the excited synchronous field and it makes difficult the comparison with available models assuming $\lambda \gg L$.

The corrugated pipes under test are 97cm long and they have an internal diameter of 6mm; four of them are available: the first one is smooth, while the 2nd and the 3rd ones have $\approx$3240 regularly spaced bumps of width

t=1.2mm and respectively $h = 0.3$mm and $h = 0.6$mm (see top and medium picture in Fig. 3). The forth one has about 2900 randomly distributed large bumps (t=1.2mm and $h = 0.6$mm) as shown in the bottom picture of Fig. 3. In order to suppress transverse wake-field effects the electron beam was centered to within 50μm using an alignment laser and optical survey. The initial measurement of the smooth tube gave an energy spread close to the intrinsic one, meaning that the wake-fields from the resistive wall and the from the two end transitions are negligible.

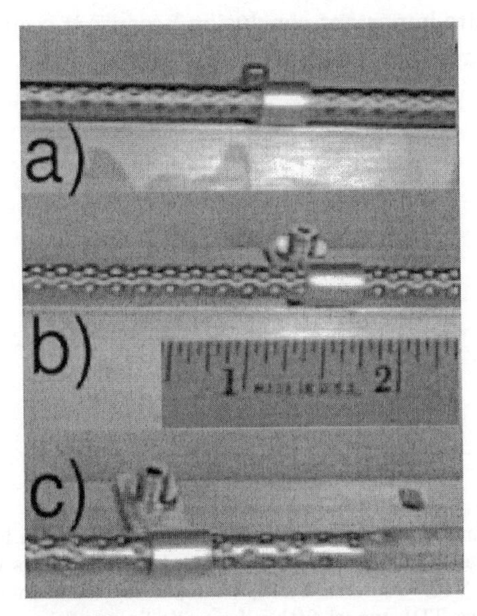

Figure 3. (Color) Three different pipes used in the BNL-ATF experiment. The top picture a) shows the tube with regularly spaced small bumps (h =0.3mm). The pipe with large periodical bumps is shown in picture b), while the third pipe c) has random large corrugations (h=0.6mm). The internal diameter is 6mm and the whole pipe is 97cm long (courtesy of F. Zhou, UCLA/BNL).

Energy loss and energy spread measurement as a function of the pulse length are reported and compared to theoretical model in Refs. 42–44 where they show these quantities normalized to the bunch charge and to the particle energy facilitating the comparisons. Good fit of the experimental data are obtained using the energy spread computed as in Eq. (9). The quantity $\mathcal{L}$ is calculated as in Ref. 17 for the randomly distributed bumps (that is considering each bump acting separately), while the effect of the synchronous mode on the imaginary part is added to fit the data of the regularly spaced bumps; in fact the contribution of the synchronous mode dominates the energy spread for the large bumps corrugated tube.

The energy loss deserves a more detailed discussion since it is due only to the synchronous mode. Figure 4 shows the energy loss after normalization as a function of the bunch length. A significant energy loss is observed in the regularly corrugated beam tubes while in the pipe with random bumps the loss is lower than expected as if the random location of the bumps destroys the coherent build up of the radiated wave reducing the induced field. This is a quite important conclusion because it seem to disagree with previous measurements (see Sec. 5.1).

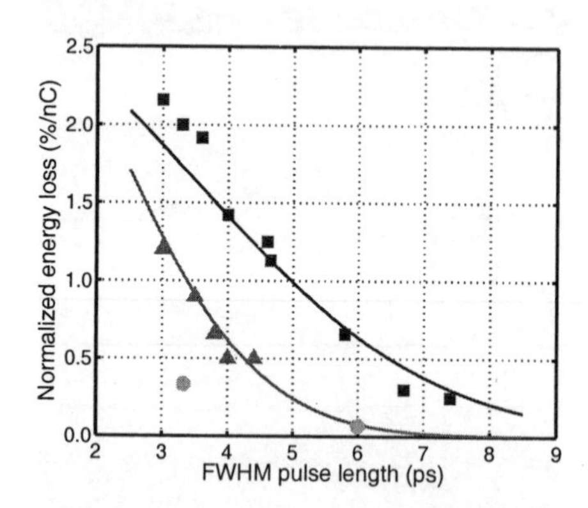

Figure 4. (Color) Energy loss of rough beam tubes versus pulse length from Ref. 42. Blue squares show data for periodic large bumps; blue line shows thin dielectric layer (TDL) model fitting with Gaussian beam; triangles represent data for periodic small bumps; red line shows TDL model fitting for periodic small bumps; green circles are data for random large bumps. The fit parameters are the same as in Ref. 42.

The authors of Ref. 42 used a "modified" thin dielectric layer model to fit the data; the particle wake function amplitude is assumed as in Eq. (5) with

$$w_0 = \frac{Z_0 c}{\pi r_0^2} F_0 \tag{29}$$

where F_0 is a so called coupling coefficient and the frequency ω_0 chosen to best fit the data. The angular frequencies 0.76 and 0.47 Trad/s fit well both the additional energy spread and the energy loss for the small and large bump pipe respectively; it means that, according to Eq. (14), the equivalent dielectric constant had to be chosen equal to 1.82 for large bumps and

1.53 for the small bumps tube. Also the introduction of a coupling factor depending on the height of the bumps can not be justified in the framework of the thin dielectric model. The numerical values for F_0 proposed in Ref. 42 are F_0 =0.552 for the tube with large bumps and F_0 =0.676 for the one with small bumps. It should be stressed that the assumptions of the thin dielectric layer model are not fulfilled by the geometry (in particular λ is not $\gg L$).

The normalized energy loss E_{loss} curves in Fig. 4 assume a Gaussian bunch of $Q = 10^{-9}e$ charge of FWHM size σ_{FWHM}, that is

$$E_{loss} = \frac{w_0 Q L_{tot}}{2E} \exp\left(-\omega_{sc}^2 \sigma_\tau^2\right) \quad \text{with} \quad \sigma_\tau = \frac{\sigma_{FWHM}}{2\sqrt{2\ln(2)}}. \tag{30}$$

E is the particle energy (E=40eV), L_{tot} is the corrugated beam pipe length (L_{tot}=0.97m) the wake function amplitude is given by Eq. (29) with the pipe radius r_0=3mm. Similar fit using parabolic bunch shape is proposed in Ref. 43 giving very close results.

The data reported in Refs. 42–44 for the periodically spaced bumps can also be analyzed with the corrugated waveguide model discussed in Sec. 4.2, approximating the pipes both with circular geometry and with square one. Some approximations are necessary because the geometry does not fulfill exactly the theoretical assumptions. Fist of all the width of the bumps t is not negligible with respect to the period L. Therefore from Eqs. (17, A.2) the synchronous frequencies for circular cross section (radius r_0) and square cross section (side a) are modified as

$$\omega_{sc}^{square} \approx c\sqrt{\frac{\pi}{ah}\coth\left(\frac{\pi}{2}\right)\frac{L}{L-t}} \quad \text{and} \quad \omega_{sc}^{circ} \approx c\sqrt{\frac{2}{r_0 h}\left(1+\frac{h}{2r_0}\right)\frac{L}{L-t}}, \tag{31}$$

coherently with Eq. (20). Such expression for ω_{sc}^{square} is also demonstrated in Ref. 36. To estimate the value of L we remind that the total number of bumps is about 3240 over L_{tot} = 970mm. If we consider 8 bumps per each cross section and two consecutive rings interleaved (see Fig. 3), two bumps on the same azimuth are separated by $970/(3240/16) = 4.8$mm. Since the synchronous mode is constant with the azimuth, to apply the corrugated waveguide model one has to take $L = 4.8/2 = 2.4$mm. Such dimensions are compatible with Fig. 3. Since $t = 1.2$mm for both kinds of bumps (and $r_0 = 3$mm), inserting numbers in Eq. (31) one gets a synchronous angular frequency of 0.47 Trad/s for large bumps and 0.65 Trad/s for small bumps, considering the circular cross section case. The corresponding wavelength, about 4mm (3mm) for large (small) bumps, doesn't fulfill the condition $L \gg$

λ. Similar comparison can be done assuming a square cross section of radius equal to the diameter ($a = 2r_0 = 6$mm). The synchronous frequencies are similar to the circular case, 0.44 Trad/s for large bumps and 0.61 Trad/s for small bumps. To get the energy loss one has to multiply by a factor two the w_0^{rect} given in Eq. (18) which assumes the corrugations to be only on two opposite sides and not on the four of them. By using Eq. (18, 31) in Eq. (30), we get the curves in Fig. 5.

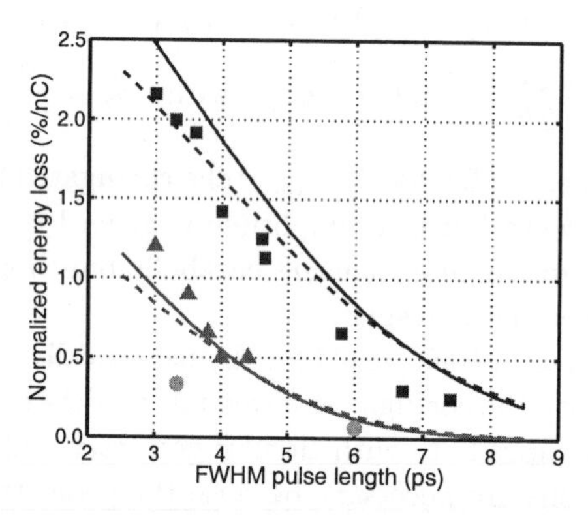

Figure 5. (Color) Energy loss of rough beam tubes versus pulse length. Blue squares (red triangles) show data for periodic large (small) bumps from Ref. 42 ; blue (red) solid line shows corrugated waveguide model for a circular pipe of $r_0 = 3$mm, $L = 2.4$mm and $h = 0.6$mm ($h = 0.3$mm). The dashed lines are predictions of the corrugated waveguide model for a square pipe of side $a = 2r_0 = 6$mm. Green circles are experimental data for random large bumps.

The agreement is surprising considering the difference of the real geometry from the model assumed in the theoretical predictions. In fact the wavelength of the synchronous mode can not be considered much bigger than the period L and the round shape of the bumps creates some uncertainty on the value of h to use in the analytical predictions. Anyway it is worth noting that the results of the BNL-ATF experiment seems to indicate clearly a dependency on the height of the bump not only of the frequency of the synchronous mode, but also of the wake function amplitude.

In conclusion, this experiment shows that a synchronous mode exists in pipes with periodically spaced bumps. It interacts with the beam affecting considerably the energy spread and causing an energy loss. In the case of randomly spaced bumps, the measured energy spread agrees with a purely imaginary impedance model and no significant energy loss is measured;

therefore they conclude that the synchronous mode decays significantly due to the randomization of the corrugations.[43] The resistive wall effect is negligible (according to smooth pipe measurements) and it can not explain the decay of the synchronous mode.

5.3. *Discussion*

The DESY-TTF and BNL-ATF experiments arrives to different conclusions: the first one shows that synchronous modes (leading to relevant energy loss) can exist also on random bumps, while the second one concludes that the synchronous mode decades on such a surface. It is worth observing that the experiments are exploring two different parameter regimes. In the DESY-TTF experiment the λ of the mode varies roughly from 0.4mm to 0.6mm (pipes with different radius are measured) while the bump height is in the 8μm range, that is much smaller. The average distance between the bumps is not mentioned explicitly, but it can be guessed to be of the same order of the height, since that is the case usually studied by the authors of Ref. 39. On the contrary the BNL-ATF experiment investigates the regime when the wavelength of the mode in the 3-4mm range (pipes with different corrugation height are measured) is of the same order of the period L =2.4mm. The aspect ratio of the corrugation is of the order of ten and the authors of Ref. 43 claim that this may be the reason of such a different behavior. In surface roughness of realistic cases (e.g. LCLS), aspect ratios are larger than 100. Some theoretical results in Sec. 4.3 suggest a possible dependency of the physical behavior on the aspect ratio (or on t/L): a more detailed experimental investigation is surely advisable.

In conclusion, a surface wave can be generated at wavelengths much larger than the (average) distance between the corrugation even for random distribution, while it is not the case for wavelength close to the bumps separation. The main question to be addressed is still the presence of synchronous modes in realistic (i.e. statistically) surfaces with big aspect ratios and small roughness height.

The purely inductive impedance, affecting only the energy spread, seems to be well understood as shown from the agreement between measurements and predictions of the energy spread in the randomly corrugated pipe of BNL-ATF experiment.

Even from an accelerator point of view the comparison of the two experiment is very interesting. They adopt different (and complementary) methods to measure the corrugation effect on the beam. The BNL-ATF set-up allows to see beam spectrometer images and to measure directly energy loss and energy spread which are important parameters in the specifications of future light sources. The DESY-TTF set-up uses a particular bunch and measures directly the frequency of the excited synchronous mode via the beam tail energy modulation; they profit from phase space tomography as well.[40]

6. Conclusions

The e.m. interaction of a beam with a pipe with rough walls has become of extreme importance for the dynamics of ultrashort bunches in small radius long beam pipes inside the undulators of the proposed X-ray SASE-FEL. This interaction is well understood at low frequency, i.e. for frequencies of bunch spectrum well below the pipe cut-off and unable to excite the synchronous modes that can travel in the pipe. In this particular case each bump of the rough surface behaves independently. The situation is well described by an imaginary coupling impedance and the beam traveling in such a pipe is subject to an increase of energy spread only. Such inductive effects due to the typical size of roughness are well under control with modern technology available for beam pipes machinery (see conclusions of Ref. 18). It has been shown by several authors that the roughness at the pipe walls can deform the modes of the smooth pipe, generating slow waves that can become synchronous with the charge, and exchange energy with it. This effect allows an energy loss and an energy spread in the beam. This effect, normal for periodic corrugations, seems to be induced as well by random corrugations provided that the wavelength of the slow wave is much larger than typical corrugation geometrical parameters.

For the most studied case, i.e. periodic corrugations, all the authors agree on the expression of the synchronous mode frequency, while there is a disagreement on the amplitude of the mode excited by the charge.

The main difference concerns the dependence of wake field and the loss factor on the corrugation height. Some models predict that these quantities are independent from this geometrical parameter, while others predict a wake and a loss factor proportional to h, which disappear for vanishing corrugation.

Looking at the methods of solution adopted, one finds that some models disagree since they use a different expression of the loss factor, while other models agree despite the fact that they are not using the same loss factor definition.

Therefore, a clear picture of this effect is still missing and some theoretical questions are still hanging:

a) The standard procedure of considering the spectrum of infinite charge waves (point charge spectrum) as source of fields in an infinite corrugate structure, selects only one space harmonics of the fields, and therefore only the phase velocity plays in this case a role. Where does the group velocity correction come from in the case of an infinite geometry?
b) When the structure length $L_s >> \lambda$ (typically $10^6\lambda$ for 100 m long undulator beam tube), can the infinite length model be used?
c) Some models do not predict free loss motion of the point charge for vanishing roughness. It this physically acceptable? Why the classical results are not recovered?
d) How the synchronous mode amplitude depends on the aspect ratio of the roughness?

Unfortunately, the unique experiments performed so far don't help since, for the parameter chosen, the models are able to reasonably fit the measured data, within the experimental uncertainty. About the statistical roughness case, which is closer to reality, a detailed theoretical analysis is still missing and desirable, at least concerning the build up of the synchronous mode in that case.

Acknowledgments

The author acknowledge Y. Ranguin (CERN) for interesting discussion about the results of Smirnov (e.g. Ref. 11) and F. Zhou (UCLA/BNL) for explanation on the BNL-ATF experiment.

References

1. O. Brüning, F. Caspers, I.R. Collins et al., in *Proc. IEEE Particle Accelerator Conference*, 2629 (1999).
2. The LCLS Design Study Group, *SLAC Report No. SLAC-R-521*, SLAC (unpublished).
3. G. Stupakov, in *Proc. 19th ICFA Advanced Beam Dynamics Workshop on Future Light Sources* (2000). See also *SLAC-PUB-8743*, SLAC, (2000, unpublished).
4. M. Chatard-Moulin and A. Papiernik, *IEEE Trans. Nucl. Science*, **NS–26**, 3523 (1979).
5. R.L. Gluckstern and S. Krinsky, *IEEE Trans. Nucl. Science*, **NS–28**, 2621 (1981).

6. L. Palumbo V.G. Vaccaro and M. Zobov, in *Proc. CERN Accelerator School: Advanced Accelerator Physics Course, Rhodes, 1993, CERN Yellow Report*, **95-06**, 331 (1995).
7. K.L.F. Bane and B. Zotter, in *Proc. of the 11th Int. Conf. on High Energy Accelerators*, 581 (1980).
8. E. Chojnacki, R. Konecny, M. Rosing and J. Simposon, in *Proc. Particle Accelerator Conference* (1993).
9. A. Millich, L. Thorndahl, *CLIC-Note*, **366**, CERN, (1999, unpublished).
10. L.A. Vainshtein, Electromagnetic waves, *Izd. Sov. Radio* (1957).
11. A. Smirnov, *Nucl. Instr. and Method*, **A480**, 387 (2002).
12. S. Kurennoy, *Part.Accel.*, **39**, 1 (1992).
13. R. Gluckstern, *Phys. Rev.*, **A46**, 1106 (1992).
14. S. Kurennoy, *Phys. Rev.*, **E55**, 3529 (1997).
15. K.L.F. Bane, G. Stupakov, in *Proc. International Computational Accelerator Physics Conference*(1998). See also *SLAC-PUB-8023*, SLAC, (1998, unpublished).
16. K.L.F. Bane, C.K. Ng, A.W. Chao, *SLAC-PUB-7514*, SLAC, (1997, unpublished).
17. G. Stupakov, *Phys. Rev. ST Accel. Beams*, **1**, 064401 (1998).
18. R.E. Thomson, R. Carr, G. Stupakov and D. Walz, *Phys. Rev. ST Accel. Beams*, **2**, 060701 (1999).
19. G. Di Massa, *SPS/ARF Note 85-8*, CERN, (1985, unpublished).
20. A. Mosnier, A. Novokhatski, in *Proc. Particle Accelerator Conference*, 1661 (1997).
21. K.L.F. Bane and A. Novokhatski, *SLAC Report No. SLAC-AP-117 (and also LCLS-TN-99-1)*, (1999, unpublished).
22. K.Y. Ng, *Phys. Rev.*, **D42–5**, 1819 (1990).
23. M. Timm, A. Novokhatski and T. Weinland, in *Proc. European Particle Accelerator Conference*, 1351 (1998).
24. M. Timm, H. Schlarb, A. Novokhatski and T. Weinland, in *Proc. IEEE Particle Accelerator Conference*, 2879 (1999).
25. M. Timm, A. Novokhatski and T. Weinland, in *Proc. European Particle Accelerator Conference*, 1441 (2000).
26. M. Timm, S. Ratschow, T. Weiland, in *Proc. IEEE Particle Accelerator Conference* (2001).
27. M. Timm, *Wake Fields of Short Ultra-relativistic Elecron Bunches*, Ph.D. thesis, Universität Darmstadt, (Der Andere Verlag, Osnabrück, 2001).
28. M. Timm, A. Novokhatski and T. Weinland, in *Proc. International Computational Accelerator Physics Conference* (1998).
29. A. Mostacci, F. Ruggiero, M. Angelici, M. Migliorati, L. Palumbo, S. Ugoli, *Phys. Rev. ST Accel. Beams*, **5**, 044401 (2002).
30. M. Angelici, F. Frezza, A. Mostacci, L. Palumbo, *Nucl. Instr. and Method*, **A489**, 10 (2002).
31. G. H. Bryant, *Proc. IEE*, **116**, 203 (1969).
32. R. Collin, *Field Theory of Guided Waves*, 2nd ed. (Oxford Univ. Press, Oxford, 1995).

33. K.L.F. Bane, G. Stupakov, in *Proc. International Linac Conference* (2000). See also *SLAC-PUB-8599*, SLAC, (2000, unpublished).
34. G. Stupakov, in *Proc. Workshop on Instabilities of High Intensity Hadron Beams in Rings*, Brookhaven National Laboratory, USA (1999). See also *SLAC-PUB-8208*, SLAC, (1999, unpublished).
35. R.L. Gluckstern, *Phys. Rev.*, **D39**, 2780 (1989).
36. K.L.F. Bane, G. Stupakov, *Phys. Rev. ST Accel. Beams*, **6** 024401 (2003). Work presented at the Joint ICFA Advanced Accelerator and Beam Dynamics Workshop, Chia Laguna, Sardinia (July 2002).
37. A.V. Agafonov, A.N. Lebedev, in *Proc. 19th ICFA Advanced Beam Dynamics Workshop on Future Light Sources* (2000).
38. E. Di Liberto, F. Frezza, L. Palumbo, in *Proc. 19th ICFA Advanced Beam Dynamics Workshop on Future Light Sources* (2000).
39. M. Hüning, H. Schlarb, P. Schmüser, M. Timm, *Phys. Rev. Lett.*, **88–7**, 074802 (2002).
40. M. Hüning, *Analysis of Surface Roughness Wake Fields and Longitudinal Phase Space in Linear Electron Accelerator*, Ph.D. thesis, Universität Hamburg, (2002) (to be published).
41. M. Hüning, P. Schmüser, *Nucl. Instr. and Method*, **A483**, 336 (2002).
42. F. Zhou *et al.*, in *Proc. European Particle Accelerator Conference*, 1598 (2002).
43. F. Zhou *et al.*, *Phys. Rev. Lett.*, **89–17**, 174801 (2002).
44. F. Zhou, *11th CAP Steering Committee and ATF Users Meeting* (2002, unplublished). See also World Wide Web address "http://www.bnl.gov/atf/Meetings/ATF2002/Feng.pdf".

Appendix A. Corrugated waveguide model in circular pipes

The derivation of the field generated by a beam in a corrugated waveguide is conceptually identical for circular and rectangular cross section waveguides. The rectangular case has been discussed in Ref. 29 while the circular one in Ref. 30. This last one contains some imprecisions that should be cured to get the results reported in Sec. 4.2. Thus we repeat the main steps of the derivation, referring to Ref. 30 for details in the field expression. Although there is a misprint in Eqs. (16, 15) of Ref. 30 ($\sin(n\phi)$ is missing), the dispersion equation is correct; in particular it reads (for the first mode)

$$\frac{J_1(k_0 r_0)Y_0(k_0 r_1) - J_0(k_0 r_1)Y_1(k_0 r_0)}{J_0(k_0 r_0)Y_0(k_0 r_1) - J_0(k_0 r_1)Y_0(k_0 r_0)} = \frac{k_0}{k_t}\frac{J_1(k_t r_0)}{J_0(k_t r_0)}, \tag{A.1}$$

being $r_1 = r_0 + h$ and $k_0^2 = (\omega/c)^2 = k_t^2 + \beta^2$ (k_t is the transverse propagation constant and β is the longitudinal one). The field generated by the charge (longitudinal propagation constant β) is synchronous with the charge itself (longitudinal propagation constant k_0), thus the synchronous frequencies are the frequencies such that $\beta = k_0$ (i.e. $k_t = 0$). Expanding the right hand term of Eq. (A.1) for small k_t and left hand one for small h/r_0, one gets (for $k_t = 0$)

$$\beta_{sc} = \frac{\omega_{sc}}{c} \approx \sqrt{\frac{2}{r_0 h}\left(1 + \frac{h}{2r_0}\right)} \approx \sqrt{\frac{2}{r_0 h}}, \tag{A.2}$$

which differs from what derived in Ref. 30, but it is coherent with other author results (e.g. Ref. 21).

The amplitude of the field generated by the charge can be derived from the Lorenz reciprocity principle, following the same steps as in Ref. 29. The longitudinal electric field is

$$E_z(r, z, \omega) = [c_0 \exp(-j\beta_0 z) - d_0 \exp(j\beta_0 z)]\, J_0(k_t r) \tag{A.3}$$

considering only the fundamental mode (labeled with the index 0). The coefficient c_0, d_0 are derived by the reciprocity principle, that is

$$c_0 = -\frac{qZ_0 c}{\omega\beta_0 r_0^4}\frac{k_t^2 r_0^2}{F(k_t r_0)}\delta\left(\frac{\omega}{c} - \beta_0\right) \quad \text{and} \quad d_0 = \frac{qZ_0 c}{\omega\beta_0 r_0^4}\frac{k_t^2 r_0^2}{F(k_t r_0)}\delta\left(\frac{\omega}{c} + \beta_0\right) \tag{A.4}$$

where q is the particle charge, Z_0 the vacuum impedance and

$$F(k_t r_0) = J_1^2(k_t r_0) - J_0(k_t r_0)\, J_2(k_t r_0) \longrightarrow \frac{k_t^2 r_0^2}{8} \quad \text{if} \quad k_t \to 0.$$

Since $\omega/c = \beta_0$ if $k_t = 0$, thus that ω equal to ω_{sc} found in Eq. (A.2), the longitudinal electric field is constant with the radius:

$$E_z(r, z, \omega) = -\frac{8qZ_0 c^3}{\omega_{sc}^2 r_0^4}[\delta(\omega - \omega_{sc}) + \delta(\omega + \omega_{sc})]\exp\left(-j\frac{\omega_{sc}}{c}z\right). \tag{A.5}$$

Therefore, it is straightforward to get the coupling impedance and then the wake-function per unit length as discussed in Sec. 2. In particular using Eq. (A.5) in Eq. (1) and comparing the result to Eq. (5), the amplitude of the wake function per unit length reads

$$w_0^{circ} = \frac{16}{\pi}\frac{Z_0 c^3}{\omega_{sc}^2 r_0^4}. \tag{A.6}$$

Equation (19) compares the result for a circular (radius r_0) and a square cross (side a) section beam pipe. It is interesting to see that ratio in two particular (arbitrary) cases. The first one when the side a and the diameter $2r_0$

are chosen such that the synchronous frequencies are equal $\left(\omega_{sc}^{rect} = \omega_{sc}^{circ}\right)$:

$$\frac{a}{2r_0} = \frac{\pi}{4}\coth\left(\frac{\pi}{2}\right) \approx 0.86 \quad \text{and}$$

$$\frac{w_0^{circ}}{2w_0^{rect}} = \frac{\pi}{16}\coth\left(\frac{\pi}{2}\right)^4\left[\frac{\sinh(\pi)}{\pi} - 1\right] \approx 0.74. \tag{A.7}$$

The second one when the side a is chosen equal to the diameter $2r_0$ (the circular cross section is inscribed in the rectangular one):

$$\frac{\omega_{sc}^{rect}}{\omega_{sc}^{circ}} = \sqrt{\frac{\pi}{4}\coth\left(\frac{\pi}{2}\right)} \approx 0.92 \quad \text{and}$$

$$\frac{w_0^{circ}}{2w_0^{rect}} = \frac{4}{\pi^2}\coth\left(\frac{\pi}{2}\right)\left[\frac{\sinh(\pi)}{\pi} - 1\right] \approx 1.18. \tag{A.8}$$

This last case was used to analyze the results of BNL-ATF experiment in Sec. 5.2, since the two cross sections are equal where the beam field is maximum (i.e. closer to the beam axis).

HIGH BRIGHTNESS BEAMS – APPLICATIONS TO FREE-ELECTRON LASERS

SVEN REICHE

University of California, Los Angeles
Deptartment of Physics & Astronomy
Los Angeles, CA 90095-1547

Free-Electron Lasers as high-brilliance radiation sources, rely on a high quality of the electron beam driving the FEL process. The amount of energy, transferred from the electrons to the radiation field, and thus the efficiency of the FEL depends on the provided beam parameters. The presentation discusses the impact of various beam parameter and how current designs of FEL injector try to accomplish the demands on the beam quality for reaching saturation.

1. Introduction

A characteristic parameter for any kind of light source is the brilliance, which is defined as the number of photons per second, frequency interval, area and divergence angle and typically classifies the generation of the light source. The third generation is well established and there exist several facilities around the world, based on storage ring synchrotron radiation. The 4th generation light source was proclaimed[1] as a significant improvement in the brilliance, with radiation pulses in the femtosecond regime instead of the 3rd generation picosecond pulses. In addition, the light source should be transversely and longitudinally coherent.

Linac based Free-Electron Lasers are a possible candidate for fulfilling the criteria of a 4th generation light source. The key is that Free-Electron Lasers can saturation within a single pass of an electron bunch, starting from the spontaneous emission of that bunch. This defines the operation principle of an Self-Amplifying Spontaneous Emission (SASE) FEL. Because the SASE FEL does not require seeding nor an oscillator set-up to accumulate gain, the radiation wavelength can be extended to the X-ray regime.

Ongoing research since the nineties reduced the radiation wavelength of the SASE FEL down to less than 100 nm, with the latest three experiments

(VISA[2], LEUTL[3] and TTF[4]) reaching saturation. The trend is towards even shorter wavelengths with two projects – LCLS[5] and TESLA-FEL[6] –, planning on radiating on an Ångstrom level.

All single-pass FELs rely on a high-brightness beam to enhance the efficiency of the FEL, transferring energy from the electron bunch to the radiation field. The electron beam parameters of several running or planned FELs are listed in Tab. 1. This paper discusses the importance of the beam brightness to the Free-Electron Laser. The next section gives an overview of the FEL principle and how beam parameters such as current, energy spread or emittance effect the FEL performance. Certain conditions are derived, which have to be fulfilled in order to achieve lasing in the FEL. The remaining sections deal with how the electron bunches are produced and transport from the electron source to the FEL.

Table 1. Beam parameters of recent and planned FELs

	VISA	LEUTL	TTF	LCLS	TESLA
Energy [MeV]	71	220	250	14500	25000
Energy Spread [%]	0.1	0.1	0.06	0.01	0.05
Emittance [mm·mrad]	2.3	8.5	6	1.2	1.6
Peak Current [kA]	0.25	0.27	1.3	3.4	5
Charge [nQ]	0.17	0.2	2	1	1
Wavelength [nm]	850	530	100	0.15	0.1

2. FEL principle

This section gives an overview of the FEL physics with a focus on the dependance of electron beam parameters. An electron, injected into an alternating magnetic field of an undulator or wiggler, emits radiation. The slippage of the radiation field over one undulator period λ_u with respect to the electron determines the resonant radiation wavelength

$$\lambda = \frac{\lambda_u}{2\gamma^2}\left(1 + a_u^2\right) \quad , \tag{1}$$

where γ is the Lorentz factor of the electron energy and $a_u = (e/2\pi mc)B_u\lambda_u$ is the rms undulator parameter, with B_u the on-axis field, c the speed of light and e and m the electron charge and mass.

The emission on a spontaneous, incoherent level is called undulator radiation. FEL radiation is an enhancement of the undulator radiation by increasing the coherence level due to the formation of microbunches. The spacing of the microbunches is the resonant radiation wavelength. Bunching is caused by the collective instability of the FEL, where the emitted radiation acts back on the electrons and modulates their energy. Because the electron motion in the undulator is dispersive, the energy modulation is transformed into a bunching in the current. The emission level of the undulator radiation is enhanced. The physics is similar to the motion in an rf bucket, but with the difference that the size of the FEL bucket grows with the accumulated radiation and its phase adjusts automatically to the positions of the microbunches.

The 'time'-scale of the collective instability of the FEL is given by the gain length

$$L_g = \frac{\lambda}{4\pi\rho} \quad , \tag{2}$$

where the FEL parameter ρ[7] is mainly a measure for the brightness of the driving electron beam. In the 1D limit, where diffraction effects of the radiation field are negligible, the FEL parameter is given by

$$\rho = \frac{1}{\gamma}\left[\frac{I}{I_A \Sigma_b}\frac{(\lambda_u f_c a_u)^2}{16}\right] \quad , \tag{3}$$

with I and Σ_b the current and transverse beam size of the electron beam, respectively, and $I_A \approx 17$ kA is the Alvén current. The correction factor f_c arises due to the reduced coupling of the electron beam to the radiation field in a planar undulator. Typical values for ρ are well below unity.

The FEL amplification stops at saturation, when maximum current modulation is achieved. In the longitudinal phase space, the initially continuous distribution has been rotated to an upright position in each bucket. From this picture it can easily be seen that a larger initial energy spread degrades the FEL performance, because the bunching is more spread out, reducing the coherence level of the FEL radiation. Because the maximum relative width in energy of the bucket is ρ[7], the energy spread σ_γ has to fulfill the constraint

$$\frac{\sigma_\gamma}{\gamma} \ll \rho \quad . \tag{4}$$

Otherwise the bucket is smoothly filled and rotation provides nothing.

Emittance effects are similar to those of the energy spread. The magnetic field of the undulator couples velocity spread in the transverse direction to a spread in the longitudinal direction[8]. The constraint for the normalized emittance ϵ_n is

$$\epsilon_n \ll \frac{4\lambda\beta}{\lambda_u}\rho \quad , \tag{5}$$

where the average beta-function β is a measure for the focusing strength within the undulator. Although a stronger focusing reduces the beam size Σ_b and, thus, increases the ρ parameter, it enhances the emittance effect. For any given emittance value there is an optimum value for the β-function, which balances out both effects. As an example, Fig. 1 shows the dependence of the saturation length and power on the value of the β-function for the case of the LCLS FEL. The optimum value for an emittance of 1.2 mm·mrad is 18 m.

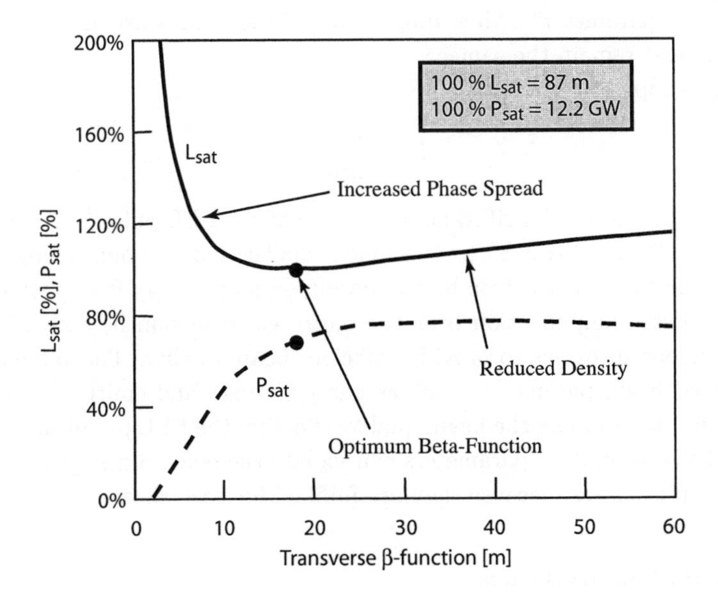

Figure 1. Saturation power and length as a function of the focusing. Optimum value is 18 m.

Because the radiation field is emitted within the finite area of the electron beam, the emitted radiation diffracts. The field amplitude at the location of the electron beam is reduced and, thus, the FEL amplification is inhibited. On the other hand diffraction spreads out the phase and amplitude information in the transverse direction, which is essential to achieving transverse coherence. For best coupling, the electron beam envelope should match that of the radiation beam, resulting in a second condition –

$$\frac{\epsilon_N}{\gamma} \approx \frac{\lambda}{4\pi} \tag{6}$$

– for the normalized emittance. However this condition is not as crucial as the first and is indeed violated for the planned X-ray FELs LCLS and TESLA.

For an FEL with strong diffraction effects, where the Rayleigh length $z_r = \Sigma_b^2 k/2$ is shorter than the gain length, the definition of the 1D ρ parameter looses its validity. The FEL performance depends rather on the beam current than on the electron density ($\propto I/\Sigma_b$). The FEL theory can be extended to include diffraction effects, which allows one to define an effective ρ-parameter[9]. All scaling laws (gain length, constraints for beam parameters) remain the same.

The cooperation length

$$l_c = \frac{\lambda}{4\pi\rho} \tag{7}$$

is a measure for the longitudinal coherence of the FEL pulse[10]. In particular, for VUV and X-ray FELs l_c is much smaller than the bunch length and various parts of the electron bunch amplify independently from each other. To guarantee amplification over the entire electron bunch, each of these independent slices has to provide sufficient beam quality. The value of the projected beam parameters such as energy spread and emittance are less useful for determining the beam quality effecting the FEL performance. Indeed, these secondary parameters can violate the constraints, given above, but the pulse lases, because they are fulfilled for each slice.

3. Beam Quality Issues

A high electron beam quality is essential for a successful operation of an Free-Electron Laser. All requirements, given in the section above, can be combined in the demand for a high density of the 6D phase space distribution. Because a linear accelerator does not allow for beam cooling as is the case for a storage ring, a sufficient phase space density has to be generated at the electron beam source. The phase space density is correlated to the beam brightness and, thus, Free Electron Lasers are commonly based on high-brightness electron beam sources, which typically excludes the use of thermionic guns. A common choice is an rf photo-electron gun. Typical parameters are a charge of around 1 nQ, rms bunch length of around 3 ps and transverse, normalized emittances of 1 mm·mrad or above.

Three effects contribute to the transverse emittance[11]: space-charge, rf curvature and initial thermal velocity spread. The first two contributions scale monotonically with the emitted charge. Lowering the charge reduces the transverse emittance and initial bunch length. Because the beam brightness is actually increase by this, the FEL performance improves[12]. The choice of operation at 1 nQ was set, when Free-Electron Lasers were discusses as the 4th generation light sources[1]. Indeed, recent experiments[2] uses a much lower charge around a few hundreds of pC to shorten the FEL saturation length. The lower limit is given by the resolution of the beam diagnostics along the beam line.

Emittance compensation is essential to achieve small values for the transverse emittances. Using this scheme a focussing solenoid compensates the space charged field so that the electrons have minimal transverse momenta, when the acceleration diminishes the space charge effects. For best performance the charge distribution should be homogeneously, so that the transverse space charge field is predominantly linear in the transverse position. Otherwise the fluctuation in the applied kicks by the field increases the emittance, which are not compensated by the solenoid field. This effect puts a stringent requirement on the transverse uniformity of the laser profile[13] as well as for the cathode itself.

The transverse emittance is the most crucial parameter, when designing and building an FEL. Only for FELs with a wavelength of 10 microns or longer, a larger emittance would actually enhanced the FEL performance, because diffraction effects are reduced and the beam couples better to the fundamental mode, while higher modes are suppressed. Phase spread effects due to the transverse emittance dominate for UV and X-ray FELs. The uncorrelated betatron oscillation of all electron yields a longitudinal spread, as discussed above. The change in the longitudinal velocity is constant for natural focusing undulators[14] as well as for FODO lattices[15], where the cell length is shorter than the gain length, and does not depend on the betatron phase of the electrons. Theoretically, it can be compensated by a correlation between energy and the amplitude of the betatron oscillation[16]. For X-ray FELs, where the emittance effects are the strongest, the required conditioning line with higher-order mode cavities are even longer than the main acceleration section due to the poor efficiency of the conditioning.

The FEL gain is less sensitive to the energy spread than to the emittance. The required energy spread can easily be achieved by the current generation of photo injectors. Even the growth of the energy spread within the undulator due to the quantum fluctuation of the spontaneous radiations[17] does not effect the FEL performance significantly. It also allows to operated at a larger energy spread than provided by the rf photo electron guns to suppress the microbunch instability in the bunch compressor (see below).

Longitudinal homogeneity of the beam parameters are not essential for the FEL process if the cooperation length is smaller than the bunch length, but can affect the output properties of the FEL. If the variation in the slice beam energy is larger than the FEL parameter ρ the spectrum becomes broader. Experiments, based on the spectral purity of the radiation, place a monochromator prior to the experimental station. Due to the wider spectrum less photons will pass the acceptance of the monochromator, thus reducing the available flux for the experiment. Another issue is that most diagnostics are based on projected values, in particular the matching to the focusing structure. With variation in the slice centroid position and angle as well as in the divergence/convergence of the beam envelope, the individual slices are not matched to the undulator focusing lattice. Mismatch typically results in a shift in the wavelength, reduced coupling between the slice and the radiation field and thus reduces the gain for the slice. This is most critical if only a small part of the bunch is lasing, such as in the TTF experiment, where less than 20% contributes to the FEL amplification[4]. Fig. 2 shows the simulated current profile for this experiment. Experiments and simulation confirm that only the bunch part around the high current spikes lases.

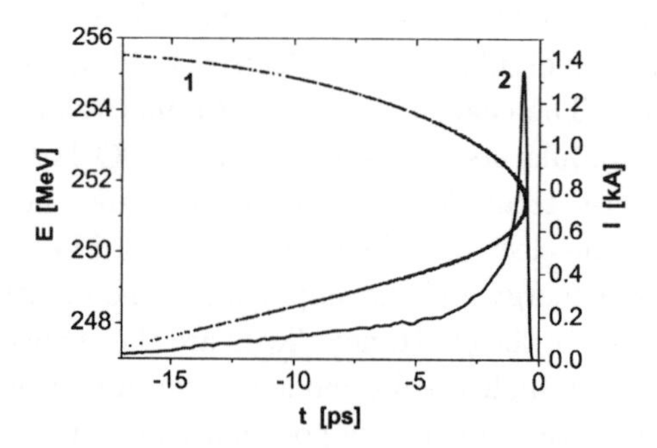

Figure 2. Phase space distribution (1) and current profile (2) of the start-end simulation for the TTF-FEL

4. Beam Generation and Transport

For all current single-pass high gain Free-Electron Lasers the driving beam is created with an rf photo-electron gun, where a laser pulse illuminates the cathode within an rf cavity. The electrons are generated by the photo-electric effect and accelerated by the electric field in the cavity. Typical design of the rf guns uses standing wave structures with a one and a half cell layout[18] and an rf frequency in the S- or L-band regime. A solenoid focuses the electron beam, compensating the defocusing effects of the space charge field. Emittance compensation[19] balances the solenoid field with the space charge field, so that the transverse emittance is minimized when further acceleration reduces the sensitivity to the space charge forces. Because the head and the tail has less charge than the core, the solenoid overcompensates the space charge. As a result the local, longitudinal sliced phase space distributions in the transverse plane are not aligned to each other. The projected emittance is larger than the individual slice emittance. As already mentioned above, projected beam properties are not useful for describing FEL performance.

The main acceleration is accomplished by a series of rf cavities. Due to the rf curvature as well as wakefields within the rf structure the longitudinal energy profile becomes modulated. For a given charge and a certain bunch charge both effects can cancel each other but only in the linear term of the effective energy change along the bunch. Higher order modulation remains. However it has been shown that a cavity at a high frequency can compensate the second order effects as well[20] and the bunch compression is more homogeneous along the bunch. Because the operation of the machines requires the cancellation of rf-curvature and wakefields, it implies tolerances on the fluctuation of charge and rf phases. It also limits the charge range in order to control saturation power and length without changing the machine settings.

The bunch compressor is the most crucial part in the beam manipulation, because there are two effects, which can spoil the beam quality. Both are based on the coherent spontaneous radiation (CSR)[21] within the magnetic chicane. The CSR causes that the emitted radiation of the tail of the bunch acts back on the head of the bunch, because the radiation travels on a straight path and can catch up with the bunch, propagating a long a curved trajectory. The first effect is caused by the fact that a particle, which changes its energy within the chicane, has a residual dispersive effect at the end of the chicane. Because the CSR effect has different amplitude along the bunch, the main effect is a mismatch of the slice phase space distribution and, thus, a growth in the projected emittance. In addition there is also a slight increase in the slice emittance itself due to the non-uniformity of the effective CSR field in the transverse direction. The second effect is a collective instability[22] very similar to that driving the FEL. Any modulation in the beam current modulates the beam energy. Due to the dispersive effects the energy modulation enhances the current modulation, and thus the emission of the coherent synchrotron radiation. For the planned X-ray FELs LCLS and TESLA the overall gain of this microbunch instability is small[23] and can be further reduced if a super-conducting, planar undulator increases the energy spread prior compression[24].

Within the undulator itself wakefields can further effect the FEL amplification. This effect is distinct from the effect of the wakefields of the linear accelerator. The latter result in a variation in the initial energy of each slice and, thus, the resonant wavelength. This effect can be regarded as static while the undulator wakefields acts dynamically during the amplification. Once the electron enters the undulator, the slice energy defines the radiation wavelength. Once defined, the slice has to stay within the FEL bandwidth, thus requiring that the energy change by the undulator wakefields are fulfilling the constraint[25]

$$\frac{d\gamma}{dz} < \frac{\rho^2 \gamma}{\lambda_u} \quad . \tag{8}$$

If it is not fulfilled the slice stops amplifying before reaching saturation and the emission is shifted to different wavelengths.

All together – the undulator wakefields, the variation in the slice emittance and energy spread as well as the offset and mismatch of each slice – can yield a strong modulation of the FEL radiation pulse[26]. Fig. 3 shows the expected profile of the LCLS FEL. The main features are the two gaps at the head and tail of the bunch. They are caused by wakefields within the undulator as well as strong degradation of slice emittance and energy spread in the preceeding bunch compressors. On the other hand, the enhanced power level at the head of the pulse arises from non-linear terms in the compression scheme, which locally increases the current up to 15 kA.

A more technical aspect is the control on the FEL pulse length. It is determined by the bunch charge and degree of compression. For X-ray FELs the bunch length will be much longer than the cooperation length. Experiments, relying on longitudinal coherence, have to place a monochromator before their experimental station. However there are classes of experiments, which requires FEL pulses no longer than 50 fs [27]. Common solutions use

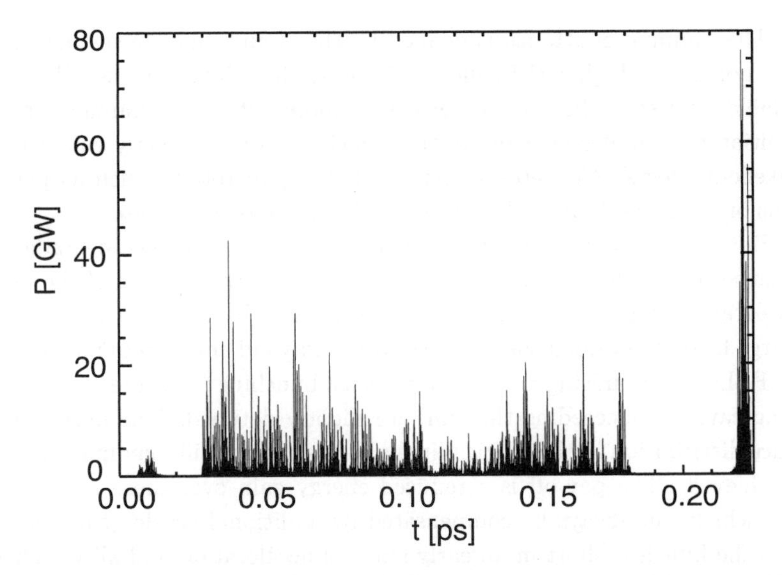

Figure 3. Simulated power profile of the LCLS FEL pulse at the undulator exit. The initial pulse (right) exceeds the expected power level due to a high local current of about 14 kA. The gaps arise due to wakefields in the undulator and insufficient slice emittances.

a monochromator to select a subsection of a chirped FEL pulse, which is produced by a chirped electron beam. The energy chirp of the bunch must be larger than the FEL bandwidth and is of the order 0.5-2%. This set-up is more feasible than compressing the bunch down to 50 fs while preserving the beam quality.

The duty cycle of the FEL depends on the acceleration technology. Normal conducting cavities have large thermal losses if a high repetition rate is desired. Also wakefields will affect succeeding bunches in a pulse train with short separation between each pulse. Superconducting cavities are much less effected by losses and beam loading, but comes with the cost of the cryogenic system. However super conducting rf guns[28] are still in its first steps of development, because their usage prohibits solenoids for emittance compensations.

5. New Concepts

The primary demand on the electron beam for a Free-Electron Laser is a high current and a low emittance. In addition a high repetition rate is desirable if the FEL operates as a user facility. Currently this is limited by the transverse emittance and bunch compression using a magnet chicane.

The minimal emittance is defined by the rf gun and the scaling shows that going to a higher rf frequency decrease the minimal value. However, reliable cavities at shorter wavelength are more difficult to manufacture. In addition the smaller aperture of these cavities would strongly enhance the wakefield effects. Concepts to replace the rf gun totally, such as plasma guns or injectors[29], are still in their initial development phase.

Bunch compressors are the strongest source of emittance degradation, because they deflect the electron beam from the primary axis of the linear accelerator. High current rf guns are not feasible because the strong space charge field of non-relativistic electrons would yield emittances not suitable for FELs. A promising concept is velocity bunching[30], where the accelerating cavities, succeeding the gun, are slightly detuned. The initial phase space distribution in the longitudinal plane is rotated till maximum current is achieved. The pay-off is a reduced energy gain over these initial cavities, which can always be compensated by additional accelerating section. Also the bunch is short in an early stage of acceleration and all wakefields adds up more coherently. On the other hand the transverse emittance is unaffected and preserved.

An alternative concept to FELs based on linear accelerator are energy recovery linacs (ERL)[31], based on super-conduction rf-technology. An electron beam is injected into a circular ring, where it is first accelerated and then injected into an FEL. Finally the bunch reenters the rf cavities but with a 180 degree phase difference. The electron energy is transferred back to the rf field, before the low-energy beam is dumped. This set-up combines the high repetition rate and energy stability of a storage ring FEL with the high beam brightness of a linac based FEL to achieve saturation within a single pass.

References

1. J. Als-Nielsen, "Scientific Opportunities for the 4th generation Light Sources, Hard X-ray", Proc. of Workshop on 4th Generation Light Sources, ESRF Report, 1996, Grenoble
2. A. Tremaine *et al.*, Phys. Rev. Lett. **88** (2002) 204801
3. S.V. Milton *et al.*, Science **292** (2001) 2037
4. V. Ayvazyan *et al.*, Phys. Rev. Lett. **88** (2002) 104802
5. *Linac Coherent Light Source (LCLS) Design Study Report*, SLAC-R-521, UC-414 (1998)
6. *TESLA - Technical Design Report* DESY 2001-011, ECFA 2001-209, TESLA Report 2001-23, TESLA-FEL 2001-05, Deutsches Elektronen-Synchrotron, Hamburg, Germany (2001)

7. R. Bonifacio, C. Pellegrini and L.M. Narducci, Opt. Comm. **50** (1984) 373
8. J. Rossbach *et al.*, TESLA-FEL 95-06, Deutsches Elektronen Synchrotron, Hamburg, Germany
9. E.L. Saldin, E.A. Schneidmiller and M.V. Yurkov, Opt. Comm. **97** (1993) 272
10. R. Bonifacio and F. Casagrande, Nucl. Inst. & Meth. **A 239** (1985) 29
11. J.B. Rosenzweig, E. Colby, Proc. Conf. Adv. Accel. Conf. **AIP 335** (1995) 724
12. C. Pellegrini *et al.*, Nucl. Inst. & Meth. **A 475** (2001) 328
13. F. Zhou *et al.*, Proc. of the European Part. Accel. Conf., 2002, Paris
14. E.T. Scharlemann, J. appl. Phys. **58** (1985) 2154
15. S. Reiche, Nucl. Inst. & Meth. **A 445** (2000) 90
16. A.N. Sessler, D.H. Whittum, L.H. Yu, Phys. Rev. Lett. **68** (1992) 309
17. M. Sands, Phys. Rev. **97** (195) 470
18. D.T. Palmer *et al.*, Proc. of the Part. Accel. Conf., Vancouver (1997) 2843
19. B.E. Carlsten, IEEE Catalog no. 89CH2669-0
20. M. Borland *et al.*, Nucl. Inst. & Meth. **A 483** (2002) 268
21. Y.S. Derbenev, J. Rossbach, E.L. Saldin and V.D. Shiltsev, TESLA-FEL 95-05, Deutsches Elektronen Synchrotron, Hamburg, Germany
22. S. Heifits,S. Krinsky and G. Stupakov, SLAC-PUB-9165, SLAC, Stanford, USA (2002)
23. Z. Huang, M. Borland, P. Emma, K.-J. Kim, Proc. of the 24th FEL Conf., 2002, Argonne
24. Z. Huang, K.-J. Kim, Phys. Rev. ST Accel. Beams 5, 074401 (2002)
25. S. Reiche and P. Emma, Proc. of the 24th FEL Conf., 2002, Argonne
26. S. Reiche *et al.*, Nucl. Inst. & Meth. **A 483** (2002) 70
27. G.K. Shenoy, J. Stoehr (eds.),*LCLS - The First Experiments*, Stanford (2000)
28. D. Janssen *et al.*, Proc. of the Part. Accel. Conf., 1999, New York
29. D. Umstadter, J.K. Kim, E. Dodd, Phys. Rev. Lett. **76** (1996) 2073
30. L. Serafini *et al.*, Proc. of the Part. Accel. Conf., 2001, Chicago
31. Maury Tigner, Nuovo Cimento **37** (1965) 1228

NEW PERSPECTIVES AND PROGRAMS IN ITALY FOR ADVANCED APPLICATIONS OF HIGH BRIGHTNESS BEAMS

L. SERAFINI

Istituto Nazionale di Fisica Nucleare
Sezione di Milano
Via Celoria 16, 20133 Milano, Italy

At the end of 2001 the Italian Government launched a call for proposals to the national research institutions for the design and construction of an Ultra-Brilliant X-ray Laser, with a dedicated funding up to 94 million € for this initiative. The Italian community responded by submitting two proposals, both envisaging a Linac based SASE-FEL as a source of coherent X-ray radiation. The two anticipated machines are quite similar to each other, as they basically consist of a few GeV Linac driving a SASE-FEL operated at a minimum wavelength around 15 Å. Here we describe the essential features of the two proposals: SPARX has been prepared by a collaboration among CNR-ENEA-INFN and Universita' di Roma "Tor Vergata", while FERMI@ELETTRA by INFM and Sincrotrone Trieste. We will also illustrate the status of the R&D project SPARC, aiming at the design and construction of an advanced 150 MeV photo-injector for generating a high brightness electron beam to drive a SASE-FEL in the optical range. This project has been approved by the Italian Government to conduct an R&D activity aimed to be strategic on the way to the coherent X-ray source; it is pursued by a CNR-ENEA-INFN-Universita' Tor Vergata-INFM-ST collaboration and will be located in the INFN National Laboratory at Frascati.

1 Introduction

Driven by the large interest that 4th generation light sources, *i.e.* X-ray SASE FEL's, have raised world-wide in the synchrotron light user community, as well as in the particle accelerator community, and following solicitations arising from several Italian national research institutions, in the year 2001 the Italian Government launched a long-term initiative devoted to the realization of a large scale ultra-brilliant and coherent X-ray source in Italy. The initiative was modulated into two phases, with anticipated budgets of 11 M€ and 96 M€ respectively: the first phase is meant to be a 3 year R&D program strategically oriented to explore the feasibility and the most crucial issues of the system which is expected to be designed and built in the second phase, aimed at the construction of the radiation source in a 5-6 year time scale. To pursue this program, the Italian Government published two calls for proposals, in March 2001 (named FISR) and in December 2001 (named FIRB), for the two phases respectively. In March 2002 the proposal SPARC, here described, was approved, among others, to be funded with 9.5 M€ over the available total budget in the FISR call (11 M€): funding should be delivered soon, allowing a prompt start-up of the project. In the meanwhile, two proposals, submitted in February 2002 at the FIRB call, are waiting a final decision of approval: one of these, SPARX, is tightly correlated to the approved R&D project SPARC, as explained in the following sections, the other one,

FERMI@ELETTRA, is competing with SPARX to obtain the 96 M€ funding from our Government.

The scientific case for these initiatives is recognized to be quite deep by the whole community of synchrotron light users. X-rays from synchrotron light sources are today widely used in atomic physics, plasma and warm dense matter, femto-second chemistry, life science, single biological molecules and clusters, imaging/holography, micro and nano lithography. The big step in the peak brilliance, several orders of magnitude, expected with the SASE-FEL sources will open new frontiers of research. New techniques in X-imaging, time resolved spectroscopy can be applied in the field of material science, biology, non linear optics. Of particular relevance are the diffractive techniques with coherent radiation on biologic tissues that allow the single-pulse crystallography of macro-molecules.

In the following we will present through different sections the status of the R&D project SPARC (Sec.2), the proposals SPARX (Sec.3) and FERMI@ELETTRA (Sec.4).

2 The R&D Project SPARC

The overall SPARC project consists of 4 lines of activity aiming at several goals: their common denominator is to explore the scientific and technological issues that set up the most crucial challenges on the way to the realization of a SASE-FEL based X-ray source. These are:

1) Advanced Photo-Injector at 150 MeV

Since the performances of X-ray SASE-FEL's are critically dependent on the peak brightness of the electron beam delivered at the undulator entrance, we want to investigate two main issues - generation of the electron beam and bunch compression via magnetic and/or RF velocity bunching - by means of an advanced system delivering 150 MeV electrons, the minimum energy to avoid further emittance dilutions due to time-dependent space charge effects [1].

2) SASE-FEL Visible-VUV Experiment

In order to investigate the problems related to matching the beam into an undulator and keeping it aligned to the radiation beam, as well as the generation of non-linear coherent higher harmonics, we want to perform a SASE FEL experiment with the 150 MeV beam, using a segmented undulator with additional strong focusing, to observe FEL radiation at 500 nm and below.

3) X-ray Optics / Mono-chromators

The X-ray FEL radiation will provide unique radiation beams to users in terms of peak brightness and pulse time duration (100 fs), posing at the same time severe challenges to the optics necessary to guide and handle such radiation. This project will pursue also a vigorous R&D activity on the analysis of radiation-matter interactions in the spectral range typical of SASE X-ray FEL's (from 0.1 to 10 nm), as well as the design of new optics and mono-chromators compatible with these beams.

4) Soft X-ray table-top Source

In order to test these optics and to start the R&D on applications, the project will undertake an upgrade of the presently operated table-top source of X-rays at INFM-Politecnico Milano, delivering 10^7 soft X-ray photons in 10-20 fs pulses by means of high harmonic generation in a gas. This will be a very useful bench-test for the activities performed in item 3 above.

In the following, the lay-out and planned activities for items 1 and 2 will be presented in more details, being these more related to the particle accelerator field.

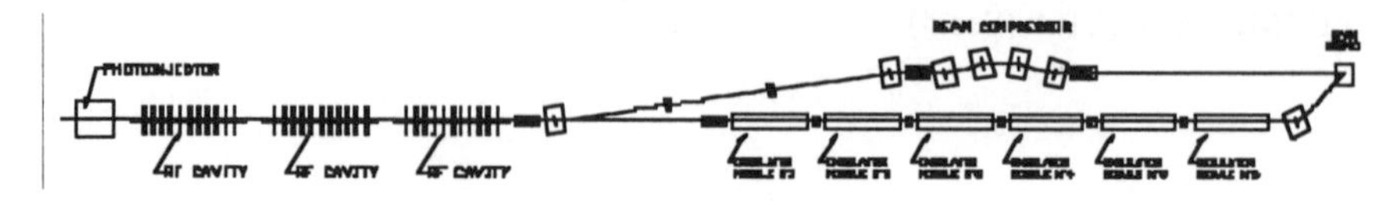

Figure 1. Lay-out of the SPARC system

2.1. Advanced Photo-Injectors

Two are the main goals of this activity in the context of the SPARC project: a) acquiring an expertise in the construction, commissioning and characterization of an advanced photo-injector system and b) the experimental investigation of two theoretical predictions that have been recently conceived and presented by members of the SPARC group. These are: the so-called Ferrario's working point[1] for high brightness RF photo-injectors and the velocity bunching technique to apply RF bunch compression[2] through the photo-injector, with emittance preservation.

The 150 MeV injector will be built inside an available bunker of the Frascati INFN National Lab: the general lay-out of the system is shown in Figure 1. It will consists of: a 1.6 cell RF gun operated at S-band (2.856 GHz, BNL/UCLA/SLAC type [3]) with high peak field on the cathode (120-140 MV/m) and incorporated metallic photo-cathode (Cu or Mg), generating a 6 MeV beam which is properly focused and matched into 2 SLAC accelerating sections.

Our simulations[5] using PARMELA indicate that we can generate with this system a beam like that needed by the FEL experiment at 150 MeV: this requires a 150 A peak current in the bunch with rms normalized emittance lower than 2 μm and energy spread below 0.1 %. Scaling the Ferrario's working point up to 1.6 nC bunch charge (instead of the nominal 1 nC) we were able to achieve in simulations a rms normalized emittance of 1.2 μm at the Linac exit, with the requested peak current and a rms correlated energy spread over the bunch equal to 0.14% .

However, the slice energy spread, calculated over a 300 μm slice length (comparable to the anticipated slippage length), is well below 0.05 % all over the bunch.

2.2. SASE-FEL Experiment

This will be conducted using a permanent magnet undulator made of 6 sections, each 2.5 m long, separated by 0.3 m gaps hosting single quadrupoles focusing in the horizontal plane. The undulator period is 3.3 cm, with an undulator parameter k_w= 1.88. Simulations performed with GENESIS show an exponential growth of the radiation power along the undulator: almost 10^8 Watts at saturation can be reached after 14 m of total undulator length, on the fundamental harmonic at 530 nm. Preliminary evaluations of the radiation power generated into the non-linear coherent odd higher harmonics show that 10^7 and $7x10^5$ W can be reached on the third and fifth harmonics, respectively.

2.3. Further Experiments

As shown in Figure 1, the SPARC lay-out anticipates two main upgrades that will be implemented in a second phase of the project: a third accelerating section, inserted between the RF gun and the 2 previous sections, and a parallel beam line containing a magnetic compressor.

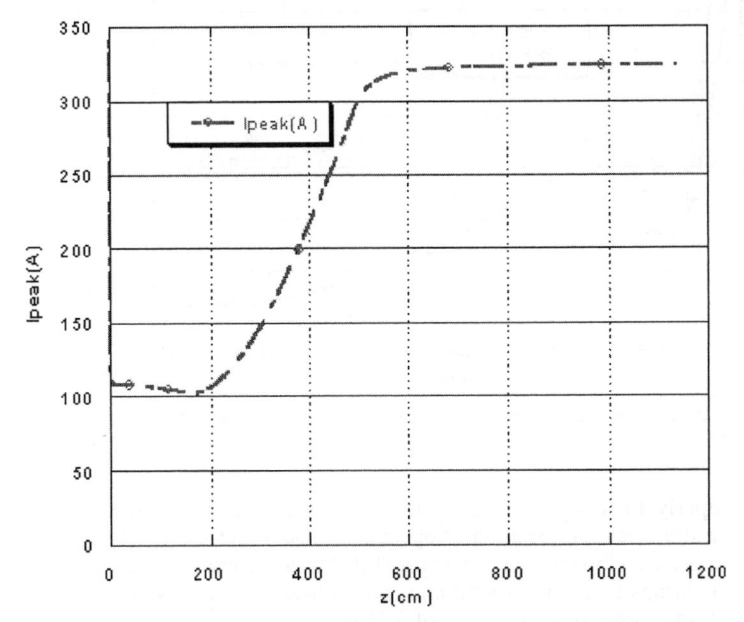

Figure 2. Rms Bunch current evolution along the photo-injector with RF Compression.

The new section will be designed to study RF compression: it will support travelling waves at an adjustable phase velocity in order to exploit the full potentialities of the velocity bunching technique[2]. Its design and construction will proceed in parallel to the commissioning of the initial SPARC injector system. These tests of RF compression assume great relevance in our R&D program[4] since the SPARX Linac foresees the use of a mixed compression scheme, as illustrated below.

Recent results[6] obtained in simulations with PARMELA show the possibility to reach bunch peak currents in excess of 500 A at the exit of the photo-injector with normalized emittances below 1 μm. The rms current carried by the bunch is plotted in Fig.2 vs. the distance along the photoinjector (taking the SPARC lay-out as in Fig.1): although it reaches 325 A, the current distribution within the bunch, as shown in Fig.4, displays a peak value close to 600 A.

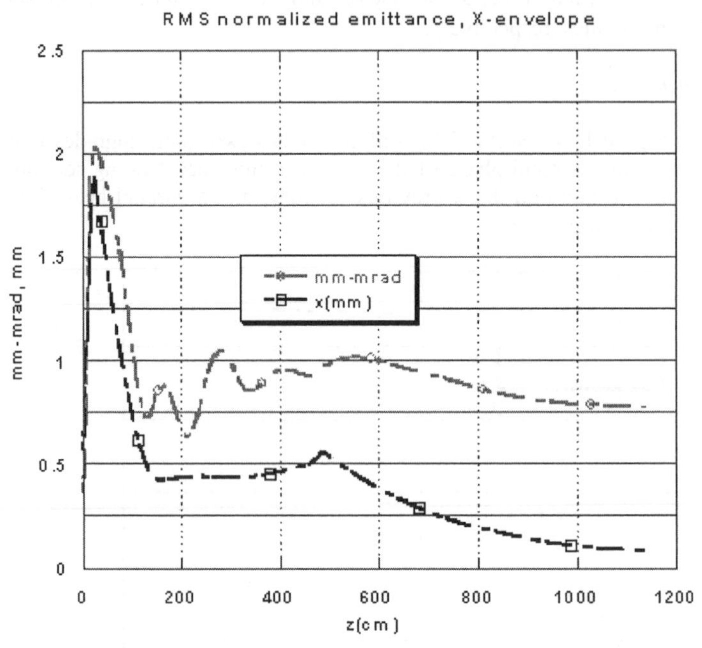

Figure 3. Emittance (red line) and envelope (black line) evolution along the photo-injector with RF Compression.

By properly focusing the beam with additional solenoids placed around the accelerating structures, the beam envelope can be taken under control as prescribed by the invariant envelope model (generalized to the RF compressor as described in Ref.4). This brings to a correct emittance compensation effect as shown in Fig.3, with a final rms normalized emittance of 0.75 μm.

The second beam line shown in Fig.1 will allow to conduct experiments on magnetic compression: we want to experimentally investigate CSR induced effects on emittance degradation and surface roughness wake-field effects, without interfering with the FEL experiment.

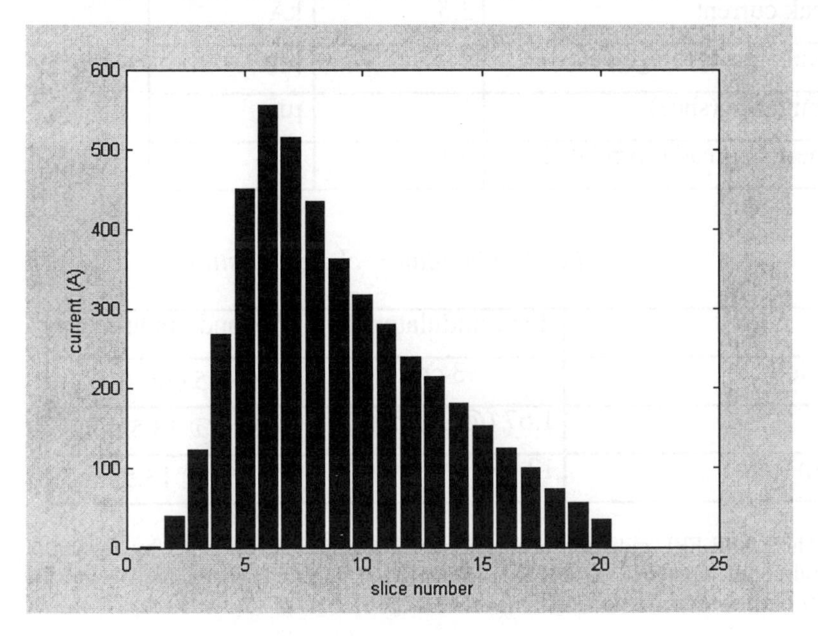

Figure 4. Bunch current distribution after RF Compression.

3 The Proposal SPARX

Two spectral complementary regions, around 13.5 nm and 1.5 nm, have been considered for this source: the first has many applications requested by future users, the second is the minimum achievable with the anticipated budget.

In order to drive a SASE-FEL at these wavelengths, it is necessary to produce a very high brightness beam, as shown by the list of parameters reported in Table 1.

We envisage the use of the same beam to feed two undulators whose characteristics are listed in Table 2: both are anticipated to be of Hallbach type (1.25 T residual field).

Table 1: Beam parameters

Beam Energy	2.5	GeV
Peak current	2.5	kA
Emittance (average)	2	μm
Emittance (slice)	1	μm
Energy spread (correlated)	0.1	%

Table 2: Undulator characteristics

	First undulator	Second undulator
Period	3 cm	5 cm
K	1.67 (@ 1.5 nm)	4.88 (@ 13.5 nm)
Gap (mm)	12.67 (@ 1.5 nm)	12.16 (@ 13.5 nm)

The SASE-FEL radiation characteristics have been investigated by means of several codes: GINGER, GENESIS, MEDUSA, PROMETEO, PERSEO, and the results are shown in Table 3.

Table 3: FEL-SASE expected performances

Wavelength (λ)	1.5 nm	13.5 nm
Saturation length	24.5 m	14.5 m
Peak Power	10^{10} W	4 10^{10} W
Peak Power 3° harm.	2 10^{8} W	5 10^{9} W
Brilliance (standard units)	1.8 10^{31}	2 10^{32}
Brilliance 3° harm.	10^{29}	10^{31}

Using two undulators allows to cover a bandwidth from 1.2nm to 13.5nm on the fundamental, while from 0.4nm to 4nm using the 3rd harmonic, which exhibits still a considerable peak power.

The SPARX Linac layout is shown in Fig.5. After the SPARC injector, a first 60m Linac section accelerates the beam up to 1 GeV, where the magnetic bunch compressor is located. The last 90 m Linac section boosts the beam up to 2.5 GeV before injection into the undulators. Adiabatic damping and longitudinal wake-fields are crucial in this Linac section to correct the correlated energy chirp applied to the bunch before the magnetic compression.

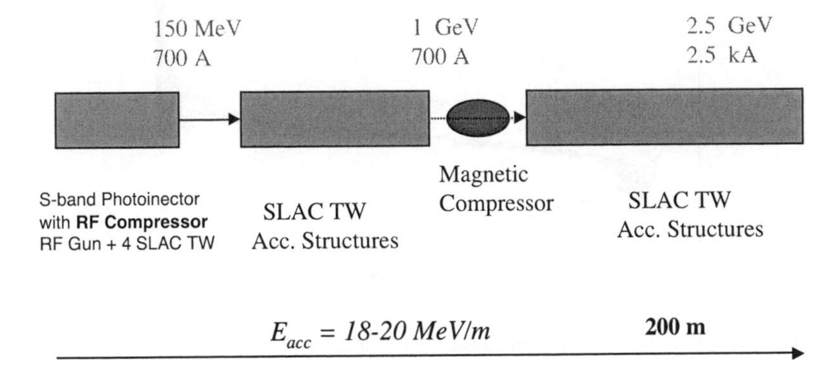

Figure 5. Linac scheme for the SPARX proposal.

Start-to-end simulations[7] performed using PARMELA and ELEGANT show the possibility to reach the anticipated beam performances (listed in Table 1) at least in the central slices of the bunch (about 60% of bunch charge). The longitudinal phase space distribution of the bunch at the Linac exit is plotted in Fig.6, while the current distribution of different bunch slices is reported in Fig.7: the normalized rms slice emittance is lower than 2 μm.

The slice energy spread is well below 0.1 %, except for the bunch tail which exhibits a huge current spike accompanied by large energy spread and slice emittance.

These preliminary results, clearly not yet optimized, show for the first time the possibility to operate a Linac in a mixed compression scheme: velocity bunching performing RF compression in the photo-injector combined to a single stage magnetic compression at intermediate energy (1 GeV).

The anticipated budget is: 34 M€ are estimated for the Linac, 10 M€ for the undulators, 10 M€ for the radiation beam lines and 13 M€ for contingency. In conclusion, SPARX is proposing an innovative solution for the Linac which relaxes the criticality of the beam compression process and seems more reliable for the achievement of the SASE-FEL beam brightness requirements. The Universita' di Roma "Tor Vergata", member of the collaboration, is ready to donate the land for the construction of a 1.5 Km tunnel hosting the Linac and radiation beam lines in case of approval of the SPARX proposal. This is shown in Fig.8, illustrating the map of the campus with the anticipated location for the tunnel.

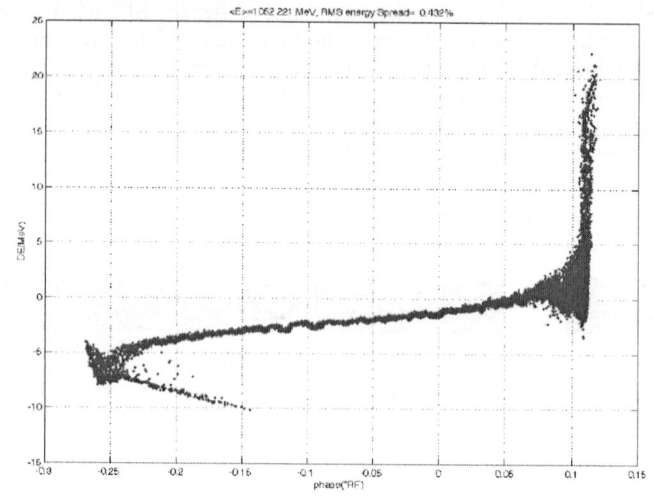

Figure 6. Longitudinal phase space distribution of the bunch at Linac exit.

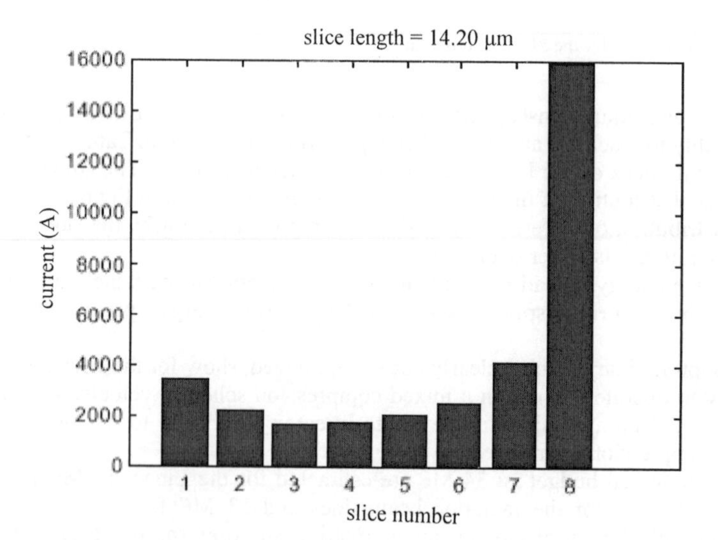

Figure 7. Current distribution of bunch slices at Linac exit.

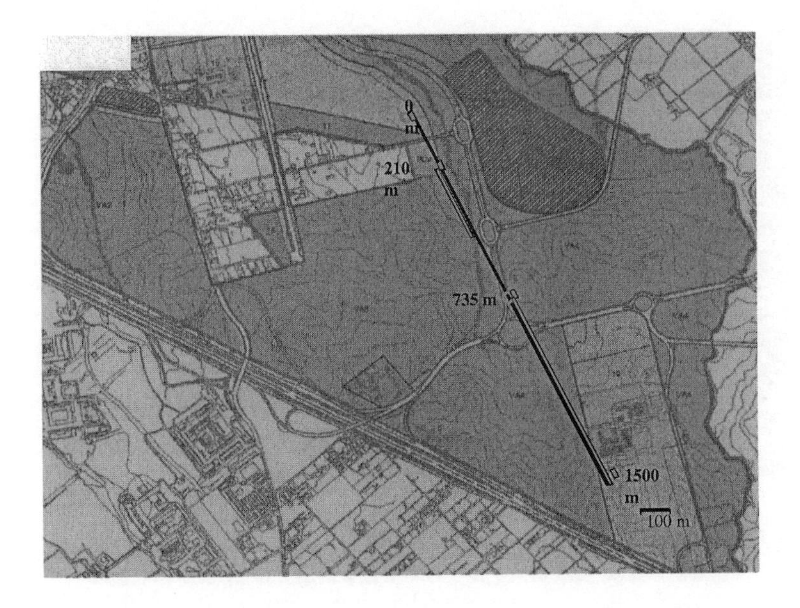

Figure 8. Anticipated location of the SPARX-tunnel on the campus map of "Tor Vergata" University.

4 The Proposal FERMI@ELETTRA

The proposal is articulated along three lines of development allowing gradual improvements and consolidation of technologies. These are schematically:

1) Use of the existing 1 GeV Linac with a new photo-injector and bunch compressors for the production of 40 nm radiation. Commissioning of 40 nm beamline in 2.5 years. Open to Users after 3.5 years.
2) Use of the Linac with increased beam quality for a second beamline at 10 nm. Commissioning of beamline in 3.5 years. Open to Users after 4.5 years.
3) Extension of the Linac up to 3 GeV and improvement of beam quality for operation at 1.2 nm (in parallel with other developments). Commissioning in 5.5 years and open to Users after 6.5 years.

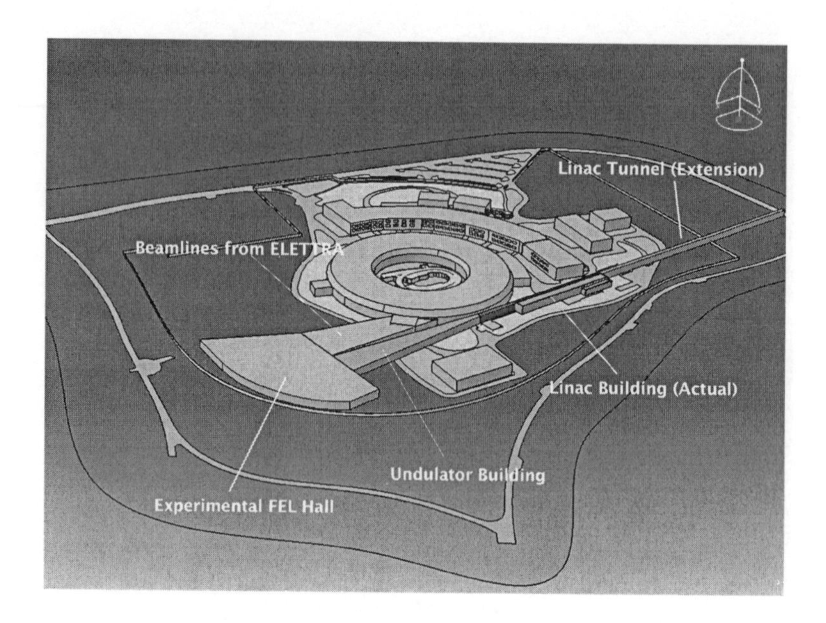

Figure 9. FERMI@ELETTRA anticipated lay-out.

A bird-view of the designed lay-out is presented in Fig.9, showing the X-ra FEL system in connection to the presently operated third generation synchrotro light source facility. The main electron beam parameters are listed in Table 4 fo different phases of the project.

Table 4: Electron Beam Parameters

Wavelength (nm)	100,40,10	1.2
Beam Energy (GeV)	1.0	3.0
Bunch Charge (nC)	0.38	1.0
Peak Current (kA)	0.6	2.5
Norm. Emittance (μm)	2.0	2.0
Energy Spread (%)	0.05	0.05

Validation of these values is based on simulations[8] performed with ELEGANT. The anticipated performances of the radiation source at shorter wavelengths are reported in Table 5.

Table 5: Radiation Beam Parameters

Wavelength (nm)	10	1.2
Undulator Period (cm)	3.19	3.26
Undulator Parameter	1.18	1.24
Gain Length (m)	1.3	2.6
Peak Brightness (st. un.)	9.4×10^{29}	5.5×10^{31}
Average Bright. (st. un.)	2.9×10^{19}	1.1×10^{21}

In order to improve stability, reproducibility and temporal control of the radiation pulses, FEL seeding based on High Gain Harmonic Generation is anticipated at 40 and 10 nm. At 1.2 nm the development of a seeding scheme is under study: if not possible, SASE operation will be pursued. Major details on this proposal can be found in Ref.8.

5 Conclusions

The Italian Government is expected to nominate soon an international committee to review the two proposals presented here, with final decision foreseen by this year. The design of an Italian Coherent X-ray Source should start in 2003 with the aim to deliver a TDR by the end of 2004.

6 Acknowledgements

We are indebted to many colleagues for the material provided and for many useful discussions: C. Bocchetta, P. Dattoli, P. Emma, M. Ferrario, L. Giannessi, L. Palumbo, A. Renieri, C. Ronsivalle, C. Sanelli and C. Vaccarezza .

References

1. M. Ferrario et al., *Homdyn Study for the LCLS RF Photoinjector* , in The Physics of High Brightness Beams, World Sci., ISBN 981-02-4422-3, June 2000
2. L. Serafini and M. Ferrario, AIP CP 581, 2001, pag.87
3. D.T. Palmer, PhD. Thesis, Stanford University
4. M. Boscolo et al., *Beam Dynamics Study of RF Bunch Compressors for High Brightness Beam Injectors* , EPAC-2002 Proceedings, Paris, June 2002
5. L. Serafini et al., *A R&D Program for a High Bright. El. Beam Source at LNF* , EPAC-2002 Proceedings, Paris, June 2002
6. C. Ronsivalle, *private communication*
7. C. Vaccarezza, *private communication*
8. G. D'Auria et al., *The FERMI Project at Trieste: a Linac Based Bright Photon Source* , Linac-2002 Conference proceedings

WORKING GROUP D

Advanced Accelerators

APPLICATION TO ADVANCED ACCELERATORS WORKING GROUP D SUMMARY

Mitsuru Uesaka[1], Patrick Muggli[2]

[1]*Nuclear Engineering Research Laboratory School of Engineering, University of Tokyo, 22-2 Shirane-shirakata, Tokai, Naka, Ibaraki, Japan*

[2]*University of Southern California, Dept.of EE-Electrophysics Los Angeles, CA 90089-0484*

E-mail: uesaka@utnl.jp

The work group D in application to advanced accelerators at the Sardinia workshop in "The Physics and Applications of High Brightness Electron Beams" focused on plasma acceleration this time. It held a series of meetings during the workshop. Presentations and discussion were performed on electron beam driven plasma wakefield acceleration, laser plasma wakefield acceleration and laser plasma X-ray generation. Finally, It discussed the integration to accelerator systems.

1. Introduction

Advanced accelerators mean the integration of higher RF technologies, table-top high-powered ultrashort lasers, plasma technology science and computer simulation, which enables their downsizing. Based on the pre-communication with the chair-persons, we decided to focus on plasma acceleration and its feasibility for future accelerator system. We formed the three topics and six sessions as shown in Table 1, including 7 speakers and 8 talks. We also merged to the working groups A (High brightness beam production) and C (Application to FELs) and the two talks were done there in Thursday morning.

The topics of this working group are organized into the three as illustrated in Table 2. 8 talks are devoted to updated achievement and progress in electron beam driver plasma wakefield acceleration, laser plasma wakefield acceleration and X-ray generation. Finally, we discussed the current status of those beam generation concepts and the feasibility and breakthrough forward to future accelerator systems.

A complete list of presentations and its their highlights are included in the Appendix.

Table 1: Schedule-Working Group D Application to Advanced Accelerators

Monday, July 01	Tuesday, July 02	Wednesday, July 03	Thursday, July 04	Friday, July 05
		11:15~12:30 Electron Beam Driven Plasma Wakefield Acceleration	8:30~12:30 Laser Plasma Accelerator and X-ray Generation, Joint Sessions with A and C	11:15~12:00 Report on working group D
18:30~19:30 Working Group Organization	17:00~19:30 Electron Beam Driven Plasma Wakefield Acceleration		17:00~19:00 Integration to Accelerator Systems	

Table 2: Topics/ Issuers
Application to Advanced Accelerator.

Topics	Number of talks
Electron Beam Driven Plasma Wakefield Accelerator	5
Laser Plasma Wakefield Accelerator and X-ray Generation	3
Integration to Accelerator Systems	Discussion

2. Electron Beam Driven Plasma Wakefield Acceleration

P.Muggli of USC presented energy gain and loss in E162. A new imaging spectrometer to measure the energy gain eliminating the effect of beam divergence was developed and used at SLAC wakefield experimental line (E-162) under collaboration of USC/UCLA/SLAC/etc. Here, the Cherenkov radiation by the electro beam in air after the achromatic optics was imaged by the CCD camera. As results the energy loss/gain of 15 GeV beam with several charges were successfully obtained as 159+/-40, 156+/-40 MeV with the resolution of ~10MeV, respectively.

M.Thompson of UCLA presented plasma density transition trapping as a possible high brightness electron beam sources plasma density transition trapping is a recently purposed self-injection scheme for plasma wake-field accelerators. This technique uses a sharp downward plasma density transition to trap and accelerate background plasma electron in a plasma wake-field. They examine the quality of electron beams captured using this scheme in terms of emittance, energy spread, and brightness. Two-dimensional Particle-In-Cell (PIC) simulations show that these parameters can be optimized by manipulating the plasma density profile. We also develop, and support with simulations, a set of scaling laws that predict how the brightness of transition trapping beams scales with the plasma density of the system. These scaling laws indicate that transition trapping can produce beams with brightness > 5x10^{14} Amp/(m-rad)^{2}. A proof-of-principle transition trapping experiment is planned for the UCLA Neptune Laboratory in the near future.

R.Yoder of UCLA presented resonant, THz slab-symmetric dielectric-based accelerator. Slab-symmetric dielectric-loaded structures, consisting of a vacuum gap between dielectric-lined conducting walls, have become a subject of interest for short-wavelength acceleration due to their simplicity; relatively low power density, and advantageous beam dynamics. Such a structure can be resonantly excited by an external power source and is known to strongly suppress transverse wakefields. Motivated by the prospect of a high-power FIR radiation source, currently under construction at UCLA, we investigate a high-gradient slab-symmetric accelerator powered by up to 100 MW of laser power at 340μm, with a predicted gradient near 100 MeV/m. Three-dimensional simulation studies of the structure fields and wakes were presented and compared with theory, and a future experiment discussed.

J.Rosenzweig of UCLA talked on energy loss of ultrahigh charge electron bunches in plasma. There has been much experimental and theoretical interest in blowout regime of plasma wakefield acceleration (PWFA), which features ultra-high accelerating fields, linear transverse focusing forces, and nonlinear plasma motion. Using an exact analysis, we examine here a fundamental limit of nonlinear PWFA excitation, by an infinitesimally short, relativistic electron beam. The beam energy loss in this case is shown to be linear in charge even for nonlinear plasma response, where a normalized, unitless charge exceeds unity, and relativistic plasma effects become important or dominant. The physical bases for this persistence of linear response are pointed out. As a byproduct of our analysis, we re-examine the issue of field divergence as the point-charge limit is approached, suggesting an important modification of commonly held views of evading unphysical energy loss. Deviations from linear behavior are investigated using simulations with finite length beams. The peak accelerating field in the plasma wave excited behind a finite-length beam is also examined, with the artifact of wave spiking adding to the apparent persistence of linear scaling of the peak field amplitude well into the nonlinear regime. On the other hand, at large enough normalized charge, linear scaling of fields collapses, with serious consequences for plasma wave excitation efficiency. The dramatic implications of these results for observing the collapse of linear scaling in planned experiments were discussed.

3. Laser Plasma Wakefield Acceleration and X-Ray Generation

D.Umstadter of Univ.Michigan talked on updated results on the original plasma gun, LILAC, a stationary optical trap for relativistic electrons by colliding two 400fs intense lasers and laser plasma higher harmonic X-ray generation. They did the LILAC experiments using the two long 400fs orthogonal laser pulses. More collimated electrons were observed than in the case of one laser. Interference fingers were observed by the two orthogonal colliding laser (400fs, 1.053 μ m) pulses (4×10^{17}W/cm^2). Maximum $\delta n / n$ and electric field are 12 and 3GeV/cm, respectively. It is expected to be used for the stationary optical trap for relativistic electrons and a micro-wiggler.

M.Uesaka of Univ. Tokyo explained generation of relativistic electrons by interaction between a high intensity ultra-short laser pulse (Ti: Sapphire, 12 TW, 50 fs, λ=790 nm) and gas jet. In the experiment, spatial and energy distribution of energetic electrons produced by an ultra-short, intense laser pulse in a He gas jet are measured. They depend strongly on the contrast ratio and shape of the laser prepulse. In the case of a proper prepulse the electrons are injected at the shock front produced by the prepulse and accelerated by consequent plasma wake-field up to tens MeV forming a narrow-coned ejection angle. In the case of non-monotonic prepulse, hydrodynamic instability leads to a broader, spotted spatial distribution. The numerical analysis based on a 2D hydrodynamics (for the laser prepulse) and 2D particle-in-cell simulations justify the mechanism of electron injection and acceleration.

K.Floettmann of DESY reported numerical comparison between RF photoinjector and plasma gun (one laser and wavebreaking) with respect to emmitance done by Michael Geissler and Thomas Brabec of Tech. U. Vienna. Emittance of the electrons from a plasma gun via wavebreaking is about 10 times more than that from a photoinjector. However, multi-laser scheme, plasma density transition trapping or control of laser group velocity is going to give better beam quality for the plasma guns in near future. Several promising beam parameters (charge, emittance, energy spread, divergence, etc.) have begun to be demonstrated in the world.

4. Integration to Accelerator Systems

4.1 *Electron Beam Driven Plasma Wakefield Accelerator*

It is still in the stage of proof-of-principle. More new ideas (plasma density transition trapping, THz acceleration, etc), numerical simulations, experiments and new diagnostics (imaging spectrometer, etc) are needed. The linear scaling law should be checked and up-graded.

4.2 *Laser plasma wakefield accelerator*

Although the plasma gun is still in the stage of proof-of-principle, several promising beam parameters (charge, emittance, energy spread, divergence, etc.) have begun to be demonstrated. New diagnostic methods for prepulse, plasma channel, etc., should be developed. Concerning the stability of laser, environments (room temperature, dusts, humidity, etc) should be strictly controlled. Furthermore, feedback control to adjust input RF phase, laser timing and optical alignment should be implemented.

4.3 *Laser plasma X-ray source*

Among the quality parameters, the aspects of short-pulse (sub-pico~10ps) and compactness can be emphasized. Several pump (laser)-and-probe (X-rays) experiments are under way and planned in the world. It is complementary to existing light sources and XFELs.

5. Appendix: Presentations and Highlights in the Working Group D

Electron Beam Driven Plasma Wakefield Acceleration

5.1. *P.Muggli (USC)*

"Energy gain and loss in E162"

A new imaging spectrometer to measure the energy gain eliminating the effect of beam divergence was developed and used at SLAC wakefield experimental line (E-162) under collaboration of USC/UCLA/SLAC/etc. Here, the Cherenkov radiation by the electro beam in air after the achromatic optics was imaged by the CCD camera. As results the energy loss/gain of 15 GeV beam with several charges were successfully obtained as 159+/-40, 156+/-40 MeV with the resolution of ~10MeV, respectively.

5.2. *M.C. Thompson (UCLA)*

"Plasma density transition trapping as a possible high brightness electron beam sources"

PIC simulation was performed for investigation of electron infection and trapping in density transition region. 14MeV, 5.9nC, 6ps drive beam induces 1.2MeV, 120pC, 1ps bunch. To verify this, the experiment with the new plasma shaping designs is going to be done soon.

5.3. *M.C. Thompson (UCLA)*

"Emittance growth in intense beam due to collective effects at metallic boundaries"

Theoretical analysis based on an expansion of current work in our group on coherent transition radiation produced from rough surfaces. Improved simulations and experiments are under consideration.

5.4. *R. Yoder (UCLA)*

"A resonant, THz slab-symmetric dielectric-based accelerator"

Advantages are that transverse wakefields is strongly suppressed, that planar structure is easy to build and tune and that dielectric breakdown limit is potentially easier. Numerical simulation is under way and experiment is planned.

5.5. *J.B. Rosenzweig (UCLA)*

"Energy loss of ultrahigh charge electron bunches in plasma"

The linear scaling of energy loss and accelerating field vs electron beam parameters and plasma density should be carefully re-checked for recent and next PWFA experiments.

6. Laser Plasma Wakefield Acceleration and X-Ray Generation

6.1. *D.Umstadter (Univ. Michigan)*

"Laser plasma guns and wigglers"

Experiments on long-pulsed LILAC demonstrated that two pulses scheme gave better electron beam generation. Higher harmonics by relativistic electron motion have been observed.

6.2. *M.Uesaka (Univ.Tokyo)*

"Numerical analysis and experiment on plasma cathode by self-injection"

New scheme to generate a bunched ~10fs beam with a single TW laser by tuning the laser group velocity is theoretically proposed. 40MeV at maximum electrons were observed by using a newly designed gas jet nozzle and 5TW laser pulse.

6.3. *K.Floettmann (DESY)*

"Simulation of a plasma beam source"

Numerical comparison between RF photoinjector and plasma gun (one laser and wavebreaking) with respect to emittance done by Tech. U. Vienna is reported. Emittance of the electrons from a plasma gun via wavebreaking is about 10 times more than that from a photoinjector.

ENERGY LOSS OF A HIGH CHARGE BUNCHED ELECTRON BEAM IN PLASMA: ANALYSIS

N. BAROV

Department of Physics
Northern Illinois University, DeKalb, IL 60115

J.B. ROSENZWEIG, M.C. THOMPSON, AND R. YODER

UCLA Department of Physics and Astronomy
405 Hilgard Ave., Los Angeles, CA 90095-1547

There has been much experimental and theoretical interest in blowout regime of plasma wakefield acceleration (PWFA), which features ultra-high accelerating fields, linear transverse focusing forces, and nonlinear plasma motion. Using an exact analysis, we examine here a fundamental limit of nonlinear PWFA excitation, by an infinitesimally short, relativistic electron beam. The beam energy loss in this case is shown to be linear in charge even for nonlinear plasma response, where a normalized, unitless charge exceeds unity, and relativistic plasma effects become important or dominant. The physical bases for this persistence of linear response are pointed out.

1. Introduction

The transfer of energy from short, intense electron beams to collective electron plasma oscillations is a critical component of the advanced, high-gradient acceleration scheme known as the plasma wakefield accelerator (PWFA)[1,2,3,4]. While the original proposal for the PWFA and related concepts was in the linear regime[1,2], where the plasma oscillations can be considered small perturbations about an equilibrium, highly nonlinear regimes have recently been favored[3]. For example, in the highly nonlinear "blow-out" regime[4], the plasma electrons are ejected from the channel of the intense driving electron beam, resulting in an electron-rarefied region with excellent quality electrostatic focusing fields, as well as longitudinal electromagnetic fields, which can, in tandem, stably accelerate and contain a trailing electron beam. While many aspects of the beam dynamics in this regime are linear and analytically tractable, the plasma dynamics are not.

Most quantitative predictions concerning plasma behavior in the blow-out regime have been deduced from numerical simulations.

This paper represents a step in the direction of an analytical understanding of the plasma response in the blow-out regime - we shall see that new, surprising physical aspects of the beam-plasma interaction become apparent from the analysis presented. The analytical results deduced are exact in the limit of an infinitesimally short beam. Particle-in-cell simulations that both support the results given in this paper, and extend the analysis of energy loss and gain in the nonlinear PWFA to finite-length beams, are presented in an accompanying paper.

We now review some relevant background concerning the blow-out regime of the PWFA. Despite the lack of analytical models for the nonlinear plasma response, it has been noted in a number of studies [5,6,7,8] that the beam energy loss rate in the PWFA blow-out regime obeys a scaling law usually associated with the interaction of charged particles with linear media [10] (*i.e.* Cerenkov radiation). This scaling, which persists even when the beam is much denser than the plasma, and the concomitant plasma response is nonlinear and relativistic, predicts that the energy loss rate is proportional to the square of the plasma frequency [10], ω_p. Since the efficient excitation of an oscillatory system by a pulse occurs when the pulse is short compared with the oscillator period [5,6,7,8], this scaling further implies that the PWFA driving beam's energy loss rate is proportional to the inverse square of the achievable driving beam's rms pulse length, σ_z. This prediction has led to a number of experiments that employ bunch compressors in order to decrease σ_z, thus dramatically increasing the transfer of beam energy to the plasma. In recent measurements with compressed beam at FNAL [11], the trailing portion of a 5 nC, 14 MeV, $\sigma_z = 1.2$ mm, beam pulse was nearly stopped in 8 cm of $n_0 \simeq 10^{14}$ cm^{-3} plasma, a deceleration rate of over 150 MeV/m.

The large collective field observed in this as well as other recent PWFA experiments [11,12], was obtained the context of nonlinear plasma electron motion. Because of the onset of experiments in the nonlinear blow-out regime, the issue of wakefield scaling validity has taken on new urgency. In addition, it has recently been a proposed (in the SLAC E-164 experiment) to use ultra-short, high charge beams to a drive PWFA in the tens of GeV/m range, for creation of an ultra-high energy plasma accelerator, the so-termed "afterburner" concept[8,13,14]. As we shall see below, in the limit of an infinitesimally short beam, when either the plasma density or the beam charge is increased, the response of the plasma to the beam eventually becomes nonlinear. Nevertheless, the energy loss rate, somewhat surprisingly, still scales linearly with the charge in the short beam limit, even for nonlinear plasma motion. This paper is primarily intended to address the new aspects of the underlying physics of the linear-like wakefield scaling of relativistic beam energy loss in plasma. Deviations from this scaling will be studied in the accompanying simulation-based paper[9].

2. Dimensionless Cold Fluid Analysis

To examine the physics relevant to ultra-short electron-beam energy loss in a plasma, we perform an analysis of the motion of cold plasma electrons having an initial ambient density n_0 (equal and opposite in charge density to the ions, which are assumed stationary) as they are excited by the passage of the ultra-relativistic ($v_b \simeq c$) beam. The state of plasma motion is described in terms of the velocity $\vec{v}$ and related momentum $\vec{p} = \gamma m_e \vec{v}$, where the Lorentz factor $\gamma = \left[1 - (\vec{v}/c)^2\right]^{\frac{1}{2}}$. The necessary relations for describing the cold plasma electrodynamic response are the Maxwell equations

$$\vec{\nabla} \times \vec{H} = \frac{4\pi}{c}\vec{J} + \frac{1}{c}\frac{\partial \vec{E}}{\partial t}, \text{ and } \vec{\nabla} \times \vec{E} = -\frac{1}{c}\frac{\partial \vec{H}}{\partial t} \tag{1}$$

the Lorentz force equation in convective form,

$$\frac{\partial \vec{p}}{\partial t} + (\vec{v}\cdot\vec{\nabla})\vec{p} = -e\left[\vec{E} + \frac{1}{c}\vec{v} \times \vec{H}\right] \tag{2}$$

and the equation of continuity for charge and current density

$$\frac{\partial n}{\partial t} + \vec{\nabla} \cdot (n\vec{v}) = 0. \tag{3}$$

The results of our analysis will be made more transparent by the adoption of unitless variables. The natural variables used in discussing a cold plasma problem parameterize time and space in terms of the plasma frequency $\omega_p = \sqrt{4\pi e^2 n_0 / m_e}$, and wave-number $k_p = \omega_p / c$, respectively, densities in terms of n_0, and the amplitudes of the and fields in terms of the commonly termed "wave-breaking limit", $E_{WB} = m_e c \omega_p / e$. In addition, all velocities and momenta are normalized to c and $m_e c$, respectively. We thus write the spatio-temporal variables, charge and current density, and electromagnetic field components in our analysis as

$$\tilde{r} = k_p r, \; \tau = \omega_p (t - z/c), \; \tilde{v}_i = v_i / c, \; \tilde{p}_i = p_i / m_e c, \tag{4}$$

$$\tilde{n} = n_e / n_0, \; \tilde{J}_i = J_i / n_0 c \tag{5}$$

$$\tilde{E}_i = E_i/E_{WB},\ \tilde{H}_i = H_i/E_{WB}. \tag{6}$$

Note that use of Eqs. 4-6 implies that we are assuming a steady-state response (wave *ansatz*), where t and z occur only in the combination $t-z/c$. As we are eventually only interested in the region directly in contact with the relativistic beam, this steady-state assumption is entirely reasonable – it implies that we expect the energy loss rate in the plasma to be constant after the passage of a short (k_p) transient region associated with the entrance into the plasma, or some other boundary. This assumption is validated by simulations.

Using the variables in Eqs. 4-6, we may write a general equation for the azimuthal component of $\tilde{H}$

$$\frac{\partial^2 \tilde{H}_\phi}{\partial \tilde{r}^2} + \frac{1}{r}\frac{\partial \tilde{H}_\phi}{\partial \tilde{r}} - \frac{\tilde{H}_\phi}{\tilde{r}^2} = \frac{\partial \tilde{J}_r}{\partial \tau} + \frac{\partial \tilde{J}_z}{\partial \tilde{r}}. \tag{7}$$

In addition to the governing equation for $\tilde{H}$, we will need relationships between fields and current sources,

$$\frac{\partial \tilde{E}_z}{\partial \tilde{r}} = \tilde{J}_r \text{ and } \frac{\partial}{\partial \tau}\left(\tilde{E}_r - \tilde{H}_\phi\right) = -\tilde{J}_r. \tag{8}$$

In this analysis the induced is found most directly by determining the transverse current, as is customary in media-stimulated radiation calculations (*cf.* Jackson, Ref. [10]).

3. Linear Plasma Response

Equation 7 is nonlinear, but may be simplified by assuming small amplitude response, in which the $|\tilde{n}|$, $\left|\tilde{E}_i\right|$ and $\left|\tilde{H}_i\right|$ are small compared to unity. In fact, to place our results in perspective, we must begin with a review of previous work in the linear regime[1,2], which, because they are models that simplify the physical scenario, did not need to employ the induced magnetic field as the initially-solved variable.

From the viewpoint of the fluid equations, linearity importantly implies that the plasma electron response is nonrelativistic, *i.e.* $\vec{v} = \vec{p}/m_e$. In addition, the plasma electrons can then be assumed to be unaffected by magnetic fields, and the resultant fields excited behind the beam are approximately electrostatic. Note, in contrast, that in the nonlinear blowout regime, the excited fields in the rarefied region behind the beam head are qualitatively different than those in the linear, electrostatic regime. In the blowout regime, the excited fields can be described as a superposition of a radial electrostatic field due to the ions, and a TM electromagnetic wave arising from the plasma electron motion[4]. In addition, it should be noted that in deriving Eq. 7, it is also implicitly assumed that the beam density is smaller than the plasma density, or the magnetic force will not be ignorable in determining the plasma electron response (the right-hand-side of Eq. 7).

In order to illustrate the dependence of $\tilde{E}_z$ on transverse beam size, and to examine the limit of a point charge, we assume a disk-like beam, uniform up to radius $\tilde{a} = k_p a \ll 1$, and δ-function in τ. We are interested in the instantaneous response of the plasma directly upon beam passage, and integrate over the δ-function in Eq. 8 to obtain $\tilde{H}_\phi = \tilde{E}_r$, as a valid condition before, during, and immediately after beam passage. Further, we use this condition to find

$$\frac{\partial^2 \tilde{H}_\phi}{\partial \tilde{r}^2} + \frac{1}{\tilde{r}}\frac{\partial \tilde{H}_\phi}{\partial \tilde{r}} - \frac{\tilde{H}_\phi}{\tilde{r}^2} - \tilde{H}_\phi = \frac{\tilde{Q}}{\pi \tilde{a}^2}\delta(\tau)\delta(\tilde{r} - \tilde{a}). \tag{9}$$

Note that $\tilde{H}_\phi = \tilde{E}_r = 0$ just in front and behind the beam, while at $\tau = 0$ these fields are singular (they are δ-functions, while the plasma $\tilde{J}_r$ remains finite in Eqs. 8) in the ultra-relativistic $v_b \Rightarrow c$ limit. These comments are equally valid when the assumptions needed for a linearized, small amplitude analysis are violated. Note also that per Eqs. 8, the condition $\tilde{H}_\phi = \tilde{E}_r$ does not generally hold further than an infinitesimal distance behind the beam - it does not apply to the wake-region that is not in direct contact with the disk-like beam.

In Eq. 9 we have introduced a fundamental quantity that controls the scale of the beam-plasma interaction, the normalized beam charge

$$\tilde{Q} = 4\pi k_p r_e N_b. \tag{10}$$

When $\tilde{Q} \ll 1$, this indicates that the response of the system is linear. Note also that $\tilde{Q}$ can be written as $\tilde{Q} = N_b k_p^3 / n_0$, which is the ratio of the number of beam electrons to plasma electrons within a cubic plasma skin-depth k_p^{-3}. Thus we may also write the underdense condition as $n_b/n_0 = \tilde{Q}/(2\pi)^{3/2} k_p \sigma_z (k_p \sigma_r)^2$, and when is near unity, and (as is typical in previous experiments) $k_p \sigma_r \ll 1$ then $\tilde{Q} \simeq 1$ implies that the beam is denser than the plasma. It should be noted in this regard that the experiments[11,12] that are performed in the blowout regime (n_b/n_0 well in excess of unity) have beam-plasma systems yielding $\tilde{Q}$ values between 1.5 and 4.

Equation 9 has a temporal δ-function which we again integrate over, to obtain an inhomogeneous modified Bessel equation in $\tilde{r}$

$$\frac{\partial^2 \mathbf{H}}{\partial \tilde{r}^2} + \frac{1}{\tilde{r}}\frac{\partial \mathbf{H}}{\partial \tilde{r}} - \frac{\mathbf{H}}{\tilde{r}^2} - \mathbf{H} = \frac{\tilde{Q}}{\pi \tilde{a}^2}\delta(\tilde{r} - \tilde{a}), \tag{11}$$

where $H = \int_{\epsilon-}^{\epsilon+} \tilde{H}_\phi d\tau = \int_{\epsilon-}^{\epsilon+} \tilde{E}_r d\tau$. In the linear, non-relativistic limit, We interpret $\mathbf{H}$ as the total radial momentum impulse $\tilde{p}_r$, which, also in this limit, is approximately equal to $\tilde{J}_r$ immediately behind the beam. The solution to Eq. 11 is given by

$$\mathbf{H}(\tilde{r}) = \frac{\tilde{Q}}{\pi\tilde{a}} \begin{cases} K_1(\tilde{a})I_1(\tilde{r}) \; (\tilde{r} < \tilde{a}) \\ K_1(\tilde{r})I_1(\tilde{a}) \; (\tilde{r} > \tilde{a}) \end{cases} \tag{12}$$

where I_1 and K_1 are modified Bessel functions. We are interested in $\tilde{E}_z$ directly behind the beam, which is found by integrating Eq. 12

$$\tilde{E}_z(\tilde{r})\Big|_{\tau=\epsilon+} = \int_\infty^{\tilde{r}} \mathbf{H}(\tilde{r}')d\tilde{r}' = \frac{\tilde{Q}}{\pi\tilde{a}^2} \begin{cases} 1 - \tilde{a}K_1(\tilde{a})I_0(\tilde{r}) \\ \tilde{a}I_1(\tilde{a})K_0(\tilde{r}) \end{cases} \tag{13}$$

over the radial coordinate.

In the limit that $\tilde{a} \ll 1$, the field inside of the disk region is nearly constant, and given by

$$\tilde{E}_z(\tilde{r})\Big|_{\tau=\epsilon+} \simeq \frac{\tilde{Q}}{\pi\tilde{a}^2}[1 - \tilde{a}K_1(\tilde{a})] \simeq \frac{\tilde{Q}}{2\pi}\left[\ln\left(\frac{2}{\tilde{a}}\right) - 0.577\ldots\right] \tag{14}$$

which is to leading order proportional to $\tilde{Q}/2\pi$. In physical units we may write Eq. 14 as

$$eE_z|_{\tau=\epsilon+} \simeq 2e^2k_p^2N_b \ln\left(\frac{1.123}{k_pa}\right). \tag{15}$$

Several comments arise from inspection of Eq. 15. The first is that the scaling of E_z with respect to wavenumber k is dominated by the factor k_p^2 that is typical of Cerenkov radiation [10], if we interpret k_p as the maximum allowable value of k that is radiated (in the linear regime it is the only value). The second comment is that the linear result is ill-behaved in the limit of $k_pa \ll 1$, as Eq. 15 predicts a logarithmic divergence in E_z. This pathology is a result of allowing $\tilde{J}_r$ (through $\mathbf{H}$) to diverge as r^{-1}.

In the limit of $a \to 0$ (the point-charge limit), this divergence is mitigated by the replacement of a in Eq. 15 with the minimum impact parameter, b_{min}, using either the classical ($b_{min} = e^2/\gamma m_e$) or quantum mechanical (see discussion at end of next section) expression for this parameter. In this limit, a significant protion of the energy loss predicted by equation 15 is due to small impact parameter collisions with a correspondingly large energy transfer. These particles are energetic enough to simply leave the plasma region, and are thus unable to fully couple their energy back to the plasma wave.

4. Nonlinear Plasma Response

As $\tilde{Q}$ is raised, we must consider the plasma electrons' relativistic response to large amplitude fields, under the general condition that $\tilde{H}_\phi = \tilde{E}_r$. The problem of charges moving in perpendicular electric and magnetic fields of equal magnitude has been solved by Landau[16], and we base our analysis on this solution. In this analysis, the relation between the transverse and longitudinal momentum (in our units) is,

$$p_z = -\frac{\alpha}{2} + \frac{p_z^2 + \epsilon^2}{2\alpha}, \tag{16}$$

where α and ϵ are two constants which are both equal to unity for our initial conditions where the particle starts from rest. Equation 16 then becomes,

$$p_z = \frac{p_r^2}{2}. \tag{17}$$

This result is not dependent on the fields being constant in r or z, but only on them being orthogonal and equal.

The passage of the beam induces not only an impulsive change in the radial momentum, but a longitudinal impulse in the positive (beam motion) direction. Thus, for large **H**, the plasma electrons experience a large forward momentum impulse, and may have a relativistic v_z just after passage of the beam.

The equation of motion for p_r is,

$$\frac{dp_r}{dt} = (1 - v_z)H \tag{18}$$

where the time derivative can be transformed to the beam longitudinal coordinate $\xi = z - t$ with the relation $d/dt = -(1 - v_z)d/d\xi$. Cancellation of the $1 - v_z$ terms and integration over the beam extent leads to the result

$$\tilde{p}_r = \mathbf{H}, \tag{19}$$

which is identical to the result for the linear case.

With the use of Eqs. 19 and 17 the plasma electron's transverse velocity becomes

$$\tilde{v}_r = \frac{\mathbf{H}}{\sqrt{1 + \mathbf{H}^2 + \frac{1}{4}\mathbf{H}^4}} = \frac{\mathbf{H}}{1 + \frac{1}{2}\mathbf{H}^2} \tag{20}$$

In order to relate this result to $\tilde{J}_r$ we must multiply by $\tilde{n}$, which due to the change in $\tilde{v}$ directly after passage of the beam, is predicted with the aid of the continuity relation (Eq. 3), and Eqs.19 and 17 to be

$$\tilde{n} = (1 - \tilde{v}_z)^{-1} = 1 + \frac{1}{2}\mathbf{H}^2. \tag{21}$$

Thus, we are led to the remarkable result that the relativistically correct induced radial current is identical to the approximate, linear, non-relativistic expression,

$$\tilde{J}_r = \tilde{n}\tilde{v}_r = \left(1 + \frac{1}{2}\mathbf{H}^2\right) \cdot \frac{\mathbf{H}}{1 + \frac{1}{2}\mathbf{H}^2} = \mathbf{H} \tag{22}$$

Since the induced $\tilde{J}_r$ is unchanged from the linear case, the analysis of the decelerating field $\tilde{E}_z$ leading to Eq. 13 remains valid. Therefore we see that the "linear" scaling observed in simulations of short pulse beam-excited wake-fields may be have some basis in this on an analytically predicted effect.

The result in Eq. 22 arises from two effects which cancel each other: the induced $\tilde{v}_r$ saturates (at a value well below 1), yet the density enhancement due to longitudinal motion - a "snow-plowing" of the plasma electrons by the electromagnetic pressure - exactly makes up for this saturation, and the induced $\tilde{J}_r$ remains linear in $\tilde{Q}$. This snow-plowing is analogous to the scenario from laser wake-field acceleration, where the electromagnetic pressure in gradient of a short, intense laser gives rise to a density enhancement in the laser's leading edge.

In order for $\tilde{n}$ to be enhanced directly after the beam passage, no net longitudinal displacement of the plasma electrons while the beam interaction must occur – the density enhancement is caused only after the beam by particles with relativistic longitudinal velocity "catching up" to the beam. In fact, when one takes the limit of a δ-function in ζ in the integrals of the motion, one finds that the displacements in both r and z vanish in this limit. This is reassuring, as it validates our use of the Landau result. Numerical integration of the equations of motion also confirm our result. In this case if one performs the integration over ζ as the independent variable, care must be taken to expand the differential interaction period by the factor $(1 - \tilde{v_z})^{-1}$ as described in obtaining Eq. 19.

The effects described by Eqs. 17 and 19 can be alternatively be confirmed by performing an analysis in the rest frame of the beam and Lorentz transforming to the laboratory frame. It should be noted that the vanishing of the radial displacement of the plasma electrons during the passage the δ-function beam is necessary for self-consistency with the assumptions of the derivation leading to Eqs.17 and 19. If the plasma electrons experience non-negligible displacement, then Eq. 19 and must be revisited. It is because we can neglect radial displacement in the δ-function beam limit that our analysis can rigorously account for radial variations in the plasma response –it is manifestly a two-dimensional axisymmetric result. In the δ-function beam limit, one may also self-consistently ignore effects of the induced longitudinal force from the plasma electrons themselves during their infinitesimally short interaction time with the beam, as this force remains finite and its integrated effect tends to zero.

The results obtained here are similar in appearance to those that have been derived in the context of the one-dimensional laser-plasma interaction by Sprangle, *et al.*[18] Because of some considerable confusion in formal discussions of the present results, we must remark on this situation. This similarity is not a result of the physical scenario, but arises because our 2D time-dependent results (r, z, t) bear a mathematical resemblance to the 1D fluid analysis undertaken in Ref.[18]. In our case and in Ref.[18], the analysis is begun with the wave, or steady state, assumption – all time dependence enters into the problem through the variable $\tau = t - z/c$. However, our analysis is explicitly 2D, in (r, τ), while in the 1D analysis laser-plasma analysis, the system reduces to simply τ. In our derivation, we reduce Eq. 7 to Eq. 11 by looking only in an infinitesimal region in around the beam, and only the radial coordinate remains as a variable. By integrating over the δ-function in τ, we finally obtain a 1D equation,but it is not in τ, it is in r. The coincidental similarities between our Eqs. 17-22 and the analogous expressions in Ref.[18] arise because in our case, during the passage of the beam, the effect of the δ-function integration on the equation of continuity enters purely through v_z and not v_r. Thus the powerful relation $\tilde{n} = (1 - \tilde{v_z})^{-1}$ is valid in both cases. In addition, because in both our case and the 1-D laser-plasma case, the phase velocity of the excitation is relativistic, the relationship $\tilde{p}_z = \frac{1}{2}\tilde{p_r}^2$ holds, and several of the relations listed in Sprangle, *et al.*, reduce to ours – assuming that the transverse momenta discussed in both cases are equivalent, which they are not, as we discuss below. These factors, and these along, are all there is to the formal similarity of our results to those of Ref.[18].

Beyond these similarities, the present analysis and that of Sprangle, *et al.*, are completely divergent, as a cursory examination indicates. The analysis we have presented is of an electron beam excitation of a plasma with the radial variation of the induced longitudinal field that arises from plasma currents driven by the beam. Sprangle's analysis, on the other hand, concerns the plasma response to a transverse wave – a laser beam, with only longitudinal variation, and which has no free charge. This is a critical point, because in Ref.[18] the transverse component of the vector potential $\vec{A}$ is introduced, whereas it is well-known that in beam-driven wake-fields, one must solve for the longitudinal component of the vector potential A_z, as is done in Refs.[1,19]. Further, we note that the equations governing the potentials in Ref.[18] and the integrated field in our case are completely different, as one would expect because they are completely different physical scenarios. One may not, except by error, deduce the present results from those of Ref.[18].

As a way of further investigating the nonlinear response of the plasma to the beam charge, we now examine the energy content of the excitation left in the beam's wake. The energy per unit length which must be supplied by the beam is found found by integrating the differential energy density in the plasma motion and field just behind the beam;

$$\begin{aligned} \frac{d\tilde{U}}{d\tilde{z}} &= 2\pi \int_0^\infty \left[\sqrt{1+\tilde{p}_r^2+\tilde{p}_z^2}-1\right]\tilde{n}(1-\tilde{v}_z)\tilde{r}d\tilde{r} + 2\pi\int_0^\infty \tfrac{1}{2}\tilde{E}_z^2(\tilde{r})\tilde{r}d\tilde{r} \\ &= \pi\left[\int_0^\infty H^2\tilde{r}d\tilde{r} + \int_0^\infty \tilde{E}_z^2(\tilde{r})\tilde{r}d\tilde{r}\right] \end{aligned} \tag{23}$$

Several comments are in order at this point: the first is that just behind the beam $\tilde{H}_\phi = \tilde{E}_r = 0$, so there is no contribution to the field energy density from these field components. The second is that in order to find the differential spatial rate of mechanical energy deposition by the beam, one must take into account that the plasma electrons may be traveling in z, which means that the normalized mechanical energy density $\tilde{n}(\gamma-1)$ must be multiplied by $(1-\tilde{v}_z)$, as shown in in the first integral above. This factor, which is may be understood by analogy with a familiar effect found in the study of electromagnetic wake-fields in accelerators (although not as familiar as it perhaps should be to all, see the discussion in Ref.[17]) removes the dependence of the energy loss on powers $\tilde{Q}$ of larger than 2. Evaluation of the integrals given above yields

$$\begin{aligned} \frac{d\tilde{U}}{d\tilde{z}} &\simeq \frac{\tilde{Q}^2}{2\pi\tilde{a}^2}\int_a^\infty \left[K_0^2(\tilde{r})+K_1^2(\tilde{r})\right]\tilde{r}d\tilde{r} \\ &= \frac{\tilde{Q}^2}{2\pi\tilde{a}^2}\left[1-\tilde{a}K_1(\tilde{a})I_0(\tilde{a})\right] \simeq \tfrac{1}{2}\tilde{Q}\tilde{E}_z(\tilde{r})\Big|_{\tau=\epsilon+} \end{aligned} \tag{24}$$

as expected. The factor of one-half is also familiar from the study of both plasma[2] and electromagnetic wake-fields – it arises from the averaging of the force over the bunch (zero at the front, maximum at the back), and taking the limit as the bunch length goes to zero.

As the results of Eqs. 17-20 concern beams of negligible length, they are applicable, in the limit $\tilde{a} \to 0$, to the case of a single particle. The effects of nonlinear plasma electron response do not, as might have been hoped, remove the logarithmic divergence seen in Eq. 14. Note that the logarithmic term in Eq. 15 corresponds to the familiar Coulomb logarithm[10], with an argument that is the ratio of the maximum to minimum impact parameter b, $\ln(b_{\max}/b_{\min})$. We deduce that the upper limit $b_{\max} = 2/k_p$, while the lower limit in the analysis is a. The value of a in Eqs. 13-15 cannot be drawn towards zero without violating several assumptions of our analysis, however. The fluid assumption is fine; modeling the plasma electrons as a continuous fluid introduces errors not in the average energy loss, but in the fluctuations of this quantity. For ultra-relativistic particles, quantum mechanical effects constrain the minimum impact parameter[10] to $b_{\min} \simeq (\hbar/m_e c)\sqrt{2/\gamma}$ through the uncertainty principle, however. Thus we write the energy loss rate for a point particle of charge q as

$$\frac{dU}{dz} \simeq q^2 k_p^2 \ln\left(0.794\sqrt{\gamma}\frac{m_e c}{k_p \hbar}\right) \simeq q^2 k_p^2 \ln\left(5\sqrt{\gamma}\frac{\lambda_p}{\lambda_c}\right) \tag{25}$$

where $\lambda_p = 2\pi/k_p$ and λ_c are the plasma and Compton wavelengths, respectively. Note that both limits in the Coulomb logarithm can be viewed quantum mechanically, as the minimum quantum of energy loss (emission of a plasmon) in the plasma is in fact $\hbar\omega_p$, as has been verified experimentally for very thin foils[20].

5. Numerical integration of the axisymmetric fluid equations

It is important to validate the infinitesimal length analysis of the previous sections, which by allowing an exact solution of the energy loss problem gives insight into the microscopic processes which are present in the nonlinear PWFA. Our analysis has been checked with numerical integrations of the two dimensional fluid equations for finite length beams, having a longitudinal charge distribution, $\rho_b \sim \exp(-z^2/2\sigma_z^2)$, and exploring the limit that $k_p\sigma_z \to 0$. In order to connect with the point beam limit, and to accurately quantify the energy imparted to the plasma, we compare the average on-axis beam energy loss rate, $(2\pi\sigma_z)^{-1}\int eE_z(0,\zeta)\exp(\zeta^2/2\sigma_z^2)d\zeta$ (where $\zeta = -c\tau$) for these cases with linear theory. The predictions of linear theory are obtained from using Eq. 14 to give the Green function (δ-function response), and performing a convolution integral[2] of the over the Gaussian pulse, to give an average energy loss rate of $\frac{\tilde{Q}}{2\pi\tilde{a}^2}[1-\tilde{a}K_1(\tilde{a})]\exp(-k_p^2\sigma_z^2)$.

The results of these simulations are shown in Fig. 1, which displays the average energy loss of a beam in the linear regime ($\tilde{Q} = 0.002$), a comparison to linear analytical theory, and the nonlinear regime ($\tilde{Q} = 2$). In the $\tilde{Q} = 0.002$ case, the fluid simulations agree extremely well with analytical predictions. For the case with $\tilde{Q} = 2$, the simulations disagree with linear theory over a broad range of pulse lengths, but converge to the linear theory in the limit that , as expected from the conclusions we have drawn from Eqs. 13 and 19-22. Note that the numerical integration of the fluid equations is not easily stabilized when $\tilde{Q} > 2$, and thus to perform further numerical investigations another tool must be adopted. The investigations using such a tool, electromagnetic particle-in-cell (PIC) simulation, are discussed in the companion work to this paper[9]. We note their conceptual importance is found in that the results of fluid integrations rely on a model that is simply connected to our analytical results (they are only a check on the validity of the derivation), while the PIC simulations are an entirely different, more complete model of the beam-plasma interaction. The PIC codes thus give a check on the physics of our analysis results, not solely the mathematics.

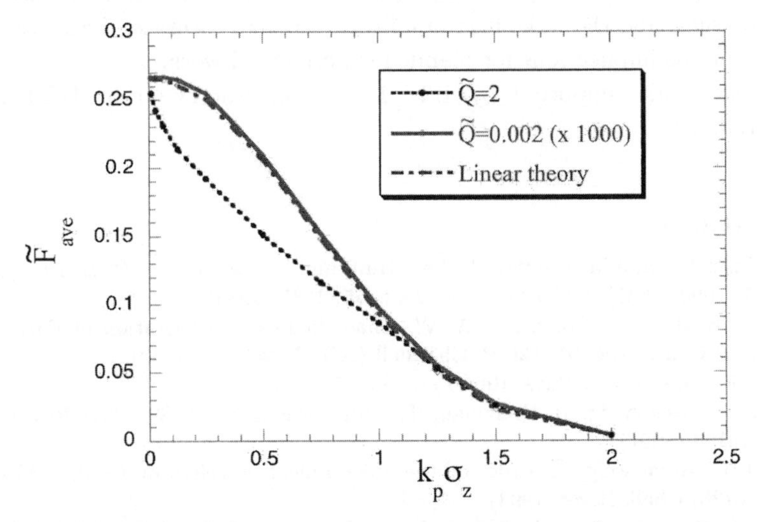

Figure 1. The average normalized energy loss rate $\tilde{F}_{dec} = e\bar{E}_z/m_e c\omega_p$ of an electron beam with $k_p a = 0.2$, as a function of $k_p\sigma_z$, for $\tilde{Q} = 0.002$ (diamonds, solid line) and $\tilde{Q} = 2$ (circles, dotted line) from cylindrically symmetric fluid simulation, and linear theory (diamonds, dashed line).

6. Conclusions

In conclusion, we restate the most surprising of our results, that the fully relativistic response of a plasma to the passage of an ultra-short beam gives an induced electric field that is identical to the linear result. We have in the process identified a single parameter, the normalized charge $\tilde{Q}$, which identifies when a bunched beam may be expected to give rise to nonlinear motion in the plasma. Further, this parameter may be used to predict the maximum physically achievable energy loss of a beam in plasma, Eq. 15. The interplay between the nonlinear effects - relativistic saturation of transverse velocity, and snow-plowing of plasma density - which cancel for infinitesimal, but not for finite, length beams must be studied in more detail by simulation. Such a study is now actively under way[9].

In addition to the verification of the relevant physical processes at play in the plasma response in finite-length beams, PIC simulations in the companion paper delve more deeply into the question of the scaling of both decelerating and accelerating wake-fields with charge ($\tilde{Q}$) in the nonlinear limit. In agreement with the results of the present analysis, the beam energy loss found in simulation scales nearly linearly with charge until $\tilde{Q}$ well in excess of unity. At high enough $\tilde{Q}$ notable deviations from linear scaling are present, and the wake-field amplitude tends to saturate. These results have serious implications for planned experimental work.

This work supported by U.S. Dept. of Energy grant DE-FG03-92ER40693.

References

1. Pisin Chen, J.M. Dawson, Robert Huff, and T. Katsouleas, *Phys. Rev. Lett.* **54**, 693 (1985), and *Phys. Rev. Lett.* **55**, 1537 (1985).
2. R. Ruth, R. D. Ruth and A. W. Chao, in Laser Acceleration of Particles. AIP Conf. Proc. 91, Ed. P. Channell (AIP, New York, 1982).
3. J.B. Rosenzweig, Phys. Rev. Lett. 58, 555 (1987).
4. J.B. Rosenzweig, B. Breizman, T. Katsouleas and J.J. Su, Phys.Rev.A 44, R6189 (1991).
5. J.B. Rosenzweig, in Proc. of the 1992 Linear Accelerator Conf., (AECL-10728, Chalk River, 1993).
6. J.B. Rosenzweig, et al., Nuclear Instruments and Methods A **410** 532 (1998).
7. N. Barov, J.B. Rosenzweig, M.E. Conde, W. Gai, and J.G. Power, *Phys. Rev. Special Topics - Accel. Beams* **3** 011301 (2000).
8. S. Lee, T. Katsouleas, R. Hemker and W. Mori, *Phys. Rev. E* **61**, 7012 (2000).
9. J.B. Rosenzweig, M.C. Thompson, R. Yoder and N. Barov, “Energy Loss of a

High Charge Bunched Electron Beam in Plasma: Simulations, Acceleration and Linear Scaling", these proceedings.

10. J.D. Jackson, *Classical Electrodynamics*, 2nd Ed, (Wiley, New York, 1975).
11. N. Barov, *et al., Proc. 2001 Part. Accel. Conf.* Ed. P. Lucas and S. Webber (IEEE, 2002) 126.
12. M. Hogan, *et al.*, Phys. Plasmas **7**, 224 (2000).
13. S. Lee, et al., *Phys. Rev. Special Topics - Accel. Beams* **4** 011011 (2002).
14. D. Bruhwiler, *et al., Proc. 2000 European Part. Accel. Conf.*, 877 (Austrian Acad. Sci. Press, 2000).
15. N. Barov, J.B. Rosenzweig, M.C. Thompson, submitted to *Physical Review Letters.*
16. L.D.Landau and E.M. Lifschitz, *The Classical Theory of Fields*, 4th ed., p. 58 Pergamon Press, 1975.
17. E. Chojnacki, *et al,. Proceedings of the 1993 IEEE Particle Accelerator Conference*, 2, 815 (Ed. S.T. Corneliussen, IEEE, New York, 1993)
18. P. Sprangle, E. Esarey, and A. Ting, *Phys. Rev. A* **41**, 4463 (1990).
19. P. Chen, Particle Accelerators, 20 171 (1985).
20. H. Raether, *Springer Tracts in Modern Physics*, **38**, Ed. G. Hohler (Springer-Verlag, Berlin, 1965), pp. 84-157
21. D. Smithe, *et al.*, Comp. Phys. Commun. **78**, 54 (1995).

PLASMA WAKEFIELD EXPERIMENTS WITH 28.5 GeV ELECTRON AND POSITRON BEAMS

P. MUGGLI[*]
[*]*University of Southern California*
University Park Campus
Los Angeles, CA 90089, USA
E-mail: muggli@usc.edu

B.E. BLUE[Δ], C.E. CLAYTON[Δ], F.-J. DECKER[†], S. DENG, E.S. DODD[Δ], M.J. HOGAN[†], C. HUANG[Δ], R. IVERSON[†], C. JOSHI[Δ], T.C. KATSOULEAS[*], S. LEE[*], K.A. MARSH[Δ], W.B. MORI[Δ], C.L. O'CONNELL[†], P. RAIMONDI[†], R. SIEMANN[†], D. WALZ[†], S. WANG[Δ]

[Δ]*University of California*
405 Hilgard Avenue
Los Angeles, CA 90095, USA

[†]*Stanford Linear Accelerator Center*
2575 Sand Hill Road
Stanford, CA 94309, USA

Plasma wakefield experiments are performed with the 28.5 GeV electron and positron beams of the Stanford Linear Accelerator Center. The lithium plasma is 1.4 m long and a density in the 0-2×10^{14} cm^{-3}. Experimental results are presented which include the focusing of the electron and positron beams, the emission of x-ray radiation by the beam electrons resulting from their betatron motion in the plasma, the refraction of the electron beam at the plasma boundary, and the energy loss and gain by electrons and positrons. Finally, a brief look is given at the future of these experiments.

1. Introduction

High brightness beams are produced in conventional accelerators where accelerating cavities are driven by high power microwave sources. In these radio-frequency structures peak fields may exceed 100 MV/m, however in present high-energy accelerators average accelerating gradients are in the 20 MeV/m range. For example the accelerator at the Stanford Linear Accelerator Center is ≈3 km long for a maximum energy of ≈50 GeV.

Plasmas can support relativistic plasma waves with very large amplitudes. The amplitude of these waves can be estimated by using Gauss' law in the case of full charge separation over a plasma wavelength $\lambda_p=2\pi c/\omega_p=2\pi c/(n_e e^2/\varepsilon_0 m_e)^{1/2}$:

$$E_{ES}=\left(\frac{m_e c^2 n_e}{\varepsilon_0}\right)^{1/2} \tag{1}$$

which corresponds to ≈1 GV/m in a plasma of density $n_e=10^{14}$ cm^{-3}. At this density the plasma frequency $\omega_p/2\pi$ is ≈100 GHz. Short, relativistic particle bunches available from present accelerators can drive these large amplitude plasma waves. In this scheme called the plasma wake field accelerator (PWFA) [1], the space charge field of the particle bunch displaces the plasma electron away, in the case of an electron bunch, or toward the bunch volume in the case of a positron bunch. The net plasma charge left behind the head of the particle bunch partially neutralizes the bunch, which is therefore focused. As the plasma electrons cross the beam axis, after overshooting in the case of an electron bunch, they create an excess of charge which accelerates the particles in the back of the same bunch, or in a witness bunch following a driver bunch. Note that the particles in the head and core of the bunch loose energy as they work on the plasma to displace the electrons. The PWFA is therefore an energy transformer where the energy if transferred from the head of the bunch, or a driver bunch, to the back of the same bunch, or a witness bunch. In the linear theory of the PWFA describing the case of a bunch with N particles and a density $n_b=N/(2\pi)^{3/2}\sigma_r^2\sigma_z$ smaller than the homogeneous electron plasma density n_e, the peak amplitude of the accelerating field is given by [2]:

$$E_z[V/cm]=\sqrt{n_e}\,\frac{n_b}{n_e}\,\frac{\sqrt{2\pi}\,k_p\sigma_z e^{-k_p^2\sigma_z^2/2}}{1+1/k_p^2\sigma_r^2} \tag{2}$$

where $k_p=\omega_p/c=(n_e e^2/\varepsilon_0 m_e)^{1/2}/c$ is the wave number of the relativistic electron plasma wave, and σ_r and σ_z are the radial and transverse and longitudinal sizes of a bi-Gaussian bunch. For the case of $k_p\sigma_r$<<1, this field is maximum a plasma density such that $k_p\sigma_z\approx\sqrt{2}$, and has the value [2]:

$$E_z \cong 88MV/m\left(\frac{N}{2\times10^{10}}\right)\left(\frac{700\mu m}{\sigma_z}\right)^2 \quad (3)$$

where N and σ_z (in μm) are referred to the typical parameters (Table 1) for the experiment described here after. Even though these experiments are conducted in the nonlinear regime of the PWFA in order to achieve high gradient acceleration, the scaling of the linear theory can be used as a guideline. Note that both electrons and positrons can be accelerated by the PWFA. In the nonlinear case where $n_b > n_e$, the field of an electron bunch expels all the plasma electron from the bunch volume leaving behind the bunch head a pure ion column and reaching the blow-out regime. The ion column with $n_i = n_e < n_b$ partially neutralizes the bunch and the corresponding focusing field and strength are given by:

$$E_r[V/m] = \frac{1}{2}\frac{n_e e}{\varepsilon_0} r, \quad \frac{B}{r}[T/m] = \frac{1}{2}\frac{n_e e}{\varepsilon_0 c} \quad (4)$$

The focusing field is linear with r and the ion column acts on the core of the bunch as an aberration-free plasma focusing element. Note that for $n_e = 10^{14}$ cm^{-3} the focusing strength is ≈3 kT/m and $E_r(r=\sigma_r=25\ \mu m) \approx 2.3\times10^7$ V/m.

In the case of a positron bunch the plasma electrons are continuously attracted towards the bunch volume and there is no blow-out regime. As a result the focusing force varies with r and z over the bunch volume, which leads to focusing aberrations and emittance dilution of the beam [3].

Preliminary numerical simulations [4] show that the energy of the 50 GeV electron and positron beams of the Stanford Linear Accelerator Center (SLAC) collider could be doubled right before the interaction point in ≈30 m long sections of plasma in an "after-burner" PWFA concept. In this concept plasma lenses are used to compensate for the charge lost in the driver bunch.

An experiment using single electron or positron bunches was set up at SLAC to study the issues related to a meter-long PWFA module in the context of an actual high-energy collider. These issues include: the creation of meter-long, homogeneous plasmas with densities $>10^{14}$ cm^{-3}; the stable propagation of electron and positron beams in meter-long, high-density plasmas; the emission of radiation by the bunch particles; and the demonstration of large energy gain by particles in a high-gradient PWFA modules. Experimental results obtained so far with relatively long, single bunches ($\sigma_z \approx 700\ \mu$m) are presented here. A brief look is given at future experiments with shorter bunches ($\sigma_z \approx 100\ \mu$m->12 μm) for which an energy gain >1 GeV over

<1 m is expected. Plasma Wakefield experiments with low energy electron beams are also performed in other laboratories [5].

2. Experimental Set Up

The experimental set up is shown on Fig. 1. The 1.4 m long plasma is created by single-photon, laser photo-ionization of a lithium vapor contained in a heat-pipe oven [6]. The ionizing ultra violet (uv) laser pulse, and thus the plasma, is made collinear with the particle beam by reflection off a 250 μm thick fused silica pellicle. The plasma density n_e is obtained from uv energy absorption and from the variations of the beam spot size as a function of the uv energy incident upon the lithium vapor (see Sect. 3.1.1). Typical plasma parameters are given in Table 1. The 28.5 GeV electron or positron beam from the FFTB line [7] with the typical parameters given in Table 1 is sent into the plasma. The beam transverse size (σ_x, σ_y) is monitored by imaging the optical transition radiation (OTR) emitted by the particles when traversing thin (12.5 μm at 45°) titanium foils located approximately one meter from the plasma entrance and exit. After exiting the plasma the beam is transported through a set of quadrupole and dipole magnets arranged in a imaging spectrometer configuration. The object plane of the spectrometer can be set at the plasma entrance or at the plasma exit. A thin piece of aerogel is placed in the image plane of the spectrometer ≈25 m downstream from the plasma exit. The magnification of the spectrometer M is ≈3 in both x-, and y-plane. The beam is dispersed in energy in the vertical y-plane with a dispersion η=10 cm at the aerogel location. The visible Cerenkov radiation emitted by the particles in the aerogel is split and a fraction is imaged onto a CCD camera to obtain a time-integrated image of the beam, with the beam size in the non-dispersive x-plane, and the beam energy spectrum in the dispersive y-plane. Another fraction of the Cerenkov light is split again, delayed, and imaged onto the slit of a streak camera to obtain the time-dispersed beam size $\sigma_x(t)$ and beam energy $E(t)$. The temporal resolution of the streak camera is ≥1 ps. The spatial resolution of imaging system is <100 μm. For plasma densities larger than $\approx 0.5\times10^{14}$ cm^{-3} the beam size at the plasma exit is smaller or equal to that at the plasma exit, and the energy resolution is thus given by $\Delta E \leq M\sigma_{y,ex}E/\eta<28$ MeV, where $\sigma_{y,ex}$ is the beam size at the plasma exit. The imaging property of the magnetic spectrometer is essential for the unambiguous measurement of the particles energy change resulting from the PWFA action. It removes the contribution to the beam size and apparent energy resulting from the transverse momentum of the incoming beam and imparted to a non-perfectly cylindrical-symmetric beam by the strong focusing plasma forces. Note that with the parameters of Table 1 the experiment is performed in the nonlinear blow-out regime in which $n_b>n_e$, and $k_p\sigma_{x,y}<<1$.

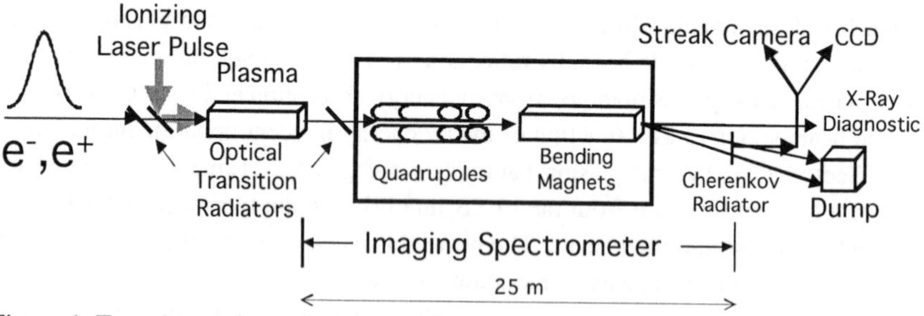

Figure 1: Experimental set up, not to scale.

Table 1: typical plasma and beam parameters.

	Symbol	Value
Plasma Length	L	1.4 m
Plasma Density	n_e	0-2×10^{14} cm^{-3}
Beam Energy	E	28.5 GeV
Beam Relativistic Factor	γ	55686
Number of Particles per Bunch	N	1-2×10^{10}
Bunch Length	σ_z	700 μm
Bunch Transverse Dimension	σ_x, σ_y	25-70 μm
Average Current	I_b	1.3 kA (for N=2×10^{10})
Average Beam Density	n_b	>1.9×10^{-14} cm^{-3} (for N=1×10^{10}, σ_x=σ_y=70 μm)
Normalized Transverse Emittance*	ε_{Nx}, ε_{Ny}	5×10^{-5}, 0.5×10^{-5} m-rad

* Before scattering through the various pellicles and foils used in the experiment.

3. Experimental Results

3.1 Electrons

3.1.1 Electron Beam Focusing

The incoming beam is focused near the plasma entrance and in absence of plasma diverges there after because of its finite emittance. As the plasma density is increased from zero, the beam size first decreases at the downstream OTR location, x1 m downstream from the plasma exit. The plasma acts as an extending focusing element or plasma lens, and a minimum size is reached at $n_e \approx 3\times10^{-12}$ cm^{-3} (see Fig. 2). As n_e is further increased, the focusing force becomes so strong that the beam waist moves into the plasma. The beam size increases on Fig. 2, and oscillates as a function of n_e as more foci move into the plasma. At the highest plasma density the beam has three foci within the plasma, i. e., the beam envelope experiences three betatron oscillations over the plasma length. At this density $n_e \approx 1.8\times10^{14}$ cm^{-3} the focusing strength is ≈5.4 kT/m. The beam size as a function of n_e is obtained from a simple envelope equation model [8] in which the plasma is modeled as an extended focusing element with a restoring term given by $K=(eE_r/\gamma mc^2)^{1/2}=\omega_p/(2\gamma)^{1/2}c$ [9]. Figure 2 shows that the measured beam size follows that model for beam parameters including the increase of beam emittance resulting from scattering through the various pellicles. It also provides a measurement of the plasma density at the time when the beam traverses the plasma. A single emittance value is used in the model which shows that the beam suffers no measurable emittance dilution from the plasma focusing. This is expected in the blow-out regime. Note that, in the case of the left hand side plot of Fig. 2, the beam beta function at the plasma entrance $\beta_i=\gamma\sigma_i^2/\varepsilon_N \approx 1.16$ m is comparable to the plasma length. The minimum beam size at the downstream OTR location is ≈50 μm, ≈2.5 times smaller than without plasma, but approximately equal to the spot size before the plasma. However, the minimum spot size within the plasma can be calculated from the envelope equation for the case of the beam focused at the plasma entrance as:

$$\sigma_{extremum} = \left(\frac{1}{K\beta_i}\right)\sigma_i \qquad (5)$$

For the parameters of Fig. 2 a ($K>1/\beta_i$) the minimum spot size in the plasma given by Eq. 5 is ≈7 μm at $n_e=1.8\times10^{14}$ cm^{-3}. Such small spot sizes could be measured at distance $<\lambda_\beta/4$ downstream from the plasma exit, where $\lambda_\beta=\omega_p/(2\gamma)^{1/2}c$ is the particles betatron wavelength. Figure 2 also shows that the beam propagation is stable over 1.4 m and up to a density of $\approx 2\times10^{14}$ cm^{-3}. In particular, no sign of the beam disrupting hose instability [10] is observed in the experiment. The propagation of a bunch creating its own channel in a homogeneous plasma is more stable that that of a bunch in a preformed plasma channel [11]. Single images of the unfocused and focused beam are shown on Fig. 4.

Figure 2 b show the beam size as a function of n_e for the case of a beam with $\beta\approx0.1$ m$<<L$ and for beam parameters closer to the condition for matching to the plasma ($\beta=1/K$). As a result the σ_x does not appear to grow at the downstream OTR location. Matching of the beam to the plasma is desirable for long-, dense plasma in which small variation of the plasma density from bunch to bunch could lead to large variation of the beam size as a result of the large number of betatron experienced by the beam. In the plasma density range of Fig. 2 b $K>1/\beta$ and the beam size at the plasma exit is smaller or equal to that at the plasma entrance. The beam is therefore channeled over more that 13 beam beta functions.

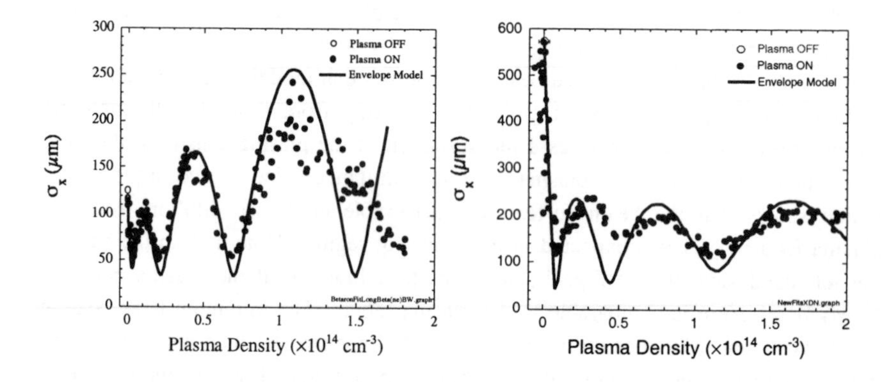

Figure 2: Beam transverse size σ_x measured $\approx$1 m downstream from the plasma as a function of the plasma density for a long (left hand side), and a short (right hand side) β electron beam. The solid curves are the result of a simple beam envelope model for the focusing of the beam by the plasma. The beam parameters at the plasma entrance for the envelope model are: a) σ_{xi}=39 μm, $\varepsilon_{Nx}=8\times10^{-5}$ m-rad, α_{xi}=-0.5, so that β_{xi}=1.06 m, and b) σ_{xi}=24 μm, $\varepsilon_{Nx}=12\times10^{-5}$ m-rad, α_{xi}=0.8, so that β_{xi}=0.11 m.

In the experiment the beam enters the plasma with a correlated energy spread (Fig. 3) resulting from the wakefields along the accelerating linac. The correlated energy spread is linear over ≈5 ps near the bunch center and is ≈86 MeV/ps. After dispersion in energy at the aerogel location the beam is therefore "stood up", allowing for the observation of the dynamics of the beam focusing ($\sigma_x(E)$ or $\sigma_x(t)$) from single time integrated images of the beam [12]. These measurements show that the focusing force increases from the beam head, and reaches a constant value in the bunch core, demonstrating the access to the blow-out regime. Correspondingly, images of the beam at high plasma density show that different z or t beam slices experience a different number of betatron oscillations over the plasma length.

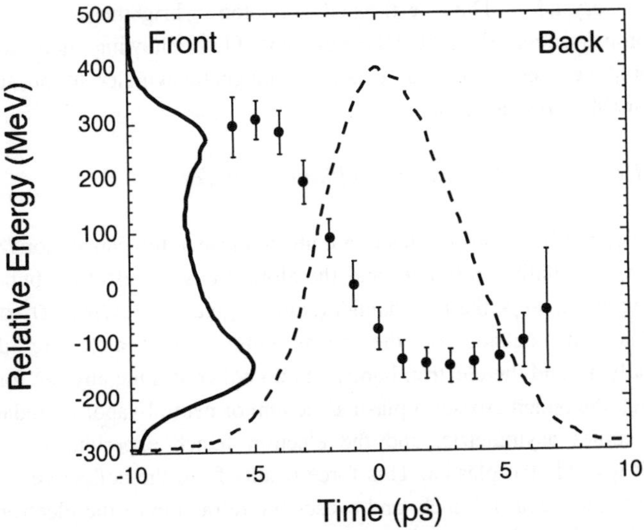

Figure 3: Measured incoming beam energy (solid circles, with standard deviation as error bars), beam charge profile (doted line, in arbitrary units), and energy spectrum obtained by summation over time (solid line). The beam has a correlated energy spread with the head at higher energy and a linear chirp of 86 MeV/ps over ≈5 ps.

3.1.2 *Emission of X-Ray Betatron Radiation*

As the beam is focused, the beam particles experience betatron oscillation and radiate at a frequencies given by:

$$\omega_r = \frac{2m_h\omega_\beta}{1+K_0^2+(\gamma\Omega)^2} \quad (6)$$

where m_h=1,2,... is the harmonic number, $K_0=\gamma\omega_\beta r_0/c$ is the wiggler strength, $\omega_\beta=\omega_p/(2\gamma)^{1/2}$ is the particles betatron frequency, and $\Omega<<1$ is the observation angle measured from the beam axis. Since K_0 depends on the particle initial radius r_0 in the bunch the radiation is emitted in a broadband spectrum. For the parameters of this experiment (Table 1), the radiation spectrum is in the x-ray range, between 6 and 30 keV. The radiation is emitted in a narrow cone with a half angle of $\approx K/\gamma$, and the angle measured ≈10 m from the plasma is 1-3 mrad. The number of photons measured at 14.2±0.002 keV is $\approx 6\times10^5$, in good agreement with the number of estimated photons 30×10^5 obtained using a unity reflection coefficient for the Bragg diffracting crystal. The estimated photon brightness is therefore 7×10^{18} photons/(sec·mrad²·mm²·0.1%bandwidth) [13], showing that sections of plasma could be used instead of expensive magnetic wiggler to produce bright bursts of tunable x-ray radiation.

3.1.3 *Electron Beam Refraction at a Plasma Boundary*

When the electron beam propagates in a uniform plasma, the plasma ions within the bunch ($r<\sigma_{x,y}$) partially neutralize, and therefore focus the electron bunch. In the blow-out regime ($n_b>n_e$), the ion channel radius is given by $r_c=(N/(2\pi)^{3/2}\sigma_z n_e)^{1/2}$ [2]. The plasma ions outside of the beam radius ($\sigma_{x,y}<r<r_c$), are distributed symmetrically around the electron bunch and do not contribute any net force on the bunch. When the bunch crosses a plasma/vacuum or neutral vapor boundary the ion channel becomes asymmetric, and the electron bunch experiences a net force attracting it towards the plasma. This force results from the collective response of the plasma to the electron bunch, and causes the refraction of the electron beam at the plasma boundary, similar to the refraction of a light beam crossing the interface between two dielectric media. However, the beam creates its own ion channel, and the particles in the bunch head are not refracted at the boundary. The beam head and core are therefore split when crossing the plasma boundary. The head/core splitting and the beam refraction were observed experimentally [14] and are in good agreement with a simple impulse model and with full tri-dimensional numerical simulations for the refraction process.

3.1.4 Energy Loss and Gain by Electrons

Numerical simulations performed with the experimental parameters indicate that a peak energy loss of ≈66 MeV/m can be expected in the core of the bunch. The highest energy gain averaged over the particles in an 1 ps beam slice, a time comparable to the time resolution in the experiment, is ≈260 MeV, while the highest gain by a simulation particle is ≈334 MeV, both after propagation through the 1.4 m long plasma at a density of 1.5×10^{14} cm^{-3}. These energy gains are smaller than the incoming bunch correlated energy spread (Fig. 3), and the Cerenkov light emitted by the beam in the aerogel needs to be dispersed in time by a streak camera to observe them. The streak camera images are analyzed by dividing the image in ≈1 ps slices and by fitting a Gaussian profile to each slice to obtain its average energy. Such an analysis reveals energy loss and gain by ≈150 MeV over the ≈1.4 m-long plasma with $n_e \approx 1.7\times10^{14}$ cm^{-3}. Detailed analysis of the images further reveals the gain of >300 MeV by bunch particles. These results represent the first demonstration of high energy gain by particles in a high-gradient PWFA mode and will be published soon [15].

3.2 Positrons

When traveling in a plasma a positron bunch attracts the plasma electrons toward the beam axis, in contrast with an electron bunch that expels the plasma electrons from the beam volume. As a result there is no equivalent to the blow-out regime for positrons, and the density of the neutralizing plasma electrons within the bunch volume varies bot along r and z with a large compression on axis. Numerical simulations show that depending on the beam parameters, focusing of the positron beam can be achieved downstream from the plasma. Previous experiments have shown evidence of focusing of positrons by a short, high-density plasma [16]. In the linear theory for the PWFA the amplitude of the wake driven by a positron or an electron bunch are identical. However, numerical simulations in the non-linear regime show that the wake of a positron beam is smaller that that of an electron beam [3].

3.2.1 Positron Beam Focusing

Experimental studies of the propagation of a positron in a long, low density plasma ($k_p\sigma_z<<\sqrt{2}$) show clear focusing of the beam at a distance of one meter downstream from the plasma exit. Time dispersed images of the beam at a distance of 10 meter downstream from the plasma and without the imaging spectrometer show the focusing dynamics in single bunches. At very low density the back of the beam is first focused, and as the density is increased the core of the bunch is focused. Detailed results will be published elsewhere [17]. At much higher densities ($k_p\sigma_z\approx\sqrt{2}$), the focusing of the positron beam depends on the beam parameters. A sample result is shown on Fig. 4 for a beam focused to a round spot of $\sigma_x\approx\sigma_y\approx 25$ μm in radius at the plasma entrance. Whereas the electron beam is focused to a small spot in both directions, the positron beam shows focusing in the horizontal x-plane by a factor of ≈ 3, and no focusing in the vertical y-plane. In both planes a halo is created which is smaller or larger than the initial spot size in the respective planes. Note that the emittance of the incoming beam is different in both planes. Ongoing data analysis is looking at the beam emittance dilution created by the r- and z-varying plasma focusing force as a function of the beam emittance, size and charge, as well as plasma density. This will be the topic of a later publication.

3.2 Energy Loss and Gain by Positron Bunches

In the nonlinear regime, but for identical beam and plasma parameters, the amplitude of the wake driven by a positron beam is smaller than that driven by an electron beam [3]. However it may be possible to increase the amplitude of the positron wake by using a hollow plasma channel. The amplitude of the wake driven by a particle bunch is proportional to the number of particles in the bunch or the bunch charge. Although lowering the number of particles per bunch decreases the wake amplitude, lowering the bunch charge reduces the effect of the wakefields along the 3 km long linac. In particular, the beam correlated energy is reduced below the measurement resolution with $N=10^{10}$ positrons per bunch, and the bunch propagation through the plasma has less shot-to-shot variations, which makes the time-dispersed beam energy measurement much easier. The measured energy loss and gain are in good agreement with numerical simulations and are the subject of a future publication [18].

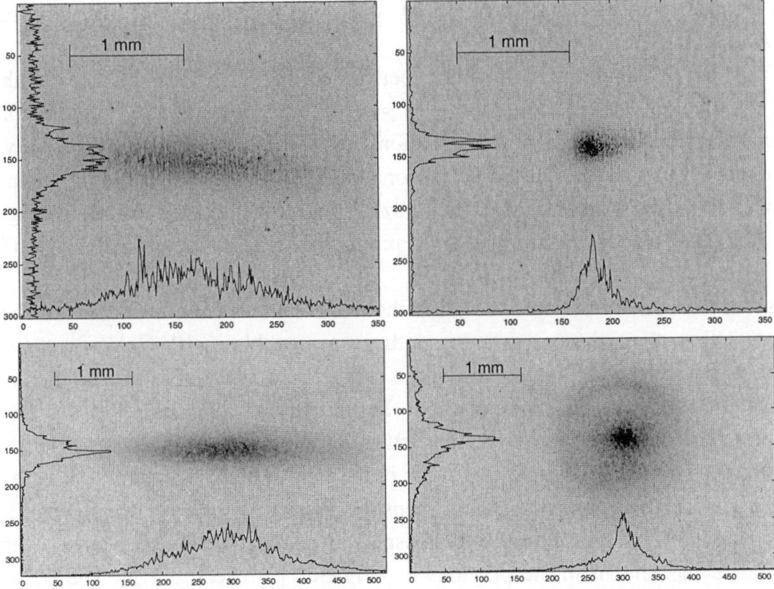

Figure 4: OTR images of the electron (top images) and positron (bottom images) beam approximately 1 m downstream of the plasma, without plasma (left hand side images), and with $n_e \approx 10^{14}$ cm^{-3} (right hand side). The beams are focused to a round spot at the plasma entrance. The beam without the plasma is elongated because the beam emittances are different in the horizontal and vertical plane. The electron beam is focused in both planes. The positron beam is focused in the horizontal plane only, and a halo is created in both planes.

4. Future Experiments

Previous experiments were performed with relatively long particle bunches: σ_z=700 μm. The linear theory for the electron PWFA predicts that the accelerated gradient scales as $1/\sigma_z^2$. Although this scaling may not be rigorously applicable to the nonlinear regime reached with the experimental parameters, numerical simulations show that the wake amplitude increases drastically with shorter bunches.

Ultra-short electron bunches with $\sigma_z \approx 12\ \mu m$ will soon be available at the Stanford Linear Accelerator Center [19].

A first set of experiments will be performed in 2003 with bunches with $\sigma_z \approx 100\ \mu m$ for which a plasma density of $\approx 6\times10^{15}$ cm^{-3} ($k_p\sigma_z \approx \sqrt{2}$) is required for maximum acceleration. Numerical simulations show that an energy gain >1 GeV can be expected in a plasma 30 cm long, corresponding to an accelerating gradient of ≈4 GeV/m. The length of the pre-ionized plasma is limited by the energy depletion of the ionization laser pulse at the required neutral density of $\approx 2\times10^{16}$ cm^{-3} [20].

A second set of experiments using the shortest electron bunch available will be proposed. Other than the very large gradient achievable (>10 GeV/m), these experiments are interesting because the space charge field of the ultra-short bunch is large enough to field-ionize a lithium vapor. Preliminary calculations show that full ionization of the lithium vapor to the first ionization level is reached over a volume large enough so that the driving of the wake in the field-ionized plasma is similar to that in the pre-ionized plasma [21]. This opens the possibility for a very large energy gain in a single PWFA module since the plasma can channel the electron beam over long distances, as shown in the previous experiments [9] (Sect. 3.1.1). The field-ionized plasma density is very reproducible form shot-to-shot since the vapor neutral density determines it. Lithium is chosen because of the very different ionization potentials for the first and second electron: 5.4 eV and 75 eV, respectively. In this scheme scattering is reduced because the amount of scattering atoms and ions is minimum at any given density and because no pellicle is required to in-couple the ionizing laser pulse, thereby also greatly simplifying the experimental set up.

5. Conclusions

The results presented here show very encouraging progresses toward a high-gradient, very high-energy gain PWFA module, and toward a PWFA after-burner. The experimental results presented here are well understood, and in agreement with models and corresponding numerical simulations. The main issues to be addressed by numerical simulations and in future experiments include: possible beam erosion associated with the propagation over very long plasmas; the preservation of the beam emittance in a plasma wake with typical dimension comparable to those of the bunch $1/k_p \approx \sigma_r \approx \sigma_z$ and in the presence of a scattering plasma background; and the loading of the wake by a short witness bunch in order to achieve a small energy spread after acceleration.

Acknowledgments

This work is supported by US DoE #DE-FG03-92ER40745, DE-AC03-76SF00515, #DE-FG03-98DP00211, #DE-FG03-92ER40727, NSF #ECS-9632735, NSF #DMS-9722121. We would like to thank Dr. Peter Tsou of JPL for providing the aerogel.

References

1. P. Chen *et al.*, *Bull Am. Phys. Soc.* **29**, 1355 (1984).
   P. Chen *et al.*, *Phys. Rev. Lett.* **54**, 693 (1984).
2. S. Lee *et al.*, *Phys. Rev. E* **61**(6), 7014 (2000).
3. S. Lee *et al.*, positron **64**(4), 045501/1 (2001).
4. S. Lee *et al.*, PRST-AB **5**(1), (2002).
5. N. Barov *et al.*, *PRST-AB* **3**, 011301 (2000).
   S. Russel and B. Carlsten, private communication.
6. C.R. Vidal and J. Cooper, *J. Appl. Phys.* **40**(8), 3370 (1960).
   P. Muggli *et al.*, *IEEE Trans. Plasma Sci.* **27**, 791 (1999).
7. V. Balakin *et al.*, *Phys. Rev. Lett.* **74**, 2479 (1995).
8. K.G. Steffen, in *High Energy Beam Optics*, John Willey, NY, 173, (1965).
9. C.E. Clayton *et al.*, *Phys. Rev. Lett.* **88**(15), 154801-1 (2002).
10. D.H. Whittum *et al.*, *Phys. Rev. Lett.* **67**, 991 (1991).
11. E.S. Dodd *et al., Phys. Rev. Lett.* **88**(12), 125001/1 (2002).
12. C. O'Connell *et al.*, submitted to *PRST-AB*, September 2002.
13. S. Wang *et al., Phys. Rev. Lett.* **88**(13), 135004-1 (2002).
14. P. Muggli *et al.*, *Nature* 411, 43 (2001).
    P. Muggli *et al.*, *PRST-AB* **4**(9), 091301 (2001).
15. P. Muggli *et al.*, in preparation.
16. J.S.T Ng *et al.*, *Phys. Rev. Lett.* **87**, 244801-1 (2001).
17. M.J. Hogan *et al.*, submitted to *Phys. Rev. Lett.*, November (2002).
18. B.E. Blue *et al.*, in preparation.
19. P. Emma *et al.*, SLAC-PUB-8850 (2001).
20. K.A. Marsh and P. Muggli, Proceedings of the 10th Advanced Accelerator Workshop, Mandalay Beach, California, 22-28 June 2002, AIP 647, 614 (2002).
21. P. Muggli *et al.*, Proceedings of the 10th Advanced Accelerator Workshop, Mandalay Beach, California, 22-28 June 2002, AIP 647, 620 (2002).

ENERGY LOSS OF A HIGH CHARGE BUNCHED ELECTRON BEAM IN PLASMA: SIMULATIONS, SCALING, AND ACCELERATING WAKE-FIELDS

J.B. ROSENZWEIG, N. BAROV[1], M.C. THOMPSON, AND R.B. YODER

UCLA Department of Physics and Astronomy,
405 Hilgard Ave., Los Angeles, CA 90095-1547

The energy loss and gain of a beam in the nonlinear, "blowout" regime of the plasma wakefield accelerator (PWFA), which features ultra-high accelerating fields, linear transverse focusing forces, and nonlinear plasma motion, has been asserted, through previous observations in simulations, to scale linearly with beam charge. In a new analysis that is the companion to this article[1], it has been shown that for an infinitesimally short beam, the energy loss is indeed predicted to scale linearly with beam charge for arbitrarily large beam charge. This scaling holds despite the onset of a relativistic, nonlinear response by the plasma, when the number of beam particles occupying a cubic plasma skin-depth exceeds that of plasma electrons within the same volume. This paper is intended to explore the deviations from linear energy loss using 2D particle-in-cell (PIC) simulations that arise in the case of finite length beams. The peak accelerating field in the plasma wave excited behind the finite-length beam is also examined, with the artifact of wave spiking adding to the apparent persistence of linear scaling of the peak field amplitude well into the nonlinear regime. At large enough normalized charge, the linear scaling of both decelerating and accelerating fields collapses, with serious consequences for plasma wave excitation efficiency. Using the results of parametic PIC studies, the implications of these results for observing the collapse of linear scaling in planned experiments are discussed.

1. Introduction

The scheme of using of electron beam-excited plasma waves, or plasma wakefields, to generate ultra-high accelerating gradients for future linear accelerators, is known as the plasma wakefield accelerator (PWFA)[2]. In the past decade, work on the PFWA has concentrated on extending the PWFA from the linear regime[1,2], where the plasma oscillations can be considered small perturbations about an equilibrium, to the highly nonlinear "blow-out" regime[3]. In the blow-out regime, the plasma response to a beam that is much denser than the ambient plasma is violently nonlinear, as the plasma electrons are ejected from the path of the intense driving electron beam, resulting in an electron-rarefied region. This region contains only (nearly stationary) ions, and thus

[1] Dept. of Physics, Northern Illinois University, DeKalb, IL 60115.

possesses linear electrostatic focusing fields that allow high quality propagation of both the driving[4,5,6] and accelerating beams. In addition, this region has superimposed upon it (TM) longitudinal electromagnetic fields, which, because the phase velocity of the wake wave is axisymmetric and nearly the speed of light, are independent of radial offset from the axis. Thus this wake may accelerate a trailing electron beam just as a traveling wave linac, with strong transverse focusing conveniently supplied by the plasma ions[3].

The beam dynamics in this scenario are termed *linear*, and are conceptually easy to understand using common tools. On the other hand, until recently, the plasma dynamics of the PWFA blow-out regime, with their extreme nonlinearity, have been only qualitatively understood, mainly through both fluid and particle-in-cell simulations[3,7,8]. The companion work[1] to this paper has made progress in moving toward an analytical understanding of the plasma response to very large beam charges.

Despite the lack of analytical models for the nonlinear plasma response, it has been noted in a variety of studies that the accelerating and decelerating fields associated with the blow-out regime obey a Cerenkov-like scaling[7,9,10,11]. This scaling[12] predicts that the fields which produce energy loss and gain in the system are proportional to the square of the characteristic maximum frequency in the system (the plasma frequency, ω_p^2). The efficient excitation of an oscillatory system by a pulse occurs when the pulse is short compared with the oscillator period, requiring the rms pulse length σ_z to obey the constraint $k_p\sigma_z \leq 2$. Thus linear scaling implies that the PWFA decelerating and (assumed proportional) acceleration fields should be related to the pulse length by $E_z \propto \sigma_z^{-2}$.

This prediction provides motivation for recent experiments that employ bunch compressors to reduce σ_z. In recent measurements with compressed beam at FNAL[13], the trailing portion of a 5 nC, 14 MeV, σ_z=1.2 mm, beam pulse was nearly stopped in 8 cm of $n_0 \cong 10^{14}$ cm^{-3} plasma, a deceleration rate of over 150 MeV/m, obtained in the PWFA blow-out regime. Despite the nonlinearity of the plasma motion in this experiment, as well as other recent measurements at SLAC, the linear scaling of wakefields with respect to charge seems to have been well upheld[14].

On the other hand, in the future, beams compressed to ever shorter lengths will be employed in experimental scenarios. For example, in the context of SLAC E-164 experiment[15] it has been proposed to use a beam that is compressed to as short as 12 μm in rms length, resulting in over 10 GeV/m acceleration gradients. This experimental test is quite important, as it is a milestone on the road to the so-called “after-burner” concept[16], in which one may use a very high energy drive beam to double the energy of a trailing beam population, *e.g* converting the SLC at SLAC from 50 GeV per beam to 100 GeV. The use of such short drive beams, and verification of scaling, could thus help realize such an ambitious goal.

The present work is concerned with two aspects of the problem of plasma wake-field acceleration and its scaling to ultra-high fields. The first is to show that the assertion of a linear relationship between the exciting beam charge and the amplitude of the excited wake-fields previously deduced from simulations does not hold when the charge is high enough. The second is to explore the implications of the approach to understanding the physics of the PWFA in the very nonlinear regime suggested by the accompanying[1] work that analyzes an interesting and illuminating limit of this regime, that of an infinitesimally short (rms bunch length $\sigma_z \to 0$) driving beam. This exploration is accomplished through particle-in-cell simulations that serve to verify the analytical results, and extend them beyond the strict applicability of the analysis.

This companion analysis paper is concerned only with the energy loss of an infinitesimally short beam. We now recapitulate the main results of this exact (in the limit that $\sigma_z = 0$, and further assuming an ultra-relativistic beam velocity $v_b \cong c$) analysis. It was found in this limiting case that the induced decelerating field at the beam is strictly proportional to the induced charge. This is despite the fact that the plasma response changes qualitatively as the normalized charge

$$\tilde{Q} = 4\pi k_p r_e N_b = \frac{N_b k_p^3}{n_0} \qquad (1)$$

is raised. Equation 1 indicates that $\tilde{Q}$ is the ratio of the beam charge to the plasma electron charge located within a volume of a cubic plasma skin-depth, k_p^{-3}. Here, the plasma skin depth is defined as $k_p = c/\omega_p$, where the electron plasma frequency is $\omega_p = \sqrt{4\pi e^2 n_0 / m_e}$, and n_0 is the ambient plasma electron density.

In the analysis of our companion paper, all time intervals are normalized to ω_p^{-1} and spatial distances to k_p^{-1}. This analysis further normalizes all densities to n_0, velocities to *c*, momenta to $m_e c$ all current densities to $en_0 c$, and all fields to $m_e c\omega_p / e$. All normalized variables are indicated by the tilde symbol, *e.g.* $\tilde{H}_\phi \equiv eH_\phi / m_e c\omega_p$, and $\tilde{Q}$. In this regard, it should be noted that if $\tilde{Q}$<<1, the plasma response should be linear — all other normalized variables are small compared to unity. Nonlinear features appear in the response when $\tilde{Q}$ approaches or exceeds one.

Some of these nonlinear features can be anticipated easily; as noted before, the plasma electrons become rarefied from the beam channel. We can comment in this regard that it is assumed for the moment that the beam is both radially narrow $k_p\sigma_r << 1$ and short $k_p\sigma_z < 2$, and thus if $\tilde{Q}$ exceeds unity then the ratio of the beam-to-plasma density, $n_b / n_0 = \tilde{Q}/(2\pi)^{3/2} k_p\sigma_z (k_p\sigma_r)^2$, is greater than one. Under these assumptions, which are most often, but not always, obeyed in experimental conditions, the statement that $\tilde{Q}$ is much greater than unity implies blow-out regime conditions, where the beam is much denser than the plasma. Under such conditions, the plasma electrons are then ejected from the beam channel. For present experiments in the blow-out regime[13,14], the value of $\tilde{Q}$ is in the range of 1.5-4.

It may also be anticipated that accompanying this large amplitude plasma density modulation implies a large velocity response in the plasma electrons themselves. In fact, it has been observed in simulations that the plasma electrons attain relativistic velocities as they are ejected from the beam channel[3,8]. What was not appreciated before the analysis presented in our companion paper is that a strong component of the imparted relativistic momentum impulse is predicted to be in the forward longitudinal ($+z$) direction. For infinitesimally short (longitudinal δ-function) beams, the longitudinal momentum impulse is predicted to be[1] increasingly dominant; the radial and longitudinal components are related by

$$\Delta\tilde{p}_z = \tfrac{1}{2}\Delta\tilde{p}_r^2, \tag{2}$$

where $\tilde{p} = p / m_e c$ is the normalized momentum. In contrast, in the case of excitation of small amplitude plasma waves, the initial longitudinal acceleration of the plasma electrons due to the introduction of the beam pulse is always in the negative direction ($-z$).

The analysis presented in Ref. 1 quantitatively predicts the state of the plasma and fields directly behind the δ-function beam (located at the zero of the variable $\tau = \omega_p(t - z / v_b)$), having uniform surface charge density up to a normalized radius $\tilde{a} = k_p a$, The longitudinal electric field is obtained through a process that initially requires calculation of integral of the magnetic field $\mathbf{H} = \int_{\varepsilon-}^{\varepsilon+} \tilde{H}_\phi d\tau$ during beam passage;

$$\mathbf{H}(\tilde{r}) = \frac{\tilde{Q}}{\pi\tilde{a}} \begin{cases} K_1(\tilde{a}) I_1(\tilde{r}) & (\tilde{r} < \tilde{a}) \\ K_1(\tilde{r}) I_1(\tilde{a}) & (\tilde{r} > \tilde{a}), \end{cases} \tag{3}$$

where I_l and K_l are modified Bessel functions.

Noting further that the radial momentum impulse is simply $\Delta\tilde{p}_r = \mathbf{H}$, one has $\Delta\tilde{p}_z = \frac{1}{2}\mathbf{H}^2$; the normalized velocity components associated with these momenta are $\tilde{v}_r = \mathbf{H}/\left(1+\frac{1}{2}\mathbf{H}^2\right)$ and $\tilde{v}_z = \frac{1}{2}\mathbf{H}^2/\left(1+\frac{1}{2}\mathbf{H}^2\right)$. The equation of continuity for the plasma electrons further gives an enhancement of the plasma electron density immediately behing the driving beam, $\tilde{n} = \left(1-\tilde{v}_z\right)^{-1} = 1+\frac{1}{2}\mathbf{H}^2$. The combined effects of the crossed electric and magnetic fields causes a "snow-plowing" of the plasma electrons, resulting in enhanced density.

Thus the normalized transverse current density is given by $\tilde{J}_r = \tilde{n}\tilde{v}_r = \mathbf{H}$, which is identical to the linear (non-relativistic) analysis. This linear-like scaling is caused by the fortuitous cancellation of two effects: the radial velocity is limited by relativistic effects, while the density grows via the snow-plow effect, exactly compensating for the reduced radial velocity response. Further, in the case of the δ-function beam, we have for the field directly behind the disk-like beam, in the limit $\tilde{a}$<<1,

$$\begin{aligned}\tilde{E}_z(\tilde{r})\Big|_{\tau=\varepsilon+} &= \int_{\infty}^{\tilde{r}} \mathbf{H}(\tilde{r}')d\tilde{r}' = \frac{\tilde{Q}}{\pi\tilde{a}^2}\left[1-\tilde{a}K_1(\tilde{a})I_0(\tilde{r})\right], \quad \tilde{r}<\tilde{a} \\ &\cong \frac{\tilde{Q}}{\pi\tilde{a}^2}\left[1-\tilde{a}K_1(\tilde{a})\right] \cong \frac{\tilde{Q}}{2\pi}\ln\left(\frac{1.123}{\tilde{a}}\right).\end{aligned} \tag{4}$$

The energy loss gradient associated with this field is given, for a δ-function beam, by one-half of the longitudinal force directly behind the beam[1,17], $\tilde{F} = \tilde{E}_z/2$. It is predicted to be linear in charge regardless of the size of $\tilde{Q}$.

This scaling in energy loss for the limiting case of the δ-function beam has been suggested as an illuminating example that helps explain the persistence of linear-like scaling of both the deceleration of a drive beam in the blow-out regime of the PWFA, and in the subsequent available acceleration fields. In this paper, we quantitatively explore, with particle-in-cell (PIC) simulations using the 2D axi-symmetric codes OOPIC[8] and MAGIC[18], the dependence of these fields on $\tilde{Q}$.

In order to perform such an analysis, several definitions associated with the simulations must be introduced. The typical on-axis longitudinal field profile excited in the blow-out regime is shown in Fig. 1. Three measures of the field amplitudes are given in this figure: the well-behaved decelerating field inside of the driving beam[2], the peak accelerating field, which is characterized by a narrow spike, and "useful" field, that which directly precedes the spike. The decelerating field is investigated in detail in this paper to make a strong connection to the analysis in Ref. 1. The second measure of acceleration given is termed "useful" because the acceleration associated with the spike is an extremely narrow region, with negligible stored energy, and therefore of very limiting use for efficiently accelerating a real beam. It will be seen that the scaling of plasma wake-field amplitudes, as measured in particular by the peak accelerating field spike, follows linear-like behavior well into the nonlinear regime. The mechanisms behind this anomalous scaling are explored, as are the ways in which they fail in the extremely nonlinear limit.

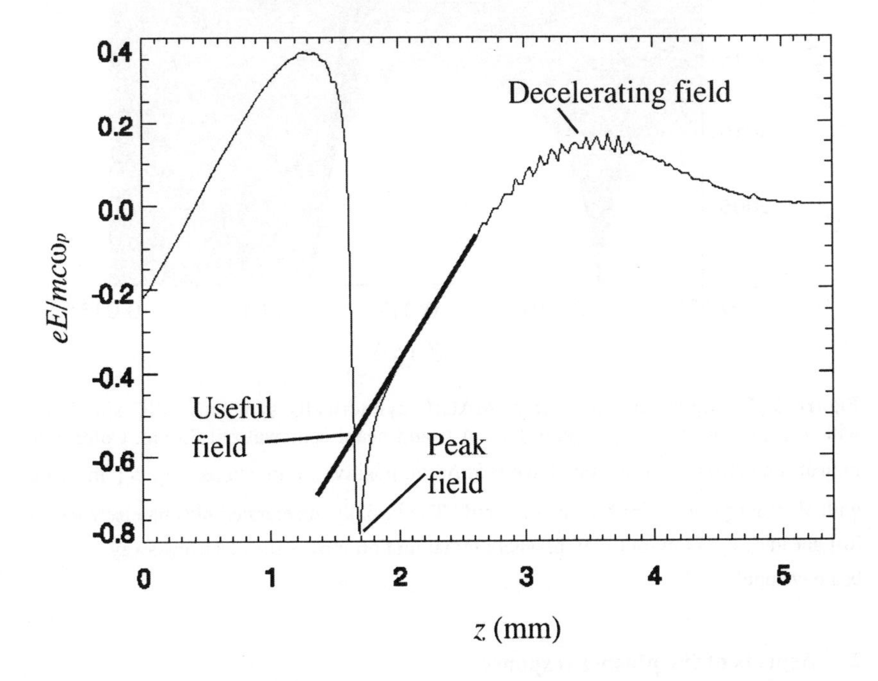

Figure 1. The longitudinal field profile given by PIC simulation for PWFA excitation in the blowout regime. The decelerating field, peak accelerating field, and a defined "useful field", which avoids the narrow spike region, are indicated in the drawing.

[2] One must take care that the PIC simulation parameters (mesh size and simulation particle number) are chosen to give well-behaved decelerating fields. Simulation noise problems become more serious when the beam charge is raised to $\tilde{Q}$ larger than unity.

Before examining the scaling of the longitudinal fields in the nonlinear regime of the PWFA, we begin our discussion of simulation results by looking at the qualitative aspects of the plasma electron response. In this way, we can verify aspects of the predictions of the nonlinear theory given in Ref. 1 using a method that is independent of the fluid analysis employed therein.

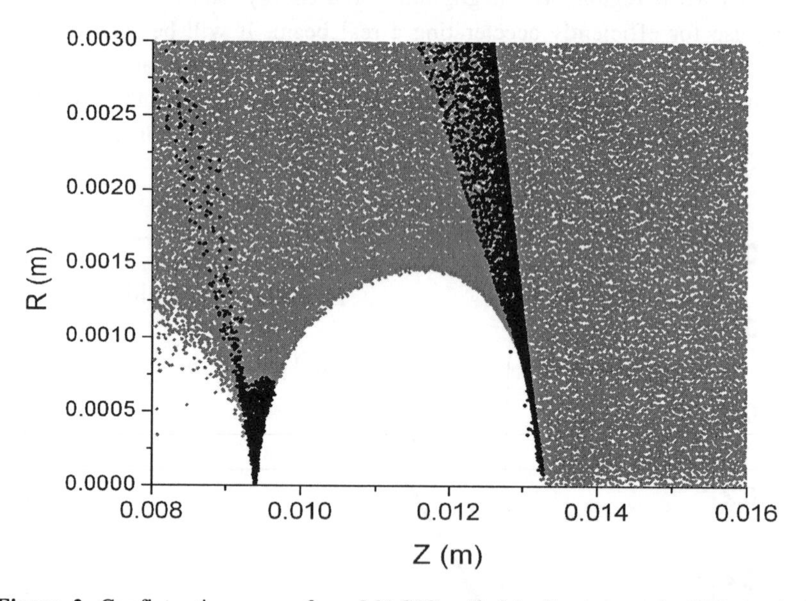

Figure 2. Configuration space from MAGIC cylindrically-symmetric PIC simulation with $\tilde{Q} \cong 20$, and $k_p\sigma_z = 0.11$, and $k_p a = 0.2$, and beam center at z=1.33 cm. Color code indicates electron positions that have *relativistic* positive momenta $p_z/m_e c > 1$ in black, with all other plasma electrons colored red. The initially accelerated plasma electrons are just ahead of the blow-out region, where radial motion moves the electrons away from the beam channel.

2. Aspects of the plasma response

In order to explore the predictions concerning the nonlinear plasma response from the analytical $k_p\sigma_z << 1$ result, we have performed a series of simulations using the fully relativistic PIC codes, MAGIC and OOPIC. Two codes were used initially to check consistency; both codes gave essentially the same answers for all comparisons. Both codes were run with 15 GeV initial beam energy, to guarantee that the beam is ultra-relativistic, and to suppress transverse evolution of the beam distribution. The first investigation undertaken using MAGIC concerned the validity of the physical model we have deduced from the analysis of Ref 1. In particular, as one never expects the snow-plow effect from linear theory, it is important this is observed. The two main characteristics of snow-plow are: 1) a forward velocity component, and 2) a plasma electron density increase, both occurring in the region directly behind the beam. These effects are noted in beams of moderate length ($k_p\sigma_z \sim 1$), but in order to observe this effect most strongly, we next display the result of a short (to approach the δ-function limit) beam simulation.

Both of the qualitative predictions are dramatically verified in Fig. 2, where we display a simulation with high charge, $\tilde{Q} \cong 20$, that is very short, $k_p\sigma_z = 0.11$, and is also narrow, $k_p a = 0.2$. Note that the "shock front" shown in this case, which consists of electrons moving both forward and radially outward at relativistic speeds (the selected electrons must have $\tilde{p}_z > 1$) is *not* a representation of the initial disturbance, which is localized around the longitudinal position of the beam. The front is canted because many of the electrons in it that are located far from the axis originated quite close to the axis. In fact, one may expect that the leading edge of this front consists of particles that are ejected with relativistic transverse velocity, and thus the edge must have roughly a 45-degree angle with respect to the axis (recall that the electrons are launched by a source moving at nearly light speed). This angle may be verified from inspection of Fig. 2; note the difference in radial and longitudinal scales. The ultimate trajectory of these ejected electrons impacts the possible acceleration available in the wake-field behind the beam; we will return to this subject below.

In order to more quantitatively explore the predictions of the analysis given in Ref. 1, a series of additional simulations were undertaken with OOPIC, that had the following cylindrically-uniform beam shape: flat-top radial distribution of width $k_p a = 0.2$ and flat-top longitudinal of length $k_p l_z = 0.1$ (effectively much shorter than even in the example of Fig 2). This type of beam, which is as close as possible that we could obtain (given the constraints of numerical stability in the simulation) to the ideal δ-function length used in the analysis of Ref. 1, was then scaled in charge $\tilde{Q}$ upwards from 0.2, to 2, 20 and 200. To illustrate the relevant physical processes, we begin by the plotting plasma electron density as well the plasma longitudinal current density, for the $\tilde{Q}$=200 case, in Figs. 3. It can be seen from Fig. 3(a) that the plasma density is strongly snow-plowed in the vicinity of the driving beam, with a strong component of the longitudinal current density located there. Within this high forward-current region, the plasma density (Fig. 3(b)) is roughly 6 times the ambient density, indicating the severity of the snow-plow in this highly nonlinear case.

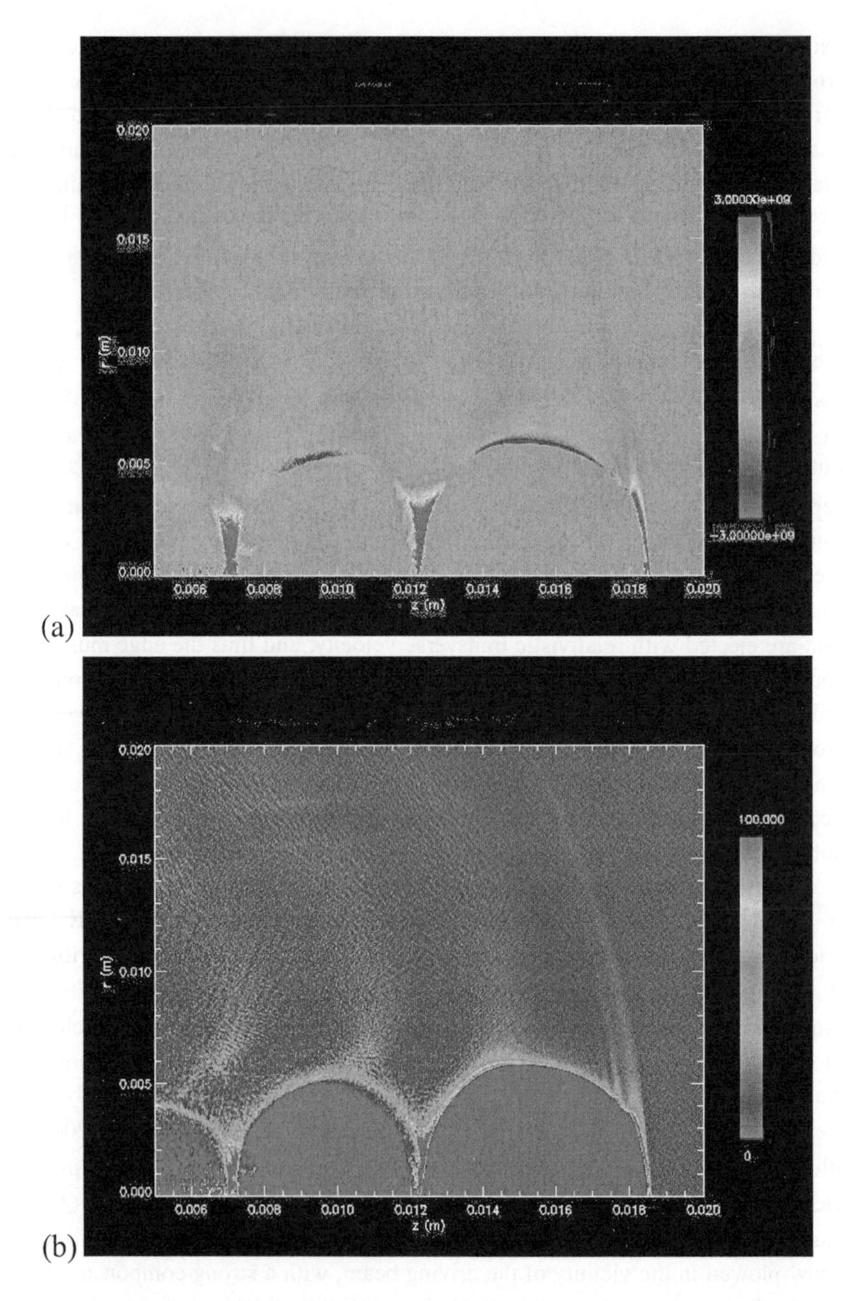

Figure 3. OOPIC simulation of $\tilde{Q}$=200 case, drive beam having uniform distribution of width $\tilde{a} = 0.2$ and length $\tilde{l}_z = 0.1$. False-color (a) Plasma longitudinal current density and (b) plasma electron density (ambient level is at 15).

Both the plasma electron density and the longitudinal currents show a strong localization around the rarefied region in Figs. 3. It can be seen that the plasma disturbance is surprisingly well-behaved even during such large amplitude motion. The nonlinear plasma motion coheres fairly well until the strong wave-breaking event located where the plasma electrons return to the axis. Note that portions of the disturbance propagate to large radial offset; these artifacts are associated with the electrons that are strongly ejected by the beam, and also with other regions having large longitudinal current density (both positive and negative).

In order examine the details of the immediate plasma response to the beam, we plot the current densities directly behind the driving beam that resulted from these simulations. This macroscopic quantity reflects both the density and velocity state of the plasma electrons, and in the analytical case $\tilde{J}_r$ also represented the magnetic field response.

It can be seen in Fig. 4 that the $\tilde{Q}$=0.2 case shows a small amount of longitudinal current density directly behind the beam, even though the beam-plasma system may be naively thought to be in the linear response regime. Inspection of our expressions for the velocity and current density indicate that this is not a completely linear system; the peak velocity induced at the beam edge is expected to be $0.15c$ even in this case. The nonlinearity arises even with a relatively small charge, because the peak beam density is 1.6 times n_0.

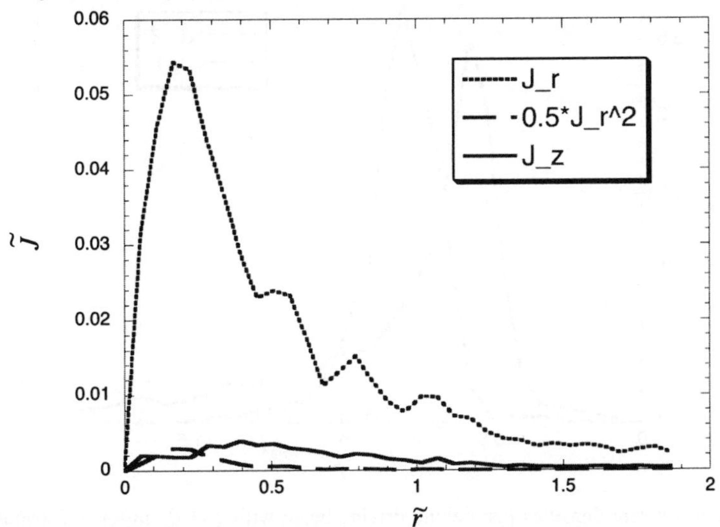

Figure 4. Current densities just behind driving beam with $\tilde{Q}$=0.2, uniform distribution of width $\tilde{a}=0.2$ and length $\tilde{l}_z=0.1$. Analytical prediction $\tilde{J}_z=\frac{1}{2}\tilde{J}_r^2$ from theory also shown.

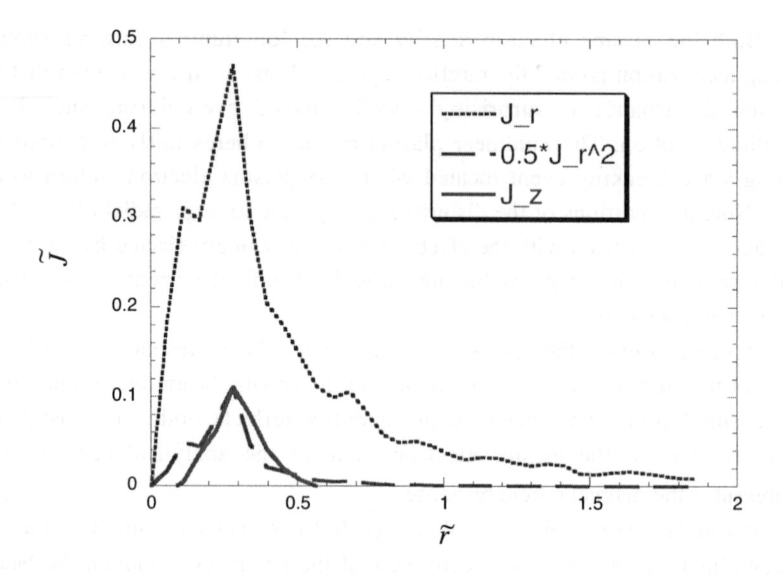

Figure 5. Current densities just behind driving beam with $\tilde{Q}$=2, uniform distribution of width $\tilde{a}=0.2$ and length $\tilde{l}_z=0.1$. Analytical prediction $\tilde{J}_z=\frac{1}{2}\tilde{J}_r^2$ from theory also shown.

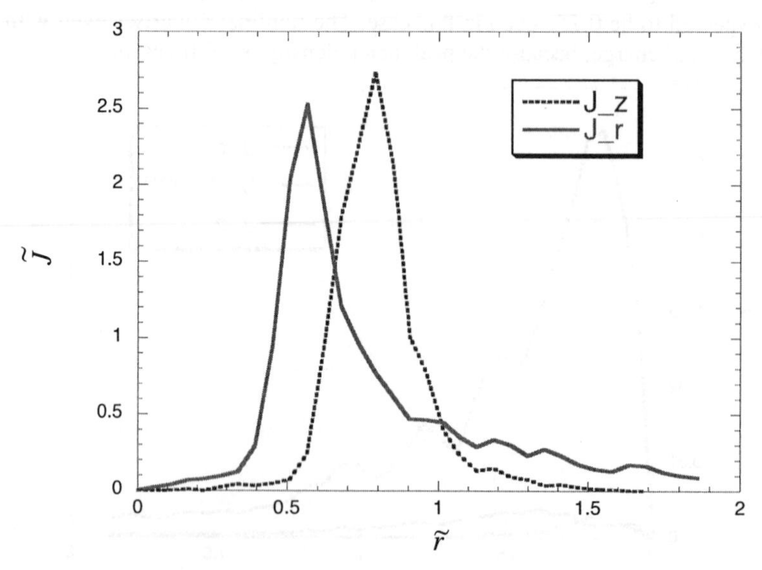

Figure 6. Current densities just behind driving beam with $\tilde{Q}$=20, uniform distribution of width $\tilde{a}=0.2$ and length $\tilde{l}_z=0.1$. Note that the currents have moved away from the beam distribution ($\tilde{r}>\tilde{a}=0.2$) significantly during the beam passage.

In the higher $\tilde{Q}$ cases, however, nonlinear effects are even more pronounced. In Figure 5, we plot, along with $\tilde{J}_z$ and $\tilde{J}_r$, the quantity which relates the two in the δ-function beam limit, $\tilde{J}_z = \frac{1}{2}\tilde{J}_r^2$; the comparison is quite good. Note that in Fig. 4, the longitudinal current density $\tilde{J}_z$ was small enough that its amplitude was dominated by noise, and the comparison is not as good.

Finally, we show the current densities associated the very nonlinear $\tilde{Q}$=20 case in Fig. 6. Note that the current densities have moved away from the beam distribution ($\tilde{r} > \tilde{a} = 0.2$) significantly during the beam passage. This further indicates that the induced decelerating electric field should be smaller, as the radial currents associated with the induced $\tilde{E}_z$ are reduced by the diminishing of the coupling (by simple proximity arguments) to the driving electron beam charge. It is also striking to note that normalized current densities $\tilde{J}$ are well larger than unity — without the snow-plow enhancement of the plasma electron density, this is strictly forbidden. For the $\tilde{Q}$=200 case, $\tilde{J}_z$ is, as expected, even larger (over 11), and exceeds $\tilde{J}_r$ by a factor of two. Thus, as predicted by analytical results of Ref. 1 (see also our Eq. 2), the longitudinal current density eventually exceeds the radial current density when $\tilde{Q}$ becomes very large.

3. Deceleration and acceleration scaling studies: ultra-short beam

In order to evaluate the characteristics of the induced electric field driven by an ultra-short beam, we summarize the results of the parametric scan in $\tilde{Q}$ (from 0.02 to 200) in Fig. 7. This scan is an example of ideal scaling, in which the beam geometry and plasma density are held constant, while the charge is increased. In this figure, we plot the average on-axis deceleration experienced by the driving beam, and the associated prediction of linear theory, which in the short beam limit is given by $\tilde{F}_{dec} \cong \left(\tilde{Q}/2\pi\tilde{a}^2\right)\left[1-\tilde{a}K_1(\tilde{a})\right]$. We choose the average deceleration as a relevant measure to accurately quantify the energy imparted to the plasma by the beam passage, and to connect with both the analytical δ-function beam limit and also the case of longer, Gaussian beams discussed below.

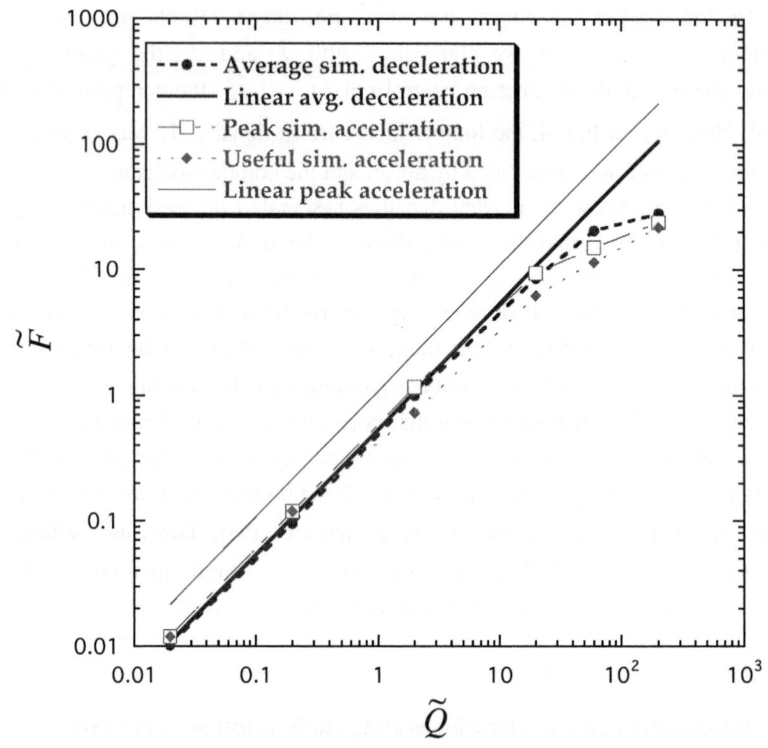

Figure 7. The average normalized energy loss rate of $\tilde{F}_{dec} = e|\langle E_z\rangle|/m_e c\omega_p$ of an electron beam with $\tilde{l}_z = 0.1$, $\tilde{a} = 0.2$, as a function of $\tilde{Q}$, from linear theory (solid bold line) and self-consistent PIC simulation (circles); the peak accelerating field behind the beam, $\tilde{F}_{\max} = e|E_{z,\max}|/m_e c\omega_p$, from linear theory (solid fine line) and PIC simulation (squares); also the useful field for acceleration (diamonds), defined by the geometry in Fig. 1.

In addition we also plot three measures of the accelerating field for each case: two from the simulation, the peak acceleration, and the "useful" acceleration; also, the peak acceleration from the predictions of linear theory, obtained (as was also done for the average deceleration) by performing a convolution integration[17] over the drive beam, using the result of Eq. 4 as the Green function in the convolution, $\tilde{F}_{\max} \cong \left(\tilde{Q}/\pi\tilde{a}^2\right)\left[1-\tilde{a}K_1(\tilde{a})\right] \cong 2\tilde{F}_{dec}$. This procedure is of course not valid for the nonlinear response, but is only employed to extrapolate the predictions of linear theory.

Figure 7 shows some expected and some unexpected behavior. First, we note with satisfaction that until $\tilde{Q}$>20, the average deceleration observed in simulation is very close to that predicted by linear theory. This is a direct verification of the extrapolation of the linear response of the decelerating field predicted by the analysis of the δ-function beam. Above $\tilde{Q}$=20, the decelerating field is smaller than one expects from this limiting case, because (as already noted in the current response shown in Fig. 5) the plasma electrons may already be rarefying the near-beam region during the passage of the beam. This effect is not possible within the analytical model — the electrons do not notably change position within the time of the beam passage.

The peak, as well as the useful, accelerating field observed in the simulations is, according to Fig. 7, well below (nearly a factor of 2) that expected from linear theory, even for small $\tilde{Q}$. This effect is due to nonlinear response having much to do with the very high density (n_b=1.6 n_0 even for $\tilde{Q}$=0.02) in such a short beam. Aspects of the microscopic mechanisms for diminishing of the acceleration field are evident in Figs. 3, and are discussed in conjunction with the results shown in Fig. 8. This type of nonlinearity is not observed for small $\tilde{Q}$ cases in the longer beam simulations discussed in the following sections — such cases agree quite well with linear theory. We now turn to our examination of these simulation results.

4. Ideal scaling with a Gaussian beam-plasma system

While the δ-function beam limit is relevant to verification of the theoretical analysis, it is not of highest practical interest in bunched beams, which generally have a Gaussian current distribution, $\rho_b(z) \propto \exp\left(-z^2/2\sigma_z^2\right)$. Further, it has often been argued that one should choose the plasma density such that $k_p\sigma_z \cong 1$ to optimize drive beam energy loss and accelerating beam energy gain in a PWFA[2-5,9-11]. In order to explore the deviation in plasma response from the analytical ($k_p\sigma_z \to 0$) result, therefore, we have performed a series of OOPIC simulations. We again take the beam of radius $\tilde{a} = 0.2$ (again keeping the transverse beam profile uniform, to compare with the extrapolations of linear theory), and Gaussian current profile with $k_p\sigma_z = 1.1$. We have scanned the charge from $\tilde{Q} = 0.02$ to 200, values indicating linear to very nonlinear cases.

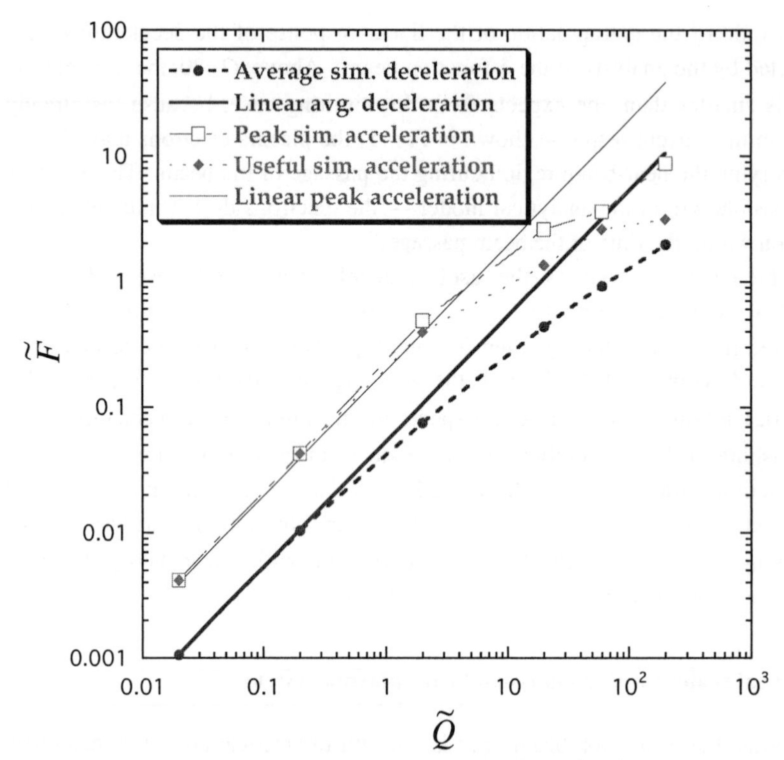

Figure 8. The average normalized energy loss rate of $\tilde{F}_{dec} = e|\langle E_z \rangle| / m_e c\omega_p$ of a Gaussian-current electron beam with $k_p\sigma_z = 1.1$, $\tilde{a} = 0.2$, as a function of $\tilde{Q}$, from linear theory (solid bold line) and self-consistent PIC simulation (circles); the peak accelerating field behind the beam, $\tilde{F}_{\max} = e|E_{z,\max}| / m_e c\omega_p$, from linear theory (solid fine line) and PIC simulation (squares); also the useful field for acceleration (diamonds).

In Fig. 8, the average on-axis decelerating field, as calculated from the simulation through $(2\pi\sigma_z)^{-1}\int eE_z(z)\big|_{r=0}\exp(-z^2/2\sigma_z^2)dz$, again compares well with the linear theory prediction of $(\tilde{Q}/2\pi\tilde{a}^2)[1-\tilde{a}K_1(\tilde{a})]\exp(-k_p^2\sigma_z^2)$, until $\tilde{Q}$ exceeds 2. It should be noted that $\tilde{Q}$=2 is just into the blow-out regime, as the beam is 2.3 times denser than the plasma. One in fact would expect that the linear prediction would begin to fail for $\tilde{Q}$ an order of magnitude smaller, but it does not, because of the snow-plow effect. The enhancement of the coupling due to snow-plowing of the plasma electrons is a reflection of the a longer interaction time, since the plasma electron are traveling longitudinally at relativistic speed, and thus stay in contact with the beam longer. Thus, even though the plasma electrons may move radially outward away from the beam, the beam-plasma coupling stays anomalously strong for moderately large $\tilde{Q}$ (less than 10). At very large $\tilde{Q}$, this coupling, as measured by the deceleration, diminishes notably — it is an order of magnitude smaller than predicted by extrapolated linear theory.

Examination of the acceleration amplitudes in the simulation shows several interesting features. The first is that if one relies on the peak as a measure of the acceleration, the spike that occurs at the back of the accelerating region misleadingly indicates linear-like response until $\tilde{Q}$=20. In fact, the spike in the peak field that we have discussed above is magnified in the more nonlinear cases, causing a field *enhancement* relative to linear theory for $\tilde{Q} \cong 1$. This phenomenon partly explains why field saturation was not noted in previous simulation scans. Even with this masking effect, however, the accelerating peak still displays saturation when $\tilde{Q} >> 1$, with increasing severity for $\tilde{Q} > 100$. Examination of the useful acceleration amplitude, however, indicates that the accelerating field response is diminished above $\tilde{Q}$=2, just as is found for the average deceleration. For $\tilde{Q} > 100$, the efficiency of exciting the acceleration field is very low. In fact, the useful acceleration diminishes more rapidly than the drive beam deceleration for large $\tilde{Q}$. This effect is more noticeable in the ultra-short beam simulation shown in Fig. 7, where the useful acceleration is *smaller* than the deceleration for even moderate $\tilde{Q}$.

There is, in the linear regime, a fixed relationship between acceleration and deceleration that is dependent only on beam geometry, not charge. The question then arises: where is the energy lost by the drive beam going, if not into acceleration? The answer is hinted at by the enhancement of this effect in the ultra-short, dense beam case — much of the energy that is deposited into the plasma by the beam does not, for plasma electrons initially close to the axis, go into generation of simple wave motion, but into very large amplitude scattered motion. While such electrons may seem, if one concentrates only on smaller radial regions, to be ejected in a near-ballistic manner from the beam region, in fact, they eventually lose energy, but at a different, much longer time-scale than the main oscillation. These electrons, having a tight time-profile, and a non-trivial density, may be considerd to make up subsidiary "beams", which produce their own decelerating wakes. Since these wakes are far from the axis, and proceed with a long periodicity, such electrons do not contribute to building the initial accelerating wake-field, and their energy is effectively lost from the main component of wave system.

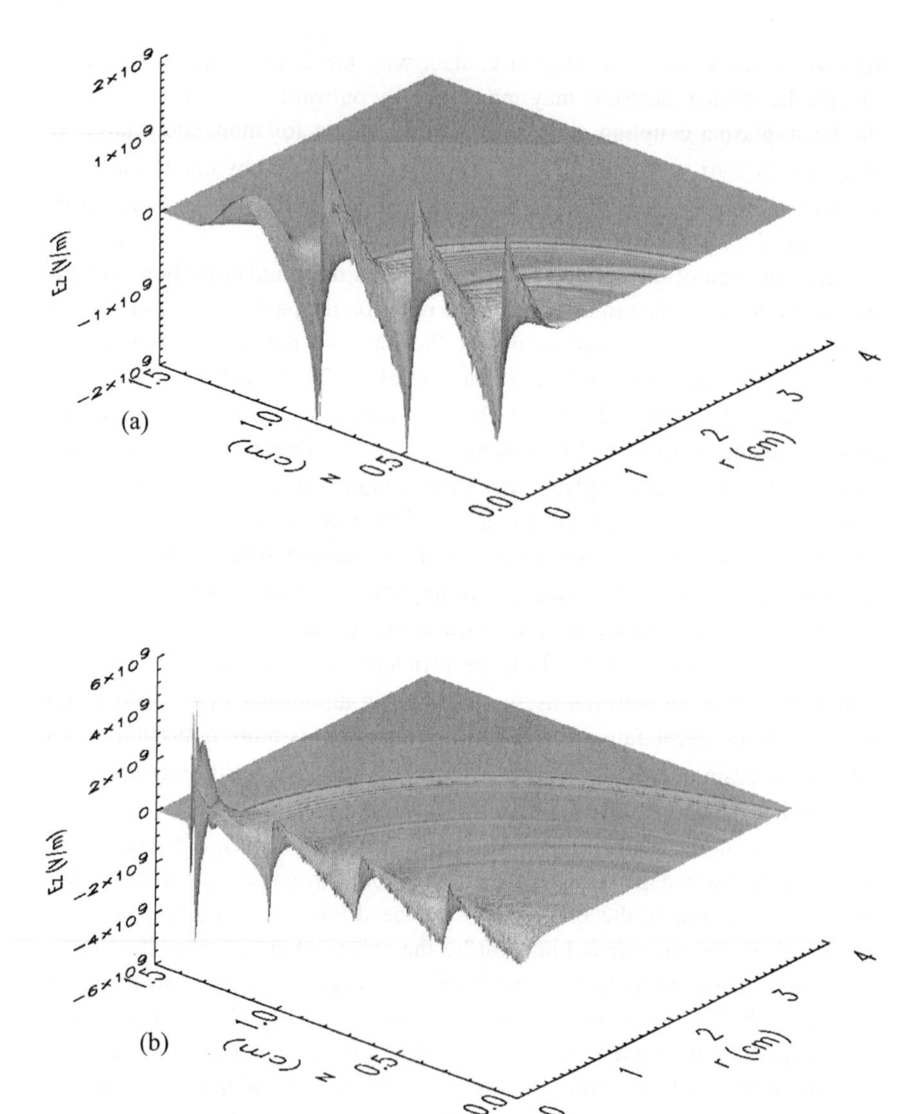

Figure 9. Surface plot of E_z in OOPIC PWFA simulation with $\tilde{Q}$=20, (a) long-beam geometry as in Fig. 8, and (b) ultra-short beam geometry, as in Fig. 7.

The generation of large radial-amplitude particles can be noted in the momentum-coded configuration space plot of Fig. 2, and the density/current profiles of Fig. 3. Other aspects of this phenomenon are shown in Fig. 9, which displays the longitudinal electric field associated with $\tilde{Q}$=20 cases, for both the long-beam geometry (Fig. 9(a)) as in the simulations of Fig. 8 (similar to Fig. 2), and for the ultra-short beam (Fig. 9(b)) as in the simulations of Fig. 7. In the long-beam case, while there is negligible field disturbance electromagnetic energy density lost to large radial amplitudes emanating from the drive-beam region, there is a pronounced effect arising from the regions immediately following the field spikes that terminate the accelerating phases of the wake. These spikes are caused by electrons having very large momentum as they return to the axis, and this momentum is not removed during the dissolution of the spike. Thus many relativistic electrons are lost to subsequently large radial amplitudes. Associated with the wake of these electrons, which have a much shorter temporal length than the original drive beam, we again see significant levels of E_z.

The ejection of plasma electrons from the drive beam region to large radial amplitude is more severe for larger $\tilde{Q}$ cases. It is also a dominant effect in the ultra-short beam response, as seen in Fig. 9(b). A very large field response is observed propagating to large radial amplitudes in this case, one which clearly carries with it a significant fraction of the total electromagnetic energy deposited in the plasma (along with the mechanical energy associated with the motion). The severity of this effect for the ultra-short beam case illustrates the underlying mechanism behind the observation of accelerating field response that is lower than linear predictions, even for relatively small $\tilde{Q}$.

5. Experimental scaling with a Gaussian beam-plasma system

The type of beam-plasma system scaling that has been explored in the simulations of Figs. 7 and 8 is what has been termed ideal. In performing these calculations, the plasma density and beam geometry were kept constant, while the charge was varied. This is of course exactly what one wants to do to conceptually understand the scaling of the system. It is not, however, what is strictly relevant to experiments. In accessing higher values of $\tilde{Q}$ in experiment, the beam is compressed longitudinally[3], and one typically then raises the plasma density to keep the condition $k_p\sigma_z \approx 1$. Thus $\tilde{Q} = 4\pi k_p r_e N_b$ is raised by

[3] In many electron sources (*e.g.* rf photoinjectors with or without compressors) when one raises the charge, the bunch length also increases.

increasing k_p with N_b constant. In ideal scaling, the beam density increases as k_p^3, while n_0 increases only as k_p^2, thus leading to higher values of n_b/n_0.

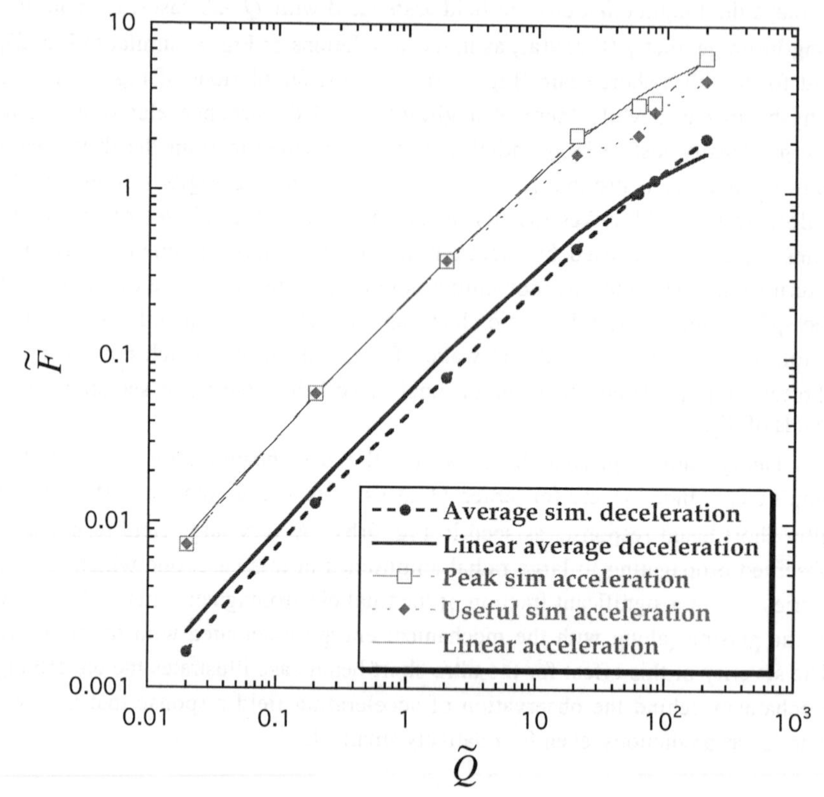

Figure 10. The average normalized energy loss rate of $\tilde{F}_{dec} = e|\langle E_z \rangle|/m_e c\omega_p$ of a Gaussian-current electron beam with $k_p\sigma_z = 0.1$, using experimental scaling, $\tilde{a} = 0.2\sqrt{\tilde{Q}/2}$, as a function of $\tilde{Q}$, from linear theory (solid bold line) and self-consistent PIC simulation (circles); the peak accelerating field behind the beam, $\tilde{F}_{\max} = e|E_{z,\max}|/m_e c\omega_p$, from linear theory (solid fine line) and PIC simulation (squares); also the useful field for acceleration (diamonds).

In order to accomplish ideal scaling[19], it is implied that the beam's matched beta-function β_{eq} scales as k_p^{-1}, which is in fact the case[3,4], as $\beta_{eq} = \sqrt{2\gamma}k_p^{-1}$. On the other hand, it is clear that the beam emittance ε does not decrease during compression; it can be at best constant, and may indeed increase due to collective effects[20,21] such as coherent synchrotron radiation. If one assumes that the beam emittance is held constant during compression, and that the beam is injected in a matched fashion to the ion focusing, then the normalized transverse beam size scales $\tilde{\sigma}_r = k_p\sqrt{\beta_{eq}\varepsilon} \propto \sqrt{k_p}$.

The results of a parametric study using what may be therefore termed "experimental" scaling are shown in Fig. 10. In these simulations, the normalized bunch length is held constant, $k_p\sigma_z = 1.1$, while the normalized transverse (uniform density) beam size is given by $\tilde{a} = 0.2\sqrt{\tilde{Q}/2}$, as dictated by the constant emittance hypothesis. Note that because $\tilde{a}$ is now a function of $\tilde{Q}$, the extrapolated linear average deceleration and peak acceleration are no longer straight lines on the log-log plot. As the beam spot becomes larger with increasing $\tilde{Q}$, the electromagnetic coupling of the beam to the plasma decreases, as could have been anticipated by inspection of Eq. 4.

One interesting point in this regard is that the beam no longer becomes relatively denser than the plasma as $\tilde{Q}$ is increased — $n_b/n_0 \propto \tilde{Q}^0$, and the relative density is stationary. In the example of Fig. 10, this ratio is 2.3, at the lower end of the blowout regime, and strong blowout is never accessed. As $\tilde{Q}$ is raised, however, the plasma response to the beam under experimental scaling also displays qualitatively different behavior — the plasma motion becomes increasingly longitudinal as $\tilde{a}$ approaches and exceeds unity, with the plasma return currents running inside of the beam. When $\tilde{a}>>1$, one may expect that there is no relative reduction of beam coupling to the plasma with increasing charge, as beam does not eject the plasma electrons radially from its path. As such, the response in this limit does not resemble the blow-out regime, but a ID nonlinear system that was discussed previous to the proposal of the blow-out regime[22].

These considerations shed light on the most striking aspect of the experimental scaling results of Fig. 10; the simulated deceleration and acceleration amplitudes are not notably different than those found from extrapolation of linear theory. Even the useful acceleration fields do not degrade from linear expectations by more than a factor of two. The electric field response in this study is well approximated by the linear prediction, since the beam is not too much denser than the plasma, and the plasma response is not radial motion-dominated as the normalized charge increases to nominally "nonlinear" levels.

This study illustrates one of the expected laboratory pitfalls of experimental scaling — the plasma does not become more underdense when the beam is compressed. Thus one may expect the longitudinal fields to increase with compression, and according to linear expectations. But one may not, as indicated by simple arguments proposed by many previous authors, scale the expected field by σ_z^{-2}.

To illustrate this point, we take the case of the E164 experiment[15], now proposed for SLAC, as an extension to the E162 measurements[14]. In the first stage of E164, the beam will be compressed from ~700 μm to ~100 μm, with the plasma density raised to keep $k_p\sigma_z$ approximately constant. This jump in density represents a scaling in $\tilde{Q}$ from roughly 1.5 to near 10. An additional stage of E164 is more audacious, with the beam compressed to ~12.5 μm, and a concomitant $\tilde{Q}$ of ~80. The parameters chosen for the radial beam size in our experimental scaling scan also correspond well to those achieved at the SLAC FFTB ($k_p\sigma_r \approx 2\tilde{a} \approx 0.2$ at $\tilde{Q} \approx 2$), so this scan may be taken as a fairly accurate guideline for anticipating experimental scenarios. In fact, we have explicitly included the $\tilde{Q}$=80 point in order to compare the simulation to the approximate experimental conditions. Note that the scaling of $\tilde{Q}$ from 1.5 to 80 implies, using the linear extrapolation of field gradients (ideal scaling), that the acceleration should go from around 300 MeV/m, as is predicted by the results in Fig. 10, and was recently observed[15], to nearly 850 GV/m (!).

For the first stage of E164, experimental scaling the results of Fig. 10 gives a maximum acceleration gradient of 5 GV/m, which is slightly larger than claimed in Ref. 15. For the second stage of the E164 experiment, our scaling study predicts that the useful accelerating field is near 50 GV/m. This is an impressive number, but is over an order of magnitude shy of the prediction given by the σ_z^{-2} extrapolation of linear theory. It also interesting to note in this context that if one could perform ideal scaling (by reducing the emittance along with bunch length), the nonlinearity of the plasma response, dominated by the radial ejection of the plasma electrons and concomitant loss of coupling, produces a useful acceleration which is nearly the same as in the experimental scaling case. Thus the nonlinear saturation of the plasma wave response with such a high value of $\tilde{Q}$ negates the advantage of using a narrower beam that is indicated by Eq. 4.

6. Conclusions

In conclusion, we have examined, through PIC simulation, aspects of the physics of the plasma electrons as they responds to a very high charge beam. We have concentrated on two areas of investigation: verification of the analytical results concerning the fluid response derived in Ref. 1, and the scaling of the beam-plasma interaction in both ideal and experimental scenarios.

The simulations that most directly dealt with the predictions of Ref. 1 were performed with an ultra-short ($k_p l_z$=0.1) beam. The most important qualititave analytical predictions, those of a strong initial forward component of the plasma electrons, along with the associated increase in density (and current density) excited by a large $\tilde{Q}$ beam, were indeed verified by these simulations. In addition, the linear-like scaling of the deceleration fields driven by the beam was observed occur up to extremely large $\tilde{Q}$, with significant deviations entering in only when $\tilde{Q}$ exceeded 20. This is again is accordance with the most striking of the analytical results given in Ref. 1. This linear-like response in the fields stands in stark contrast to the nonlinear aspects (phenomena *not* associated with the linear theory) of the current density response that was observed even for $\tilde{Q}$ significantly smaller than unity. It is noted that the current density disturbance even for such a short beam leaves the immediate radial vicinity of the beam for $\tilde{Q}$>>1. This fast ejection of the plasma electrons results eventually in a loss of coupling between beam and plasma, and thus in lower wake-field amplitudes.

In order to connect the nonlinear physics observed in analysis and theory for the ultra-short beam case to more experimentally relevant scenarios, a series of studies were undertaken with beam-plasma systems that have $k_p \sigma_z$=1.1. To explore the physics most characteristic of the PWFA blow-out regime, an ideal scaling study, where both $k_p a$ and $k_p \sigma_z$ were held constant while $\tilde{Q}$ was varied, was performed. In this case, the decelerating fields inside of the beam and the accelerating fields behind the beam displayed deviations from linear behavior. In particular, while the peak acceleration that is dominated by the spike at the tail of the accelerating portion of the wave scales nearly with linear expectations until $\tilde{Q}$>20, the useful field for acceleration is strongly degraded from the expectations of linear theory above $\tilde{Q}$=2. The masking of this degradation by reliance on measurement of the acceleration amplitude through the spike is noted as a probable cause for missing the deviations from linear scaling in previous studies.

The ideal scaling study definitively showed that while the previously proposed scaling of wake-field amplitudes as linear with k_p^2 (or σ_z^{-2} for constant $k_p a$ and $k_p \sigma_z$) is remarkably persistent until $\tilde{Q}$ well exceeds unity for finite length beams, when $\tilde{Q}$ is large enough, the coupling of the beam to the plasma becomes much less efficient, and field amplitudes do not grow as expected. Further, for very high $\tilde{Q}$, plasma electrons that are strongly ejected in the radial direction, and therefore do not contribute to creation of a useful initial accelerating portion of the wake wave, form an increasing component of the energy lost by the drive beam to the plasma. For these reasons, it may be unwise to adopt this regime of the PWFA. We thus suggest that a practical limit of useful $\tilde{Q}$ for operation in the blow-out regime is around 10.

On the other hand, for the presently proposed experiments that may access very large values of $\tilde{Q}$, these results may not apply. This is because when one compresses a beam of constant charge and emittance, the transverse beam size decreases only as $k_p^{-1/2}$, not as k_p. In such a case, while $\tilde{Q}$ increases due to the increase in k_p, the relative beam density does not. For proposed experiments at SLAC, this means that scaling to larger $\tilde{Q}$ through beam compression and use of a denser plasma does not imply moving into a more complete blow-out scenario. In fact, since $k_p\sigma_r$ steadily increases under these conditions, for the most ambitious planned experiment ($\tilde{Q}$=80), $k_p\sigma_r$ is near 1, and the plasma response is more longitudinal than radial. In the experimental scaling study we have performed, where we take into account the growth in $k_p\sigma_r$ (which may be taken as approximately $\tilde{a}/\sqrt{2}$ for the sake of comparison between theory, simulation and experiment) with increasing $\tilde{Q}$, it is found that all of these effects conspire to produce wake-field amplitudes that are very close to those predicted by linear theory, even for $\tilde{Q}$=200. It should be emphasized that this experimental scaling does not produce wake-fields that are proportional to k_p^2 (or σ_z^{-2}), as previously proposed, because the beam-plasma coupling decreases with normalized beam radius $k_p\sigma_r$.

It should be noted that experimental scaling is dependent on the available beam emittance. As future high-brightness electron sources[23] should have emittances a factor of 10 better than those found currently at the SLAC FFTB, for the same parameters as the second phase of E164 proposes, $k_p\sigma_r$ would still be much less than one. In addition, the beam would be more than an order of magnitude denser than the plasma, and this experiment would be well into the blow-out regime. In this ($\tilde{Q}$=80) case, one may expect that the plasma wake-fields would display significant degradation in amplitude due to nonlinear effects.

Finally, we end our discussion by noting that the plots of very nonlinear plasma disturbances shown in Figs. 3 and 9 emphasize that, while a notable amount of the energy deposited by the beam into the plasma is lost to creation of large amplitude particles that do not contribute to the wave, the majority of the plasma electrons taking part in the oscillation (especially in the more realistic long-beam case), engage in coherent motion. The degree to which these electrons stay localized together in both radial and longitudinal dimensions is remarkable. We may speculate that a more detailed examination of the microscopic physics of the plasma motion accompanying the formation of the accelerating phase of the wake may yield an explanation for such coherence.

Acknowledgments

This work supported by U.S. Dept. of Energy grant DE-FG03-92ER40693.

References

1. N. Barov, J.B. Rosenzweig, M.C. Thompson, and R.Yoder, "Energy loss of a high charge bunched electron beam in plasma: simulation, acceleration and linear scaling", these proceedings.
2. Pisin Chen, J.M. Dawson, Robert Huff, and T. Katsouleas, *Phys. Rev. Lett.* **54**, 693 (1985), and *Phys. Rev. Lett.* **55**, 1537 (1985).
3. J.B. Rosenzweig, B. Breizman, T. Katsouleas and J.J. Su, *Phys.Rev.A* **44**, R6189 (1991).
4. N. Barov and J.B. Rosenzweig, *Phys. Rev. E* **49** 4407(1994).
5. N. Barov, M.E. Conde, W. Gai, and J.B. Rosenzweig, *Physical Review Letters* **80**, 81 (1998).
6. C. Clayton, *et al., Phys. Rev. Lett.* **88**, 154801 (2002)
7. S. Lee, T. Katsouleas, R. Hemker and W. Mori, *Phys. Rev. E* **61,** 7012 (2000).
8. D. L. Bruhwiler, et al. *Phys. Rev. ST Accel. Beams* **4**, 101302 (2001).
9. J.B. Rosenzweig, in *Proc. of the 1992 Linear Accelerator Conf.*, (AECL-10728, Chalk River, 1993).
10. J.B. Rosenzweig, *et al., Nuclear Instruments and Methods A* **410** 532 (1998).
11. N. Barov, *et al., Phys. Rev. Special Topics – Accel. Beams* **3** 011301 (2000).
12. J.D. Jackson, *Classical Electrodynamics*, 2nd Ed, (Wiley, New York, 1975).
13. N. Barov, *et al., Proc. 2001 Part. Accel. Conf.* Ed. P. Lucas and S. Webber (IEEE, 2002) 126.
14. M. Hogan, *et al.*, Phys. Plasmas 7, 224 (2000).
15. P. Muggli, *et al.,* "Plasma wakefield experiments with 28.5 GeV electron and positron beams", these proceedings.
16. S. Lee, et al., *Phys. Rev. Special Topics – Accel. Beams* **4** 011011 (2002).
17. R. Ruth, R. D. Ruth and A. W. Chao, in *Laser Acceleration of Particles.* AIP Conf. Proc. **91**, Ed. P. Channell (AIP, New York, 1982).
18. D. Smithe, *et al., Comp. Phys. Commun.* **78,** 54 (1995).
19. J.B. Rosenzweig and E. Colby, in *Advanced Accelerator Concepts*, p. 724
20. (AIP Conf. Proc. 335, 1995).

21. H. Braun, Phys. Rev. Lett. 84, 658 (2000).
22. S. Heifets, G. Stupakov, and S. Krinsky, Phys. Rev. ST Accel. Beams **5**, 064401 (2002)
23. J.B. Rosenzweig, *Phys. Rev. Lett.* **58**, 555 (1987).
24. M. Ferrario, et al. New Design Study and Related Experimental Program for the LCLS RF Photoinjector", in *Proc.EPAC 2000*, 1642 (Austrian Acad. Sci. Press, 2000).

PLASMA DENSITY TRANSITION TRAPPING AS A POSSIBLE HIGH-BRIGHTNESS ELECTRON BEAM SOURCE

M.C. THOMPSON AND J.B. ROSENZWEIG

Department of Physics and Astronomy,
University of California, Los Angeles, CA 90095

H. SUK

Korea Electrotechnology Research Institute,
Changwon 641-120, Republic of Korea

Plasma density transition trapping is a recently purposed self-injection scheme for plasma wake-field accelerators. This technique uses a sharp downward plasma density transition to trap and accelerate background plasma electron in a plasma wake-field. This paper examines the quality of electron beams captured using this scheme in terms of emittance, energy spread, and brightness. Two-dimensional Particle-In-Cell (PIC) simulations show that these parameters can be optimized by manipulating the plasma density profile. We also develop, and support with simulations, a set of scaling laws that predict how the brightness of transition trapping beams scales with the plasma density of the system. These scaling laws indicate that transition trapping can produce beams with brightness $\geq 5x10^{14}$Amp/(m-rad)^2. A proof-of-principle transition trapping experiment is planned for the UCLA Neptune Laboratory in the near future. The proposed experiment and its status are described in detail.

1. Introduction

In a plasma wake field accelerator (PWFA) a short, high density electron beam is used to drive large amplitude plasma waves. Accelerating gradients in these systems scale with the non-relativistic plasma frequency $\omega_p = (4\pi n_0 e^2/m_e)^{1/2}$, where n_0 is the plasma density, e is the electron charge, and m_e is the electron mass. It follows that high gradient PWFAs have very short period waves. Accelerating a second beam in such a system and maintaining its energy spread and emittance requires injecting a sub-picosecond beam into the drive beam's wake with well sub-picosecond timing accuracy. This is often referred to as witness beam injection, which has never been fully achieved experimentally. All experiments to date that have injected external electrons into accelerating plasma waves have used either continuous electron beams or beam pulses that were long compared to the plasma wave [1,2,3,4,5]. As a result the accelerated electrons had induced energy spread equivalent to the acceleration, which would eventually result in 100% energy spread.

The difficulty of witness beam injection makes it desirable to develop a system in which charge is automatically loaded into the accelerating portion of the wake by the drive beam's interaction with its environment. Suk et al. [6] recently proposed a new self-trapping system for the use in the blow out regime of PWFA where $n_b > n_0$ (underdense condition). In this scheme the beam passes though a sharp drop in plasma density where the length of the transition between the high density in region one (1) and the lower density in region two (2) is smaller than the plasma skin depth $k_p^{-1} = v_b/w_p$, where $v_b \cong c$ the driving pulse's velocity. As the drive beam's wake passes the sudden transition there is a period of time in which it spans both regions. The portion of the wake in region 2 has lower fields and a longer wavelength than the portion in region 1. This means that a certain population of the plasma electrons at the boundary will suddenly find themselves rephased into an accelerating portion of the region 2 wake. When the parameters are correctly set, these rephased electrons are inserted far enough into the accelerating region to be trapped and subsequently accelerated to high energy.

The scheme originally proposed by Suk, et al., like all plasma injection systems [7,8], provides very short injection pulses that are phase locked to the plasma wave, but suffer from a lack of beam quality, as defined by energy spread and transverse emittance. In this paper, we expand on the original proposed transition trapping system, examining in greater detail the issues of trapped beam quality and scaling of the system to higher plasma density. We also present a detailed plan for a plasma density transition trapping proof-of-principle experiment and report on substantial progress towards realizing this experiment.

2. Trapping Scenarios

The current development of the idea of plasma density transition trapping centers around the detailed study of two particular scenarios. The first case uses a high charge beam to create a very strong blowout [9] of plasma electrons in a plasma with a simple step function longitudinal plasma density profile. This is the original case proposed for transition trapping [6]. The

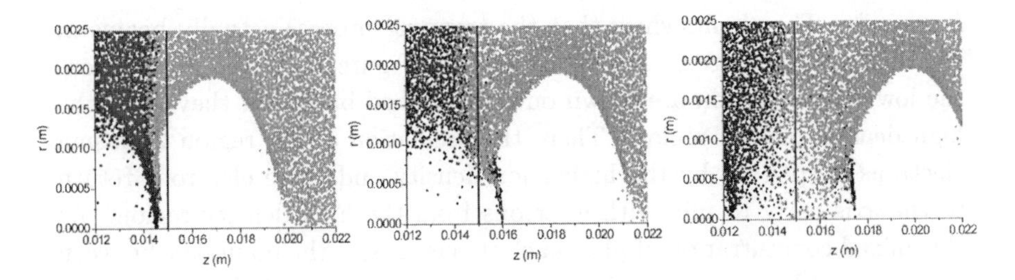

Figure 1. Illustration of particle trapping in the strong blowout case. The vertical black line indicates the original position of the density transition. Plasma electron particles originating in the high density region are colored black while particles originating in the low density region are colored grey.

second case is optimized for a proof-of-principle trapping experiment. This case uses a beam of much more modest charge to create a weak blowout in the high density region, and uses a sloped plasma density profile to enhance charge capture and reduce energy spread.

2.1. *Strong Blowout Scenario*

The strong blowout scenario uses the parameters presented in Table 1 and illustrated in Figure 1. The plasma density profile is a simple step function with a constant density of $n_{region1} = 5\text{x}10^{13}\text{cm}^{-3}$ in the high density region and a constant density of $n_{region2} = 3.5\text{x}10^{13}\text{cm}^{-3}$ in the low density region. The high charge driver produces a very strong blowout, which in turn results in a clear picture of the trapping process.

In order to increase our understanding of the trapping mechanism we preformed a series of simulation with the 2D Particle-In-Cell code MAGIC [10] in which the high and low density plasma electron populations are tracked

Table 1. Drive and Captured Beam Parameters in the Strong Blowout Case. Figures for the captured beam are for the core of the captured beam, which is about 20% of the captured particles, after 12 cm of acceleration.

Drive Beam		Captured Beam	
Beam Energy	50 MeV	Beam Energy	56 MeV
Beam Charge	63 nC	Beam Charge	5.9 nC
Beam Duration σ_t	3 psec	Beam Duration σ_t	161 fsec
Beam Radius σ_r	500 μm	Beam Radius σ_r	112 μm
Peak Beam Density	$1.2x10^{14}\text{cm}^{-3}$	Normalized Emittance ε_x	155 mm-mrad
		Total Energy Spread	13%

separately. The results show that the trapping process actually begins in the high density region, as can be seen in Figure 1. As electrons from the low density region are blown out and pushed backward they enter the high density plasma region. There the oscillation of the region 2 plasma electrons is sped up by the higher ion density and these electrons return to the axis early to mix with electrons from the high density region. As this mixed concentration of plasma electrons crosses the boundary between the high and low density regions many of the electrons find themselves in an accelerating phase of the low density plasma wake and are trapped and accelerated.

The properties of the beam captured in this scenario are listed in the second column of Table 1. The captured beam is very short and has a small radius, both of which originate from the small accelerating volume of the accelerating plasma wave. The beam also has a high charge that results from the very high concentration of electrons in the oscillation density spike that are injected. Unfortunately, the captured beam has a significant energy spread that results from the fast variation in the plasma wake field accelerating gradient where the particles are captured. The beam also has a poor transverse emittance. This is an unavoidable consequence of trapping background plasma particles in the strong blowout regime. The large amplitudes of transverse momenta imparted to the plasma electrons as the drive beam space charge blows them out to the side remains with the particles as they are trapped and accelerated to high energy.

In addition to the undesirable emittance and energy spread properties of the captured beam, this transition trapping scenario is also impractical from an experimental stand point. The drive beam parameters listed in the first column of Table 1 are not currently achievable. For this reason we began to look at what sort of trapping experiments could be done with the more modest driver beams that are available. During this development we also found ways to improve both the emittance and energy spread of the captured beams.

2.2. *Weak Blowout Scenario*

A great deal can be learned about the mechanism and dynamics of density transition trapping by comparing the strong blowout case previously described to a case in which a weak blowout is used. Our standard example of a weak blowout case is the proof-of-principle experimental case designed for the Neptune Advanced Accelerator Laboratory at UCLA [11]. This case

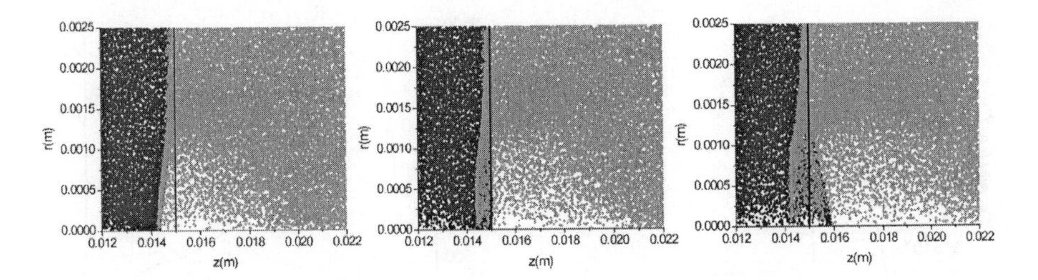

Figure 2. Illustration of particle trapping in the weak blowout case. This figure is directly comparable with Figure 1. The scale and particle coloring are identical. Note that the trapping mechanism is essentially the same except that it proceeds more slowly due to the low plasma density in the down stream region compared to strong blowout case. The weaker blowout also leads to much less transverse disturbance in the plasma, which in turn yields lower emittance.

was developed and optimized for parameters achievable at the Neptune Laboratory through extensive simulations with MAGIC. The driving beam parameters of the simulation are shown in Table 2. The driving beam has a ramped longitudinal profile as shown in Figure 3. Ramped profiles of this type maximize the transformer ratio of the wake field [12] and can be produced using a negative R_{56} magnet compressor system. We are developing such a compressor system for the Neptune Laboratory [13]. While the ramped beam profile improves performance, it is not critical to this trapping scenario.

The plasma density profile used in this case is illustrated in Figure 3. The plasma density profile is tailored to maximize the amount of charge captured while maintaining an acceptable amount of acceleration. The first cm of the profile reflects a realistic finite rise time from zero to the maximum plasma density. After 5 mm of maximum density the transition takes place and the density is reduced to 18% of the maximum. This density

Table 2. Drive and Captured Beam Parameters in the Weak Blowout Case.

Drive Beam		Captured Beam	
Beam Energy	14 MeV	Beam Energy	1.2 MeV
Beam Charge	5.9 nC	Beam Charge	120 pC
Beam Duration	6 psec	Beam Duration σ_t	1 psec
Beam Radius σ_r	540 μm	Beam Radius σ_r	380 μm
Normalized Emittance ε_x	15 mm-mrad	Normalized Emittance ε_x	15 mm-mrad
Peak Beam Density	$4x10^{13}$cm^{-3}	Total Energy Spread	11%

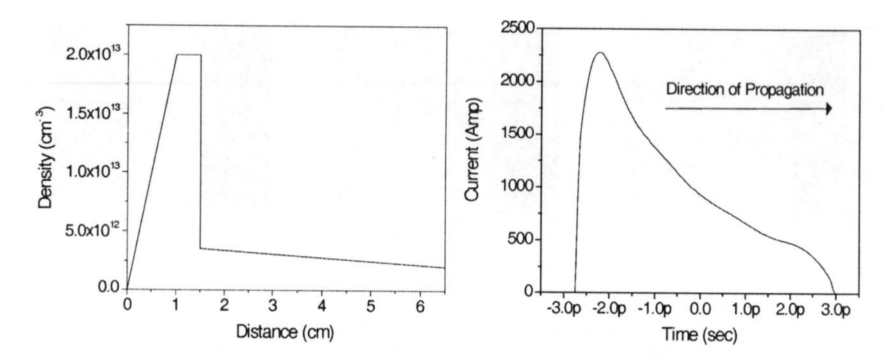

Figure 3. Plasma Density (left) and Drive Beam Current (right) Profiles.

drop is near the optimum to maximize charge capture. Decreasing the density of region 2 increases the wavelength of the accelerating plasma wave. This has the effect of enlarging the volume of the capture region and enhancing the amount of charge trapped. Lowering the plasma density also reduces the accelerating gradient, however, reducing the number of initially captured particles that ultimately achieve resonance with the accelerating wave. These two effects compete with the charge capture maximum occurring at $n_{region2} = 0.18 n_{region1}$. To quantify the degree of blowout in this case, we note that the electron beam density is 2 times larger than the peak plasma density of $2\text{x}10^{13}\text{cm}^{-3}$, as can been seen from Table 2 and Figure 3.

In the simulation the transition is approximated as a perfect step function; the validity of this assumption will be elaborated on later in this paper. Finally, the gradual decline in plasma density after the transition slowly increases the plasma wavelength, and thus the extent of the accelerating phase of the wake field region. The growth in plasma wavelength reduces the peak gradient but rephases the captured charge forward of the peak field of the wake into a region of slightly weaker, but more uniform, acceleration. This rephasing both increases the amount of charge trapped and reduces energy spread. Gradually declining post transition plasma densities have been shown to have similar benefits in the strong blowout regime [14].

The parameters of the bunch of captured plasma electrons are given in Table 2. The captured plasma electrons form a well defined beam of substantial charge that can be propagated and detected without major difficulty. The captured beam is also well separated from the drive beam in

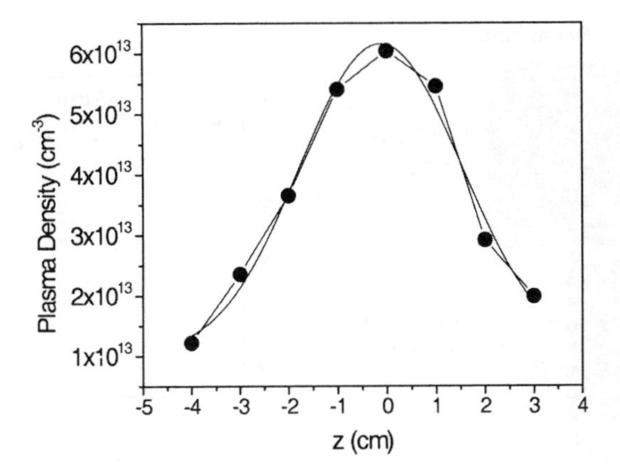

Figure 4. Measured Transverse Density Profile of the Plasma Column

energy and should be easy to isolate. If a bi-gaussian beam with $\sigma_t = 1.5$ ps is substituted for the ramped profile shown in Figure 3 the captured beam parameters remain unchanged except for a 20% loss of captured charge.

3. Experimental Plans

An experiment is planned for the Neptune laboratory at UCLA with the parameters presented in Table 2 and Figure 3. By modifying an existing pulse discharge plasma source [15] we have created a plasma column with a peak density of $6x10^{13}cm^{-3}$. As shown in Figure 4 the raw plasma column has a gaussian transverse density profile and over 6 cm of the plasma has density greater than $2x10^{13}cm^{-3}$, the required peak density for the trapping experiment.

Experimental realization of plasma density transition trapping depends on the creation of sharp density transitions. The limit on the sharpness of the transition necessary to produce trapping is set by the trapping condition

$$k_p^{region1} L_{Transition} < 1. \tag{1}$$

As can be seen from Figure 5, this is a very strict condition. The turn on of the capture in this regime is nearly a step function.

The creation of a density transition that satisfies this criteria is a interesting experimental challenge. At high plasma densities, $n \geq 10^{14}\text{cm}^{-3}$, it will probably be necessary to directly create the plasma with the required density profile already built in. This might be accomplished though photoionization using a laser with an intensity profile that matches the desired

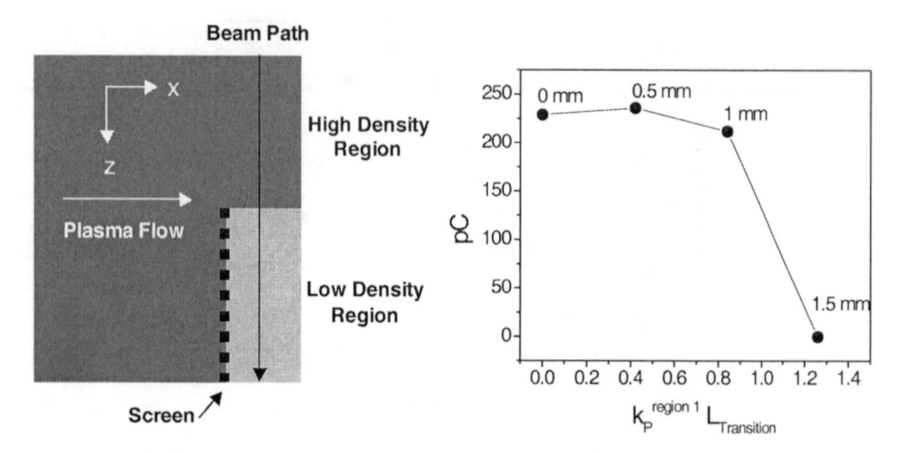

Figure 5. Left: Simplified diagram of a plasma density transition produced by a obstructing screen. Right: Simulated dependence of captured charge on transition length in the proposed experimental case. Each point is marked with the length of the transition.

plasma density profile or using a uniform laser to ionize a dual density gas jet. At lower densities such as 10^{13}cm^{-3}, which can be easily produced using discharges, it is possible to consider a simpler method using a masking screen to generate the density transition. The basic concept of operation for the masking screen is illustrated on the left of Figure 5. Consider a system in which the plasma discharge is separated from the path of the driver beam. Once the plasma is created in the discharge apparatus it will diffuse and flow towards the beam path. If a perforated metal foil or grid of wires is placed in the path of the plasma flow it will block a portion of the flow creating a low density region. Unfortunately, the plasma density transition will not remain sharp as the distance from the screen grows as portrayed in the simple picture of Figure 5. In reality the two plasma regions will diffuse into one another on the far side of the screen so that the plasma density transition will lengthen and blur as the distance from the screen edge increases. This process can be quantified using a simple model based on the velocities with which the plasma diffuses as shown in Figure 6. On the far side of the screen from the plasma source the high density plasma will continue to flow past the screen in the direction of the bulk plasma flow with a velocity $V_{||}$ and will begin flowing into the low density region with

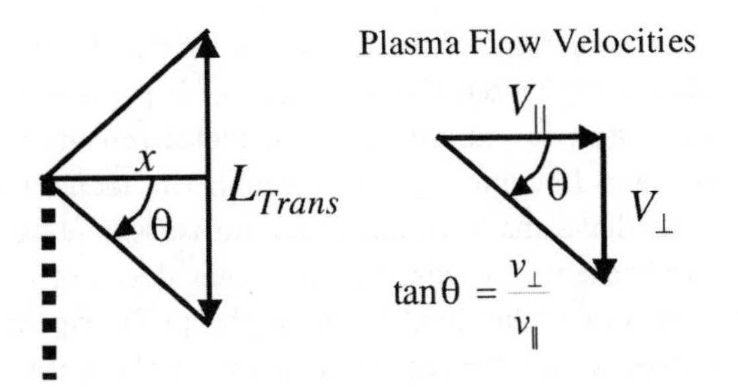

Figure 6. Illustration of the transition geometry.

a velocity $V_{\perp}$. The sum of these two vectors defines the line which marks the end of the transition into the low density plasma region. Symmetry dictates that the start of the transition in the high density region can be defined in the same way so that the total transition length is given by

$$L_{Trans} = 2x\tan\theta = 2x\frac{V_{\perp}}{V_{\parallel}}. \tag{2}$$

Since our plasma is weakly magnetized it is reasonable to assume that the parallel and perpendicular plasma flow velocities are approximately equal. This assumption leads to the conclusion

$$V_{\perp} \approx V_{\parallel} \rightarrow L_{Trans} = 2x, \tag{3}$$

which in turn leads to a new experimental constraint on achieving efficient trapping

$$x < \frac{k_p^{-1}}{2}. \tag{4}$$

This new trapping condition for obstructing screens requires that the drive beam passes within half a plasma skin depth of the boundary. For a $2x10^{13}\mathrm{cm}^{-3}$ plasma the drive beam will have to pass within 600μm of the screen. This level of pointing accuracy and stability is not difficult to achieve.

We have explored the validity of this model through simulations. We began by looking at MAGIC PIC simulations in which a neutral plasma was initialized in half the simulation volume and allowed to diffuse through a periodic series of conduction obstructions into the rest of the volume. This is essentially the situation that we wish to create in our discharge plasma source. By plotting histograms of the simulation particles contained in small bands of Δx at various distances from the edge of the obstructions we were able to make predictions of the transition length at various positions. The results of this study match the prediction of Eq. (3) almost exactly.

We have also begun to experiment with metal screens in our plasma source. By moving a Langmuir probe through the transition region behind the screen we have made rudimentary measurements of the plasma transition. Unfortunately, our existing equipment does not allow a precise distance to be set and maintained between the probe tip and the metal screen. This makes a direct comparison between the data and Eq. (3) impossible. The data do, however, appear to be in approximate agreement with Eq. (3) and we are upgrading our equipment to make more precise measurements. While giving inconclusive measurements of the plasma transition, these early screen experiments did confirm our ability to reliably set the plasma density behind the screen. In the limit of thin sheaths, the density of the plasma in the low density region behind the screen should be related to high density plasma by the relation

$$\frac{n_{low}}{n_{high}} \approx \frac{A_{ScreenOpenArea}}{A_{ScreenTotalArea}}. \tag{5}$$

The screen experiments were conducted using a micro-etch perforated stainless steel screen with 152μm diameter holes and an open area of 21%. The observed plasma density behind the screen was 19.5% of the unfiltered plasma density, which agrees well with Eq. (5). This means that we should be able to produce the density profile shown in Figure 3 from the natural gaussian profile of the plasma column by varying the open area of the screen.

Propagating a beam so close to a metallic screen leads to other difficulties. Interactions with the screen over the entire length of the low density plasma region will completely disrupt the processes of trapping and acceleration. To circumvent this problem we examined many alternative geometries and arrived at a solution based on a screen with a solid metal baffle attached to its edge. As shown in Figure 7 this baffle moves the sharp portion of the density transition away from the screen so that the beam and plasma wake will no longer interact with it. During the trapping process at the transition, however, the beam and wake still interacts with the baffle. The primary effect of the baffle is to block a portion of the particles participating in the plasma wake oscillation, as illustrated on the right in Figure 7.

Simulating the effects of the baffle on particle trapping is a complex problem. The baffle breaks the cylindrical symmetry of the problem requir-

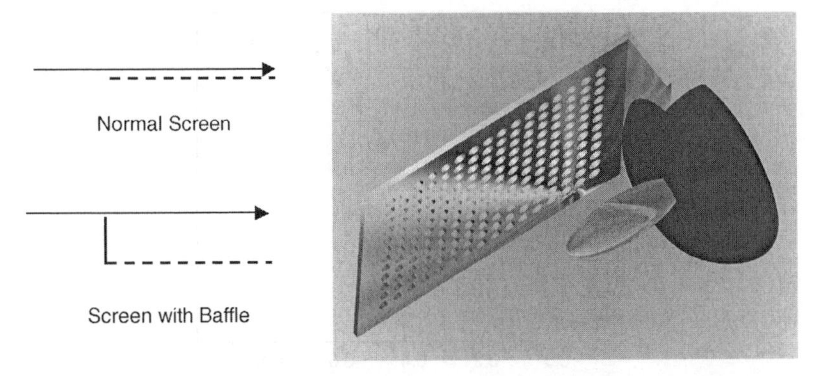

Figure 7. Artist's conception of partial blocking of wake particles by the baffle. The prolate spheroid represents the electron beam and the toroid that partially intersects the screen baffle is the plasma wake.

ing that any simulations of its effects must be done in three dimensions.

We used the three dimensional version of the PIC code MAGIC to simulate this problem. Unfortunately, MAGIC 3D is not a parallel code, which ultimately limited the accuracy we could obtain in our simulation, but it was the only three dimensional code available to us that would allow a conducting boundary to be placed near the beam path in the plasma. We were able to reproduce the results of our 2D simulations almost exactly with the 3D version of the code. We then modelled the same system with a metallic baffle at various distances from the beam center. The results of these simulations are summarized in Figure 8. The points in the graph are taken from simulations in which the simulation cells are $0.17k_p^{-1}$ on a side, which is the maximum resolution we could obtain with our computing hardware.

Since Eq. (4) indicates that the beam must pass within $k_p^{-1}/2$ of the baffle edge, the results shown in Figure 8 predict an approximate 50% loss of total captured charge. This may not translate into a 50% loss of particles in the beam core, however, since the large amplitude particles block by the baffle are not necessarily the ones that form the beam core. The 3D simulations lacked the resolution to resolve this question.

The final issue with the use of screen produced plasma density transitions is the rapid growth of the transition length with distance from the screen. The growth rate is large enough that there will be a significant transition length gradient over distance spanned by the plasma wake field. The effect of this transition length gradient is unknown, but will soon be

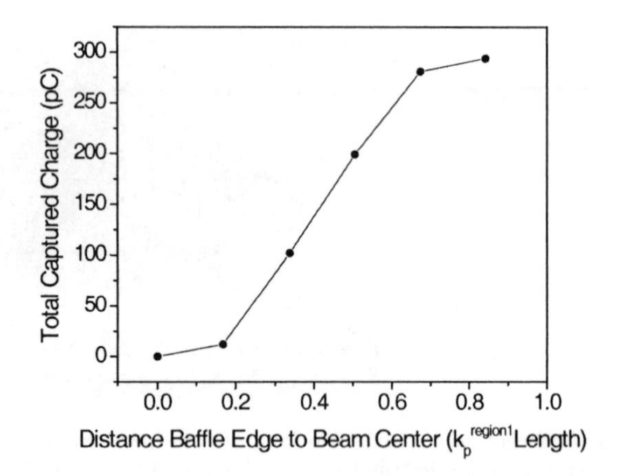

Figure 8. Effect of the Beam-Baffle Distance on Trapping

examined in simulation. We expect this effect to produce another minor but acceptable degradation of the trapping performance.

4. Driver Charge Scaling

hspace.1inWhile the captured beam parameters presented in Table 2 are adequate for a first step, proof-of-principle type experiment they are far inferior to the state of the art beams produced in modern photoinjectors. In order to find a systematic way to improve the captured beam parameters we began to simulate the effects of scaling up the drive beam charge without altering the rest of the experiment. The results of these simulation are shown in Figure 9. Increasing the driver charge increases the strength of the blowout forming a larger amplitude more non-linear plasma wave. It follows that all the accelerating fields in the problem are increased as is the size of the accelerating wave. The impact on the captured beam is clearly shown in Figure 9. The amount of charge captured, the length of the beam, and the emittance all grow as the driver charge is increased. Although it deserves more detailed study, simple scaling of the driver charge appears to lead to bigger captured beams but not higher quality ones.

5. Wavelength Scaled Sources

We have seen the performance of density transition trapping in densities $n_0 \sim 10^{13}\text{cm}^{-3}$ in the preceding sections, as well as how the performance

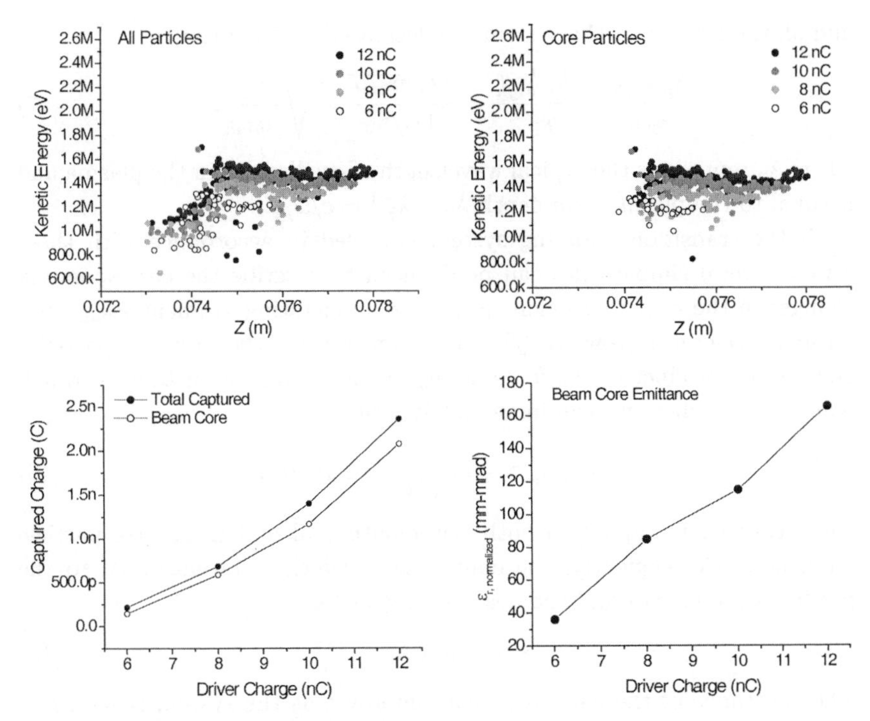

Figure 9. Variation of the captured beam parameters with scaling of the driver beam charge in the weak blowout case.

changes with driver charge scaling. From these studies it is clear that transition trapping at $n_0 \sim 10^{13}\text{cm}^{-3}$ produces beams of low brightness when compared to the benchmark of the LCLS photoinjector [16], see Table 3. It is therefore interesting to examine how the captured beam performance scales with plasma density or, equivalently, the plasma wavelength. This type of wavelength scaling, and its impact on predicting beam emittance and brightness behavior, has been previously examined in the context of rf acceleration [17] in photoinjector sources, where the beam displays palsma-type behavior. In order to scale the transition trapping system to a higher plasma density n_{high} all the charge densities in the system must be increased by the ratio,

$$n_{high}/n_0, \tag{6}$$

and all the lengths in the system are decreased by the ratio,

$$\frac{\lambda_{p\ high}}{\lambda_{p\ 0}} = \frac{k_{p\ high}^{-1}}{k_{p\ 0}^{-1}} = \frac{1/\sqrt{n_{high}}}{1/\sqrt{n_0}} = \sqrt{\frac{n_0}{n_{high}}}, \tag{7}$$

where λ_p represents the typical wavelength of oscillations in the plasma and is equal to the plasma skin depth $\lambda_p = k_p^{-1} = c/\omega_{pe}$.

If the transition trapping system is scaled in accordance with these ratios, several simple rules can be deduced to describe the corresponding changes in the captured beam. Since the accelerating gradient E_{max} in a plasma accelerator grows as $\sqrt{n_0}$ and momenta p of the captured particles goes as the product of the force acting on them and the distance through which it is applied, we can immediately write,

$$p \propto E_{max}\lambda_p \propto \sqrt{n_0}\frac{1}{\sqrt{n_0}} = Constant. \tag{8}$$

This scaling law applies to both the longitudinal and transverse particle momenta. Consequently, the emittance ε, which is proportional to the product of momenta and the beam size, goes like

$$\varepsilon \propto \lambda_p p \propto \lambda_p. \tag{9}$$

The emittance of the captured beam improves as the system is scaled to higher density as a result of the reduction in the transverse beam size.

The amount of charge captured Q depends on both the available plasma electron density n_0 and the volume of the accelerating portion of the wave, which is proportional to λ_p^3. This scaling can be written as

$$Q \propto n_0\lambda_p^3 \propto n_0(\frac{1}{\sqrt{n_0}})^2\lambda_p \propto \lambda_p. \tag{10}$$

While the captured charge goes down as the plasma wavelength is reduced, the current I remains constant since the length of the beam also goes down with the plasma wavelength,

$$I \propto \frac{Q}{\lambda_p/c} = Constant. \tag{11}$$

Finally we can combine the scaling laws for emittance and current to deduce the scaling of the beam brightness B

$$B \propto \frac{I}{\varepsilon^2} \propto \frac{1}{\lambda_p^2} \propto n_0. \tag{12}$$

Thus the brightness of electron beams produced using density transition trapping increases linearly with the density of the plasma.

Table 3. Simulations of Wavelength Scaling using MAGIC 2D.

Peak Density	$2x10^{13}$cm^{-3}	$2x10^{15}$cm^{-3}	$2x10^{17}$cm^{-3}	
$\sigma_{t,Driver}$	1.5 psec	150 fsec	15 fsec	LCLS
Q_{Driver}	10 nC	1 nC	100 pC	Photoinjector
$\sigma_{t,Trap}$	2.7 psec	270 fsec	28 fsec	Specifications
Q_{Trap}	1.2 nC	120 pC	12 pC	———
$I_{Peak,Trap}$	163 Amp	166 Amp	166 Amp	100 Amp
$\varepsilon_{x,norm,Trap}$	57 mm-mrad	5.9 mm-mrad	0.6 mm-mrad	0.6 mm-mrad
$B_{norm,Trap}$	$5x10^{10}$	$5x10^{12}$	$5x10^{14}$	$2.8x10^{14}$

These scaling laws were tested using the 2D PIC code MAGIC. The cases examined are scaled versions of the proof-of-principle experimental case with a slightly larger driver charge. The results are summarized in Table 3. The simulation results follow the scaling laws precisely in the range studied. At $2x10^{17}cm^{-3}$ transition trapping can produce an extremely short beam with excellent emittance and a brightness that exceeds state of the art photo-injectors. The drive beams needed at all densities must be of similar length and approximately one order of magnitude greater charge than the beams they capture. The emittance of the driver, however, is irrelevant as long as the driving beam can be focused sufficiently to match into the plasma. This means that plasma density transition trapping might be used as an emittance transformer to produce short, low emittance beam from short beams with high emittances that were produced using extreme magnetic compression or other techniques that produce significant emittance growth. The feasibility of this idea is still under study and may be enhanced by our effort to find new scenarios that produce low emittance trapped beams.

As described previously, plasma density transition trapping, at least in the regimes examine so far, produces beams of large emittance do to the sizeable transverse momenta the plasma particles have at capture. Scaling to higher density improves the emittance by reducing the beam size rather than reducing the transverse momentum. We are continuing to explore alternative transition trapping scenarios in an effort to reduce the transverse momentum of the beam further. This may be accomplished by using drive beams that are wide and or long compared to the plasma skin depth. The development of the technique of foil trapping, which is discussed in the next section, might also lead to lower transverse momenta.

6. Future Directions

In order to proceed beyond a proof-of-principle transition trapping experiment will necessarily require scaling to higher plasma densities. This will require improvements to both the driver beam and higher density plasma sources with sharp transitions. The production of very short, high current electron drive beams is a matter discussed at great length elsewhere. Ideas for producing plasmas with transition that satisfy Eq. (1) at high densities $n \geq 10^{14}\text{cm}^{-3}$ are still in the conceptual phase. H. Suk plans to pursue development of a laser ionized dual density gas jet for use in laser wake field driven transition trapping [14]. We have tentative plans at UCLA to pursue development of a high density source based on photo-ionization of Lithium using a laser with step function intensity profile.

In the extreme limit, one can imagine creating an ultra-sharp transition into a plasma by simply replacing the high density plasma region in a transition trapping scenario with a solid metal foil. Electrons would be provided for trapping from the foil via Fowler-Nordheim field emission [18]. Since this situation is much easier to produce experimentally then sharp plasma density drops, we have begun to look at the idea closely.

The field values necessary for significant Fowler-Nordeim emission are easy to achieve in current plasma wake field experiments. N. Barov et al. have produced wake fields $\geq 140\text{MeV/m}$ in a 10^{14}cm^{-3} plasma at FNAL [19]. In this experiment the drive beam enters the plasma through a metal foil, one side of which is immersed in the plasma and experiences the large plasma fields. Taking a reasonable value of $\beta \geq 50$ for the microscope surface field enhancement factor of the foil, Fowler-Nordheim theory predicts a large emission $J \geq 100\text{Amp/mm}^2$ under these conditions. Unfortunately, the emission of charge does not guarantee that the emitted charge will be trapped and accelerated. The charges emitted from the foil due to the plasma wake fields start essentially at rest and must be accelerated to resonance with the wave within the same period of the plasma wake. This situation is analogous to that in RF photoinjectors and the same dimensionless parameter can be used to evaluate the plasma wake's potential to capture foil electrons. This parameter, α, is the ratio of the maximum normalized energy gain per unit length and the wave number of the accelerating wave.

$$\alpha = \frac{qE_{\text{max}}}{k_z m_e c^2} = \frac{\left.\frac{d\gamma}{dz}\right|_{\text{max}}}{k_z}, \tag{13}$$

where $k_z = \omega/v_\phi$. The capture of electrons starting from rest typically

Table 4. Comparison of α parameters.

Accelerating Structure	E_{max}	*frequency*	v_ϕ	α
1.6 Cell Photoinjector	80MeV/m	2.856Ghz	c	2.6
Barov et al. Wake Field Experiment (7 nC)	300MeV/m	90Ghz	c	0.3
Experiment with High Charge Driver (70 nC)	1.5GeV/m	90Ghz	c	1.6

requires $\alpha \geq 1$. If we compare the α parameters of the Barov et al. experiment and a standard 1.6 cell photoinjector, see Table 4, we see that a plasma wake is not capable of capturing charge from a foil in this regime since its α is only 0.3. The frequency of the accelerating wave is too high in comparison to the accelerating field and the emitted particles can not achieve resonance with the wave.

The peak accelerating field can be increased by increasing the driver beam charge. If this is done while holding the plasma density constant, the plasma frequency will remain essentially unchanged and α will increase. The driver charge can be increased to the point where $\alpha > 1$ and charge is captured from the foil in the plasma wake. If the driver charge in the Barov et al. experiment is increased by a factor of ten the α of the system reaches 1.6 and charge is captured. The trapping behavior predicted by the α parameter has been verified by initial MAGIC 2D simulations. Further work needs to be done to explore the parameter space of foil trapping and characterize the captured beams.

7. Conclusions

The theoretical understanding of the basic plasma density transition trapping mechanisms are well developed. The quality of the beams produced by transition trapping is also well understood, as are a number of methods for optimizing the beam quality. We have shown, through the development of the wavelength scaling laws, that at high densities transition trapping can produce beams with brightness $\geq 5\text{x}10^{14}\text{Amp}/(\text{m-rad})^2$. This study shows, for the first time, a pathway toward ultra-high brightness, small energy spread beam production using wake field trapping that is competitive with state-of-the-art photoinjectors. Many variations on the idea of transition trapping, such as foil trapping and the use of drive beams that are long and or wide compared to the plasma skin depth, remain to be studied. In addition, many of the ideas developed in this paper, such as scaling to higher density and the use of gradually declining plasma density profiles, may be applicable to other classes of plasma-based electron beam injectors [7,8].

A detailed plan is in place for a proof-of-principle plasma density transition trapping experiment. This experiment will be conducted at low density using density modifying screens, a technique which we have made substantial progress in developing. The rest of the hardware necessary for the experiment is either already in place, or being constructed at this time. We plan to conduct this experiment in the second or third quarter of 2003.

With further research and refinement plasma density transition trapping holds promise as a future high brightness beam source. This source may take several forms such as an automatically timed PWFA injector or an "emittance transformer" used to convert short beams with poor emittance into short beams with a much lower transverse emittance.

Acknowledgments

The authors would like to thank T.C. Katsouleas and L. Ludeking for their assistance with the MAGIC PIC simulations. The authors also thank Mike Schneider, Chris Muller, Soren Telfer, and Ronald Agustsson for their technical assistance. This work was supported by the U.S. Dept. of Energy grant number DE-FG03-92ER40693.

References

1. Rosenzweig, J. B., Cline, D. B., Cole, B., Figueroa, H., Gai, W., Konecny, R., Norem, J., Schoessow, P., and Simpson, J., *Phys. Rev. Lett.*, **61**, 98 (1988).
2. Clayton, C. E., Marsh, K. A., Dyson, A., Everett, M., Lal, A., Leemans, W. P., Williams, R., and Joshi, C., *Phys. Rev. Lett.*, **70**, 37 (1993).
3. Amiranoff, F., Baton, S., Bernard, D., Cros, B., Descamps, D., Dorchies, F., Jacquet, F., Malka, V., Marques, J. R., Matthieussent, G., Mine, P., Modena, A., Mora, P., Morillo, J., and Najmudin, Z., *Phys. Rev. Lett.*, **81**, 995 (1998).
4. Barov, N., Rosenzweig, J. B., Conde, M. E., Gai, W., and Power, J. G., *Phys. Rev. Special Topics - Accel. Beams*, **3**, 011301 (2000).
5. Hogan, et al., M. J., *Physics of Plasmas*, **7**, 2241 (2000).
6. Suk, H., Barov, N., Rosenzweig, J. B., and Esarey, E., *Phys. Rev. Lett.*, **86**, 1011 (2001).
7. Umstadter, D., Kim, J. K., and Dodd, E., *Phys. Rev. Lett.*, **76**, 2073 (1996).
8. Esarey, E., Hubbard, R. F., Leemans, W. P., Ting, A., and Sprangle, P., *Phys. Rev. Lett.*, **79**, 2682 (1997).
9. Rosenzweig, J. B., Breizman, B., Katsouleas, T., and Su, J. J., *Phys. Rev. A*, **44**, R6189 (1991).
10. Goplen, B., Ludeking, L., Smithe, D., and Warren, G., *Computer Physics Communications*, **87**, 54–86 (1995).

11. Thompson, M. C., Clayton, C. E., England, J., Rosenzweig, J. B., and Suk, H., "Beam-Plasma Interaction Experiments at the UCLA Neptune Laboratory," in *Proc. PAC 2001*, IEEE, 2001, p. 4014.
12. Chen, P., Su, J. J., Dawson, J. M., Bane, K. L., and Wilson, P. B., *Phys. Rev. Lett.*, **56**, 1252–1255 (1986).
13. England, R. J., Rosenzweig, J. B., and Thompson, M. C., in *Advanced Accelerator Concepts*, 2002, vol. 647 of *AIP Conf. Proc.*, p. 884.
14. Suk, H., *Journal of Applied Physics*, **91**, 487 (2002).
15. Suk, H., Clayton, C. E., Hairapetian, G., Joshi, C., Loh, M., Muggli, P., Narang, R., Pellegrini, C., Rosenzweig, J. B., and Katsouleas, T. C., "Underdense Plasma Lens Experiment at the UCLA Neptune Laboratory," in *Proc. PAC 1999*, IEEE, 1999, p. 3708.
16. Ferrario, et al., M., "New Design Study and Related Experimental Program for the LCLS RF Photoinjector," in *Proc. EPAC 2000*, Austrain Acad. Sci. Press, 2000, p. 1642.
17. nB. Rosenzweig, J., and Colby, E., "Charge and Wavelength Scaling of RF Photoinjector Designs," in *Advanced Accelerator Concepts*, 1995, vol. 335 of *AIP Conf. Proc.*, p. 724.
18. Fowler, R. H., and Nordheim, L., *Proceedings of the Royal Society of London*, **A119**, 173–181 (1928).
19. Barov, N., Bishofberger, K., Rosenzweig, J. B., Carneiro, J. P., Colestock, P., Edwards, H., Fitch, M. J., Hartung, W., and Santucci, J., "Ultra High-Gradient Energy Loss by a Pulsed Electron Beam in a Plasma," in *Proc. PAC 2001*, IEEE, 2001, p. 126.

GENERATION OF RELATIVISTIC ELECTRONS VIA INTERACTION BETWEEN ULTRA-SHORT LASER PULSE AND SUPERSONIC GAS JET

Mitsuru Uesaka[1], Alexei Zhidkov [1,2], Tomonao Hosokai [1], Kenichi Kinoshita[1,2], Takahiro Watanabe [1], Koji Yoshii [1], Toru Ueda[1], Hideyuki Kotaki[3], Masaki Kando[3], Kazuhisa Nakajima[4]

[1]Nuclear Engineering Research Laboratory School of Engineering, University of Tokyo, 22-2 Shirane-shirakata, Tokai, Naka, Ibaraki, Japan

[2]National Institute of Radiological Sciences, 4-9-1 Anagawa Inage Chiba, Japan

[3]Advanced Photon Research Center Kansai Research Establishment Japan Atomic Energy Research Institute, 8-1 Umemi-dai Kizu Soraku Kyoto, Japan

[4] High Energy Accelerator Research Organization (KEK), Tsukuba, Ibaraki, 305-0801 Japan

E-mail: uesaka@tokai.t.u-tokyo.ac.jp

We have studied generation of relativistic electrons by interaction between a high intensity ultra-short laser pulse (Ti:Sapphire, 12 TW, 50 fs, λ=790 nm) and gas jet. In the experiment, spatial and energy distribution of energetic electrons produced by an ultra-short, intense laser pulse in a He gas jet are measured. They depend strongly on the contrast ratio and shape of the laser prepulse. In the case of a proper prepulse the electrons are injected at the shock front produced by the prepulse and accelerated by consequent plasma wake-field up to tens MeV forming a narrow-coned ejection angle. In the case of non-monotonic prepulse, hydrodynamic instability leads to a broader, spotted spatial distribution. The numerical analysis based on a 2D hydrodynamics (for the laser prepulse) and 2D particle-in-cell simulation justify the mechanism of electron injection and acceleration.

1. Introduction

Recent rapid progress in intense ultra-short pulse lasers has opened high-field sciences such as particle acceleration via laser-plasma interactions. Among a number of concepts of the particle acceleration by laser fields, the laser wake-field acceleration (LWFA) in underdense plasma [1] provides one of the most promising approaches to high performance compact electron accelerators. We have been studying numerically and experimentally generation of relativistic electrons in a plasma wake wave-breaking scheme [2-4] aiming at a compact ultra-short pulse relativistic electron accelerator / injector, which we called laser-plasma cathode. In this report we show the recent results of the numerical simulation and the experiment at NERL of Univ.of.Tokyo.

2. Theoretical Study of Self-injection Processes in LWFA

Injection of electron into the wake fields for their further acceleration is the most important process in LWFA. In the usual injection schemes, in coincidence with the laser pulse a high quality electron beam from a conventional linac is injected and then further accelerated by the wake-field [1]. Other schemes inject electrons from the background plasma via the interaction of additional injected laser pulses [5,6] or by wave-breaking [2-4]. The latter is the simplest way to generate energetic electrons and wakes produced by an intense laser pulse. However, the wave-breaking is initially a stochastic process that provokes rapid randomization in energy of accelerated electrons. This makes such kind of injection sometimes inefficient and usually very sensitive to plasma parameter changes. That means that along with study of injection originated from wave-breaking, new schemes based on self-injection processes have to be investigated.

2.1. *Effect of Laser Pre-pulse on the Wave-breaking Injection*

If intensity of the laser pulse is not very high (see Ref. [7]) for the plasma wave amplitude exceeds the wave-breaking field $E_B \sim [2(\omega/\omega_{pl}-1)]^{1/2}\ mc\omega_p/e$, where ω and ω_{pl} the laser and plasma frequency, the wave-breaking appears in a plasma with rather steep density, $\lambda_{pl}\, dN/dx \sim 1$, $\lambda = 2\pi\omega_{pl}/c$, where λ_{pl} is the plasma wave wavelength [8]. However in a gas-jet the density gradient is much smaller, since usually $N/(dN/dx) \sim$ 200/500 μm the injection originated from the wave-breaking of plasma waves hardly occur if only main laser pulse coming. At real condition, a laser prepulse, with approximately 2 ns duration, precedes the main laser pulse. Usual contrast ration varies from $1{:}10^6$ to $1{:}10^7$ for fundamental laser frequency. If the Raleigh length, L_R, is short enough, the prepulse can form a cavity with a shock wave in the front of laser propagation. In contrast to the plasma channel produced by long Raleigh length laser beam [9-12], the length of the cavity is determined by this small L_R, because the energy is deposited in the plasma mostly near the focus point x=0 as

$W(x)\sim 1/(1+(x/L_R)^2)$. For low intensity laser pulse, the electron temperature, T_e, can be estimated via the collision absorption mechanism [13], $dT_e/dt=\Delta\varepsilon\, \nu_{ei}(1eV)/T_e^{\ 3/2}$, where $\Delta\varepsilon= 2\pi e^2I/m\omega^2$ the energy acquired by an electron in a collision; ν_{ei} the frequency of electron –ion collisions. For intensity I=10^{13} W/cm^2 and ion density N_i=$3x10^{18}$ cm^{-3} (in the cavity) and pulse duration τ=2 ns, T_e=150 eV. If $X=C_s\tau> L_R$, where C_s the ion sound speed, a shock wave can be formed in the plasma. If the shock wave relaxation depth $\Delta x\sim(M/m)^{1/2}\, l_i$ (M is the ion mass, l_i the ion free path) less than the plasma wave wavelength l_{pl}, the strong wave-breaking of wake-field produced by the main pulse there can be a good source of injection. For temperature T_e~150 eV in a He gas-jet, the ion sound speed is $C_s\sim 5x10^6$ cm/s and X~100 μm so that the effect appear for the laser pulse with the Raleigh L_R< 100 μm. The shock wave can be generated in get with $\omega_{pl}l_i(M/m)^{1/2}/2\pi c>1$ that gives N_i> $5x10^{18}$ cm^{-3}.

To evaluate the effect of the laser prepulse we solve numerically hydrodynamics equation in the following form [13],

$$\frac{\partial}{\partial t}n+\frac{\partial}{\partial \vec r}(n\vec u)=0$$
$$\frac{\partial}{\partial t}\vec u+\left(\vec u\frac{\partial}{\partial \vec r}\right)\vec u=-\frac{1}{Mn}\frac{\partial}{\partial \vec r}(znT)-\eta\Delta\vec u \qquad (1)$$
$$\frac{\partial}{\partial t}(nT)+\frac{\partial}{\partial \vec r}\left(\frac{5}{3}\vec u nT\right)+n\frac{\partial}{\partial \vec r}\left(\kappa\frac{\partial}{\partial \vec r}T\right)=\frac{8\pi e^2}{3mc\omega^2}\nu_{ei}nI_0\frac{d_0^2}{d_R^2}\exp\left(-0.5r^2/d_R^2\right),$$

where n, z, M are the ion density, charge, and mass; T is the electron temperature; η,κ are the plasma viscosity and thermal conductivity, $\nu_{ei}=4\pi e^2z^2n\Lambda/m^{1/2}T^{\ 3/2}$ is the electron-ion collision frequency with Λ the Coulomb logarithm. The last term of third equation describes laser energy deposition,

$$d_R=d_0\sqrt{1+x^2/L_R^2}\ ,$$

where d_0 is the laser spot size, x is coordinate in the laser propagation direction, L_R the Raleigh length.

The system of Euler equations (1) is solved numerically by fully conservative scheme, assuming cylindrical symmetry, for a fully ionized He slab with 2 mm depth. The plasma density linearly increases from zero to maximal density n=1.5 $x10^{19}$ cm^{-3} after 300 μm, which is the focus point, and than, the density is constant. Initial temperature is uniform and equals T_0=0.1 eV, the initial velocity is zero. Calculation is performed for the prepulse intensity $I_p=10^{13}$ W/cm^2 with duration 2 ns; d_0=7 μm and L_R=50 μm. These parameters are chosen to be close to experimental ones.

The radial and longitudinal distribution of the ion density in He gas–jet after the laser prepulse are shown in Fig.1. The radial distribution of the plasma density is very close to those obtained experimentally in Ref. [9-12]. However, in the direction of laser pulse propagation there is no channel. After 2 ns of irradiation by the laser prepulse, a clearly seen shock wave is formed. The thickness of the shock wave, ~ 10 μm, is comparable with the plasma wavelength so that strong wave breaking and electron injection is expected for this condition. For the prepulse intensity I=10^{13} W/cm^2, the shock wave is formed after 1 ns of irradiation. That means if the laser prepulse is shortened two times, there should not be the injection. Since wave-breaking for the condition is strongly dominated by plasma dynamics we expect that an instability in the shock front may strongly affects the process if the laser prepulse is not monotonic.

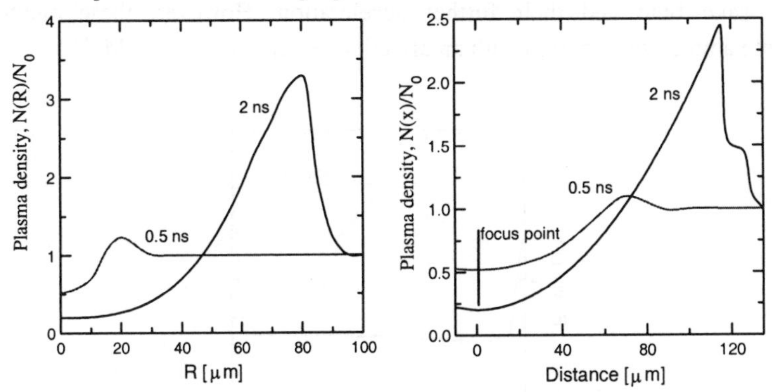

Fig.1 Radial (x=0) and longitudinal (R=0) distribution of He-jet density after the laser prepulse.

Longitudinal density distribution obtained from the hydrodynamics simulation is used for two dimensional particle-in-cell simulation employing moving window technique [8]. However, we neglect the effect of the transverse density distribution on the laser propagation assuming that the cavity width is much larger than that of the laser spot size. The size of the window is 250 μm x 40 μm. The computational grid is 5000 x 960. We use 4 particle per cell, there is no ionization included for this simulation. The initial laser field is chosen as a plane wave with E_Z and H_Y component with distribution as following,

$$E_Y(x,y,t) = a_0 \exp\left(-i\omega t + ik\left[x + y^2/(2xf^2)\right] - i\arctan(x/L_R) - y^2/(d_R)^2\right)/f,$$

where $a_0=eE_0/mc$, x, y the longitudinal and transverse coordinates, $f=\sqrt{1+x^2/L_R^2}$; $k=c/\omega$. The maximal laser intensity is $2x10^{19}$ W/cm^2, λ=0.8 μm (a_0=2.4). The pulse duration (FWHM) is 53 fs; the focus spot size at $1/e^2$ intensity is 7 μm so that the Raleigh length is L_R=50 μm.

The evolution of laser spot size during pulse propagation in the jet with the shock wave given by 2D PIC calculation is presented in Fig.2. Since the critical power for self-focusing for the plasma density $N_e=3x10^{19}$ cm^{-3} is $P_{cr}=1.7x10^{-2}$ n_{cr}/n TW=0.9 TW, the laser pulse with P=4 TW forms a laser channel. Passing distance L~0.8 mm, the pulse diffracts because its power becomes smaller than critical one. However, the laser energy absorbed by electrons is only 58% so that approximately 25% of the laser pulse energy is lost for diffraction during its propagation. The electron momentum in the direction of laser propagation is shown in Fig.3. According to calculation there is effective injection of electron to the wake field and their further acceleration. However, these electron constitute a broad distribution with an effective temperature T_e~10 MeV.

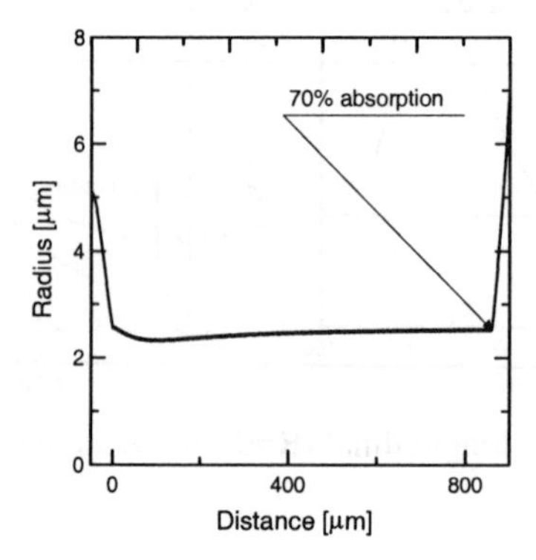

Fig. 2 The radius of the laser pulse (at 50% of the total pulse energy) with propagation distance.

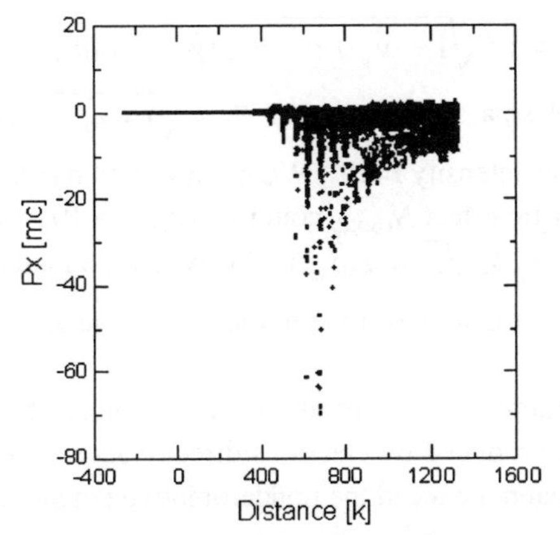

Fig. 3 Electron momentum distribution in the direction of laser pulse propagation after injection due to wave breaking.

2.2. *Effect of self-injection*

Along with the common wake-field, a relativistically intense laser pulse moving in an under-dense plasma with group velocity less than the light speed generates two additional electrostatic waves, which have group velocity close to the group velocity of the laser pulse from the linear theory. The first wave comes from electrons accelerated directly by the laser pulse forming a bunch at the front of the laser pulse. This bunching of electrons creates a potential difference between the front and back of the laser pulse due to the evacuation of electrons. Electrons accelerated by this potential difference form the second wave. Electron repulsion from the second bunch can provide efficient injection of electrons for wake-field acceleration. This injection, which can be considered as a self-injection, contends with injection after wave-breaking. At low plasma density, the self-injection produces efficient acceleration with a low energy spread while at high density, wave-breaking dominates producing a Maxwellian distribution of energetic electrons with an effective temperature [4].

Propagating in under-dense plasma with the group velocity, v_g, an intense laser pulse can accelerate plasma electrons up to the energy $\varepsilon_{e\,max} = mc^2 a_0^2 / 2$, where $a_0 = eE/mc\omega$ with E the laser electric field, and ω the laser frequency. If the velocity of such electrons exceeds the group velocity of the laser pulse, these electrons can be trapped and move with the pulse forming an electrostatic wave. The matching condition can be written in the following form,

$$\gamma_{e\max} = a_0^2/2 = \gamma_g = 1/\sqrt{1-v_g^2/c^2} \approx \omega\tilde{\gamma}_e^{\,1/2}/\omega_{pl} \tag{2}$$

where ω_{pl} is the plasma frequency and $\tilde{\gamma}_e \sim \sqrt{1+a_0^2/2}$ is the electron quiver energy. (For a laser intensity I=10^{20} W/cm^2, λ=1 μm: a_0=8, N_{ematch}=$5x10^{18}$ cm^{-3}; without the relativistic effect N_{ematch} could be equal to 10^{18} cm^{-3}; for a_0>>1 the matching occurs for $a_0 \geq \sqrt{2}(\omega/\omega_{pl})^{2/3}$). When the matching condition is met, the electron spends the longest time in the laser wave and, therefore, gets the largest acceleration.

In plasma, the number of trapped electrons cannot be infinite. The potential difference, $\Delta\phi$, produced by electrons directly accelerated by the laser at the front of the pulse cannot exceed the ponderomotive potential $mc^2a_0^2/2$. As a result we get,

$$e\,|\Delta\phi| \approx 2\pi z e^2 N_i d^2 = mc^2 a_0^2/2\,, \tag{3}$$

where z is the ion charge, N_i is the ion density, and d is the length of a cavity behind the laser pulse. (We assume that the length, d_e, of the electron cloud generated at the front of the pulse is such that d_e<<d).

Calculated absorption rate of the laser pulse with the intensity I=10^{20} W/cm^2 is shown in Fig.4 for different plasma densities. The absorption rate increases non-linearly with the plasma density, ~$N_e^{\,3/2}$, though energy losses for the wake-field excitation and electron acceleration by the potential difference given in Eq. (3) linearly depends on the plasma density. This non-linear dependence could be explained by the frequency down shift absorption mechanism given in Ref. [14].

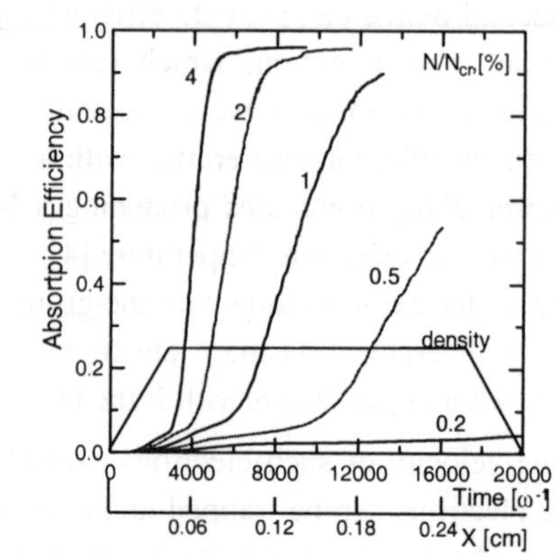

Fig. 4 The temporal evolution of the absorption rate of the laser pulse with intensity I=10^{20} W/cm^2 for different

The spatial distribution of the longitudinal momentum of electrons in the plasma is shown in Fig.5 and Fig.7.

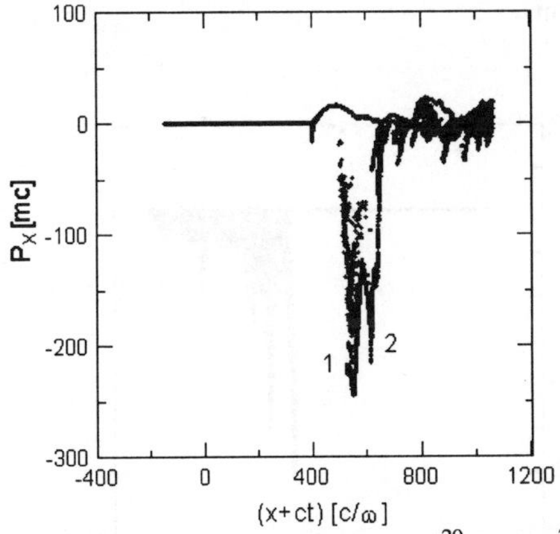

Fig. 5 Spatial distribution of P_x at ωt=6000 I=10^{20} W/cm^{2}and plasma density N_e=2x10^{19} cm$^{-3.}$

For the lower density plasma, the mechanism of electron injection is totally different from that in the higher density plasma, where the wave-breaking process is dominant. As seen in Fig.5 in lower density plasmas, electrons accelerated by the laser pulse are further trapped by the wake-waves. When the maximum charge that can be sustained by the laser pulse is reached, the second electrostatic wave is produced. Then, electrons repelled forward from this wave are further accelerated forming almost mono-energetic bunches with a maximal energy of ε~400 MeV forming a peaky distribution as seen in Fig. 6

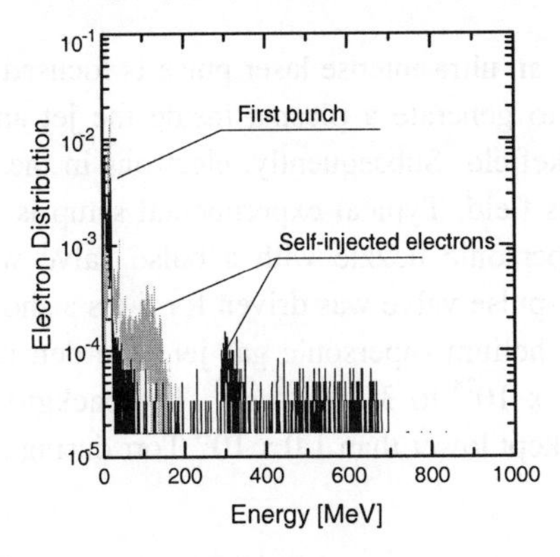

Fig.6 Distribution function of electrons self-injected at lower gas density as shown in Fig. 5

In higher density plasmas, the wave-breaking process becomes important as presented in Fig.6. One can see two bunches originating from different processes: the first bunch is due to self-injection and the second is due to the wave-breaking injection.

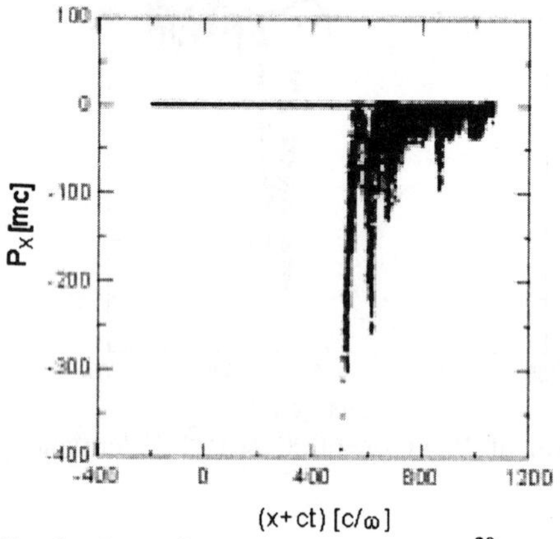

Fig. 7 Spatial distribution of P_x at ωt=6000 I=10^{20} W/cm^{2}and plasma density N_e=5x10^{19} cm$^{-3.}$

We also believe that this effect can be found in any plasma system, such as cluster jet or capillary discharge and so on, where the group velocity of the laser pulse can be considerably less than the light speed.

3. Plasma Cathode Experiment

In the experiment an ultra-intense laser pulse is focused on a gas jet, which is sufficiently strong to generate a plasma inside the jet and to excite a high-amplitude plasma wakefield. Subsequently, electrons in the plasma are trapped and accelerated in this field. Typical experimental setup is shown in Fig.8. An axially symmetric supersonic nozzle with a pulse valve was fixed inside the vacuum chamber. The pulse valve was driven for 5 ms a shot at a repetition rate of 0.2 Hz to generate helium supersonic gas jet. The density at the exit of the nozzle ranged from 7 x 10^{18} to 3 x 10^{19}cm^{-3}. The background pressure of the vacuum chamber was kept lower than 1.0 x 10^{-4} Torr during the gas jet operation.

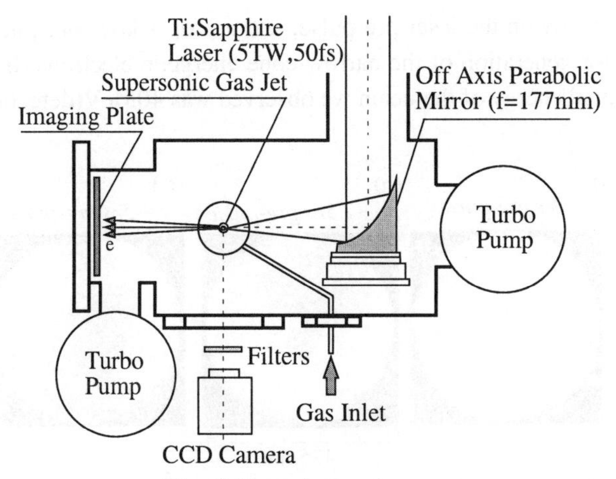

Fig. 8 Experimental setup.

The 12 TW Ti:Sapphire laser system based on CPA technique gives up to 600 mJ, 50 fs laser pulses at the wavelength of 790 nm at the repetition rate of 10 Hz. In this study, available laser power at the target in the vacuum chamber was up to 5 TW. The *p*-polarized laser pulse with diameter of 50mm was delivered into the vacuum chamber and focused on the front edge of the helium gas jet column at the height of 1.3 mm from nozzle exit with an *f*/3.5 off-axis parabolic mirror. The maximum laser intensity on the target was estimated to be approximately 1.0 x 10^{19}W/cm^2. In order to investigate correlation between the electron generation and laser pre-pulse, ns-order laser pre-pulse was produced before the main pulse by detuning of the pockels cell of the Ti:Sapphire laser system.

The electron emission from the gas jet was directly measured by the imaging plates (I.P.). They were laminated with an aluminum foil in 12μm thickness on the surface to avoid exposure to X-rays or laser pulses. The electron signals were accumulated for 300 shots. The energy spectra of the generated electrons were obtained by a compact magnetic electron spectrometer set on the laser axis behind the jet.

Figure 9(a)-(c) shows typical images of beam spot on the bottom plate. In case of a laser pulse with 1ns duration of the pre-pulse just before the main pulse there was no electron beam generation as shown in (a). In the second case of a 2.5ns pre-pulse, the narrow cone energetic beam was generated and an enhanced spot was observed as shown in (b). In the third case of a 5ns non-monotonic pre-pulse, the beam was exploded to pieces and smaller spots were observed as shown in (c). The experimental results clearly show that the beam generation depended strongly on the laser pre-pulse, and a proper laser pre-pulse condition is essential for generation of the narrow-cone energetic electrons. In the case of (b) the maximal energy of the beam we observed was 40MeV(detection limit).

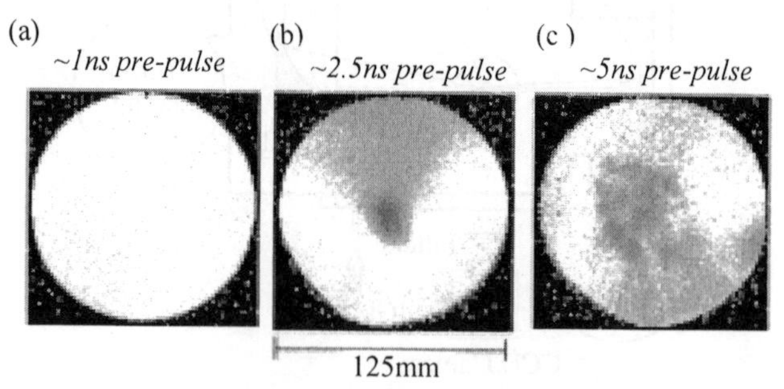

Fig. 9 Typical images of the bottom plate of the cup-shaped I.P. detector. A laser pulse with 1ns duration of the pre-pulse just before the main pulse(a), 2.5ns (b) and 5ns (c).

As a future plan, we will try further acceleration of the electron bunch from the laser-plasma cathode using channel-guided laser wakefield acceleration [15]. Figure 10 shows a setup of the channel-guided laser wakefield acceleration experiment. A capillary discharge plasma will be used for optical guiding channel [16]. The electron bunch from the gas jet is accelerated to higher energy through the capillary.

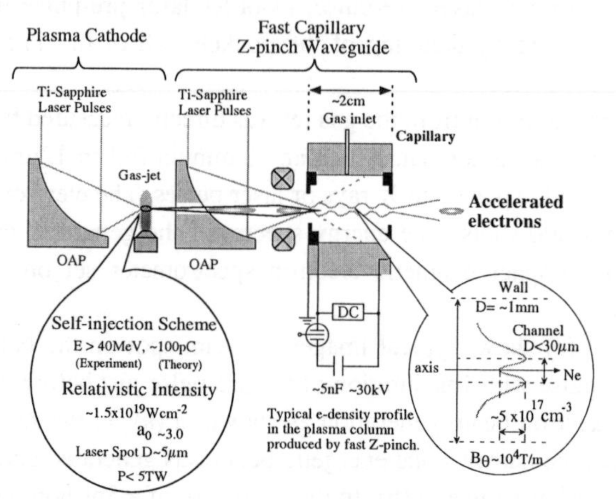

Fig. 10 Experimental setup of the channel-guided laser wakefield acceleration (Future plan).

3. Summary

We have generated the relativistic electrons via laser plasma interaction. The spatial and energy distribution of electrons accelerated by ultra-short laser pulse depending on its pre-pulse were measured. The narrow-coned energetic electrons (up to 40MeV) were observed in the forward direction. We have carry out 2D hydrodynamic and PIC simulations to discuss the injection and acceleration process by laser plasma interaction in details. The experimental results were well explained by the simulations over the entire parameter range, which means the electron bunch is shown to appear due to the injection by wave-breaking in the front of the shock wave which formation depends on the laser pre-pulse parameters.

References

1. 1.T. Tajima and J. Dawson, Phys. Rev. Lett. 43, 267 (1979)
2. D. Umstadter, J.M. Kim, and E. Dodd, Phys. Rev. Lett. 76, 2073 (1996)
3. E. Esarey, R.F. Hubbard, W.P. Leemans, A. Ting, and P. Sprangle, Phys. Rev. Lett. 79, 2682 (1997)
4. S. Bulanov, N. Naumova, F. Pegoraro et al., Phys. Rev. E58, R5257 (1998)
5. K. Nakajima et al., Phys. Rev. Lett. 74, 4428(1995); F. Amiranoff et al., Phys. Rev. Lett. 81, 995 (1998); F. Dorchies *et al.*, Phys. Plasmas 6, 2903 (1999).
6. D. Umstadter *et al.*, Science 273, 472 (1996)
7. S.V. Bulanov, V.I. Kirsanov, and A.S. Sakharov, JETP Lett. 53, 565 (1991)
8. R.G. Hemker, N.M. Hafz, M. Uesaka, Phys. Rev. ST5, 041301 (2002)
9. J. Faure, V. Malka, J.-R. Marques et al., Phys. Plasmas 7, 3009 (2000)
10. V. Malka, J. Faure, J.R. Marques et al., Phys. Plasmas 8, 2605 (2001)
11. C.G. Durfee, J, Lynch, H.M. Milchberg, Phys. Rev E51,2368 (1995)
12. L.C. Jonson, T.K. Chu, Phys. Rev. Lett. 32 ,517 (1974)
13. Ya. B. Zel'dovich, Yu.P. Raizer, Physics of shock waves and high-temperature hydrodynamics phenomena, v. 1, p.342;v.2 p.515 (Academic Press, New York, 1967).
14. S.V. Bulanov, I.N. Inovenkov, V.I.Kirsanov, N.M. Naumova, A.S.Sakharov, Phys. Fluids B4, 1935 (1992)
15. 8. A.J.Reitsma, W.P.Leemans, E.Esarey, C.B. Schroeder, L.P.J. Kamp, and T.J.Schep. Phys. Rev. ST-AB, **5**, 05131 (2002)
16. 9. T.Hosokai, M.Kando, H.Dewa, H.Kotaki, S.Kondo, K.Horioka, and K.Nakajima, Opt.Lett. **25**,10 (2000)

A LASER-DRIVEN ACCELERATOR AND THOMSON X-RAY SOURCE*

DONALD UMSTADTER, SUDEEP BANERJEE, FEI HE, YY LAU, RAHUL SHAH, TREVOR STRICKLER AND TONY VALENZUEALA

University of Michigan,
Ann Arbor, MI 48109-2099
E-mail: dpu@umich.edu

With the development of table-top high-peak-power optical laser systems, it has become feasible to generate---for the first time---ultrashort-pulse-duration x-ray probes. Laser produced x-ray sources are promising on account of their compact size, reduced complexity and significant reduction in costs. There is currently a worldwide effort to produce a high brightness x-ray beam using an ultrafast optical pump, for myriad applications in chemistry and biology [1-8].

1. Introduction

Developments over the last decade have led to high power, ultra-short duration, laser systems which can routinely provide focused intensities in excess of 10^{18} W cm^{-2}. At the FOCUS Center, University of Michigan, there now exist an array of laser systems capable of producing multi-Terawatt power per pulse with pulse durations ranging from 20-400 femtoseconds (fs) and repetition rates ranging from single shot to 10Hz. In such high intensity laser fields, all matter is ionized [9] to form a high-density plasma composed of quasi-free electrons and ions. The free electrons liberated by the ionization process interact directly with the electric field of the light pulse and are accelerated to velocities close to that of the speed of light in each cycle of the light pulse [10, 11]. Collective processes in the form of laser wakefields also accelerate electrons, resulting in a well collimated, highly energetic beam of electrons from a laser produced plasma [12]. Such laser produced electron beams have extremely small transverse emittance, high brightness and are produced in a spatial region extending over only one mm [13, 14]. Experiments carried out over the last couple of years have shown that such a relativistically moving electron beam can be used to generate short wavelength light extending into the x-ray region of the electromagnetic spectrum [15]. A high-power, short-pulse, Petawatt-class laser

* This work is supported by the Chemical Sciences, Geosciences and Biosciences Division of the Office of Basic Energy Sciences, Office of Science, U.S. Department of Energy and the National Science Foundation.

is currently under development which would provide focused intensities in excess of 10^{22} W cm^{-2} [16]. Under such extreme conditions, the ultra-relativistic motion of electrons should lead to a high brightness beam in the hard x-ray (1-10 keV) spectral region.

The availability of a table-top hard x-ray source will have significant impact in the chemical and biological sciences. Experiments currently done with third generation synchrotrons could be done on laser based systems once the delivered x-ray flux is comparable. These would include most studies in the currently burgeoning field of proteomics [17]. Since the produced x-rays will be of ultrafast duration, it will lead to the attainment of submicron spatial resolution in studies of femtosecond time resolved chemical dynamics. Current systems based on optical wavelengths can reach only micron resolution, the large wavelength of visible light being the limiting factor. Hard x-ray based systems will also not suffer from the large absorption of optical wavelengths in most media. One of the most challenging goals in imaging technology is to develop radiation sources that can safely probe biological systems. One possible solution is to create holographic images of living cells and micro-organisms by using coherent X-rays with wavelengths in the so-called water window. This spectral region provides the best contrast between carbon-based biological structures and water. Since the laser and the x-ray pulse will be absolutely synchronized in time, the complexities of studying in real time with high spatial resolution, photo-initiated processes [18] will be significantly reduced.

There is a significant effort worldwide to produce a laser-based high quality x-ray source. Amongst the promising candidates are line [19] and continuum emission [20, 21] from laser heated solids. While these produce x-rays in the keV region, the emission is isotropic [22] leading to small x-ray flux on target. The x-ray pulse is also long, typically of the order of a picosecond or more [23] even when a sub-100 fs pump laser is used to generate the highly overdense plasma. Thus, x-rays produced from solid targets are nearly impossible to transport over large distances and have poor spatial coherence [24]. Another promising area of research is that based on high-harmonic generation from atoms. In this case, harmonics are generated from electrons bound to atoms [25, 26]. It has been shown that this mechanism can produce large photon flux in a well defined cone [27, 28] and the emission has large spatial coherence, which can be improved by guiding the laser beam through a capillary [29]. The disadvantage is that this scheme cannot be extended beyond the XUV region [30] since at high enough intensities atoms ionize and this leads to a reduction in the efficiency of harmonic generation from this process [31]. It is difficult to compare various laser based x-ray sources on account of the widely differing laser and material parameters. As such, this will not be attempted here. A comprehensive review of current x-ray sources and the current status of subpicosecond x-ray experiments is given by Rousse et al. [32]. Instead, in a later section, x-ray yields from the mechanism proposed here will be compared to the brightness of the best available synchrotrons.

It is, therefore, necessary to obtain a laser based source which can be extended into the keV region and has beam-like characteristics and good spatial coherence. A laser based source will also have distinct economic advantages; namely compact size and a significant reduction is installation and maintenance costs. Such a source would be easily accessible to university researchers and lead to a significant increase in experiments which currently are hampered by limited access to large synchrotron facilities. Time resolved x-ray diffraction experiments to study laser initiated process like non-thermal melting [33], shock wave propagation [34] and phase transitions [35] could also be done in greatly simplified fashion. There exist techniques to generate femtosecond pulses from a synchrotron using an optical pulse from an ultrashort laser as a gate subdividing the much longer x-ray pulse [36]. This suffers from the incommensurability between the synchrotron pulse rate and the repetition rate of a high-power laser leading to a dramatic reduction in the x-ray flux by at least six orders of magnitude [37]. The LCLS and TESLA are expected to overcome these challenges, but they require accelerators that are several kilometers in length and wigglers that are almost 100 meters long.

We are studying a new mechanism of x-ray generation using the interaction of free electrons with ultraintense light pulses. Based on our recent work, we have shown that electrons accelerated to relativistic energies by ultra-intense laser pulses can radiate in the XUV region [38]. We now propose to extend this study to obtain a high- energy, well-collimated, ultrashort beam of hard x-rays. While preliminary results have shown that the current theoretical understanding of the process is on the whole a good approximation of the actual interaction, significant modification in our understanding will happen as the interaction process becomes more extreme due to the use of significantly higher laser powers. Current research is focused on the improvement of electron beam quality using various injection schemes will also impact x-ray source development [39]. Specifically, a high energy beam with small longitudinal and transverse emittance is desired for optimal x-ray generation [40]. Thus, developments in x-ray generation will go hand in hand with efforts to improve the characteristics of electron beams produced from ultra-intense laser plasma interaction.

Based on current projections, it is reasonable to expect that such a laser based free electron x-ray source would be able to compete with third generation synchrotrons [41]. The delivery system will be largely based on what is used currently in synchrotrons [42] except for the fact that the x-ray pulses will be of femtosecond duration. Thus, it will be possible to follow the dynamics of processes that would otherwise have to be studied in time integrated manner. It is also proposed to investigate novel collective effects due to the trap formed by interfering intense laser pulses in a plasma. Such traps have been shown to bunch electrons [43] and may lead to significantly improved coherence of the emitted x-ray beam. Self amplified spontaneous emission (SASE) effects [44] are not expected due to the large longitudinal emittance of current laser produced electron beams. Here again, success of injection schemes leading to significantly smaller electron beam emittance may alter the picture and lead to dramatic improvements in the x-ray emission process.

2. Hard X-ray Sources

In this section, we will describe in detail the approach being used to obtain a beam of hard x-rays from a laser based source. Our approach primarily relies on the process of nonlinear Thomson scattering using a laser generated relativistic electron beam scattering off an ultraintense laser pulse. It is to be noted that this process is distinct from previous work on x-ray generation from laser-electron scattering carried out with high energy electron beams produced in accelerators. The crucial distinction is the fact that, in conventional accelerators, electrons interact with the light pulses with $a_0 << 1$, while the mechanism we are using depends critically on the nonlinear interaction between light and electrons which occurs for the case $a_0 \geq 1$ [45]. This enables access to x-ray wavelengths with electron beams of lower energy. While the mechanism of nonlinear Thomson scattering will be the primary focus of our experimental effort, it will be also possible to investigate novel schemes when two or more laser beams are used as in the colliding pulse geometry due to coherent effects in a plasma. We, therefore, propose to investigate x-ray emission from novel optical traps which have been shown to preheat electrons, reduce the transverse emittance and could also act as laser produced wigglers. Such traps have been shown to arise by interfering high intensity laser pulses in an underdense plasma [43].

2.1. *Nonlinear Thomson scattering*

The interaction of free electrons with the electric field of light pulses has been studied for well over a century. In the limit of low fields, the electron motion is linear and along the polarization of the light field and the scattered radiation is at the frequency of the incident light. This is the process of linear Thomson scattering, known for well over a century [46]. At high intensities, the magnetic field associated with light comes into play and the electron motion becomes highly nonlinear. The intensity scale is set by the dimensionless parameter $a_0 = eE/m\omega c$, where E is the electric field of the light wave with frequency ω and m is the rest mass of the electron. This corresponds to the normalized electromagnetic field experienced by the electron. The relativistic regime is accessed when $a_0 \sim 1$ which corresponds to a peak intensity $I \sim 10^{18}$ W cm^{-2} for 1-micron wavelength radiation. The electrons move in characteristic "figure-8" orbits for linearly polarized and circular orbits for circularly polarized light and radiate at harmonics of the Doppler-shifted laser frequency [47]. The usefulness of this light, however, is severely limited, owing to the fact that the harmonics are emitted at large angles with large angular spread. Hence, the flux of x-rays would be quite small. However, at relativistic laser intensities, it has been shown that a directed beam of relativistic electrons is produced by the laser wakefield acceleration mechanism [12-14, 48-50]. This high brightness beam is in the direction of the laser light. Nonlinear scattering of the laser beam off such a co-propagating relativistic electron beam is predicted to produce a highly collimated beam of high-order harmonics extending into the x-ray region [51]. This beam is directed primarily along the direction of the electron beam due to the nature of the relativistic scattering process [52].

Generation of x-rays via Compton scattering of a laser beam from a relativistic electron beam was proposed three decades ago [53] and has been demonstrated on almost every available accelerators [2,54]. However, its low scattering efficiency (low photon number per pulse) is the major problem for applications. This arises from the large spatio-temporal mismatch between the size of the electron beam which typically has a diameter > 50 μm and the laser focus (< 10 μm), leading to an extremely small overlap region [2]. The pulse duration of the electron bunch from state of the art linear accelerators (LINAC) is > 1 ps. The most commonly implemented geometry is when the electron and laser pulses collide head on. In this case the pulse duration of the x-rays is determined by the longer of the electron or laser pulse duration. Thus, current and proposed sources which combine LINACs and lasers will have correspondingly long (> 1 ps) pulse durations. This can be improved by having the electron beam and the laser pulse intersect at right angles. In this case, the x-ray pulse duration is determined by the transit time of the laser pulse across the waist of the electron beam and subpicosecond durations can be obtained, at the expense of a loss in x-ray efficiency. For laser produced electrons beams, the spatial overlap is nearly perfect while the electron pulse duration is subpicosecond when a subpicosecond laser driver is used [55]. However the low repetition rate of the high-peak-power short-pulse lasers leads to low average power x-ray flux which baffles applications that need signal accumulation. To increase the x-ray photon number per pulse and obtain ultrashort pulse duration, it is advantageous and feasible to use the femtosecond electron pulses generated from a laser-plasma-based electron accelerator and sub-30 fs terawatt-peak-power high repetition rate laser pulses from a Titanium:Sapphire laser in a nearly counter-propagating configuration. In this way, while the x-ray pulse duration is determined by the longer of the electron pulse or the laser pulse, the scattering efficiency is increased because all of the laser photons and the electrons see each other and thus participate in the interaction. Shorter x-ray pulses can be obtained by colliding the laser and electron pulses orthogonally. However, it is also known that the scattering cross-section is largest when electrons and photons collide head-on. The actual geometry must therefore be a compromise between these conflicting requirements. In the colliding pulse geometry, access to x-ray wavelengths is easier because the Doppler shift of the emitted radiation extends it to shorter wavelengths. The additional fact that a laser based electron accelerator is on the mm scale, as compared to the km scale of conventional accelerators, leads to a similar orders-of-magnitude reduction in the cost and complexity of the system.

The number of x-ray-photons per pulse is comparable to that of synchrotron radiation source. However, the former approach provides several advantages. First, the pulse duration is about two orders of magnitude smaller than conventional synchrotron radiation, providing much higher temporal resolution for pump-probe experiments. Second, the short duration also means a much higher peak power (intensity), enhancing nonlinear processes. Third, the photon energy can be potentially high (< 10 keV), so it can cover a large spectral range and can compete with the x-ray obtained from synchrotrons and free electron lasers. For instance, the x-ray photon energy reaches 60 keV when an electron beam of 50-MeV energy is used. Fourth, the use of relativistic laser intensity leads to the occurrence of nonlinear Thomson scattering and, thus, extends the spectral range by generating harmonics of x-ray photons. Furthermore, because of the femtosecond duration of the electron pulses, collective Thomson scattering may occur or partially occur to boost the conversion efficiency by orders of magnitude. When compared with Thomson scattering x-ray sources based on RF-frequency accelerators, the laser driven source has the distinct advantage of absolute synchronization between the electron bunch and the laser pulse.

2.2. *Recent Developments*

Over the last few years we have shown for the first time generation of harmonics from relativistic laser electron scattering. Initially signatures for this process were sought in the visible regions of the spectrum. At high light intensities, electrons are predicted to quiver nonlinearly, moving in figure-eight patterns, rather than in straight lines, and thus to radiate photons at harmonics of the frequency of the incident laser light, with each harmonic having its own unique angular distribution. This is referred to as nonlinear Thomson scattering or relativistic Thomson scattering. Previously, we observed the generation of the second and third harmonics, with their unique angular patterning originating from this process [56]. We have also observed for the first time relativistic harmonics emitted along the laser axis [57]. Third harmonic light was detected and discriminated spectrally and angularly from the harmonics generated from competing atomic processes, specifically due to rescattering of bound electrons from the nucleus during each cycle of the laser pulse. Fig. 1(a) shows the spectral characteristics of the emission at the third harmonic wavelength from both bound and free electrons. The angular profile of the third harmonic is depicted in Fig. 1(b).

Based on initial experiments and theoretical understanding of the process, the outlook was not very promising. The harmonics were of low order (second and third) so that even the vacuum-ultraviolet (VUV) region of the spectrum was not reached. The emission was also over a large angle and the resultant flux was quite small. With such characteristics, the process would have remained a laboratory curiosity. Over the last year, we have extended this study to much shorter wavelengths with dramatic results. Investigations carried out by us, led to the discovery, that not only was there significant amount of light in the VUV, but at the highest intensities the short wavelength light was emitted as a well-collimated beam in the forward direction which we ascribe to scattering from a relativistic electron beam, which was accelerated by the same laser pulse [58]. It is shown that at high intensities, when the normalized electric field, a_0, approaches unity, in addition to the conventional atomic harmonics from bound electrons, there is significant contribution to the harmonic spectrum from free electrons. The characteristic signatures of this are found to be the emission of even order harmonics, linear dependence on the electron density, significant amount of harmonics even with circular polarization and a much smaller spatial region over which these harmonics are produced as compared to the atomic case. Imaging of the harmonic beam shows that it is emitted in a narrow cone with a divergence of 2-3 degrees. Besides its promise as an efficient source of XUV radiation, the generation of high-order harmonics is of interest from a purely physical point of view because of the fact that they can be produced by fewer nonlinear processes than low-order harmonics and thus provide definitive signatures of their origin.

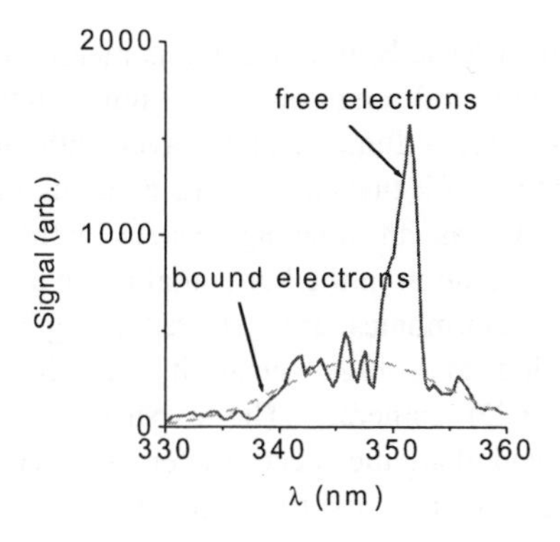

Fig. 1. (a) Spectra of the third harmonic generated in helium for 3×10^{19} cm^{-3} gas density and 2×10^{17} W cm^{-2} pump laser intensity. The dotted line is the contribution from bound electrons which disappears when a preionized plasma is used. (b) Angular pattern of the third harmonic generated under the same conditions.

These exciting results presented a new puzzle: the experimental data was contrary to theoretical predictions that the emission would be peaked off axis with a large angular spread. This was the case with the third harmonic as well, where the harmonic emission was found to be along the laser propagation direction as shown in Fig. 1(b). Clearly, a crucial piece was missing in our understanding of the process. It is well known, that under the conditions of our experiment, a high energy electron beam is produced due to acceleration by the laser wakefield [14]. The large initial velocity of the electrons significantly modifies the scattering process leading to a more efficient and better collimated emission in the short wavelength region of the spectrum. Based on our results, it is reasonable to claim that we have a high brightness beam in the soft x-ray and vacuum ultraviolet (XUV) region of the spectrum from free electrons using a tabletop setup.

3. Experiments

The experiments were performed with a hybrid Titanium:Sapphire–Neodymium:Glass laser system that produced pulses of 400 fs duration at 1.053 μm with a maximum output energy of 6 J. The maximum power delivered at the experimental setup is 5 TW. The two inch diameter laser beam is focused with an off-axis paraboloid leading to a focal spot 8-10 μm in diameter. The corresponding peak laser intensity accessed in our experiments is ~5×10^{18} W

/cm^2. High harmonic emission from the underdense Helium plasma is measured using a Seya-Namioka spectrometer with a spectral range of 30-200 nm, which corresponds to the harmonic range of 6-30 for a fundamental wavelength of 1.053 μm. It consists of a toroidal grating (1200 lines/mm) with a radius of curvature of 1 m and is configured to act as an imaging system with a magnification of 1:1 providing wavelength resolution in the horizontal plane and spatial information in the vertical plane. Harmonics are detected using an imaging multichannel plate (MCP) coupled to a high sensitivity and high dynamic range charge-coupled device (CCD) camera. The experiment is configured such that the laser beam is directed along the spectrometer axis. The spectrometer and the target chamber are separated by a slit to allow differential pumping of the system as well as to allow the laser beam to enter the spectrometer. A schematic of the setup used in these experiments is depicted in Fig. 2. Plasma conditions are monitored by imaging the channeling of the laser beam through the plasma as well as spatial diagnostic of the electron beam by use of a fluorescent screen (LANEX). Under good conditions, the laser is observed to stay self focused over the entire one mm length of the gas jet and a low divergence electron beam is produced.

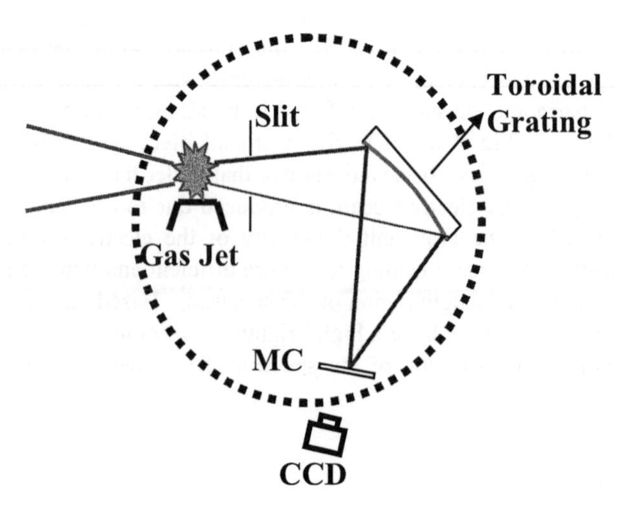

Fig. 2: Schematic of experimental setup to study nonlinear Thomson scattering. The slit separates the experimental chamber from the spectrometer and permits differential pumping of the system.

Fig. 3 (a) shows the spectrum of high order harmonics obtained from underdense helium plasma at intensity of 5×10^{16} W cm^{-2} and for linearly polarized light. It can be seen that only odd (7th and 9th) harmonics are produced as expected [59]. When the light intensity is increased to 2×10^{18} W cm^{-2} and the polarization changed from linear to circular, there is now a clear signal at the position of the 8th harmonic, as shown in Fig. 3(b). It was checked that the signal disappears at lower intensities (< 10^{17} W cm^{-2}) for all gas densities. The emission in the even order was weaker than that in the odd orders for all the orders studied (≤ 30). The presence of the even harmonics even with circularly polarized light and their absence at intensities < 10^{17} W cm^{-2} ($a_0 < 0.2$) would suggest that scattering from free electrons is the major contributing factor. It is also observed that the line width at high intensities is broadened which would arise from the Doppler shift of the emitted radiation. The largest shift possible for our case is about 4 nm and is independent of the wavelength. The overall efficiency of the process is larger by a factor of two for the case of linear polarization as compared to when circularly polarized light is used, at the highest intensity accessed in the experiment. This can be seen from the fact that, to leading order, the single electron scattering cross-section scales as a_0^2. Similar experiments were carried out in N_2 and Ar (results not shown) with linearly polarized light and it was found that the even orders are enhanced relative to the odd orders, because of the higher free electron density. The overall efficiency of the process in high-Z gases decreases because excessive ionization produces significant defocusing of the beam [60] and loss of phase matching [31].

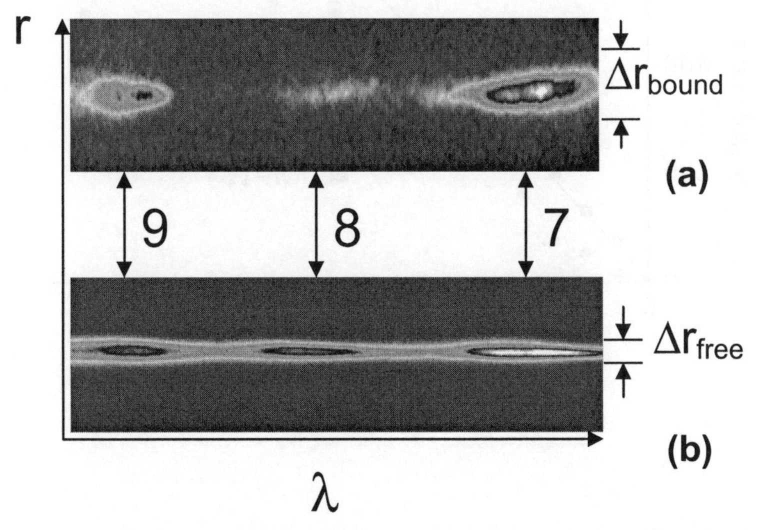

Fig. 3. Spatial profile of the high order harmonics (a) I=5×10^{16} W cm-2, n=10^{17} cm^{-3} and linear polarization. (b) I=5×10^{17} W cm-2, n=10^{18} cm^{-3} and circular polarization. The signal in (a) is 18 times smaller than that in (b).

The imaging property of the spectrometer was used to infer the source size of the harmonics. For the case of Fig. 3 (a) (linearly polarized light at low intensity and low density), the spatial extent of the odd orders (7th and 9th) is large $\approx$ 100 μm, as would be expected from the fact that bound harmonics would be produced in the focal volume where the intensity exceeds $\approx 10^{13}$ W cm^{-2}. With circularly polarized light at high intensity, only the free electron signal is expected to be present [61]. As shown in Fig. 3(b), the spatial extent of the harmonics is found to be significantly smaller $\approx$20 μm. When linear polarization is used (I=2$\times10^{18}$ W cm^{-2}), the 7th harmonic is measured to have a width (FWHM) of 70 μm, while the 8th harmonic has a spatial extent of 30 μm. The larger spatial extent of the odd orders arises because of contribution from bound electrons, while the even orders have a larger spatial extent as compared to the case for circular polarization because of the larger electric field component.

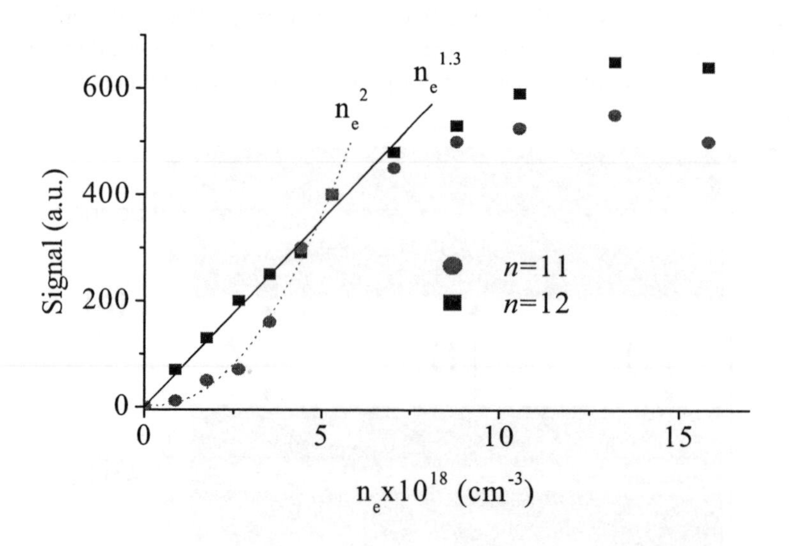

Fig. 4. Dependence of high harmonic generation on the electron density (*n*e) at an intensity of 6$\times10^{17}$ W cm^{-2}. The odd harmonic scales quadratically with the density while the even shows a dependence that is close to linear.

Since the harmonics are emitted from free electrons, the emission should be an incoherent sum from each individual scattering center. As such, it is to be expected that the radiation should scale linearly [62] in contradistinction to the case for atomic harmonics, wherein a quadratic scaling is observed on account of phase matching [63]. Fig. 4 shows the results obtained for the 11th and 12th harmonics in Ar for various gas densities at 6×10^{17} W cm^{-2} and linear polarization. In the former case, there is a quadratic dependence initially, while in the latter case, it is linear. Both signals saturate and then decay, which is expected from the known ionization-induced defocusing of the laser beam [60]. It should be noted that the dependence on the density for the even orders is not exactly linear – in fact, in our experiments, the scaling ranged from 1.2--1.4 for the various harmonic orders from 3-30. This may possibly arise from quasi-phase matching in the plasma [64] or due to coherence effects [65] and is currently being investigated. Similar results are found in other gases and for all harmonics orders studied. For the case of circular polarization, the density scaling is approximately linear for all the harmonic orders.

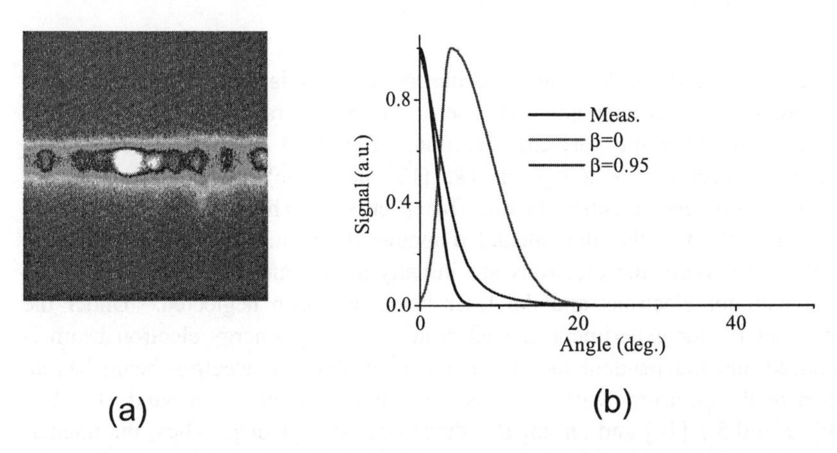

Fig 5 (a) Image of the high order harmonics after diffraction through a 200 μm slit. The horizontal axis corresponds to wavelength while the vertical axis to the spatial extent of the harmonics The measured angular divergence for the harmonics shown is about 2.5 deg (b) Calculated spatial profile of the 11th harmonic for the case where the electron starts at rest (red curve) and for an initial velocity of 0.95c (blue curve) and the measured profile (black curve).

Theoretically, the angular distribution of these harmonics is known and would provide a simple and direct comparison of experiment and theory. This is achieved by imaging the harmonics on a 200μm wide slit placed on the object plane of the spectrometer. The system was calibrated with a Helium-Neon laser such that, based on the size of the image on the MCP, the size of the beam on the slit could be inferred. Fig. 5 (a) shows the pattern obtained on the MCP in this case for the 11th and the 12th harmonics in helium. The multiple lobes correspond to single slit diffraction of the high-order harmonics. Based on the measured magnification of the system, it is found that the harmonic beam has a divergence of about 3 deg. at the 6th harmonic, which decreases to ≈ 2 deg. at the 20th harmonic. The harmonics are thus emitted as a low-divergence forward-directed beam. The angular spread of these harmonics is comparable to that obtained for bound-electron harmonics when no guiding is employed [66] and should lead to better collection and focusability for the short wavelength light.

At first sight there would appear to be a significant discrepancy between experiment and theory. Based on the exact solution of the radiation pattern for the case of an electron initially at rest in a circularly polarized light field, the angular distribution of the nth harmonic, $dP^n / d\Omega$ is given by

$$\frac{dP^n}{d\Omega} = a_0^2 \frac{2n^2}{\left[1+2b\sin^2(\theta/2)\right]^4}\left\{\frac{2\left[\cos\theta - 2b\sin^2(\theta/2)\right]^2}{a^2\sin^2\theta}\times J_n^2(n\Theta) + J'^2_n(n\Theta)\right\} \quad (2.1)$$

$$b = f(a_0), \Theta = g(\theta, \phi, a_0)$$

where θ, ϕ are the polar angles of observation. It is easy to show that the harmonics peak off-axis and that there can be no radiation other than the fundamental along the forward direction. Fig. 5(b) (red curve) shows the calculated spectrum for a typical case (12th harmonic) which is significantly different from the measured profile (black curve). The apparent discrepancy arises from the fact that the calculated angular distribution depicted in Fig. 6 is for the case where the electrons are initially at rest and the finite divergence angle both for electrons and the laser beam has been neglected. Under the conditions of our experiment, a well collimated, high-energy electron beam is produced and independent measurements show that this electron beam has an exponential spectrum with a most probable energy of around 1 MeV $(\Delta E / E \approx 0.5)$ [14] and an angular divergence of < 1 deg. Thus, the angular distribution of harmonics with a non-zero initial electron velocity needs to be considered. For this case Eq. 2.1 is modified to

$$\frac{dP^n}{d\Omega} = \left[\frac{1+\beta_0}{1-\beta_0}\right]\frac{2n^2}{\left[1+2b\sin^2(\theta/2)\right]^4}\left\{\frac{2\left[\cos\theta - 2b\sin^2(\theta/2)\right]^2}{a^2\sin^2\theta}\left[\frac{1-\beta_0}{1+\beta_0}\right]\times J_n^2(n\Theta) + J'^2_n(n\Theta)\right\} \quad (2.2)$$

$$b = f(a, \beta_0)$$

$$\Theta = g(\theta, \phi, a, \beta_0)$$

where β_0 is the initial normalized velocity of the electron [67]. Our calculation also takes into account the profile of the electron beam and the fact that the laser has a divergence angle of 5 deg. in the focal region. Fig. 6 (blue curve) shows the radiation spectrum obtained when the electron energy is 1 MeV. The calculated width is about 3.1 deg., which is in good agreement with the measured width of 2.6 deg.

The current work demonstrates a new mechanism of high-harmonic generation. The conversion efficiency integrated over all the orders observed in the experiment is estimated to be 10^{-7}. This is smaller than current capillary-based experiments which guide the laser pulse through a long length of high density gas (10 cm - 1m) with the density adjusted to optimize phase matching for a particular set of harmonic orders [68]. However, as has been noted previously harmonic generation from bound electrons drops dramatically at high intensities because of ionization of the medium. Harmonic generation from free electrons does not suffer from this drawback. In this experiment even with lower conversion efficiency, because of the higher laser intensities used, the photon yield is comparable to that obtained from bound electrons. In principle, it should be possible to get higher photon yields with higher intensities when 100 TW – 1 PW class lasers are used. This technique has several advantages over linear Compton scattering from a conventionally accelerated electron beam: (1) the acceleration length is 10,000 times shorter; (2) the conversion efficiency is much higher because the electron beam is laser-driven and self-aligned spatially and temporally with the laser focus; (3) the conversion efficiency is also higher and the scattered light is more energetic because higher laser intensity is used as a result of which nonlinear Thomson scattering dominates. Combining the best currently available laser produced electron beams with a counter propagating geometry, it should be possible to have peak spectral brightness comparable to the third generation synchrotrons at >1 keV. There is substantial current effort to improve the energy, monochromaticity and emittance of laser produced electron beams via injection and plasma guiding. Success of these schemes should lead to the attainment of > 10^{10} photons per shot in the hard x-ray region. High average brightness is also possible if a high repetition rate laser system based on Titanium: Sapphire technology is used.

4. Theory

Over the last year in collaboration with Prof. Y. Y. Lau (University of Michigan), we have begun to study, theoretically, the scattering of electrons from high-intensity laser pulses [69-71]. In the not-too-distant future, joules of laser energy will be focused down to a spot size of one wavelength (~ 1 μm), with the laser pulse length being compressed to the order of a few optical cycles. The electric field in such a laser pulse is extremely high; its normalized value, a_0 is in the tens and beyond. The phase at which the electron experiences this pulse will become important because of the shortness of the laser pulse. There are many instances in the interactions of intense lasers with matter, where there can be a large spread in the initial phase angles, θ_{in} of electrons, with respect to the light wave. For instance, electrons that are born during the ionization of a gas will do so at different phases depending on the ionization potential [26]. Finite plasma temperature will also lead to a spread in initial phase angles. Electrons will have different initial phases during the Thomson/Compton scattering of intense light from a relativistic electron beam (at all collision angles except 180° between the light beam and the electron beam). Electrons interacting with realistic three-dimensional focusing light pulses, with curved wave fronts, will also see different initial phases relative to other electrons that are at the same axial position but at a different radial position.

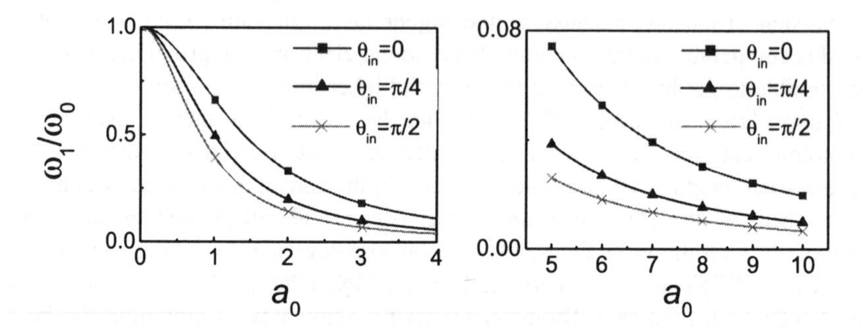

Fig. 6: Dependence of ω_1/ω_0, the normalized fundamental frequency of Thomson backscattered radiation, on the normalized laser field strength, a, for various θ_{in}

The dependence of free electron orbits on the phase of an ultra-intense light field has never before been studied, although it is critical for a thorough understanding of basic phenomena such as Thomson scattering [72]. Including this phase, we have shown that the familiar figure-8 motion of an electron that is superimposed onto the forward drift is more of an exception than the rule. Specifically, an electron tends to drift across the laser path, and this transverse drift is not due to the ponderomotive force. The orbital periodicity, and therefore its nonlinear Thomson scattering spectrum, depend critically on the amplitude of the ultra-intense laser field and on the above-mentioned phase, but are otherwise insensitive to the laser frequency [73]. Fig. 6 shows the dependence of the fundamental frequency in the back-scattered direction ($\mathbf{n} = -\mathbf{z}$) as a function of a_0, for several values of θ_{in}. The radiation spectrum extends to frequencies much lower than the laser frequency ω_0. The reason for this behavior is that, as the electron drifts longitudinally at velocities close to the speed of light, the electron moves closer in phase with the laser and the figure eight frequency (of electron oscillations) decreases.

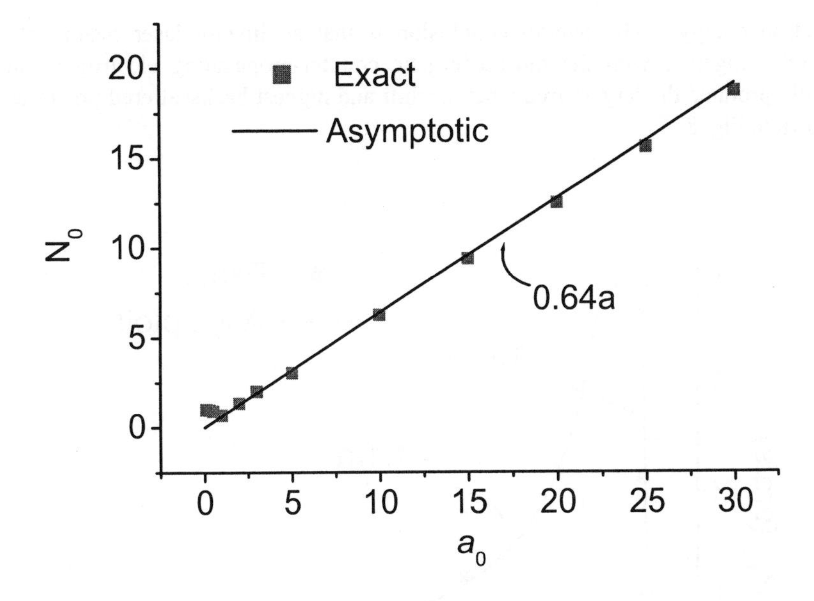

Fig 7: The harmonic number of the laser frequency, N_0, at which maximum backscatter occur for the special case when the electron starts with zero initial velocity. Note that N_0 is linearly proportional to a_0, for $a_0 >> 1$.

Recently, the classical theory for radiation from a relativistic electron beam colliding with an ultra-intense laser pulse has been revisited. While analytic and numerical solutions to this problem have been obtained many times in the past [74-76], the novel aspects of the present work include a sharp delineation of the role of the intense laser and of the electron's kinetic energy in the production of high harmonics [77]. For example, we find that (1) the relative shape of the backscatter spectrum depends only on the laser amplitude and is independent of the electron energy, (2) that a high-power laser does not necessarily produce high power radiation or overwhelmingly favorable frequency upshift, and (3) very significant frequency upshift is achieved most effectively by the use of an energetic electron beam. For radiation backscattered from the laser, the frequency upshift is proportional to a_0, as depicted in Fig. 7. These findings are consistent with some conventional notions, while contradicting others [71, 76]. They are based on the closed form solutions of the backscattered spectrum that were obtained for arbitrary laser intensity and arbitrary electron energy (including zero), together with the *simple* asymptotic expressions in the various regimes. This has led to a unified treatment for the scattering problem from low to ultrahigh laser power and from low to ultrahigh electron energy. The overall conclusion is that an intense laser beam with a_0~O(1) together with the most energetic counter-propagating electron beam would produce the largest frequency upshift and highest backscattered power as shown in Fig. 8.

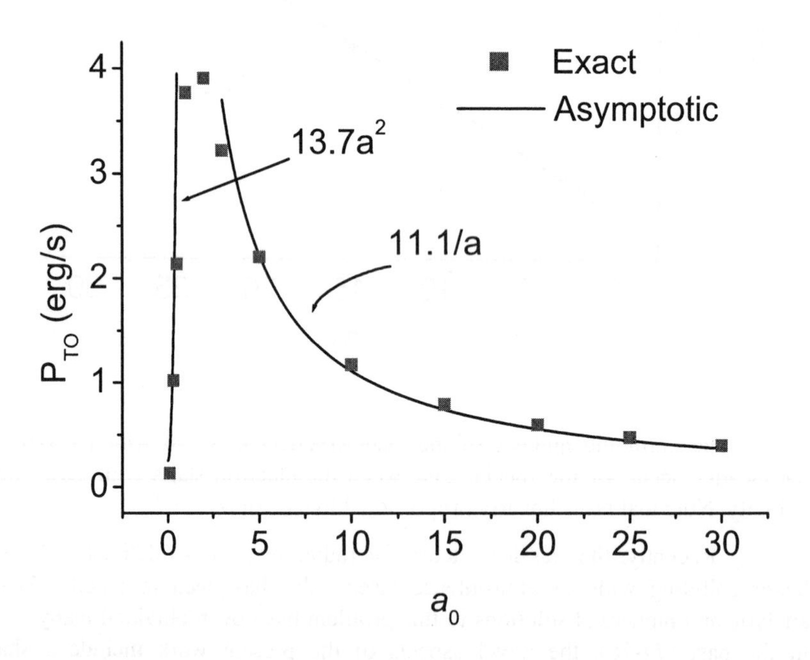

Fig. 8 The total scattered power, P_{TO}, per unit solid angle for the special case when the electron has zero initial velocity, and laser wavelength $\lambda = 1\,\mu m$.

Detailed numerical work is also in progress to elucidate the mechanism of nonlinear Thomson scattering in the presence of realistic light pulses. Electron motion and the resulting radiation patterns in the presence of low and high intensity interfering counter propagating laser fields is also being studied. This is relevant to our experiments when two ultra-intense laser beams are used in the counter-propagating geometry. As a first approximation, all collective plasma effects have been neglected so that the system is reduced to a single electron interacting with the laser fields. From this study, the trajectories of single electrons initially at rest within interfering counter propagating laser beams with intensities on the order of 10^{17} W cm^{-2} are found to be described by Lissajous patterns [78, 79]. This motion satisfies a nonlinear equation of motion that has been solved analytically near equilibrium provided the initial velocity is small [80]. The motion of an electron at a magnetic field node within the standing laser wave represents a special case that has also been found exactly for arbitrary intensity. The Thomson radiation emitted by such an electron in the presence of weak laser fields has been shown to consist of only the first few harmonics of the laser frequency. When driven by intense laser fields (I ~ 10^{19} W cm^{-2}) the electron velocity can be approximated by a step function and the radiation pattern has been computed analytically. The scattered light occurs at integral multiples, *m*, of the laser frequency falling off as m^{-2}.

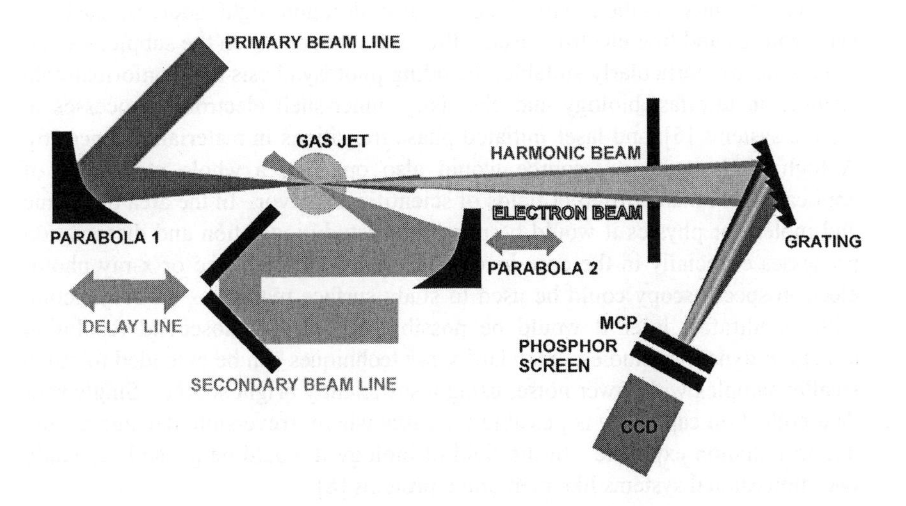

Fig. 9: Schematic of experimental setup to study nonlinear Thomson scattering in the colliding pulse geometry. The primary beam generates an electron beam which then scatters off an ultraintense laser pulse (secondary).

To improve upon this model, the laser fields have also been approximated through the addition of a triangular temporal envelope function with a total duration on the order of a few hundred femtoseconds. The motion of electrons and the resulting Thomson scattering spectra for several electron initial conditions have been computed numerically within the confluence of the two pulses. These spectra all contain a finite width peak at the laser fundamental frequency due to the quiver motion of the electrons within the laser fields and the finite duration of the pulse. As the initial electric field and, hence, the electron orbit becomes dominated by the initial magnetic field, this fundamental mode gives way to a series of sideband frequencies spread more or less uniformly about the laser frequency. This diverse set of frequencies can be understood as the intermodulation of the axial ponderomotive and transverse quiver time scales present in the electron motion: that is $\omega = \omega_0 \pm \omega_{pond}$. This results in the creation of an efficient high bandwidth light source centered about the laser frequency with approximately 30 percent of the emitted light being contained within the sidebands. Initial experiments have shown that such modulations do indeed exist and efforts are underway to have a more detailed understanding.

5. Conclusion

Ultrafast x-rays can be applied with advantage to almost any problem that is currently studied with longer pulse duration light sources, such as synchrotrons and free-electron lasers. Processes that occur on the subpicosecond time scale are particularly suitable, including photosynthesis and conformational changes in ultrafast biology and chemistry, inner-shell electronic processes in atomic systems [5], and laser initiated phase transitions in materials science [6]. A high brightness x-ray source would also open up a whole new array of applications in many different fields of scientific endeavor. In the area of atomic and molecular physics it would be possible to study ionization and dissociation processes especially in the core levels [7]. X-ray fluorescence or x-ray photo-electron spectroscopy could be used to study surface reactions. In conjunction with an ultrafast laser it would be possible to setup femtosecond resolution structural dynamics studies. Standard x-ray techniques can be extended to much smaller samples with lower noise, using a sufficiently bright source. Single shot data collection capability is possible for cases where irreversible damage occurs due to radiation exposure. In the field of biology it would be possible to study two-dimensional systems like membrane proteins [8].

The most obvious advantage of laser-produced x-ray sources is their compact size (table-top) and affordability (< $1M), features which potentially permit their operation at university, industrial and hospital labs. Ultrafast laser-driven x-ray sources are also particularly advantageous in the study of photo-initiated processes, since in this case the optical pump and x-ray probe are absolutely synchronized to each other, both being derived from the same laser. Also, in cases where there is significant absorption of the x-ray probe pulse (such as in-vitro imaging of live biological cells), the sample is often destroyed by the probe due to radiation damage. Thus, in order to acquire an image before the occurrence of blurring from heat-induced motion, a single-shot pump-probe measurement with ultrashort-duration and high-peak-power x-rays is required. Low-repetition rate, laser-produced x-ray sources are also best suited for studies of systems that are slow to relax or that involve irreversible processes, in which the sample must be moved between shots due to damage by the pump.

It is obvious that there exists a pressing need for a femtosecond, high-brightness source in the soft and hard x-ray regions. Specifically, the x-ray source should emit at wavelengths shorter than the water window region (2-4 nm) to be useful for biological studies. The source needs to have a minimum level of coherence to be useful for most experiments and the technology required should be compact, inexpensive and robust. Nonlinear Thomson scattering holds promise as a source of high brightness, ultrashort duration x-ray pulses with good coherence. Our recent work demonstrates that in principle a synchrotron like table-top x-ray source can indeed be obtained using a compact, high-power laser system.

References

1. A.H. Zewail, *Femtochemistry - Ultrafast Dynamics of the Chemical Bond*, Vol. I and II, World Scientific (20th Century Chemistry Series), Singapore (1994).
2. R.W. Schonlein, W.P. Leemans, A.H. Chin, , P. Volfbeyn, T.E. Glover, P. Balling, M. Zolotorev, K.-J. Kim, S. Chattopadhyay, and C.V. Shank, Science **274**, 236-238 (1996).
3. G. Gruebel, ed., Workshop on *Perspectives of X-Ray Photon Correlation Spectroscopy*, ESRF (1996).
4. G. Mourou and D. Umstadter, Phys. Fluids **B 4**, 2315 (1992).
5. T. Zuo and A.D. Bandrauk, Phys. Rev. A **51**, R26 (1995).
6. C. W. Siders, A. Cavalleri, K. Sokolowski-Tinten, Cs. Tóth, T. Guo, M. Kammler, M. Horn von Hoegen, K. R. Wilson, D. von der Linde, and C. P. J. Barty, Science **286**, 1340 (1999).
7. M. Gavrila and J. Shertzer, Phys. Rev. A **53**, 3431 (1996).
8. H. Luecke, B. Schobert, H. Richter, J. Cartailler, and J. K. Lanyi, Science **286**, 255 (1999).
9. M Protopapas, C H Keitel and P L Knight, Rep. Prog. Phys. **60**, 389 (1997).
10. Vachaspati, *Phys. Rev.* **128**, 664-666 (1962).
11. J.H. Eberly, *Progress in Optics* **VII**, North Holland, Amsterdam (1969).
12. D. Umstadter, Phys. Plasmas **8**, 1774 (2001) and references therein.

13. D. Umstadter, S.-Y. Chen, A. Maksimchuk, G. Mourou, and R. Wagner, Science **273**, 472 (1996).

14. S.-Y. Chen, M. Krishnan, A. Maksimchuk, R. Wagner, and D. Umstadter, Phys. Plasmas **6**, 4739 (1999).

15. S. Banerjee, A. R. Valenzuela, R. C. Shah, A. Maksimchuk, and D. Umstadter, Phys. Plasmas **9**, 2393 (2002).

16. V. Yanovsky, N. Saleh, D. Milathianaki, C. Felix, K. Flippo, A. Maksimchuk and D. Umstadter, *CLEO Technical Digest*, 288 (2000).

17. C. Ezzell, Scientific American, 42, April (2002).

18. C. Rose-Petruck, R. Jimenez, T. Guo, A. Cavalleri, C. W. Siders, F. Rksi, J. A. Squier, B. C. Walker, K. R. Wilson and C. P. J. Barty, Nature **398**, 310 (1999).

19. A. Rousse, P. Audebert, J.P. Geindre, F. Fallies, J.C. Gauthier, A. Mysyrowicz, G. Grillon, A. Antonetti, Phys. Rev. E **50**, 2200 (1994).

20. J. Workman, A. Maksimchuk, X. Liu, U. Ellenberger, J. S. Coe, C.-Y. Chien, and D. Umstadter, Phys. Rev. Lett. **75**, 2324 (1996).

21. J. Workman, M. Nantel, A. Maksimchuk and D. Umstadter, Appl. Phys. Lett. **70**, 312 (1997).

22. Y. Hironaka, K. G. Nakamura, and K. Kondo, Appl. Phys. Lett. **77**, 4110 (2000).

23. U. Andiel, K. Eidmann, and K. Witte, Phys. Rev. E **63**, 026407 (2001).

24. J. Zhang, M. Zepf, P. A. Norreys, A. E. Dangor, M. Bakarezos, C. N. Danson, A. Dyson, A. P. Fews, P. Gibbon, M. H. Key, P. Lee, P. Loukakos, S. Moustaizis, D. Neely, F. N. Walsh, and J. S. Wark, Phys. Rev. A **54**, 1597 (1996).

25. X.F. Li, A. L'Huillier, M. Ferray, L.A. Lompre, and G. Mainfray, Phys. Rev. A **39**, 5751 (1989).

26. P. Corkum, Phys. Rev. Lett., **71**, 1993 (1993).

27. Ch. Spielmann, N. H. Burnett, S. Sartania, R. Koppitsch, M. Schnürer, C. Kan, M. Lenzner, P. Wobrauschek, and F. Krausz, Science **278**, 661 (1997).

28. A. Rundquist, C. G. Durfee, Z. Chang, C. Herne, S. Backus, M. M. Murnane, and H. C. Kapteyn, Science **280**, 1412 (1998).

29. R. A. Bartels, A. Paul, H. Green, H. C. Kapteyn, M. M. Murnane, S. Backus, I. P. Christov, Y. Liu, D. Attwood, and C. Jacobsen, Science **297**, 376 (2002).

30. M. Schnürer, Ch. Spielmann, P. Wobrauschek, C. Streli, N. H. Burnett, C. Kan, K. Ferencz, R. Koppitsch, Z. Cheng, T. Brabec, and F. Krausz, Phys. Rev. Lett. **80**, 3236 (1998).

31. S.C. Rae, K. Burnett, and J. Cooper, Phys. Rev. A **50**, 3438 (1994).

32. A. Rousse, C. Rischel, and J.-C. Gauthier, Rev. Modern Phys. **73**, 17 (2001).

33. A. Rousse, C. Rischel, S. Fourmaux, I. Uschmann, S. Sebban, G. Grillon, Ph. Balcou, E. Förster, J.P. Geindre, P. Audebert, J.C. Gauthier, D. Hulin, Nature **410**, 65 (2001).

34. M. J. Edwards, A. J. MacKinnon, J. Zweiback, K. Shigemori, D. Ryutov, A. M. Rubenchik, K. A. Keilty, E. Liang, B. A. Remington, and T. Ditmire, Phys. Rev. Lett. **87**, 085004 (2001).

35. A. Rischel, A. Rousse, I. Uschmann, P. A. Albouy, J. P. Geindre, P. Audebert, J. C. Gauthier, E. Forster, J. L. Martin, A. Antonetti, Nature **390**, 490 (1997).

36. A. A. Zholents and M. S. Zolotorev, Phys. Rev. Lett. **76**, 912–915 (1996).

37. R. W. Schoenlein, S. Chattopadhyay, H. H. W. Chong, T. E. Glover, P. A. Heimann, C. V. Shank, A. A. Zholents, M. S. Zolotorev, Science **287**, 2237 (2000).

38. S. Banerjee, A. R. Valenzuela, R. C. Shah, A. Maksimchuk, and D. Umstadter, J. Opt. Soc. Am. B (In press).

39. D. Umstadter, J.K. Kirn, and E. Dodd, E., Phys. Rev. Lett. **76**, 2073 (1996).

40. F. V. Hartemann, H. A. Baldis, A. K. Kerman, A. Le Foll, N. C. Luhmann, and B. Rupp, Phys. Rev. E **64**, 016501 (2001).

41. S. Banerjee, A. R. Valenzuela, R. C. Shah, A. Maksimchuk, and D. Umstadter, CLEO Technical Digest, 78 (2002).

42. D. Atwood, *Soft X-rays and Extreme Ultraviolet Radiation*, Cambridge University Press, New York (2000).

43. P. Zhang, N. Saleh, S. Chen, Z.-M. Sheng, and D. Umstadter, Phys. Plasmas (submitted).

44. V. Ayvazyan et al., Phys. Rev. Lett. **88**, 104802 (2002).

45. W.P. Leemans, R.W. Schoenlein, P. Volfbeyn, A.H. Chin, T.E. Glover, P. Balling, M. Zolotorev, K.-J. Kim, S. Chattopadhyay, and C. V. Shank, IEEE J. Quant. Electron. **33**, 1925 (1997) and references therein; D.D. Meyerhofer, *ibid.*, 1935 (1997).

46. J.J. Thomson, *Conduction of Electricity through Gases*, Cambridge University Press, Cambridge (1906).

47. E.S. Sarachik, and G.T. Schappert, Phys. Rev. D **1**, 2738 (1970).

48. M. I. K. Santala, Z. Najmudin, E. L. Clark, M. Tatarakis, K. Krushelnick, A. E. Dangor, V. Malka, J. Faure, R. Allott, and R. J. Clarke, Phys. Rev. Lett. **86**, 1227 (2001).

49. F. Amiranoff, S. Baton, D. Bernard, B. Cros, D. Descamps, F. Dorchies, F. Jacquet, V. Malka, J. R. Marquès, G. Matthieussent, P. Miné, A. Modena, P. Mora, J. Morillo, and Z. Najmudin, Phys. Rev. Lett. **81**, 995 (1998).

50. V. Malka, J. Faure, J. R. Marquès, and F. Amiranoff J. P. Rousseau, S. Ranc, and J. P. Chambaret Z. Najmudin and B. Walton, Phys. of Plas. **8**, 2605 (2001).

51. Y. Salamin, and F.M.H. Faisal, Phys. Rev. *A* **54**, 4383 (1996).

52. J.D. Jackson, *Classical Electrodynamics*, John Wiley and Sons, New York (1999).

53. R.H. Milburn, Phys. Rev. Lett. **10**, 75 (1963); F.R. Arutyunian and V.A. Tumanian, Phys. Lett. **4**, 176 (1963).

54. A. Ting, R. Fischer, A. Fisher, K. Evans, R. Burris, J. Krall, E. Esarey, and P. Sprangle, J. Appl. Phys. **78**, 575 (1995); C. Bula et al., Phys. Rev. Lett. **76**, 3116 (1996).

55. E. Esarey, R. F. Hubbard, W. P. Leemans, A. Ting, and P. Sprangle, Phys. Rev. Lett. **79**, 2682 (1997).

56. S.-Y. Chen, A. Maksimchuk, and D. Umstadter, Nature **396**, 653 (1998).

57. S.-Y. Chen, A. Maksimchuk, and D. Umstadter, Phys. Rev. Lett. **84**, 5528 (2000).

58. S. Banerjee, A. Valenzuela, R.C. Shah, and D. Umstadter, Nature (submitted).

59. R.W. Boyd, *Nonlinear Optics*, Academic Press, San Diego (1991).

60. A. J. Mackinnon, M. Borghesi, A. Iwase, M. W. Jones, G. J. Pert, S. Rae, K. Burnett, and O. Willi, Phys. Rev. Lett. **76**, 1473 (1996).

61. J.F. Ward and G.H.C. New, Phys. Rev. **185**, 57 (1969).

62. P. Catravas, E. Esarey and, W.P. Leemans, Meas. Sci. Technol. **12**, 1828 (2001).

63. E. Constant, D. Garzella, P. Breger, E. Mével, Ch. Dorrer, C. Le Blanc, F. Salin, and P. Agostini, Phys. Rev. Lett. **82**, 1668 (1999).

64. J.M. Rax, and N.J. Fisch, IEEE Trans. Plas. Sci. **21**, 105 (1993).

65. P. Sprangle, E. Esarey, and A. Ting, Phys. Rev. Lett. **64**, 2011 (1990).

66. J. Peatross, and D.D. Meyerhofer, Phys. Rev. A **52**, 3976-3987 (1995).

67. Y. Salamin, and F.M.H. Faisal, Phys. Rev. *A* **55**, 4383 (1997).

68. J.-F. Hergott, M. Kovacev, H. Merdji, C. Hubert, Y. Mairesse, E. Jean, P. Breger, P. Agostini, B. Carré, and P. Salières, Phys. Rev. A **66**, 021801 (2002).

69. D. Umstadter and T. Norris, IEEE J. Quantum Electron. **33**, 1878 (1997).

70. C. I. Castillo-Herrera and T. W. Johnston, IEEE Trans. Plasma Sci. **21**, 125 (1993).

71. E. Esarey, S. K. Ride, and P. Sprangle, Phys. Rev. E **48**, 3003 (1993).

72. J.E. Gunn, and J.P. Ostriker, Astrophys. J. **165**, 523 (1971).

73. F. Hei, Y.Y. Lau, T. Strickler and D. Umstadter, Phys. Plas. **9**, 4325 (2002).

74. A. K. Puntajer, and C. Leubner, Phys. Rev. A **40**, 279 (1989).

75. S. K. Ride, E. Esarey, and M. Baine, Phys. Rev. E **52**, 5425 (1995).

76. P. Sprangle, A. Ting, E. Esarey and A. Fisher, J. Appl. Phys. **72**, 5032 (1992).

77. Y. Y. Lau, F. Hei, D. Umstadter and R. Kowalczyk, Phys. Rev. Lett. (submitted).

78. H. Goldstein, *Classical Mechanics*, Addison-Wesley, Boston, USA (1980).

79. R. Haberman, *Elementary Applied Partial Differential Equations*, Prentice-Hall, Englewood Cliffs (1998).

80. E.L. Ince, *Ordinary Differential Equations*, Dover Publications, New York (1956).

APPLICATIONS OF HIGH BRIGHTNESS ELECTRON BEAMS TO VACUUM LASER ACCELERATORS

A.A.VARFOLOMEEV*
Russian Research Center "Kurchatov Institute",
Kurchatov Sq.1, Moscow, 123182,
Russian Federation
E-mail: varfol@dnuc.polyn.kiae.su

Investigations on the laser acceleration in vacuum are reviewed including up-date experimental results, projects of new experiments and theoretical prospects of vacuum acceleration schemes for high energies of accelerated particles. Most attention is devoted to analysis of the IFELs with focused laser beam drivers. The free space acceleration with the axicon and the Gaussian beam are also considered as prospects for the high energy particle acceleration. It is shown that at high quality of accelerated beams the vacuum laser accelerators could provide required high acceleration gradients as well as high energy gains and can be considered as possible prototypes of the future accelerators.

1. Introduction

In accordance with the title of our review applications of high brightness electron beams to the laser acceleration in vacuum will be considered. The history of the laser acceleration as a physical problem is of the same age as the laser story itself since ideas to use laser fields for an acceleration of charged particles appeared just after the laser invention in 60-ths. Up to 80-ths this problem was a curious and interesting one from the physical point of view but practically not yet requested. It was the period of theoretical ideas accumulation, study and understanding physical aspects of the problem At the beginning of 80-ths first serious analysises of different approaches to the laser and plasma acceleration were made (see conference proceedings [1-3] and contained there references). The terminologies of acceleration mechanism as far field, near field, plasma wake field, all were introduced at that time. First experiments regarded as proof-of-principles of different types of the laser acceleration appeared the next decade. A very good analysis of the current laser acceleration research was given at Snowmass 2001 [4]. At the present time high technology of charged particle beams has been developed to the extent adequate to the challenges of the laser

* Presented at the Joint ICFA Advanced Accelerator and Beam Dynamics Workshop, Chia Laguna, July 1 - 6, 2002, Sardinia, Italy.

acceleration experiments with practically useful beam densities. This is a very important step to the experiments on construction of laser accelerator models.

In this review we discuss mainly the pure vacuum acceleration paying more attention to the physics than to technology. One of arguments for this is the fact that not many experiments were really done up to date. Further physics study is necessary since common opinion is not yet elaborated which type of the laser or the plasma acceleration schemes can really be the most efficient in providing high acceleration rates of high brightness particle beams in a wide energy range. Our review consists of three parts which include recent experimental data on the vacuum acceleration, projects on developing next stage IFEL experiments in strong laser fields and some new concepts of the vacuum acceleration respectively.

2. Research Results on Acceleration by Lasers in Vacuum

We will start with some general comments on the subject including definitions and approaches.

Acceleration in vacuum means that no media problem is concerned in this so called far field approximation. Most interesting experimental data have been obtained with the accelerator schemes based on the interaction of an electron beam with laser fields and transverse magnetic fields. Such devices are called Inverse Free Electron Lasers (IFELs). The acceleration process is described by the same as in the FEL physics well known equations. Principle differences from the FELs are the scale and the sign of the electron energy change. A relatively great total energy gain in the IFEL case makes the whole problem more complicate including the electron beam dynamics evaluation, providing stability in the nonlinear interaction processes, the radiation losses as well as some technological aspects as the beam focusing, the undulator tapering for the interaction synchronism providing and etc. First detailed theoretical analysises of this problem for the high energy electrons were given in the papers [5, 6]. The IFEL dynamics was considered in an adiabatic approach with plane wave laser fields. We will underneath return to the IFELs with a focused laser beam where a more correct approach is needed.

First experimental evidences of laser energy absorption by electron beams were obtained in the FEL experiments. It was shown that the radiation gain of the FEL can be negative as well as positive. Special experimental IFEL performances and the respective experiments were made later. Basic and most interesting experimental results are presented in Table 1. Not many experiments were done up to date. All of them are actually proof-of-principles ones. Let us analyze some key point results.

Table 1. Data on the recent IFEL experiments.

	Laser λ / P	Gradient / Gain
Experimental test of the IFEL accelerator principle. I. Wermik, T.C. Marshall [7]	1.6 mm, 5 MW	0.7 MV/m, 0.2 MeV
Observation of energy gain at the BNL IFEL accelerator. A. van Seenbergen, J. Gallardo et al. [8].	10.6 μm, 1-2 GW	2.5 MV/m, 1 MeV
Experimental observation of femtosecond e.b. microbunching by IFEL acceleration. Y. Liu, X. J-Wang, D.B. Cline et al. [9].		
Detailed experimental results for laser acceleration staging. STELLA 2001. W.D. Kimura, W.D. Cumbel, C.E. Dilley at al. [10]	10.6 μm, 0.1-0.3 GW	≤ 2 MeV
STELLA 2: Staged monoenergetic laser acceleration – experimental update; W.D. Kimura, M. Babzien, I. Ben-zvi et al. [11]	10.6 μm, 28 GW	5.9 MeV

The acceleration rates and the total energy gains given in Table1 are low (<2.5 MeV/m and <4 MeV respectively). These data are less impressive than the acceleration gradients demonstrated in Laser-induced Wake Field Accelerators (LWFA) or Plasma Wake Field Accelerators (PWFA). The ultra-high accelerator gradients, >100 GeV/m, provided by the LWFA were observed in different European, Asian and US laboratories (see published review [4] with contained there references and other reports presented at this Workshop). However, this acceleration was demonstrated only on lengths less than 1 mm and the obtained beams had a large energy spread with only very few electrons at the highest energies [4]. For the IFELs acceleration gradients can not be so high but total acceleration lengths can be much greater. Radiation energy losses of the accelerated particles can practically limit the highest electron energy of the IFEL accelerators [5, 6].

A very important evidence of the IFEL capability was achieved with the staged experiments STELLA [10, 11]. The electron beam from the ATF linac (single 6 ps, 45 MeV macropulse) was used. Two identical IFELs were utilized as a prebuncher and an accelerator respectively. The IFELs were positioned in series along the beam line and separated by 2.3 m. The pulsed ATF CO_2 laser beam was split into 2 parts one of which with the lower peak power (24 MW) was sent to the buncher (IFEL1) and the other part (100-300 MW) was sent to the prebuncher (IFEL2). Fine phase adjustment to a fraction of the laser wave-length (< 1 μm) was provided. Electrons in the macropulse left the IFEL with their energy modulated by the field inside the IFEL1. This modulation transformed into the longitudinal density by formation a train of microbunches roughly 1 μm in length separated by 10.6 μm and contained within the 6 ps macropulse envelope. This way the laser beam and the electron beam micropulses were synchronized to a precision of the laser optical wave-length order. Practical ability to maintain a stable phase control (on a few fs over many minutes) was demonstrated for 0.1 nC. Maximum energy gain of 5% of the primary energy was observed [10] corresponding to the acceleration rate 7 MeV/m. As authors have written [10] maximum gradient was not a primary goal of this STELLA experiment. The next experiment STELLAII [11] was designed with some modifications for inproving results and demonstration first of all monoenergetic laser acceleration. These modification included a new tapered undulator in the IFEL2 and the upgraded ATF laser up to several hundred GWs of peak power. Preliminary information on the result is given in Table 1.

Noteworthy results of the STELLA experiments are the possibility of a staging as well as the demonstration that a large portion of the microbunches were accelerated with the maximum gain. The general importance of the STELLA results is their applicability to other acceleration methods which are looking at the present time as more promising since they have demonstrated much higher gradients or have more favorable scaling. On other side a very strict requirement for the accelerators is maintaining of a high beam quality during the entire acceleration process. Here again the IFEL scheme appears to be more preferable since the process is going in the vacuum with much less unsertainties from nonlinearities. In what follows we will consider what can be done for the further development of the vacuum accelerator schemes with the purpose to improve their main characteristics including the acceleration rate and the total energy gain. We will not consider technical and technological aspects common for all types of the laser accelerators and will emphasize physical aspects of possible processes and schemes which could be potential for the improving of the basic vacuum acceleration schemes.

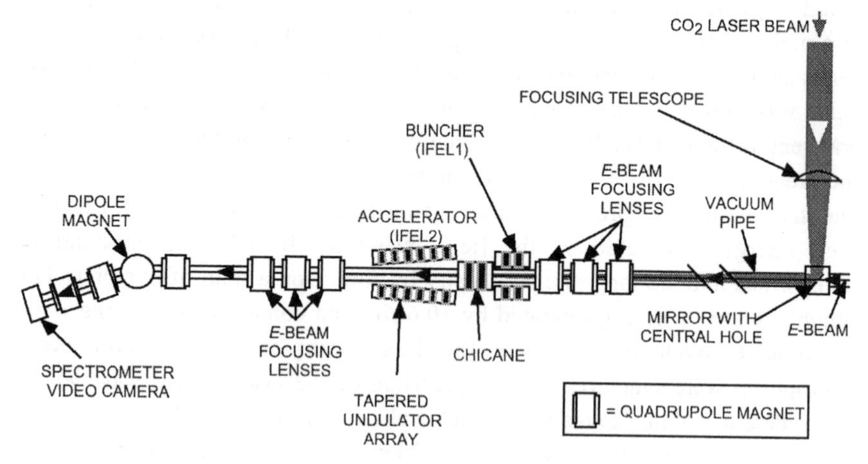

Fig. 1. Schematic layout for the STELLA-II experiment.

3. Focused Laser Beams as IFEL Drivers

The Inverse Free Electron Laser is the most developed laser acceleration concept from many points of view. Its roots are in the theory and the hardware technology of the FEL which has been developed much from 80-ths. Recent proof-of-principle experiments have successfully demonstrated ability of the IFELs to bunch electron beam along with its acceleration, to provide good acceptances of the beam and electron acceleration with a small energy spread. The most feeble feature of the recent experiments on the IFELs is a low relative energy gain demonstrated up date. The total energy gain can be encreased by staging of the accelerator system as was shown by the experiments but the acceleration rate is still low for such systems. That is why new projects on the IFELs are planning with using a higher laser power (100 GW - 1 TW) of CO_2 lasers in ATF and UCLA. At the given total laser power the most strong optical fields can be provided by focusing of the laser beams. An impressive project of this type IFEL is the project [13] to be held at Neptun Lab with the facility designed and constructed by the UCLA-KIAE group. Before describing in detail this project we will analyze common properties of the IFELs based on the laser Gaussian beams as most promising options of the high gradient laser accelerators. Reminding of some features of the laser focused fields is unavoidable.

4. Theoretical Approaches to IFEL Process in Intense Gaussian Beams

The laser fields for the fundamental TEM_{00} mode of a linearly polarized wave in the conventional units are expressed as:

$$E_x = E_0 \frac{e^{-\frac{x^2+y^2}{w(z)^2 w_0^2}}}{w(z)} \cdot \sin\left(kz - \grave{u}t + \varphi_0 + \frac{k\left(x^2+y^2\right)}{2R(z)} - \text{arctg}\frac{z}{z_R} \right) \quad (1)$$

$$E_z = E_0 \frac{e^{-\frac{x^2+y^2}{w(z)^2 w_0^2}} \cdot 2x}{k \cdot w(z)^2 \cdot w_0^2} \cdot \cos\left(kz - \grave{u}t + \varphi_0 + \frac{k\left(x^2+y^2\right)}{2R(z)} - 2 \cdot \text{arctg}\frac{z}{z_R} \right) \quad (2)$$

$$B_y = B_0 \frac{e^{-\frac{x^2+y^2}{w(z)^2 w_0^2}}}{w(z)} \cdot \sin\left(kz - \grave{u}t + \varphi_0 + \frac{k\left(x^2+y^2\right)}{2R(z)} - \text{arctg}\frac{z}{z_R} \right) \quad (3)$$

$$B_z = B_0 \frac{e^{-\frac{x^2+y^2}{w(z)^2 w_0^2}} \cdot 2y}{k \cdot w(z)^2 \cdot w_0^2} \cdot \cos\left(kz - \grave{u}t + \varphi_0 + \frac{k\left(x^2+y^2\right)}{2R(z)} - 2 \cdot \text{arctg}\frac{z}{z_R} \right) \quad (4)$$

Here z axis is the optical beam axis, $k=2\pi/\lambda$ – the carrier wave-number, $z_R=kw_0^2/2$ is the Rayleigh length, w_0 is the radius of the field waist, $w(z)=(1+z^2/z_R^2)^{1/2}$. The Gaussian beam shape in a resonator is schematically shown by Fig.2 along with an electron trajectory crossing the resonator focus. In the common case the resonator axis and the electron beam are noncollinear having a small crossing angle α between them.

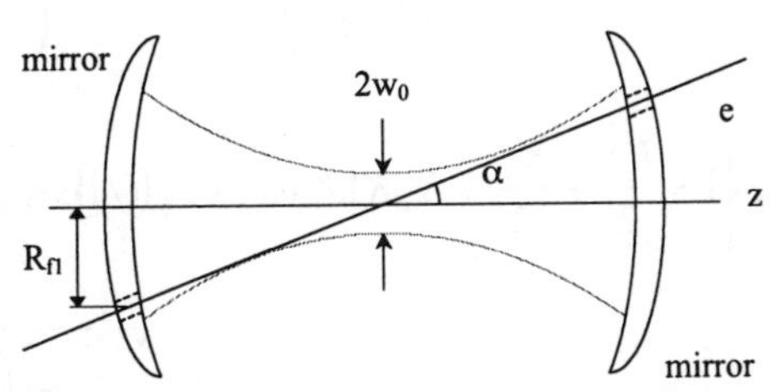

Fig. 2. Scheme of the Gaussian beam resonator with noncollinear electron beam line.

At large Rayleigh lengths z_R the well known 1D equations [5, 6] for a planar wiggler and plane wave fields can be used as a first approximation of the

IFEL equations. After some corrections made for taking into account possible undulator tapering more accurately these equations for the particle energy γ and its relative phase ψ take the form:

$$\frac{d\gamma}{dz} = \frac{eE_0}{2mc^2} \cdot \frac{K}{\gamma}\left[J_0(G) - J_1(G)\right]\sin\psi + \frac{eE_0}{2mc^2}\frac{K_L}{\gamma}\sin 2k_w z - $$
$$- \frac{2}{3} r_e \gamma^2 k_w^2 \left\{ \frac{K^2}{2} + \frac{K_L^2}{2} + K_L K\left[J_0(G) + J_1(G)\right]\cos\psi - \frac{K^2 + K_L^2}{2}\cos 2k_w z \right\} \tag{5}$$

$$\frac{d\psi}{dz} = k_w - k\frac{1}{2\gamma^2}\left\{1 + \frac{K^2}{2} + \frac{K_L^2}{2} + K_L K\left[J_0(G) - J_0(G)\right]\cos\psi\right\} \tag{6}$$

Here the same conventional notations are used:

$$G(z) = \frac{k}{k_w} \cdot \frac{K^2 + K_L^2}{8\gamma^2}; \; K = \frac{eB_{w0}}{mc^2 k_w}; \; K_L = \frac{eE_0}{mc^2 k_s}; \; k = \frac{\omega}{c}; \; k_w = \frac{2\pi}{\lambda_w};$$

$$\psi(z) = \int_0^z k_w dz - k\int_0^z \frac{dz}{2\gamma^2}\left\{1 + \frac{K^2}{2} + \frac{K_L^2}{2} + K_L K\left[J_0(G) - J_0(G)\right]\cos\psi\right\} - \psi_0$$

J_0, J_1 – Bessel functions

The used adiabatic approximation means that all parameters $k_w(z)$, $K(z)$, $K_L(z)$, $E_0(z)$, $B_{w0}(z)$ are very slowly varying with z (small relative chahges on the lengths ~2π k_w) and the parameters K and K_L are small so that the 1D approximation is still valid.

For the IFEL in the Gaussian beam with a small z_R the diffraction of the laser beam should be more accurately taken into account. In the 1D approximation which is valid for the case of small transverse electron deviations the IFEL equations take the form:

$$\frac{d\gamma}{dz} = \frac{eE_0}{2mc^2} \cdot \frac{K}{\gamma}\left[J_0(G) - J_1(G)\right]\frac{\sin\psi}{\left(1 + \frac{z^2}{z_R^2}\right)^{\frac{1}{2}}} + \frac{eE_0}{mc^2}\frac{K_L}{\gamma}\frac{\sin 2k_w z}{\left(1 + \frac{z^2}{z_R^2}\right)^{\frac{1}{2}}} \tag{7}$$

$$\frac{d\psi}{dz} = k_w - k\frac{1}{2\gamma^2}\left\{1 + \frac{K^2}{2} + \frac{K_L^2}{2} + K_L K\left[J_0(G) - J_0(G)\right]\cos\psi + \gamma^2\alpha^2\right\} - $$
$$- \frac{1}{z_R\left(1 + \frac{z^2}{z_R^2}\right)} \tag{8}$$

For very strong laser field in the focus area the approximation $r_\perp$~0 is actually never valid. So for real facilities based on using Gaussian modes of strong laser fields numerical 3D analysis has to be used. The laser field map (1)-(4) as well as a 3D undulator magnetic field map can be used this way. Solution of the system (7), (8) can be considered as a first approximation only.

5. Gaussian Mode IFEL Experiment at UCLA

The high power laser drive IFEL experiment will be held in the near future at the Neptune Laboratory under the UCLA - Kurchatov Instittute Project [13, 14]. The goal of the experiment is achievement of one order higher electron energy gain than was demonstrated before in experiments (see Table 1) along with demonstration of high acceleration rate of monochromatic electron bunches. Schematic layout of the laser line and the electron source is shown in Fig.3. The electron beam is coming from a photoinjector+booster linac system having energy 14 MeV at the entrance of an undulator. A CO_2 laser beam is focused by a lens with focal distance of 2.6 m to a tight spot of a few hundreds microns. Basic parameters of the IFEL device under construction are given in Table 2. It is seen that the Gaussian beam to be used has a short Rayleigh length. The high power laser mode with this small Rayleigh length creates a specific diffraction dominated IFEL regime absolutely new for both the experiment and the theory. Non-adiabatic approach and numerical 3D simulations are needed for the theoretical understanding of the processes. The very specific undulator construction was required with strong tapering and small magnetic field tolerances. The simulations concerning the undulator construction as well as the entire electron beam dynamics problem are now completed [13, 14]. The undulator is being manufactured and will soon be commissioned.

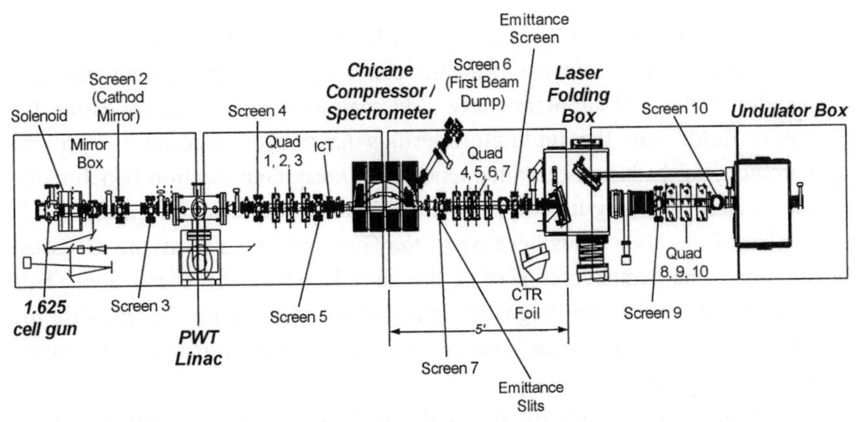

Fig. 3. Schematic layout of the IFEL experiment at UCLA on the Neptune beam line.

Table 2. Basic parameters of the IFEL device.

Laser beam parameters:	
Wave length λ	10.6 μm
Power range	0.4 – 0.8 TW
Rayleigh length z_R	3.6 cm
Waist at the focus w_0	0.35 mm
Size at the undulator entrance $w_0 w$	2.5 mm
Initial electron beam parameters:	
Initial energy	14 Mev
Energy spread	0.5%
Emittance ε_n	10 mm×mrad
Charge	300 pC
Pulse length	6 ps – 3 ps
Rms radius at the focus	0.15 mm
Rms radius at the undulator entrance	0.5 mm
Undulator parameters:	
Total undulator length	524.5 mm
Undulator period at the entrance $\lambda_w(0)$	15.16 mm
Undulator period at the exit $\lambda_w(z_{max})$	52.1 mm
Initial field strength	0.115 T
Field strength at the exit	0.626 T
Number of periods	17.5

The basic initial parameters of the IFEL undulator are given in Table 2. Fig.4 shows a schematic design of the undulator displacing permanent magnets, permendure poles and supplementary side magnets configurations. Both the magnetic fields and the undulator periods have been tapered to provide synchronicity of the laser beam interaction with a captured electron bunch along the entire undulator length. The most critical part of the electron trajectories is the region of the laser focus because of the Guoy phase shift in the Gaussian beam and a small transverse size of the laser beam at the focus area. This problem dictates very fine magnetic field shaping to be done. To provide the required shape just at the focus a special magnetic corrector was inserted between the undulator sections.

Fig.5 shows the undulator magnetic field profiles and a trajectory of a single reference electron obtained by the simulations with using the 3D map of the undulator fields. The respective electron velocities and the laser field strength experienced by the electron at position z are given by Fig.6. Synchronization of the interaction is provided at the same signs of the electron velocity and the laser field. Predicted characteristics of the output electron beam are presented in Table 3. Acceptable tolerances were also estimated (Table 4).

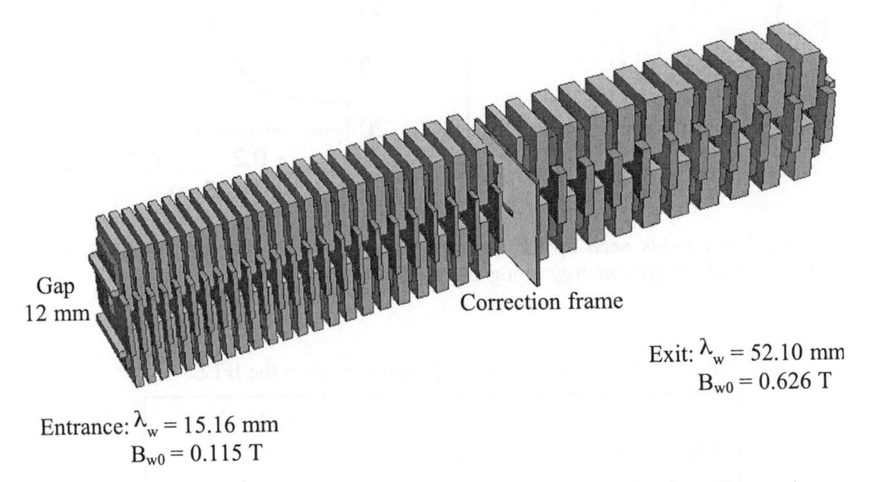

Fig. 4. Schematic construction of the planar hybrid undulator for the IFEL project

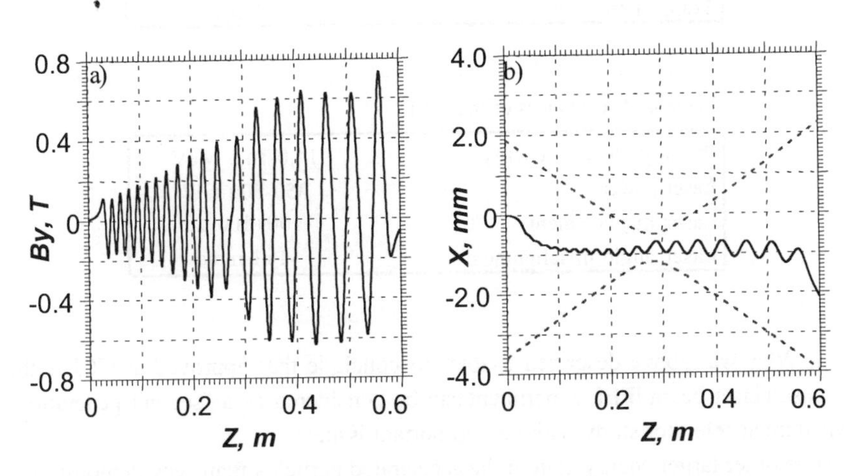

Fig. 5.Magnetic fields (a) and a trapped electron trajectory (b) in the undulator.

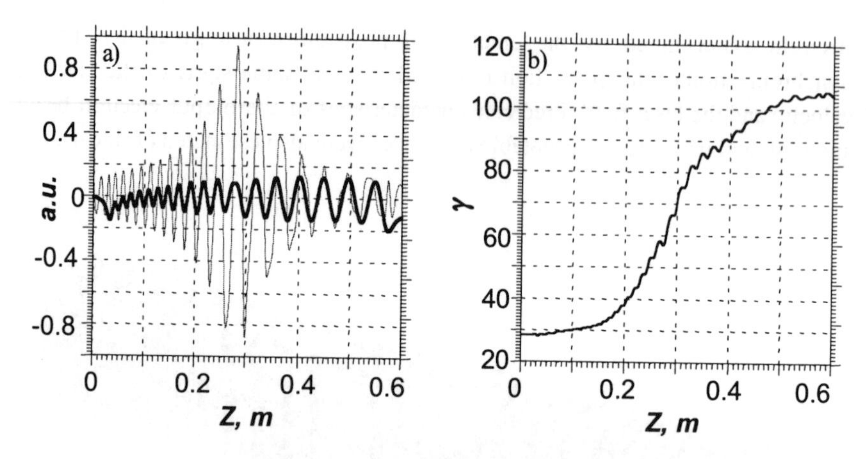

Fig. 6. (a) Laser fields seen by the electron (thin line) and electron velocity (bold). (b) Accelerated electron energy along the undulator depth.

Table 3. Output electron beam parameters from the IFEL.

Energy	55 MeV
Microbunch length	3 fs
Peak current	3 kA
Transverse emittance	10 mm×mrad
Trapped fraction	>20%

Table 4. Acceptances of the IFEL.

Primary electron energy	14.0-14.5 MeV
Laser power	350-500 GW
Laser displacement	-100÷100 μm
Laser angle misalignment	-1÷1 mrad

What was above described enables to conclude that approved at UCLA the focused laser beam IFEL experiment can be considered as a "second generation" vacuum accelerator study with very important issues:

a) one order larger energy gain of the accelerated particles than was demonstrated up date.

b) large accelerating gradient of the quasimonochromatic beam (up to 100 MeV/m).

c) appreciable fraction of the primary microbunches captured for the laser acceleration (tens %%).

So the IFEL laser accelerators with development of the staging systems are potential for the high rate acceleration in a wide energy range of the accelerated particles.

6. Perspectives of Focused Laser Beam Drivers for IFELs at Higher Electron Energies

The analysis of the current situation with the IFEL accelerators shows that the best results for relatively low electron energies (within ~100 MeV range) can be achieved by staging sections based on the using focused laser beams and particularly the Gaussian beams having small Rayleigh lengths and very strong laser fields at the focus area. The IFEL perspectives at the higher electron energies are much less evident. The key problem is how to increase the acceleration lengths over distances of many Rayleigh lengths providing strong laser fields on these distances with the existed lasers. This problem is not new. Different proposals were made on possible waveguide constructions for the laser wave propagation in a channel. One of the first descriptions of dielectric coated guides was given in [15]. Experimental measurements were made of the CO_2 laser beam attenuation. Optical guiding of the laser pulses is possible in a specially created plasma as well [16]. This process is intensively studied in connection with the laser wake field acceleration processes [4]. We will not consider the vacuum acceleration by the laser beams confined in the optical wave-guides for some reasons. First of all, these schemes are not developed enough and no experiment of this kind is actually done till now.

Another principal problem for using the conventional IFELs at the high electron energies is well known [5, 6]. The synchrotron radiation losses of the undulating electrons (see second term of eq. (5)) grow with the electron energy so that at some electron energy these losses balance the acceleration energy gain from the interaction with the laser field. It gives zero net acceleration rate. Accelerated particles (electrons) can not be accelerated beyond an upper energy limit which depends on both the IFEL and the laser parameters (see [5, 6]) but for real parameters does not exceed 200-300 GeV.

The above energy limit for the acceleration is inherent to the IFEL principal scheme but not to the vacuum acceleration in common sense. It was shown earlier [16] that the acceleration by a laser field can be provided without using the transverse deviations of the electrons induced by the IFEL undulators. This scheme does work at the high electron energies without the inherent radiation losses typical for the IFELs and consequently without the principal upper energy limit. We want again to draw attention to this free space laser acceleration scheme as a potential one for the vacuum laser acceleration of the high energy particles. In the next paragraph some modifications of the free space laser accelerator schemes will be considered.

7. Prospects of Free Space Laser Acceleration

Accelerator schemes based on the interaction of free propagating electrons with a laser beam only have been earlier proposed and analyzed. In one case [18] the laser beam is specially configurated with an optical system axicon [19], schematically shown in Fig.7. In another case [20] the Gaussian mode laser beam is used (Fig.2). The same physical principle is used in both schemes. The laser beam interaction with the particles being accelerated is provided due to two peculiar features of the schemes:

- The wave-vector **k** of the laser wave and the particle velocity **v** are not collinear ($\alpha > 0$ in Fig.2) what gives a nonzero interaction proportional to $|\mathbf{v}\times\mathbf{k}_s|$.

- The interaction is terminated by the device system so that the interaction length is limited.

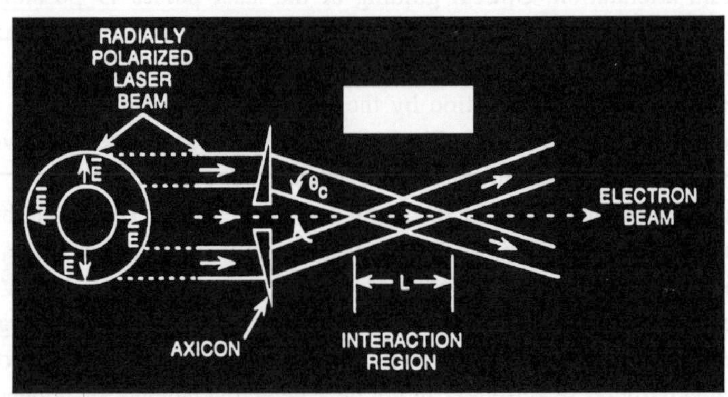

Fig. 7. Schematic of the axicon beam.

The last condition is very important for getting not zero net acceleration. It is generally recognized that if a wave overtakes a free electron the latter can gain energy from the wave only for the time being in the wave. The electron reverts to its initial energy once the wave has passed. That is why it was believed that laser field does cancel out net acceleration [21]. But this is correct for any oscillatory e.m. field only in the case of an unlimited interaction length. This condition is not so evident and was reanalyzed in many papers. In one of the latest publications [22], dedicated to this problem, it was shown that interruption of the interaction can occur due to different reasons including an electron scattering process.

Both above schemes have potential for creating the high gradient accelerators. The axicon vacuum laser acceleration is more known since the optical scheme is similar to that used in the experiments on the Inverse Cherenkov Accelerators. We will mainly consider the free space vacuum acceleration in the Gaussian beams since it can be preferable due to capability to provide the higher acceleration gradients.

7.1. *Free Space Acceleration by Gaussian Mode Laser Field*

Configuration of the rezonator with the Gaussian beam shown in Fig.2 can be considered as a schematic of the free space electron accelerator. Two resonator mirrors provide the Gaussian mode configuration of the laser beam and terminate the laser field interaction with the electron outside the resonator. Holes in the mirrors made for the electron beam propagation can be at the edge of the first Fresnel zone or at some larger distances what is practically not harmful for the Gaussian mode field configuration [23]. In what follows we will show that such system can really provide the free space acceleration since all requirements for this can be fulfilled. Characteristics of the resonator as an unit cell of the accelerator will be estimated and namely: limited slippage and synchronicity conditions, termination of the interaction length, selffocusing and limited defocusing of the accelerated beam, phase feed back in the beam dynamics. We want to show that some existent critical representation of this process are too pessimistic or not correct.

Let us consider the resonator scheme presented in Fig.2. with the linearly polarized fields given by the eqs.(1), (2). In the approximation of negligibly small transverse deviations of the electron beam (see below) the interaction energy exchange rate is proportional to the electric field projection component on the electron beam trajectory:

$$E_\alpha = E_x\,(x,y,z,t\,)\,\mathrm{Sin}\,\alpha\; - \;E_z\,(x,y,z,t,\,)\,\mathrm{Cos}\,\alpha \qquad (9)$$

where α is the noncollinearity angle shown in Fig.2, x,y,z,t are coordinates of an considered electron. The phase of the laser wave at the electron position at the time t (phase "seeing" by the electron) is defined by these coordinates.

The phase velocity of the wave differs from the electron velocity what results in some phase slippage as a functional argument of the field (9). The term arctg(z/z_R) in the argument of (1), (2) gives additional so called Guoy phase shift increasing the slippage. Some compensation of the phase slippage is provided by the term k $(x^2+y^2)/2R(z)$. Variation of the noncollinearity angle α can give at some α_{opt} a more smooth slippage variation over z near laser focus where the phase control is especially important [24]. The main factor increasing the electron phase and respectively decreasing the slippage is large γ parameters of the used electrons. At the high electron energies the phase slippage limit will be reached at rather long distances providing long interaction lengths comparable with the electron path length in the resonator. Simulation results [24] can help to understand the properties of the considering scheme more clearly.

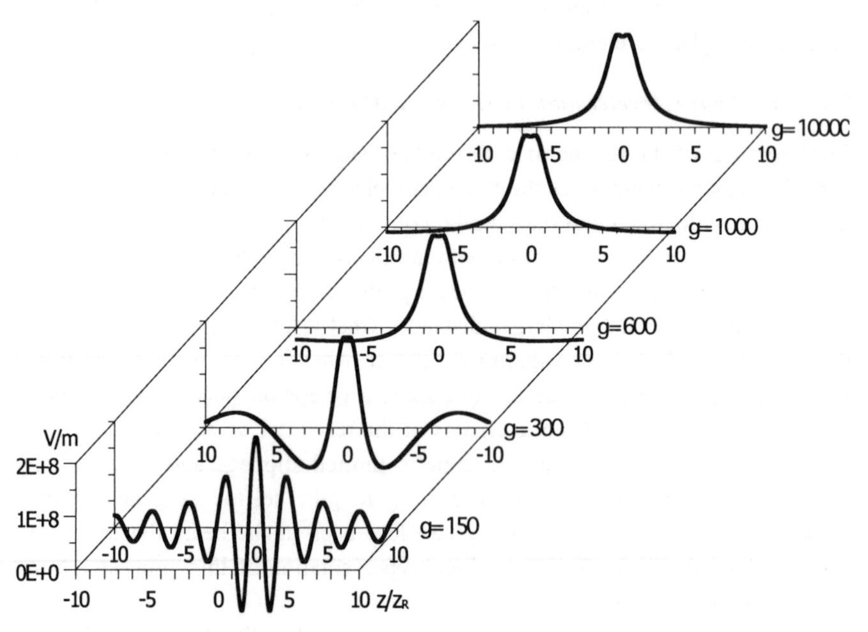

Fig. 8. Accelerating field profiles E_α of the Gaussian beam for the parameters: $\alpha = \frac{1}{2}\sqrt{\frac{\lambda_s}{2z_R}}$, $\varphi_0=\pi/2$, $\lambda=10.6$ μm, $P=10^{13}$ W, $z_R=18$ cm.

The accelerating field profiles (9) of the Gaussian mode are given in Figs.8, 9. The calculations were made with an optimal angle α for two Rayleigh lengths 3 cm and 18 cm respectively and for different energy parameters γ. Other used parameters shown in the figures are φ_0 – initial electron phase, λ - wave-length of the laser field, P – laser power. It is evident that for $\gamma > 600$ the energy exchange is positive in a wide region of the electron beam trajectory and namely in the ranges $-2z_R$ - $+2z_R$ for $z_R = 18$ cm and $-6z_R$ - $+6z_R$ for $z_R = 3$ cm respectively. The profile shapes are practically not dependent on the laser power P and on γ as well. The last can be easily explained since the used angle α_{opt} does not depend on γ.

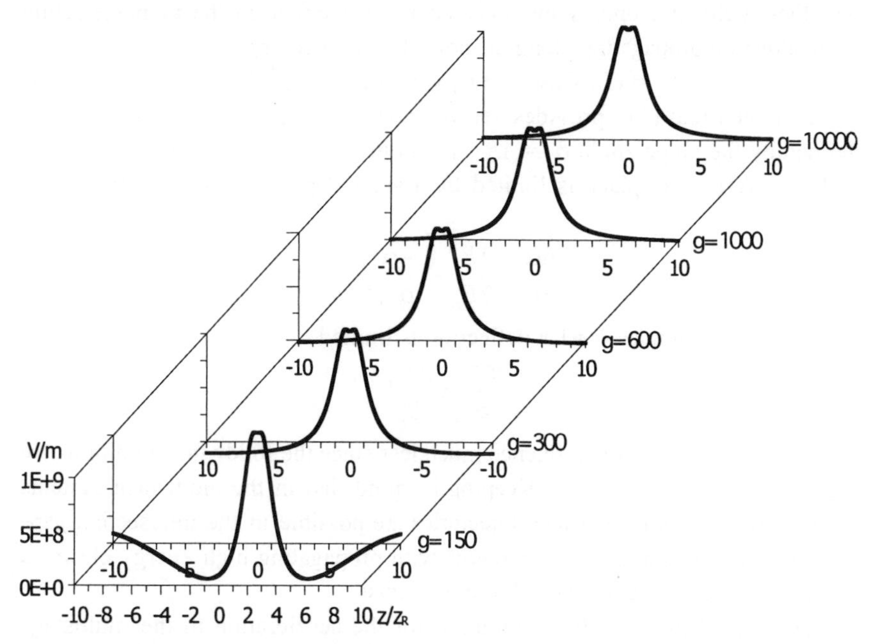

Fig. 9. Accelerating field profiles E_α of the Gaussian beam for z_R=3 cm. Other parameters are the same as given in Fig. 8.

Given by Figs.8, 9 field profiles can be used for evaluation of the acceleration gradients indused at the focus area on the path length - z_R - $+z_R$. It was found that these acceleration gradients not dependent on the electron energy are given by

$$\frac{dE_{el}}{dz}\left(\frac{MeV}{m}\right)=9.2\frac{\sqrt{P(W)}}{z_R\,(m)} \tag{10}$$

where P(W) is the laser power in W. It gives the acceleration gradients from some hundreds MeV/m to ~ 1 GeV/m. For the interaction lengths $L>z_R$ the averaged gradients can be lower in (L/z_R) times.

One of the important problem concerning realization of the free space focused laser acceleration is focusing of the accelerated beam with conservation of a high precision alignment of the electron beam. Some authors regarded it as not feasible. This opinion was based on a the confirmation that the focused laser beam has significant transverse field components deflecting particles and increasing the beam emittance [25]. Fortunately it is not true for the beams with not small γ . The transverse deviations of electrons crossing the Gaussian mode focus area can be induced mainly by the fields (1) and (3). Taking into account both these field components one finds the transverse field in the xz plane acting on an elrectron propagating along the noncollinar trajectory

$$F_{tr} = eE_x \, Cos\alpha \; - e \, B_y \, v/c = eE_x \, (1 - \alpha^2\gamma^2)/2\gamma^2 \qquad (11)$$

It is seen that (11) provides selffocusing of the electron to the trajectory defined by the slope angle $\alpha = 1/\gamma$. It can be shown that for the case $\alpha > 0$ deflection $\Delta\alpha$ in xz plane is limited by a value depending on the total energy gain ΔE_e

$$\frac{\Delta\alpha}{\alpha} = \frac{\Delta E_{el}}{2E_{el}} \frac{(1 - \alpha^2\gamma^2)}{\alpha^2\gamma^2} \qquad (12)$$

Thus for all angles $\alpha \geq 1/\gamma$ deflection is limited

$$\frac{|\Delta\alpha|}{\alpha} < \frac{\Delta E_{el}}{E_{el}}$$

The fields (2) and (4) influence much less since the following conditions are satisfied: $\alpha x \ll z_R$, $y \ll z_R$. Keeping in mind that in the multistage systems small corrections of the particle trajectories are possible in the intersection gaps we can conclude that serious problems with propagating high energy electrons along the trajectories at $\alpha \geq 1/\gamma$ should not exist.

Dynamical phase stability required for the acceleration in the multistage system can be provided as well. The acceleration rate in the fields (1) - (4) is a function of the initial electron phase φ_o with a maximum at $\pi/2$. At the range $\pi/2$ - $3\pi/2$ it has a slope admitting a negative feed back required for the beam stability providing.

The above cited simulations on the acceleration rate have shown that the gain changes linearly with the electron angular deviation $\delta\alpha$ and quadratically with its transverse trajectory displacements x_o .y_o. From these data some conclusion can be made on the beam quality requirement for the free space acceleration. It was found that at normalized emittance $\varepsilon_n \sim 1$ mm mrad a small dispersion of the gain per one accelerator section $\delta(\Delta E_{el})/\Delta E_{el} < 1$ % can be achieved.

7.2. Comparison of Free Space Accelerator schemes based on Different Configurations of Optical Beams

The last question we wish to discuss is the comparison of the Gaussian mode vacuum accelerator version having been considered in detail with the axicon vacuum accelerator mentioned above. The main difference is in the configurations of the used laser beams. In one case it is the Gaussian mode (Fig.2) with the conventional point like focus and the crossing angle α which can be $\geq 1/\gamma$. In another case it is the axicon (Fig.7) with the line focus specific to the axicon. The configuration of the axicon field is shown in Fig.10. The crossing angle of the axicon is supposed to be

$$\theta >> 1/\gamma \, . \tag{13}$$

We will use the computed characteristics of the axicon accelerator scheme given in [26] and the Gaussian beam accelerator given in [24] respectively. To make the analysis and comparison easier we introduce the following definitions: l_s is a distance over which electron propagates with a slippage phase less than π (limited slippage distance); l_a is a distance at which electron is accelerating by the laser field not changing its sign (with no deaccelerating path); l_{int} is a total path length of the electron in the laser field including in common case both accelerating and deaccelerating phases. The length l_{int} is defined by beam crossing area determined by the geometry of the device.

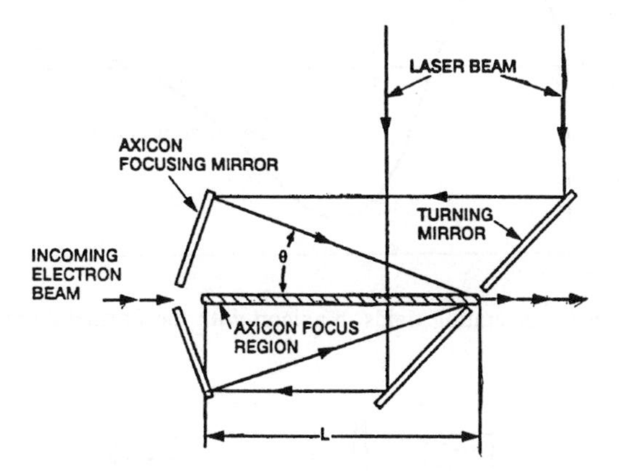

Fig. 10. Scheme of the laser accelerator with radially-polarized axisymmetric electric field.

The accelerating field profiles $E_z(z)$ from [26] are presented in Figs. 11 - 13. Here a is a transverse mirror size, L is an interaction length defined by the axicon geometry, $\theta = a/L$ is the crossing angle, $N_f = a^2/\lambda L$ is the Fresnel number of the mirror. The accelerated field profiles do not depend on γ. It is a consequence of the condition (13) so that the distance of the limited slippage equals

$$l_s = \lambda / \theta^2 \tag{14}$$

- The laser field is more concentrated at the Gaussian mode focus than at the axicon line focus. The relative field enhancement by the Gaussian mode is approximately equal to

$$\kappa = 2\,(eN_f)^{1/2} \tag{15}$$

This difference is illustrated by Fig.12 where the accelerating field profiles are given for both kinds of focusing.

- The Guoy phase jump equals π in the Gaussian mode. By using an optimal crossing angle the influence of this jump on the electron slippage can be decreased [24] so that the slippage π is reached again at the distance approximately given by

$$l_{aG} = \lambda/\alpha^2 \tag{16}$$

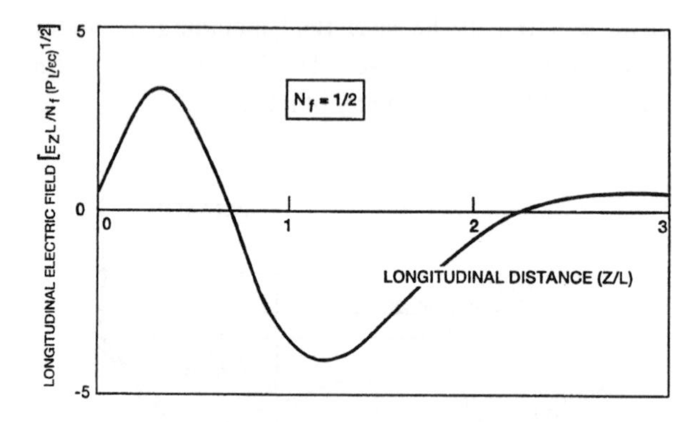

Fig. 11. Normalized accelerating fields in axicon with the Fresnel number N_f=0.5

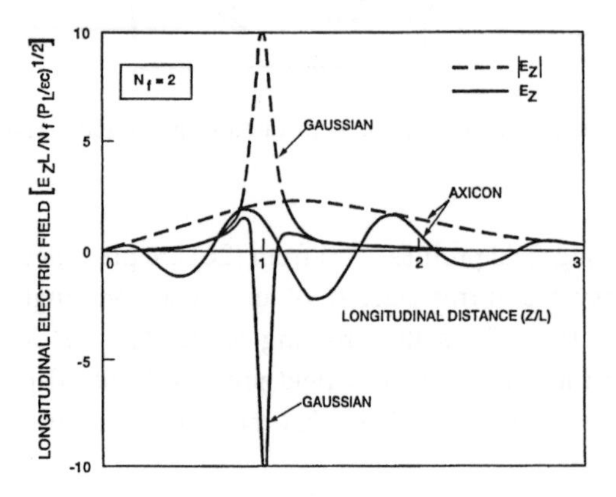

Fig.12. Normalized accelerating fields provided by axicon with N_f=2 and Gaussian beam with the same mirror size.

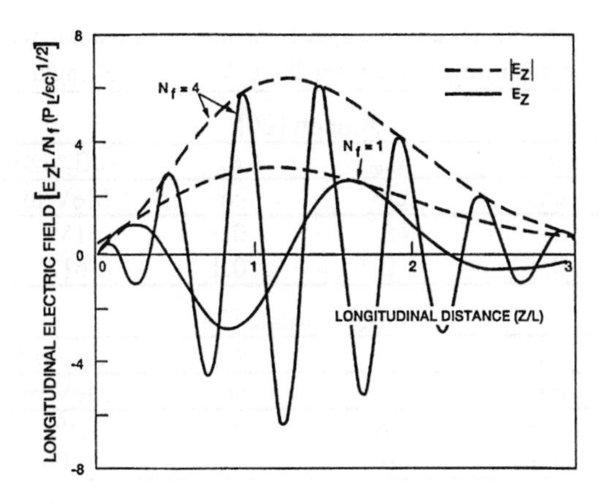

Fig. 13. Accelerating fields profiles in axicon for two different Fresnel numbers N_f.

In the axicon focused beam the respective phase jump is twice larger so that a real slippage distance is less than (14). The real slippage distance defines the above space period of the accelerating field oscillations and equals actually to the acceleration length

$$l_{aA} < \lambda/\theta^2 \qquad (17)$$

To provide larger fields E_z in the axicon [26] the crossing angle θ is made large $>> 1/\gamma$. So acceleration lengths in the considered schemes are different $l_{aA} << l_{aG}$.

- As it was shown above the interaction length for the Gaussian mode version can be chosen any in a wide range from $-z_R$ - z_R to $-10z_R$ - $10z_R$ with a single maximum of the accelerating field on these lengths. In the axicon however the interaction length L is defined by the device geometry. It contains several maxima of the accelerating field functions since $L = N_f\ l_{aA}$. These functions change sign at least N_f times (see Figs.11 - 13).

- The averaged acceleration rate of the axicon is given [26] by the formula

$$\langle dE_e/dz \rangle = 5{,}94\ [P\ \lambda\theta/a^3]\ eV \qquad (18)$$

This value is ~5 times less than an energy gain on one acceleration path length averaged over the intire interaction length or $[E_z]_{max}\ l_{aA}/L$. From the numerical simulations [24] it follows that for the considered above Gaussian beam acceleration scheme the gradients (10) are achievable. A numerical example for a direct comparison of the two schemes is given in Table 5 where the acceleration gradients as well as the total energy gains on the entire interaction lengths are presented. The calculations were made for the laser power $P = 10^{13}$ W and the wave-length $\lambda = 10{,}6\ \mu m$ for both schemes.

Table 5. Nnumerical results on the free space vacuum acceleration with the Gaussian beam and the axicon beam respectively for P=10^{13} W at wave-length λ=10.6 μm.

		Axicon [26]			
θ	4a	L_{int}	l_{aA}	$\langle dE_e/dz \rangle$	ΔE_e
mrad	cm	cm	cm	MeV/m	MeV
100	1	5	0.1	155	7.75
100	0.4	2	0.1	612	12.24
		Gaussian mode [24]			
α	w_0	$2z_R$	l_{aG}	$\langle dE_e/dz \rangle$	ΔE_e
mrad	cm	cm	cm	MeV/m	MeV
5.4	0.073	36	36.3	200	72
6.6	0.030	6	24.3	1200	72

It is seen from Table 5 that the Gaussian mode can provide 6 - 9 times larger total energy gain than the axicon along with two times higher acceleration gradients. It gives a more smooth acceleration field profile without oscillations on the interaction length in distinction from the axicon. The Gaussian accelerating envelope (with a single maximum of one sign) would give less problem with the beam dynamic stability. Of course the Gaussian beam laser acceleration is a challenge to the further improvement of the beam to be accelerated quality. Strict requirements on the small emittances follow from the small waists of the Gaussian beams used for the analysis and from the high sensitivity to the crossing angles of the scheme. With progress in the physics of high brightness electron beams perspectives of the free space vacuum acceleration by high power lasers will grow as well.

References

1. *Proceedings ICFA-RAL Meeting on Challenge of Ultra High Energies*, Oxford, England (1982).
2. *Laser Acceleration of Particles, AIP Conf. Proc.* No**91,** ed. by P.J. Channell New York (1982)
3. *Proceedings of Workshop on Laser Acceleration of Charged Particles,* Nor-Amberd, 21-24 September 1982, Yerevan Physics Institute (1983).
4. P. Sprangle, Ch. Joshi, *Final Report on the Advanced Acceleration Technique Working Group,* T8, Proceedings of Snowmass 2001.
5. E.D. Courant, C. Pellegrini, W. Zakowicz, *Phys. Rev.* **A32,** No**5**, 2813-2823 (1985)
6. A.A. Varfolomeev, Yu.Yu. Lachin. *Sov. Phys. Tech. Phys.* **31**, No11, 1273-1278 (1986)

7. I. Wernick, T.C. Marshall, *Phys. Rev.* **A 46**, 3566-3568 (1992)
8. A. Van Steenbergen, J. Gallardo, J. Sandweiss, J.-M. Fang, M. Babzien, X. Qiu, J. Scaritka, X.J. Wang, *Phys. Rev. Let.* **77**, 2690-2693 (1996)
9. Y. Liu, X.-J. Wang, D.B. Cline, M. Babzien, J.-M. Fang, J Gallardo, K. Kusche, I. Pogorelsky, J. Skaritka, A. Van Steenbergen, *Phys. Rev. Let.* **80**, 4418-4421 (1998).
10. W.D. Kimura, L.P. Campbell, C.E. Dilley, S.C. Gottschalk, D.C. Quimby, A. van Steenbergen, M. Babzien, I. Ben-Zvi, J.C. Gallardo, K.P. Kusche, I.V. Pogorelsky, J. Skaritka, V. Yakimenko, D.B.Cline, P. He, Y. Liu, L.C. Steinhauer, R.H. Pantel, *Phys. Rev. Spec. Top.* **AB 4**, 101301-1-101301-12 (2001)
11. W.D. Kimura, M. Babzien, L.C. Campbell, D.B. Cline, C.E. Dilley, J.C. Gallardo, S.C. Gottshalk, K.P. Kusche, R.H. Pantell, I.V. Pogorelsky, J.C. Quimby, J.S. Karitka, L.C. Steinhauer, V. Yakimenko, F. Zhou, *Proceedings of Advanced Accelerator Concepts,* June 23-28, 2002, Oxnard, CA.
12. Ralph W. Apmann. *Proceedings of EPAC 2002*, Paris, Francce. p.p.64-68.
13. P. Musumeci, C. Pellegrini, J.B. Rosenzweig, A. Varfolomeev, S. Tolmachev, T. Yarovoi, *Proceedings of the 2001 Particle Accelerator Conference*, June 18-22 2001, Chicago, Il., p.p. 4008-4010.
14. A.A. Varfolomeev, S.V. Tolmachev, T.V. Yarovoi, P. Musumeci, C. Pelligrini, J. Rosenzweig, *Proceedings of the 23-d International Free Electron Laser Conference,* 20-24 Aug. 2001, Darmshtadt, Germany, *Nucl. Instr. And Meth.* **A483,** 377-382 (2002).
15. A.S. Fisher, J.C. Gallardo, J. Sandweiss, A. van Steenbergen, *AIP Conference Proceeding* 279, ACC, Port Jefferson, N.Y., 1992, Ed. J.S. Wurtele, AIP, New York, (1993), p.p. 299-318.
16. J. Krall, G. Joyce, P. Sprangle, E. Esarey, *AIP Conference Proceeding* 279, ACC, Port Jefferson, N.Y., 1992, Ed. J.S. Wurtele, AIP, New York, (1993), p.p. 528-538.
17. A.A. Varfolomeev, A.H. Hairetdinov, *Proceeding of Fourth European Particle Accelerator Conference,* London, 27 June – 1 July, 1994, Ed. V. Suller, Ch. Petit-Jean-Genaz, World Scientific, Singapore, New Jersey, London, Hong Kong, p.p. 799-801.
18. L.C. Steinhauer, W.D. Kimura, *AIP Conference Proceeding* **279**, ACC, Port Jefferson, N.Y., 1992, Ed. J.S. Wurtele, AIP, New York, (1993), p.p. 539-550.
19. J.R. Fontana, R.H. Pantell, *J. Appl. Phys.* **53**, 5435 (1982).
20. A.A. Varfolomeev, A.H. Hairetdinov, ACC, Port Jefferson, N.Y., 1992, Ed. J.S. Wurtele, AIP, New York, (1993), p.p. 319-320.
21. J.A. Edighoffer, R.H. Pantell, *J. Appl. Phys.* **50**, 6120-6122 (1979)
22. Kirk T. Mc Donald, K. Shmakov, *Phys. Rev.* **ST-AB2**, 12301-1-12301-5 (1999).
23. A.A. Varfolomeev, A.H. Hairetdinov, AIP Conference Proceeding **279**, AAC, Port Jeferson, N.Y. 1992, Ed. J.S. Wurtele, AIP, New York, (1993), p.p. 319-329.

24. A.A. Varfolomeev, T.V. Yarovoi, "Prospects of free space vacuum acceleration by laser Gaussian beam.", Unpublished.
25. J.K. Mc Iver, Jr, M.J. Lubin, *J. Appl. Phys.* **45**, 1682-1687 (1974)
26. L.C. Steinhauer, W.D. Kimura, *J. of Appl. Phys.* **72**, 3237 (1992).

A RESONANT, THZ DIELECTRIC-BASED ACCELERATOR WITH SLAB SYMMETRY

R. B. YODER AND J. B. ROSENZWEIG

University of California, Los Angeles
Dept. of Physics and Astronomy
405 Hilgard Ave.
Los Angeles, CA 90095-1547, USA
E-mail: yoder@physics.ucla.edu

Slab-symmetric dielectric-loaded structures, consisting of a vacuum gap between dielectric-lined conducting walls, have become a subject of interest for high-gradient acceleration of high-charge beams due to their simplicity, relatively low power density, and advantageous beam dynamics. Such a structure can be resonantly excited by an external power source and is known to strongly suppress transverse wakefields. Motivated by the prospect of a high-power FIR radiation source, currently under construction at UCLA, we investigate a high-gradient slab-symmetric accelerator powered by up to 100 MW of laser power at 340 μm, with a predicted gradient near 100 MeV/m. Theory and simulation studies of the structure fields and wakes are presented, with an outline of a future experiment.

1. Background

The accelerator community is often reminded that the requirements for future high-energy physics experiments include the production of very high-brightness electron and positron beams. The fundamental challenges of this goal include both the realization of very high energy gain in a device of realistic size and the preservation of a high-charge, low-emittance beam through acceleration and transport. While many approaches to meeting these challenges have been proposed, the use of lasers to provide strong accelerating fields has emerged as one of the most visible areas of research, both experimental and theoretical, in high-gradient acceleration methods. However, the structures required to efficiently couple power at laser frequencies to an electron beam tend to have characteristic dimensions on the order of the laser wavelength, which leads to limitations on both usable drive power (through field-induced breakdown) and beam charge (through strong wakefields).

A variety of methods, including most prominently plasma-based mechanisms as well as optical focusing schemes, have been described which could avoid the problems associated with propagating a high-charge beam through a small structure. One concept which has received attention is that of a multi-beam or parallel-beam device, which could accelerate a number of lower-charge electron bunches in parallel structures and then combine them at some final focus. This idea applies particularly well to laser-powered accelerators, since available radiation power is not generally a limitation. A so-called "slab geometry," in which the structure and beam are both made very wide in one transverse dimension, is a variation on this theme. By spreading the beam out, in effect creating a large number of parallel accelerators, the effects of wakefields are mitigated. This paper describes a resonant slab-symmetric structure using submillimeter-wavelength drive power, development of which is in progress at the UCLA Neptune laboratory.

1.1. *The case for slab-symmetry*

Schemes for coupling laser fields to an electron bunch fall generally into two categories, known as "near-field" and "far-field" mechanisms. (Here, we disregard plasma-based schemes, in which a relativistic plasma wave provides the actual accelerating fields, though it may be excited using a laser.) Loosely speaking, these refer respectively to devices in which the fields are shaped by structures near the particle trajectory and to those in which the fields arise from free propagation of one or more lasers in the acceleration region. Near-field devices are, from the point of view of the basic physics, versions of standard radio-frequency accelerating structures which have been scaled to optical dimensions. To create a traveling-wave linac or a resonant cavity, the electric field pattern must be dominated by structures or boundaries with characteristic dimensions on the order of the radiation wavelength. The fields in far-field structures, on the other hand, are generally dominated by diffraction as laser beams propagate through vacuum or some other medium; examples include "vacuum accelerators" such as crossed-laser beam devices[1,2] as well as inverse Čerenkov accelerators.[3]

Far-field devices have two obvious advantages for acceleration: since field-shaping structures (such as mirrors) are far from regions of strong field, electric breakdown in the structure is never a limitation on attainable gradient; and by the same token, the particle beam does not pass near conducting structures and generates no deleterious wakefields, which tend to perturb or destroy the beam. On the other hand, the fields in this class of devices tend in general to have transverse components much greater than those in the longitudinal or accelerating direction. Not only are such accelerators clearly inefficient, but they also tend to suffer from a high degree of transverse instability, since small asymmetries in the fields or misalignment of the particle beam can result in severe transverse deflections or in position-dependent acceleration.

Near-field structures have highly uniform fields with small transverse components, but suffer from gradient limitations as a result of breakdown, since there will be unavoidable high power density near the field-shaping boundary. Notwithstanding this practical limit, there are advantages to boundary-dominated fields: they can be resonantly excited, giving fields in the structure larger than those of the driving laser, and it becomes possible to maintain phase synchronism with the particle beam. The other difficulty in the near-field regime, of course, is the production of strong wakefields by the beam. In a conventional linac, transverse wakefields quickly become unacceptably large as the resonant wavelength decreases below a few millimeters.

If a slab-symmetric structure with one transverse dimension much larger than the other is employed, however, this effect can be mitigated by accelerating a beam which is spread out in the wide dimension. It has been shown[4] that such a beam couples very weakly to the antisymmetric dipole modes of the structure and therefore generates almost no transverse wakefield, independent of its transverse position within the gap. To ameliorate breakdown limits when filling the structure with power, it is possible to use dielectric materials that can withstand fields of a few GV/m for pulses of a few picoseconds.[5,6,7] In the structure discussed here, combining slab-symmetry and dielectric loading, transverse wakefields are strongly suppressed, and the geometry enforces phase synchronism between particle beam and accelerating wave.

2. Theory of the Structure

2.1. *Basic principles.*

The structure geometry under discussion here is shown schematically in Figure 1. It consists of two mirror surfaces, taken to be (nearly) infinite in x and z, displaced from each other in the y direction so as to create a narrow gap. The gap is lined on each side by a layer of dielectric material having relative permittivity ϵ. For the analysis below, we take the distance between the conducting boundaries to be $2b$, with the central vacuum gap having dimension $2a$, so that the dielectric thickness on each wall is $(b-a)$. The particle beam is traveling in the positive z direction. Laser light (also polarized in the $+z$ direction) impinges on the upper surface as shown, and a series of narrow transverse coupling slots in the conductor allows transmission of light into the structure. The slot spacing must equal the free-space laser wavelength; this enforces a field in the structure which is periodic in z and can therefore accelerate particles synchronously. Coupling slots in the lower surface symmetrize the field and allow the detection of transmitted light as a diagnostic.

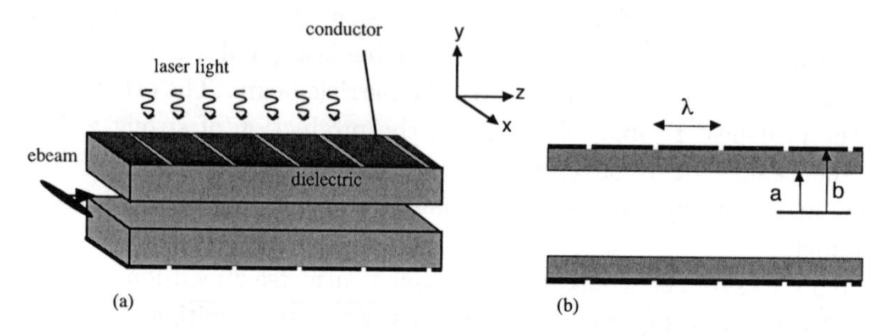

Figure 1. (a) Schematic drawing of the structure geometry. Two layers of dielectric-lined conductor surround a vacuum gap; a very wide electron beam is injected into the gap and travels in the $+z$ direction, while radiation (polarized in z) is coupled in from above through transverse slots in the conductor. (b) A cross-section in x, showing the parameters used in the analysis.

Note that in the absence of the coupling slots there would be no axial periodicity in the device at all. If the conducting boundaries were uniform partially transmitting mirrors, the structure would resemble a Fabry-Perot mirror pair, and would have resonant fields that were invariant in z. Such fields, of course, cannot accelerate particles. The coupling slots here enforce the correct field periodicity, and for a given drive frequency we can choose the structure and dielectric dimensions such that the axial wavefronts are synchronous with the particle beam, i.e. we can set the axial phase velocity equal to c.

This device continues work on slab-symmetric structures which has been underway at UCLA since 1995,[4] and is a refinement of a design for a 10.6 μm resonator[8] which performed well in simulations. However, that design had a vacuum gap spacing $2a$ equal to the laser wavelength, and due to the difficulty of producing an electron beam with such a small transverse dimension, no experiment was attempted. With a sub-millimeter laser source (discussed below) expected to be operational in the UCLA Neptune lab in the near future, the structure was redesigned to resonate near 1 THz, and the resulting increase in structure dimensions relaxes the emittance requirements on the injected beam, making experimental investigation of the slab geometry possible.

It should be noted that similar slab-symmetric dielectric-loaded structures have recently been proposed[9,10,11] for wakefield acceleration; in such a device, longitudinal wakefields excited by a drive bunch or bunches are used to accelerate a second or witness bunch following a precise distance after the driver. A theoretical analysis of wakefield acceleration shows that a very large number of structure modes are produced by the initial driving bunch, some of which combine to produce useful accelerating fields behind the bunch. In contrast, the device outlined here uses resonant coupling from an external source to drive the structure with a single dominant mode.

2.2. *Mode analysis*

The mode analysis here follows that of Tremaine et al.[12], which discusses a dielectric-loaded slab-symmetric structure with perfectly conducting boundaries, including cases with infinite as well as finite x-dimension. Recent calculations of the mode spectrum present in a slab structure with conducting boundaries at $x = \pm L$ (i.e. an asymmetric rectangular waveguide which is lined by dielectric on two sides) have also been presented by Park and Hirshfield[13] as well as Jing[14], and a planar structure without sidewalls, though with somewhat different geometry, has been investigated experimentally by Hill et al.[15] at a frequency of 91 GHz. In this section we will consider only accelerating (i.e. speed-of-light) modes in an infinite structure while neglecting the perturbation of the coupling slots. Deviations from this ideal case will be considered briefly below.

The fundamental accelerating mode must have a sinusoidal dependence on axial position in order to satisfy the wave equation; thus we solve for an axial electric field of the form $E(y)\cos(kz)\cos(\omega t)$, there being no x-dependence since the structure is translationally invariant in x. Within the vacuum gap ($|y| < a$), we further require for synchronism that the longitudinal phase velocity is equal to the speed of light, i.e. $\omega/k_z = c$. Recalling the dispersion relation

$$\omega^2/c^2 = k_x^2 + k_y^2 + k_z^2 \tag{1}$$

with $k_x = 0$, we are required to set $k_y = 0$, which implies that the axial field is constant in y:

$$E_z = E_0 \cos(k_z z) \cos(\omega t) \tag{2}$$

where E_0 is the field amplitude and $\omega k_z = c$. It is the periodicity of the structure (specifically, the coupling slots) which enforces this solution and hence causes the accelerating mode to dominate the field pattern inside the structure. Continuing with the fields, the transverse field E_y must be linear in y to keep $\mathbf{E}$ divergenceless in the gap; we therefore have

$$E_y = E_0 k_z y \sin(k_z z) \cos(\omega t) \tag{3}$$

Within the dielectric ($a < |y| < b$), there must be variation in y in order to allow E_z to decrease to zero at the conducting boundary. Here we can use an ordinary TM-mode solution as long as the boundary conditions are met. Using $\epsilon\omega^2/c^2 = k_y^2 + k_z^2$, we have

$$E_z = -AE_0 \sin[k_z\sqrt{\epsilon - 1}(b - y)] \cos(k_z z) \cos(\omega t) \tag{4}$$

and

$$E_y = -\frac{AE_0}{\sqrt{\epsilon - 1}} \cos[k_z\sqrt{\epsilon - 1}(b - y)] \sin(k_z z) \cos(\omega t) \tag{5}$$

Continuity of E_z and D_y at the dielectric boundary gives the transcendental equation

$$k_z a \frac{\sqrt{\epsilon - 1}}{\epsilon} = \cot[k_z\sqrt{\epsilon - 1}(b - a)] \tag{6}$$

for the allowed eigenvalues k_z and determines the relative amplitude A of the fields in the dielectric,

$$A = \csc[k_z\sqrt{\epsilon - 1}(b - a)]. \tag{7}$$

Consideration of Eqns. 3 and 6 shows that, for a given geometry, there is a series of eigenvalues k_z which will increase roughly linearly with mode number. It is clear that field strengths within the dielectric increase proportionately for higher mode numbers, as implied by Eqn. 7. (See Figure 2.) To avoid breakdown limitations as far as possible, the fundamental (lowest k_z) mode is obviously preferred—see Figure 3, which shows the enhanced field amplitudes within the dielectric for $n > 1$. In an experiment, naturally, k_z would be fixed by the power source, and Eqn. 6 can then be used to calculate the resonant values of a and b for a given ϵ. The fundamental mode then corresponds to taking the smallest possible solution for the gap spacing.

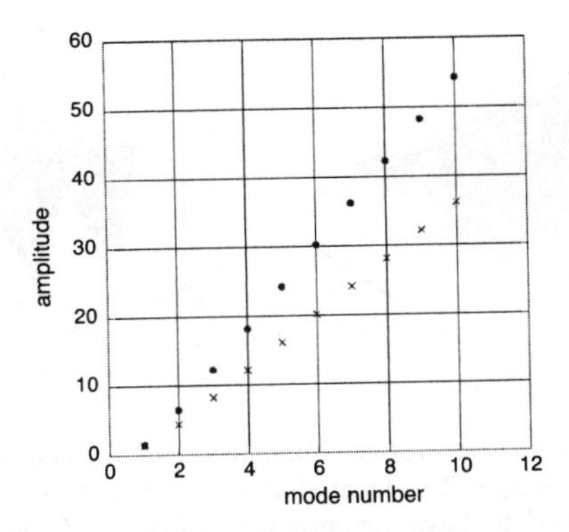

Figure 2. Plot showing values of the dielectric field amplitude A defined in Eqn. 7 ($\times$'s) and of the normalized transverse field at the dielectric boundary $E_y(a)/E_0$ (circles) for the first ten eigenmodes from Eqn. 6, for a representative geometry.

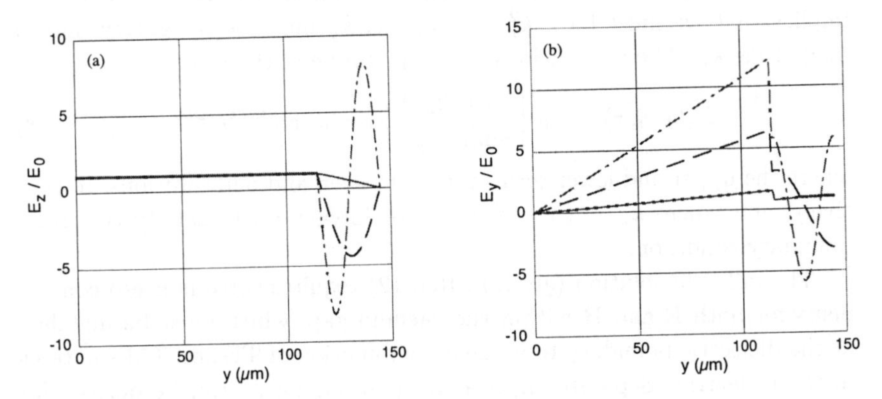

Figure 3. Field profiles for a slab-symmetric structure with $a = 115$ μm, $b = 145$ μm, $\epsilon = 3$. Longitudinal (a) and transverse (b) fields are shown, with the first three modes in each case (solid line, $n = 1$; dashed, $n = 2$, dot-dashed, $n = 3$). Large dielectric fields in both dimensions are observed for $n > 1$.

Since the speed-of-light modes supported by these structures have no y-dependence, a slab-shaped electron bunch with velocity near c will not excite transverse wakefields, even if it is displaced from the structure axis. Wakefield calculations have been carried out in Ref. 12; to demonstrate the strong suppression of transverse wakes, Figure 7 shows the results of simulation using the particle-in-cell code XOOPIC[18,19] for the fields behind a relatively long "slab beam."

To examine the non-ideal case of a slab-symmetric structure having

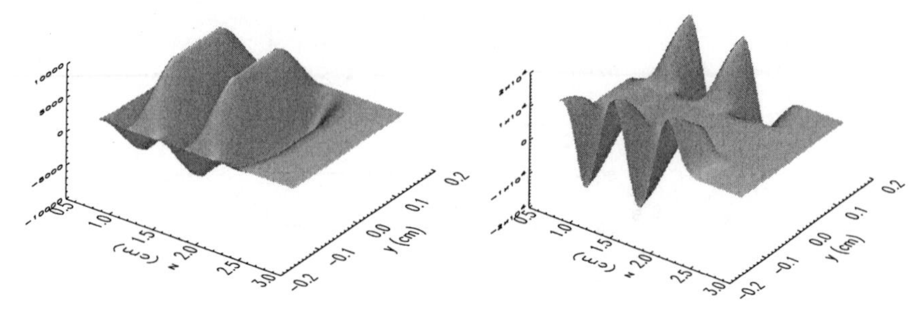

Figure 4. Surface plots (from XOOPIC simulation) of wakefields behind an infinitely-wide electron beam with $\sigma_z = 4$ ps located at $z = 2$ cm and moving in the $+z$ direction. Longitudinal wakefields (left) have the periodicity of the fundamental accelerating mode; transverse wakefields (right) are zero except within the dielectric.

finite x-width, we add conducting boundaries at $x = \pm L$, where $L \gg b$. We are interested only in modes with axial phase velocity c, and Eqn. 1 implies that we must have $k_x^2 = -k_y^2$, i.e. k_y must be imaginary and of magnitude k_x. The axial field in the gap will be of the form

$$E_z(x, y, z) = E_0 \begin{Bmatrix} \cosh(k_x y) \\ \sinh(k_x y) \end{Bmatrix} \cos(k_x x) e^{ik_z z} e^{i\omega t} \tag{8}$$

where the upper and lower terms represent even and odd solutions, respectively, and where $k_x = (m - \frac{1}{2})\pi/L$, for integer m, to satisfy the usual boundary condition.

The full field solution (given in Ref. 12) requires three nonzero components for both $\mathbf{E}$ and $\mathbf{B}$ within the vacuum gap, which must be matched at the dielectric boundary to a linear combination of TE and TM solutions in the dielectric. Separate eigenvalue equations for k_z can be derived for the odd and even solutions of Eqn. 8, meaning that a particular choice of eigenvalue k_z will correspond only to one or the other.

This approach still represents an approximation to the situation envisioned in future UCLA experiments, where a structure without sidewalls will be illuminated (as in Fig. 1) by a laser pulse with a Gaussian transverse profile in both x and z. However, since the cosine and Gaussian functions have the same limiting behavior for small argument, we expect that for x near zero Eqn. 8 will provide an approximate indication of the field profile in the accelerator structure if we take $k_x \approx 1/\sigma_x$, where σ_x is the rms width of the laser pulse. Under these conditions, the resonant values of the dimensions a and b are only slightly perturbed from the infinite-width case. Using the even solution for the gap fields, we find that the accelerating field

E_z remains nearly independent of transverse position y, with variation from flatness of less than 0.8%.

3. Experimental Design

The geometry of the experiment which is planned for the UCLA Neptune facility is dictated by available radiation wavelengths. As indicated above, the accelerator is to be driven with a novel high-power terahertz source; with a radiation wavelength of 340 μm, the slab structure will resonate in the fundamental accelerating mode when $a = 115$ μm, $b = 145$ μm, for $\epsilon = 3$. A vacuum gap of more than 0.2 mm makes it feasible to inject the 11 MeV beam from the Neptune photoinjector, with normalized transverse emittance in the range of 6–10 π mm mrad, into the structure successfully.

We expect to obtain multimegawatt laser radiation at 340 μm using a difference frequency generation scheme: two frequencies from the Neptune terawatt CO_2 laser will be mixed at high power in a gallium arsenide crystal, with conversion efficiency into the difference frequency near 1%.[16] The two input frequencies, as well as the output radiation, are non-collinear in order to maintain synchronism over a relatively large (several centimeter) interaction length. Output power levels in excess of 100 MW are projected; experimental work is currently in progress.

4. Structure Simulation

4.1. *Field Solutions*

Simulation of the slab structure was carried out using the three-dimensional finite-difference code GDFIDL.[17] For simulation purposes, magnetic boundary conditions on x were used to obtain a nearly constant field in the wide dimension without requiring a prohibitively large computational volume.

Initially, eigensolutions were found for a segment of the structure in the absence of external couplers; contour plots of the fields for the accelerating mode are shown in Figure 5. Note the flatness of the wavefronts of the accelerating component, showing that acceleration in this mode is independent of the transverse location of the beam. The fields decrease smoothly to zero in the dielectric.

Using a copper boundary with finite conductivity, the structure ohmic Q is calculated to be approximately 600, and the simulated fields give rise to a shunt impedance for the structure of 15.3 MΩ/m. This relatively low value is a consequence of the slab geometry, which may be understood in this context as a combination of a large number of cylindrically-symmetric structures in parallel. For this reason, slab structures are well suited for laser-powered accelerators, where there is effectively no limitation on available power. From this result, the accelerating field gradient can be estimated to be approximately 50–100 MV/m when the radiation power is 100 MW.

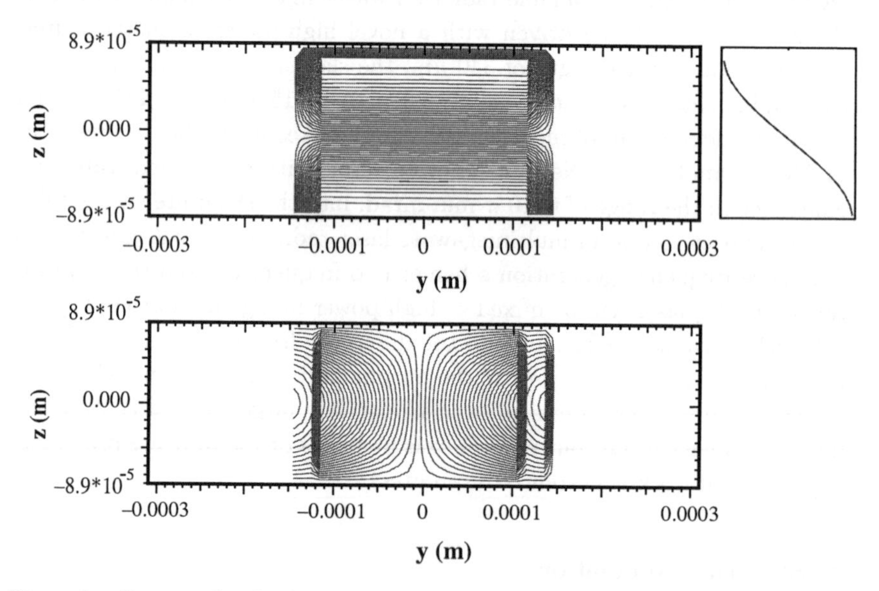

Figure 5. Contour plots in the $y - z$ plane from simulation, showing the fundamental accelerating mode over one half-period in z in the absence of coupling slots. Longitudinal field (top) extends in z from negative to positive field maxima, as shown on the lineout along $y = 0$ to the right; transverse field (bottom) is zero along $y = 0$ and discontinuous at the dielectric boundary.

To simulate the time-dependent filling of the structure when coupled to an external radiation source, the geometry was extended to include external waveguides for power input and output, coupled through narrow slots to the interior of the structure. Judicious choice of boundary conditions keeps the field polarization in the waveguide along the z direction.

The field strength in the structure as a function of time has the exponential shape associated with a resonator having finite Q, as seen in Figure 6. The filling time for fields to reach their steady-state values, roughly 400 ps, is associated with an intrinsic $Q = \omega\tau$ of about 2000, a relatively high value. Optimizing the slot width and wall thickness to obtain critical coupling has been only partially completed; best results to date, with reflected power less than 50% of input power, have been obtained with a 3 μm slot spanning the x-width of the device and a wall thickness of 90 μm.

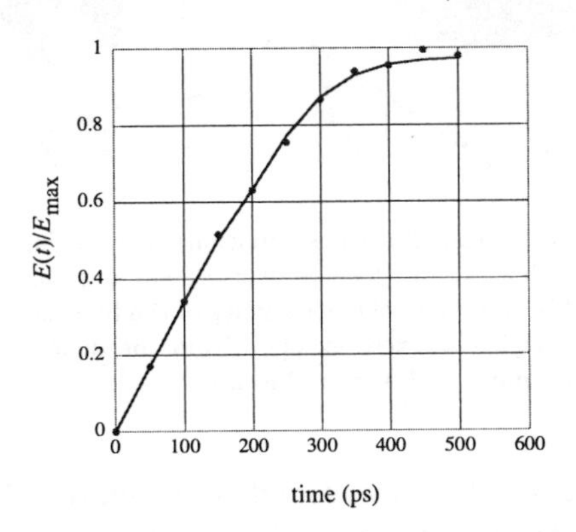

Figure 6. Axial field amplitude within the vacuum gap as a function of time, normalized to the steady-state amplitude E_{max}. The solid line is interpolated from the data points. (GDFIDL simulation)

4.2. *Wakefield Analysis*

Since the speed-of-light modes supported by a slab-symmetric structure have no y-dependence, a slab-shaped electron bunch with velocity near c will not excite transverse wakefields, even if it is displaced from the structure axis. Wakefield calculations have been carried out in Ref. 12; to demonstrate the strong suppression of transverse wakes, Figure 7 shows the results of simulation using the 2D particle-in-cell code XOOPIC[18,19] for the fields behind a relatively long "slab beam."

For the particular geometry under consideration here, the magnitudes of both transverse and longitudinal wakefields are small. Figure 8 shows the longitudinal wakefield behind the beam for two different beam lengths. When $\sigma_z = 1.2$ mm, the wakefields are almost entirely washed out. For much shorter bunches ($\sigma_z = 120$ μm), one would still expect a retarding field of only 8 to 10 kV/m to be experienced by the injected beam, so that with a structure length of a few centimeters the energy change experienced would be impossible to observe. Note in both cases that the correct wavelength (340 μm) was produced by the simulation. Fig. 8 also shows the vanishing of the wakefield after about 8 periods, which could be due to de-

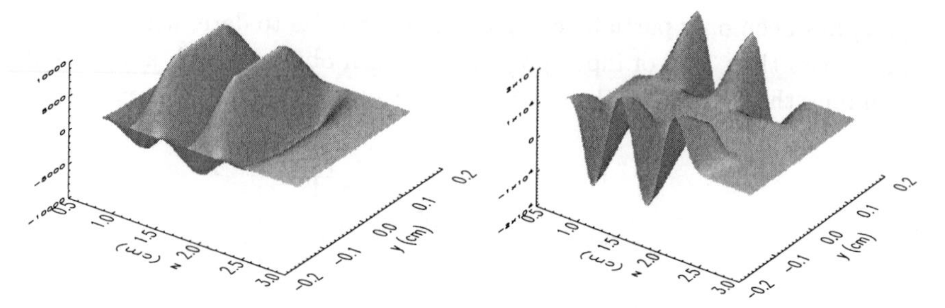

Figure 7. Surface plots (from XOOPIC simulation) of wakefields behind an infinitely-wide electron beam with $\sigma_z = 4$ ps located at $z = 2$ cm and moving in the $+z$ direction. Longitudinal wakefields (left) have the periodicity of the fundamental accelerating mode; transverse wakefields (right) are zero except within the dielectric. In this calculation, $a = 0.58$ mm, $b = 1.44$ mm, $\epsilon = 3.9$, $\sigma_r = 120$ μm.

structive interference between wake radiation and transition radiation from the beam's entry into the structure.[13]

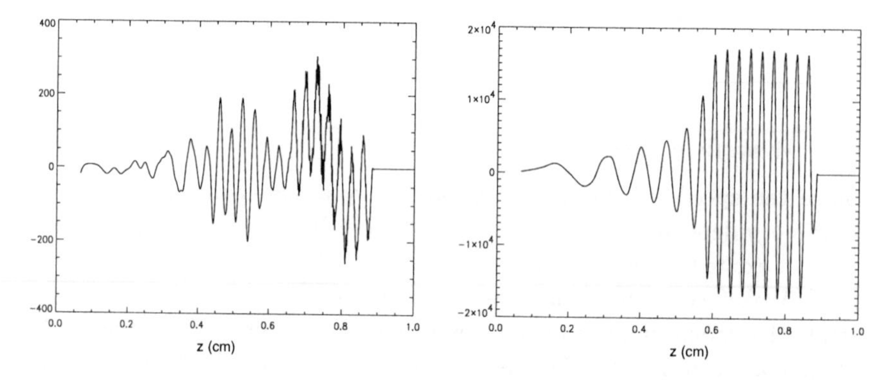

Figure 8. Lineouts of the longitudinal wakefield (in V/m) as a function of axial position for the design geometry, for a long ($\sigma_z = 1.2$ mm) bunch (left) and a short ($\sigma_z = 120$ μm) bunch (right). In both cases the leading edge of the driving beam is located at $z = 0.87$ cm, and $\sigma_y = 25$ μm.

Further simulations will be necessary to investigate wakefields in the case of beams propagating at some small transverse angle to the z-axis. Such a misalignment would break the symmetry which is responsible for the cancellation of the transverse wakes, so that it might become possible to observe some transverse effect on a test beam.

5. Summary and Future Work

In this paper, we have described a slab-symmetric dielectric-loaded structure which serves as a resonant laser-driven accelerator with advantageous transverse stability for high-charge beams. When such an accelerator is powered by a submillimeter-wave source at 340 μm, the structure dimensions become ample for acceleration of a slab electron beam with realistic transverse size.

Such structures have an intrinsically low shunt impedance, but when driven with the high power levels available from optical and submillimeter radiation sources, appreciable accelerating fields may be obtained. For the dimensions described here, with available power levels of 100 MW or more, gradients of up to 100 MeV/m are expected.

Experimental investigation of the slab-symmetric, dielectric-loaded structure is to begin in winter 2002 at the UCLA Neptune facility. Planned measurements include measurement of the wakefield radiation spectrum (which would enable checking the theoretical dependence on gap spacing); cold-testing of the structure by using the output coupling slots to verify the resonant frequency is also envisioned. Breakdown limitations are not clear at present and must also be tested experimentally by varying the input power density. A successful demonstration of energy gain by a slab beam will depend on detailed knowledge of the structure physics, as verified by experiment.

Acknowledgments

This work was supported by the US Dept. of Energy, grant number DE-FG03-92ER40693. We thank S. Tochitsky for useful discussions.

References

1. Y. C. Huang and R. L. Byer, *App. Phys. Lett.* **69**, 2175 (1996).
2. E. Esarey, P. Sprangle and J. Krall, *Phys. Rev. E* **52**, 5443 (1995).
3. J. R. Fontana and R. H. Pantell, *J. Appl. Phys.* **54**, 4285 (1983).
4. J. Rosenzweig, A. Murokh and C. Pellegrini, *Phys. Rev. Lett.* **74**, 2467 (1995).
5. D. Du, *et al.*, *Appl. Phys. Lett.* **64**, 3071 (1994).
6. M. Lenzner, *Int. J. Mod. Phys. B* **13**, 1559 (1999).
7. A. C. Tien, *et al.*, *Phys. Rev. Lett.* **82**, 3883 (1999).
8. J. B. Rosenzweig and P. V. Schoessow, "An Optimized Slab-Symmetric Dielectric-Based Laser Accelerator Structure," in *Advanced Accelerator Concepts, Eighth Workshop*, Baltimore, 1998, edited by W. Lawson *et al.*, AIP Conf. Proc. 472, pp. 693-700.

9. T. C. Marshall, C. Wang and J. L. Hirshfield, *Phys. Rev. ST Accel. Beams* **4**, 121301 (2001).
10. J. G. Power, W. Gai and P. Schoessow, *Phys. Rev. E* **60**, 6061 (1999).
11. L. Xiao, W. Gai and X. Sun, *Phys. Rev. E* **65**, 016505 (2002).
12. A. Tremaine, J. Rosenzweig and P. Schoessow, *Phys. Rev. E* **56**, 7204 (1997).
13. S. Y. Park, C. Wang and J. L. Hirshfield, "Theory for Wake Fields and Bunch Stability in Planar Dielectric Structures," presented at the 10th Advanced Accelerator Conference, Oxnard, CA, June 2002.
14. C. Jing and W. Gai, presented at the 10th Advanced Accelerator Conference, Oxnard, CA, June 2002.
15. M. E. Hill *et al.*, *Phys. Rev. Lett.* **87**, 094801 (2001).
16. S. Tochitsky, private communication.
17. W. Bruns, *IEEE Trans. Magn.* **32**, 1453 (1996).
18. J. P. Verboncoeur, A. B. Langdon and N. T. Gladd, *Comput. Phys. Commun.* **87**, 199 (1995).
19. D. L. Bruhwiler *et al.*, *Phys. Rev. ST Accel. Beams* **4**, 101302 (2001).

List of Participants

Scott Anderson
UCLA, Department of Physics & Astronomy
Los Angeles, CA 90095-1547 USA
anderson@physics.ucla.edu; Ph: 310 206 5584; fx: 310 825 8432

Sandro Angius
INFN-LNF
Via Enrico Fermi, 40, 00044 Frascati (Roma) ITALY
sandro.angius@lnf.infn.it; Ph: 39 06 9403 2580; fx: 39 06 9403 2649

Ubaldo Aresu
G. H. Chia Laguna
Localita Chia,
09010 Domus De Maria, Cagliari ITALY
ph: 39 070 92391; fx: 39 070 9239586

Francesco Ascani
G. H. Chia Laguna, Director
Localita Chia,
09010 Domus De Maria, Cagliari ITALY
ph: 39 070 92391; fx: 39 070 9239586

Alberto Bacci
INFN-Milano
Via Fratelli Cervi, 201, 20090 Milano ITALY
alberto.bacci@mi.infn.it; ph: 39 2 5031 9577; fx: 39 3 4706 26928

Carlo J. Bocchetta
Sincrotrone Trieste
Padriciano 99, 34012 Trieste ITALY
bocchetta@elettra.trieste.it; ph:: 39-40-375-8544 fx: 375 8565

Manuela Boscolo
INFN-LNF
Via E. Fermi, 40, 00044 Frascati (Roma) ITALY
manuela.boscolo@lnf.infn.it; ph: 39 06 9403 2636; fx: 39 06 9403 2256

Charles A. Brau
Vanderbilt University, Department of Physics
Box 1807 Station B, Nashville, TN 37235 USA
charles.a.brau@vanderbilt.edu; ph: 615 269 6721; fx: 343 7263

SethBrussaard
Eindhoven University of Technology
P.O. Box 513, Eindhoven 5600 mb The NETHERLANDS
g.j.h.brussaard@tue.nl; ph: 31 40 247 4359; fx: 31 40 243 8060

Rossana Centioni
INFN-LNF
Via Enrico Fermi, 40, 00044 Frascati (Roma) ITALY
rossana.centioni@lnf.infn.it; ph: 39 06 9403 2423; fx: 39 06 9403 2243

Enrica Chiadroni
University of Rome "La Sapienza"
Via Fontana Della Rosa, 81, 00049 Velletri (Roma) ITALY
enrica.chiadroni@roma1.infn.it; ph: 39 06 9633 242; fx: 39 06 4991 4386

Patrick L. Colestock
LANL
Box 1663, Los Alamos, NM 87506 USA
colestoc@lanl.gov; ph: 505 665 3565; fx: 505 667 8207

Max Cornacchia
Stanford Linear Accelerator Center, SSRL, MS-69
2575 Sand Hill Rd, P.O. Box 4349, Menlo Park, CA 94025 USA
cornacchia@slac.stanford.edu; ph: 650-926-3906 fx: 926-4100

Paolo Craievich
Sincrotrone Trieste
Trieste 34012 ITALY
paolo.craievich@elettra.trieste.it; ph: 39 040 375 8682; fx: 39 040 375 8565

Gerardo D'Auria
Sincrotrone Trieste S.C.P.A.
34012 Trieste ITALY
gerardo.dauria@elettra.trieste.it; ph: 30 040 375 8220; fx: 39 040 375 8565

Hideki Dewa
Japan Synchrotron Radiation Research Institute, Accelerator Division
1-1-1 Kouto, Mikazuki-cho, Sayo-gun, Hyogo 679-5198 JAPAN
dewa@spring8.or.jp; ph: 81 791 58 0802 (ex 3545); fx: 81 791 58 0850

Douglas Michael Dykes
Daresbury Laboratory
Warrington, Cheshire, WA4 4AD UK
d.m.dykes@dl.ac.uk; ph: 44 1925 603 142; fx: 44 1925 603 192

Paul Emma
SLAC
P.O. Box 20450, Stanford, CA 94309 USA
emma@slac.stanford.edu; ph: 650 926 2458; fx: 650 926 5368

Massimo Ferrario
INFN-LNF, Accelerator Division
Via E. Fermi 40, C.P. 13, 00044 Frascati (Roma) ITALY
massimo.ferrario@lnf.infn.it; ph: 39 06 9403 2216; fx: 39 06 9403 2256

Klaus Floettmann
DESY, MPY
Notkestr. 85, 22603 Hamburg GERMANY
klaus.floettmann@desy.de; ph: 49-40 8998-2052

Valeria Fusco
LNF-INFN
Via Enrico Fermi, 40, 00044 Trascati (Roma) ITALY
valeria.fusco@lnf.infn.it; ph: 39 06 9403 2281; fx: 39 06 9403 2565

John Galayda
SLAC, SSRL, MS-69
2575 Sand Hill Road, Menlo Park, CA 94025 USA
galayda@slac.stanford.edu; ph: 650 926 2371; fx: 650 926 4100

Luca Giannessi
ENEA
Via E. Fermi, 45, 00044 Frascati (Roma) ITALY
giannessi@frascati.enea.it; ph: 39 06 9400 5180; fx: 39 06 9400 5334

Silvia Vannucci Giromini
INFN-LNF
Via Enrico Fermi, 40, 00044 Frascati (Roma) ITALY
silvia.giromini@lnf.infn.it; ph: 39 06 9403 2643; fx: 39 06 9403 2243

Danilo Giulietti
University of Pisa, Dipartimento di Fisica E. Fermi
Via Buonarroti, n.1, 56100 Pisa ITALY
danilo.giulietti@df.unipi.it; ph: 39 050 844 840; fx: 39 050 844 333

Susanna Guiducci
LNF-INFN
Via E. Fermi, 40, 00044 Frascati (Roma) ITALY
susanna.guiducci@lnf.infn.it; ph: 39 06 9403 2221; fx: 39 06 9403 2256

Zhirong Huang
ANL, Accelerator Systems Division, Bldg 401, Rm. B2200
9700 S. Cass Ave., Argonne, IL 60439 USA
zrh@aps.anl.gov; ph: 630 252 6023; fx: 630 252 5703

F. B. Kiewiet
Eindhoven University of Technologie
P.O. Box 513, Eindhoven 5600MB The NETHERLANDS
f.b.kiewiet@tue.nl; ph: 31 40 247 4864; fx: 31 40 243 8060

Kwang-Je Kim
Argonne National Laboratory
9700 S. Cass St., Argonne, IL 60439 USA
kwangje@aps.anl.gov; ph: 630 252 4647; fx: 630 252 7369

Jim Kolonko
UCLA, Department of Physics & Astronomy
Box 951547, Los Angeles, CA 90095-1547 USA
kolonko@physics.ucla.edu; ph: 310 206 4548; fx: 310 206

Melinda Laraneta
UCLA, Department of Physics & Astronomy
Box 951547, Los Angeles, CA 90095-1547 USA
laraneta@physics.ucla.edu; ph: 310 206 2499; fx: 310 206 5251

Greg Le Sage
Lawrence Livermore National Laboratory
7000 East Avenue, L-154, Livermore, CA 94550 USA
lesage@llnl.gov; ph: 925 422 2390; fx: 925 424 4561

John W. Lewellen
ANL, APS, Bldg 401, Rm B2190
9700 S. Cass Ave., Argonne, IL 60439 USA
lewellen@aps.anl.gov; ph: 630 252 5252; fx: 630 252 4732

Steve Lidia
LBNL, MS 80-101
1 Cyclotron Road, Berkeley, CA 94720 USA
smlidia@lbl.gov; ph: 510 486 6101; fx: 510 486 4102

Torsten Limberg
DESY
Notkestrasse 85, 22603 Hamburg GERMANY
torsten.limberg@desy.de; ph: 49 40 8998 3998

Cecile Limborg
SLAC, SSRL
2575 Sand Hill Road, Menlo Park, CA 94025 USA
limborg@slac.stanford.edu; ph: 650 926 8685; fx: 650 926 4100

Dirk Lipka
DESY, PITZ
Platanenallee 6, 15738 Zeuthen GERMANY
dlipka@ifh.de; ph: 49 033 762 77280; fx: 49 033 762 77330

Vladimir Litvinenko
Duke University, FEL Lab
Box 90319, Durham, NC 27708-0319 USA
vl@phy.duke.edu; ph: 919 660 2658; fx: 919 680 2671

O.J. Luiten
Eindhoven University of Technology,
P.O. Box 513, Eindhoven 5600 MB The NETHERLANDS
o.j.luiten@tue.nl; ph: 31 40 247 4359; fx: 31 40 243 8060

Jelena Maksimovic
Universitat Rostock, Institut fur Allgemeine Elektrotechnik
Albert-Einstein-Strasse 2, Rostock, Mecklenburg 18059 GERMANY
jelena.maksimovic@stud.uni-rostock.de; ph: 49 381 498 3649;
fx: 49 381 498 3480

Viktor Maksimovic
Universitat Rostock, Institut fur Allgemeine Elektrotechnik
Albert-Einstein-Strasse 2, Rostock, Mecklenburg-Volpommein
GERMANY
viktor.maksimovic@stud.uni-rostock.de; ph: 49 381 1498 3486;
fx: 49 381 498 3480

Mario Mattioli
University of Rome "La Sapienza", Physics Department
Piazzale A. Moro, 2, 00185 Rome ITALY
mario.mattioli@roma1.infn.it; ph: 39 06 499 14202

Patrick Muggli
UCLA, Electrical Engineering, 56-125B, Eng IV
Los Angeles, CA 90095 USA
muggli@usc.edu; ph: 310 206 1913; fx: 310 206 8220

Alex Murokh
UCLA, Department of Physics & Astronomy
Los Angeles, CA 90095-1547 USA
alex@physics.ucla.edu; ph: 310 206 5584

King Y. Ng
Fermilab
P.O. Box 500, Batavia, IL 60510 USA
ng@fnal.gov; ph: 630 840 4597, fx: 630 840 6039

Heinz-Dieter Nuhn
SLAC, SSRL
2575 Sand Hill Road, MS69, Menlo Park, CA 94025 USA
nuhn@slac.stanford.edu; ph: 650 926 2275; fx: 650 926 4100

Nicola Pala
G. H. Chia Laguna, Director of Sales & Marketing
Localita Chia
09010 Domus De Maria, Cagliari ITALY
ph: 39 070 92391; fx: 39 070 9239586

Denis T. Palmer
SLAC, ARDB
2575 Sand Hill Road, Menlo Park, CA 94112 USA
dtp@slac.stanford.edu; ph: 650 926 4611; fx: 650 926 4365

Luigi Palumbo
INFN-LNF & Univ. Roma, Accelerator Division
Via Enrico Fermi, 40, 00044 Frascati (Roma) ITALY
luigi.palumbo@uniroma1.it; ph: 39 06 4976 6533; fx: 39 06 4424 0183

Claudio Pellegrini
UCLA, Department of Physics & Astronomy
Los Angeles, CA 90095-1547 USA
pellegrini@physics.ucla.edu; ph: 310 206 1677; fx: 310 206 5251

Vittoria Petrillo
Universit' di Milano, Deipartimento di Fisica
Via Celoria, 16, 20133 Milano ITALY
petrillo@mi.infn.it; ph: 39 2 5031 7345; fx: 39 2 5031 269

Philippe Piot
DESY, MPY
85 Notkestrasse, 22603 Hamburg GERMANY
piot@mail.desy.de; ph: 49 040 8998 3448

Massimo Pistoni
INFN-LNF
Via Enrico Fermi, 40, 00044 Frascati (RM) ITALY
massimo.pistoni@lnf.infn.it; ph: 39 06 9403 2372; fx: 39 06 9403 2649

Massimo Placidi
CERN, SL Division
CH-1211 Geneva 23 SWITZERLAND
massimo@mail.cern.ch; ph: 41 22 767 6638; fx: 41 22 767 7740

Marcello V. Quattromini
E.R. ENEA-Frascati
Via Enrico Fermi, 45, 00044 Frascati ITALY
quattromini@frascati.enea.it; ph: 39 06 9400 5718, fx: 39 06 9400 5334

Pantaleo Raimondi
INFN-LNF
Via Enrico Fermi, 40, 00044 Frascati (Roma) ITALY
pantaleo.raimondi@lnf.infn.it; ph: 39 069 403 2268; fx: 39 069 403 2203

Sven Reiche
UCLA, Department of Physics & Astronomy
Los Angeles, CA 90095-1547 USA
reiche@physics.ucla.edu; ph: 310 206 5584; fx: 310 206 5251

Andrew John Rollason
Keele University
Keele, Staffordshire ST5 5BG UK
pha14@phys.keele.ac.uk; ph: 01 782 583329; fx: 01 782 712378

James Rosenzweig
UCLA, Department of Physics & Astronomy
Los Angeles, CA 90095-1547 USA
rosen@physics.ucla.edu; ph: 310 206 4541; fx: 310 206 5251

Mauro Rossi
INFN-LNF
Via Enrico Fermi, 40, 00044 Frascati (Roma) ITALY
rossana.centioni@lnf.infn.it; ph: 39 06 9403 2423; fx: 39 06 9403 2243

John Schmerge
Stanford Linear Accelerator Center, SSRL, MS 69
2575 Sand Hill Road, Menlo Park, CA 94025 USA
schmerge@slac.stanford.edu; ph: 650 926 2320, fx: 650 926 4100

Siegfried Schreiber
DESY
Notkestr. 85, Hamburg 22603 GERMANY
siegfried.schreiber@desy.de; ph: 49 40 8998 4360; fx: 49 40 8998 4364

Luca Serafini
INFN Sez. Milan, University of Milan
Via Celoria 16, 20133 Milano ITALY
luca.serafini@mi.infn.it; ph: 39 02 5835 7673, fx: 39 02 5835 7687

Susan L. Smith
Daresbury Laboratory
Warrington, Cheshire WA4 4AD UK
s.l.smith@dl.ac.uk; ph: 44 1925 603 260; fx: 44 1925 603 192

Bruno Spataro
INFN-LNF, Accelerator Division
Via E. Fermi, 40, 00044 Frascati (Roma) ITALY
bruno.spataro@lnf.infn.it; ph: 39 06 9403 2253; fx: 39 06 9403 2256

Frank Stephan
DESY, PITZ
15738 Hamburg GERMANY
frank.stephan@desy.de; ph: 49 33 762 77338; fx: 49 33 762 77330

Frank Stulle
DESY
Notkestrasse, 85, 22607 Hamburg GERMANY
frank.stulle@desy.de; ph: 49 40 8998 4841; fx: 49 40 8994 4841

Gennady Stupakov
SLAC, ARD-A
2575 Sand Hill Road, Menlo Park, CA 94025-7015 USA
stupakov@slac.stanford.edu; ph: 650 926 4320

Franco Tazzioli
INFN-LNF
Via E. Fermi, 40, Frascati (Roma) 00046 ITALY
tazzioli@lnf.infn.it

Matthew Thompson
UCLA, Department of Physics & Astronomy
Los Angeles, CA 90095-1547 USA
matt@stout.physics.ucls.edu; ph: 310 206 5584

Mitsuru Uesaka
University of Tokyo, Nuclear Enegineering Research Laboratory
2-22 Shirakata-Shirane, Tokai, Naka Ibaraki 319-1188 JAPAN
uesaka@tokai.t.u-tokyo.ac.jp; ph: 81 29 287 8421, fx: 81 29 287 8488

Donald Umstadter
University of Michigan, Electrical Engineering & Computer Science
2200 Bonisteel Blvd., IST Building, Rm 1008, Ann Arbor,
MI 48109-2099 USA
dpu@umich.edu; ph: 734 763 4875; fx: 734 763-4876

Alexander Vaarfolomeev
Russian Research Center, Kurchatov Institute
1 Kurchatov Square, Moscow 123182 RUSSIA
varfol@dnuc.polyn.kiae.su; ph: 7 095 196 7764; fx: 7 095 196 77641

Cristina Vaccarezza
INFN LNF Frascati
P.O. Box 13, 00044 Frascati (RM) ITALY
cristina.vaccarezza@lnf.infn.it; ph: 39 06 9403 2517; fx: 39 06 9403 2256

Marco Venturini
SLAC
2575 Sand Hill Road, Menlo Park, CA 94025 USA
venturini@slac.stanford.edu; ph: 650 926 3365; fx: 650 926 5368

Victor Verzilov
Sincrotrone Trieste
Trieste 34012 ITALY
victor.verzilov@elettra.trieste.it; ph: 39 040 375 81; fx: 39 040 375 8565

Carlo Vicario
LNF-INFN, University "la Sapienza"
00044 Frascati (Roma) ITALY
carlo.vicario@lnf.infn.it; ph: 39 06 940 32347; fx: 39 06 940 32565

Rodney Yoder
UCLA, Department of Physics & Astronmy
Los Angeles, CA 90095-1547 USA
yoder@physics.ucla.edu; ph: 310 206 4540; fx: 310 825 8432

Alexander Zholentz
LBNL, Center for Beam Physics, MS 71-259
1 Cyclotron Road, Berkeley, CA 94720 USA
aazholents@lbl.gov; ph: 510 486 7533; fx: 510 486 6485